AF371262

The Role of Nanofluids in Renewable Energy Engineering

The Role of Nanofluids in Renewable Energy Engineering

Editors

M. M. Bhatti
Kambiz Vafai
Sara I. Abdelsalam

Basel • Beijing • Wuhan • Barcelona • Belgrade • Novi Sad • Cluj • Manchester

Editors

M. M. Bhatti
College of Mathematics and
Systems Science
Shandong University of
Science and Technology
Qingdao
China

Kambiz Vafai
Mechanical Engineering
Department
University of California
Riverside
United States

Sara I. Abdelsalam
Basic Science
The British University
in Egypt
Alshorouk City
Egypt

Editorial Office
MDPI
St. Alban-Anlage 66
4052 Basel, Switzerland

This is a reprint of articles from the Special Issue published online in the open access journal *Nanomaterials* (ISSN 2079-4991) (available at: www.mdpi.com/journal/nanomaterials/special_issues/Nanofluids_Renewable_Energy_Engineering).

For citation purposes, cite each article independently as indicated on the article page online and as indicated below:

Lastname, A.A.; Lastname, B.B. Article Title. *Journal Name* **Year**, *Volume Number*, Page Range.

ISBN 978-3-0365-9383-8 (Hbk)
ISBN 978-3-0365-9382-1 (PDF)
doi.org/10.3390/books978-3-0365-9382-1

Contents

About the Editors

M. M. Bhatti

M. M. Bhatti holds the position of Associate Professor at Shandong University of Science and Technology in China. He obtained his exceptional academic status from Shanghai University in China. Additionally, he has the position of an extraordinary associate professor at North-West University in South Africa. He has published over 210 papers in journals from various countries like the USA, Germany, China, the UK, and Canada. The number of citations for his work on Google Scholar exceeds 12,300, and his h-index is 68. He serves as a member of the editorial board for more than three international journals. In addition, he serves as a topical advisory panelist for two MDPI journals, namely *Mathematics* and *Nanomaterials*. Additionally, he has a commendable track record in instructing both undergraduate and postgraduate students, with notable accomplishments.

Kambiz Vafai

Kambiz Vafai is an internationally leading figure in the fields of heat and mass transfer. He joined the University of California, Riverside, as a Presidential Chair in 2000. He is currently a Distinguished Professor at the University of California, Riverside. Dr. Vafai has authored over 420 journal publications, book chapters, and symposium volumes and has been an invited Visiting Professor at the Technical University of Munich in Germany, the University of Bordeaux and Paul Sabatier University in France, and the Technical University of Naples in Italy. Dr. Vafai is a Fellow of the American Society of Mechanical Engineers (ASME), a Fellow of the American Association for the Advancement of Science (AAAS), and an Associate Fellow of the American Institute of Aeronautics and Astronautics (AIAA) and has been a member or the chair of numerous ASME and AIAA National committees. Dr. Vafai's works have been highly cited in various categories. Dr. Vafai is the Editor in Chief of the *Journal of Porous Media and Special Topics & Reviews in Porous Media*, an international journal; the Editor of the *International Journal of Heat and Mass Transfer*; and the Editor of the first, second, and third editions of the *Handbook of Porous Media* published by Taylor and Francis/CRC Press. His other book is entitled "Porous Media, Applications in Biological Systems and Biotechnology", which was also published by Taylor & Francis. He received the 1999 ASME Heat Transfer Classic Paper Award, the 2006 Heat Transfer Memorial Award (considered one of the highest awards in the field), and the College of Engineering Outstanding Research Award at The Ohio State University as an assistant, associate, and full professor. Dr. Vafai was given the highest award by the International Society of Porous Media (Interpore), i.e., the Honorary Member Award in 2011 for making extraordinary contributions to porous medium science and technology and being world-renowned in the porous media community.

Sara I. Abdelsalam

Dr Sara Abdelsalam is an Associate Professor of Mathematics in the Faculty of Engineering at the British University in Egypt. She earned her PhD from the Faculty of Science at Helwan University in 2013. In 2014, she was awarded The Best PhD Thesis in Applied Mathematics from the Egyptian Mathematical Society. She received the Fulbright Scholar award to conduct her research at the University of California Riverside and Caltech in 2015 and 2016, respectively.

Dr Abdelsalam was also granted The World Academy of Sciences—United Nations Educational, Scientific and Cultural Organization (TWAS-UNESCO) Associateship to visit the Instituto de Matemáticas (UNAM) in the summer of 2017 and 2019. TWAS has also nominated her to attend The Future Leaders Programme and the Science and Technology in Society (STS) forum by the Japan Society for the Promotion of Science in 2018. She has also been selected by the International Centre for Theoretical Physics (ICTP) for a fellowship at the Lebanese University in 2021 and by the Royal Academy of Engineering for the Leaders in Innovation program (LIF7-Newton Mosharafa fund). She has been selected by the Fundación de Mujeres Por Africa for a fellowship program at the Institute of Mathematical Sciences in Madrid in 2022 and 2023.

Dr Abdelsalam has been an affiliate member of the African Academy of Sciences (AAS) since 2020 and a member of the Egyptian Young Academy of Sciences (EYAS) since 2021. She has also won the State Encouragement Award in Mathematics for 2020, the Obada Award in 2021, and an honorable mention for the Bellman Prize by the *Mathematical Biosciences* journal, Elsevier, in 2021. She was also awarded the TWAS Arab Regional Award for Science Diplomacy in 2022 and the Shoman Award for Arab researchers in 2023. She is a member of the editorial board of peer-reviewed journals.

Preface

The reprint has a total of 13 chapters. A cumulative sum of 43 manuscripts has been received for potential consideration for publication. Following an extensive evaluation by peers, a mere 13 manuscripts met the criteria for publication. Each chapter has several realistic applications. In order to aid the comprehension of readers, a comprehensive compilation of references is provided at the end of each chapter to facilitate additional exploration and investigation. We express our gratitude to all the reviewers for their valuable comments and thorough assessments of the submitted publications. We were lucky to have the valuable contributions of distinguished researchers who provided their unique research. We commend all those involved for the successful culmination of the Special Issue titled "The Role of Nanofluids in Renewable Energy Engineering". Any errors or omissions should be called out, and we would be happy to include an apology in the earliest version we can publish. We kindly request valuable recommendations on potential enhancements, as well as ideas pertaining to the overall breadth and structure of the manuscript. Your input is highly valued. We would like to extend our sincere appreciation to MPDI for their publication of this reprint. We would also like to extend our appreciation to the whole editing team, as well as our family and friends, for their valuable support. It is of note that the aforementioned special edition has garnered a significant viewership of over 20,534 within a very short period. It is anticipated that this reprint will provide a comprehensive overview and the latest research outcomes from the scientific community engaged in this domain and also yield benefits for targeted market segments and end users within the industrial sectors.

M. M. Bhatti, Kambiz Vafai, and Sara I. Abdelsalam
Editors

nanomaterials

MDPI

Editorial

The Role of Nanofluids in Renewable Energy Engineering

Muhammad Mubashir Bhatti [1],* , **Kambiz Vafai [2]** and **Sara I. Abdelsalam [3]**

1 College of Mathematics and Systems Science, Shandong University of Science and Technology, Qingdao 266590, China
2 Department of Mechanical Engineering, University of California, Riverside, CA 92521, USA; vafai@engr.ucr.edu
3 Basic Science, Faculty of Engineering, The British University in Egypt, Al-Shorouk City, Cairo 11837, Egypt; sara.abdelsalam@bue.edu.eg
* Correspondence: mmbhatti@sdust.edu.cn or mubashirme@yahoo.com

Citation: Bhatti, M.M.; Vafai, K.; Abdelsalam, S.I. The Role of Nanofluids in Renewable Energy Engineering. *Nanomaterials* **2023**, *13*, 2671. https://doi.org/10.3390/nano13192671

Received: 15 September 2023
Revised: 22 September 2023
Accepted: 26 September 2023
Published: 29 September 2023

The phenomenon of nanofluid flows is intrinsically characterized by several scales and intricate physical processes. Numerous modeling and experimental methodologies have been extensively used to examine nanofluids, including both basic and applied perspectives. Significant advancements have been achieved in the field of physical understanding and research tools, with more substantial contributions anticipated in the future.

In recent times, the field of thermal technology has encountered many obstacles, with the realization of an elevated heat transfer rate emerging as a significant concern. Renewable energy often fails to attain the elevated temperatures attained by burning conventional fuels. Nanofluids have been shown to provide advantages in addressing these challenges. The fundamental principles of enhancing heat transmission through the use of nanofluids have been widely accepted and acknowledged in the academic literature. Further research is necessary to explore crucial domains that pertain to the use of nanofluids in energy-related applications. The phenomenon under consideration encompasses the interactions between nanoparticles and the base fluid, as well as the following effects these interactions have on heat convection. Moreover, the extensive use of nanofluids in renewable energy technologies represents a significantly underexplored domain. Numerous investigations have been carried out pertaining to nanofluids, specifically emphasizing their utilization in solar thermal and geothermal technologies, heat storage and networks, and other heat recovery technologies. A persistent obstacle in these domains is directed toward creating cost-effective computational tools that possess both accuracy and the ability to forecast the heat transfer characteristics of nanofluids. Consequently, it is necessary to conduct more experimental investigations on nanofluids at a systemic level, namely in applications such as solar thermal collectors, geothermal boreholes, and other renewable energy systems of significant size.

The engineering research community is confronted with a significant undertaking in the form of decarbonizing energy by shifting from fossil fuels to renewable sources. Thermal energy represents a substantial portion of the energy utilized in various regions globally and has a significant role in releasing greenhouse gases. In recent times, several nations have made the strategic choice to undertake the process of decarbonizing their respective energy sectors. Consequently, there is a growing focus on carbon-free thermal technologies, including solar thermal, geothermal, heat pumps, heat storage, and heat networks. The core issue here is the difficulty associated with transmitting thermal energy.

The primary objective of this Special Issue is to consolidate the most recent research discoveries pertaining to heat transfer facilitated by nanofluids, with a particular focus on the challenges associated with renewable energy. Next, this Special Issue collected nine selected original research papers and three comprehensive reviews on various topics in nanotechnology. A total of 68 scientists hailing from various universities and research organizations actively contributed their research efforts and knowledge, therefore playing a significant role in completing this Special Issue. Below is a summary of their scientific contributions.

The study by Ampah et al. [1] provided a comprehensive overview of the advancements in the use of nanoparticles in low-carbon fuels. The investigators noted that Asia is the primary driver of advancements in the field of biofuels, namely in the areas of biodiesel, vegetable oils, and alcohols. Nevertheless, significant emphasis has been placed on biodiesel. The findings of their study indicate that the combustion of low-carbon fuel diesel blends was boosted by the presence of nanoparticles in diesel engines. Nanoparticles function as an oxygen reservoir, supplying supplementary oxygen molecules inside the combustion chamber. This mechanism serves to reduce unburnt emissions and enhance the combustion process. The exceptional catalytic characteristics of nanoparticles are attributed to their behavior.

Nabwey et al. [2] examined the latest developments in the domain of porous media flow, including heat transfer and nanofluids. The authors set their evaluation on a curated collection of papers published within the period spanning 2018 to 2020. The participants engaged in a comprehensive discussion on several analytical methodologies for analyzing nanofluid flow inside porous media. The initial emphasis of the study was on the heat transfer phenomenon of natural convection in porous media. Subsequently, the researchers proceeded to examine heat transfer under forced convection and mixed convection scenarios. Based on their investigation, it was observed that altering the height of the medium results in a modification of the flow regime inside the chamber.

The primary emphasis of Moghaieb et al. [3] was on the advancements of nanofluids for use in direct-absorption solar collectors (DASCs). Their research primarily examined the optical properties, solar thermal conversion efficiency, production methods, and thermal and physical stability of these nanofluids. The authors also emphasized the difficulties associated with the actual use of nanofluids in direct-absorption solar collectors (DASCs) and offered recommendations for further research.

Hu et al. [4] provided an overview of the latest advancements in the development of solar thermal nanofluids capable of achieving homogeneous and stable dispersion at moderate temperatures. The authors comprehensively examined the issues associated with dispersion in several media, including ionic liquid, oil, molten salt-based, medium-temperature solar thermal nanofluids, and ethylene glycol. They conducted an in-depth investigation of the dispersion mechanism regulating these media and proposed a representative method. The benefits and application of many methods for strengthening the dispersion stability of thermal storage nanofluids were explored. These methods include electrostatic stabilization, self-dispersion stabilization, hydrogen bonding, steric stabilization, etc. Enhancing the dispersion stability of nanofluids used in medium-temperature solar thermal systems has the potential to not only encourage further investigation into the efficient absorption of solar thermal energy but also provide a promising solution to address basic challenges and limitations associated with nanofluid technology.

Zubair et al. [5] used a combination formulation of integrated fractional calculus and plasma modeling to examine the characteristics of the higher dimensional time fractional Vlasov–Maxwell system. A numerical methodology was devised for filtered Gegenbauer polynomials and temporal variables, demonstrating robustness, efficiency, speed, and accuracy. This approach combined a finite difference method with spectral approximations. Two kinds of boundary conditions were used, namely partial slip and Dirichlet.

Qureshi et al. [6] examined hybrid nanofluids, including several kinds of nanoparticles, namely Al_2O_3, Ti_2O, and Fe_3O_4. The authors investigated the morphological characteristics of these nanofluids and explored the influence of a magnetic field on their properties. The flow is presumed to be transitioning between two coaxially spinning disks. The numerical simulation used the Runge–Kutta technique in conjunction with the shooting method. It has been shown that there exists a direct relationship between the shape of nanoparticles and the heat conductivity of nanofluids.

Bhatti et al. [7] introduced a mathematical model that describes the flow of a nanofluid consisting of Magnesium oxide (MgO) and nickel (Ni) particles, taking into account the influence of magnetization. The researchers assumed a stagnation point flow occurring over an elastic porous surface. They employed homogeneous porous media with the assistance of Darcy law. The research also considers slip and thermal slip boundary conditions. They discovered that the thermal boundary layer thickness and thermal profile were greatly enhanced owing to the influence of magnesium oxide and nickel nanoparticles. The thermal profile was also found to be boosted owing to the heat production while it was suppressed due to the thermal slip effects. Furthermore, hybrid nanofluids have exhibited significant improvements in thermal enhancement.

Rehman et al. [8] employed carbon nanotubes to investigate the characteristics of nanofluids. Their research revealed that carbon nanotubes offer distinct advantages over other types of nanofluids. The authors considered a time-dependent stagnation point flow occurring on an elastic surface. To analyze the problem, they utilized an optimized version of the homotopy analysis method. The outcomes of their investigation demonstrated that nanoparticles, the unsteady parameter, and the magnetic field exhibited an opposing effect on the velocity. Conversely, the Eckert and Prandtl numbers displayed varying influences on the thermal profile.

The properties of nanofluid flows, which have been constructed based on basic principles and experimental data, were described by Rizwan et al. [9]. Various types of nanoparticles, including SiO_2, TiO_2, and MgO, were included in the study using a power law model. The researchers used a numerical shooting approach to assess the obtained findings. An observed decrease in the velocity profile has been attributed to the increasing concentration of nanoparticles, whereas a concurrent rise in the thermal profile has been noted. Nevertheless, the size of the nanoparticles has an inverse effect on both the velocity and heat profile. The findings of this study have practical implications for a range of industrial applications that need a comprehensive understanding of heat and mass transfer phenomena to optimize device design.

The study conducted by McGlynn et al. [10] focused on examining the effectiveness of hybrid liquid–plasma functionalization in enhancing the stability of carbon nanotubes while suspended in a base fluid. The objective of this research was to explore the potential uses of these stabilized carbon nanotubes in the field of solar energy conversion. The stability of nanofluids has been enhanced using surface treatments, including plasma-induced non-equilibrium electrochemistry. This allows the redispersion of these fluids under simulated circumstances. The enhanced dispersibility results in an increased absorption coefficient and better thermal profile when subjected to solar simulation.

The sorption effectiveness of biochar-mediated $ZrFe_2O_5$ nanocomposites for the treatment of Tartrazine dye-containing textile wastewater was explored by Perveen et al [11]. Batch mode experiments were used for this objective. The researchers noted a significant sorption capacity of BC-$ZrFe_2O_5$ NCs for Tartrazine dye. The use of a surface approach has been employed to optimize the parameters of the process.

Kanungo et al. [12] conducted a theoretical investigation of the electrical and electrochemical characteristics of $LiMPO_4$ materials, namely those based on iron, cobalt, manganese, vanadium, and chromium. The purpose of their research was to explore the potential of these materials for cathode design in lithium-ion batteries. The researchers used a computational method known as density functional theory, which is based on fundamental principles, to perform their calculations. The participants thoroughly examined the cathode design, including an analysis of several materials and their performances. This analysis specifically focused on key factors such as the Lithium intercalation energy inside the olivine structure, the cohesive energy of the material, and the intrinsic diffusion coefficient across the Lithium channel. The possibility of an open circuit and equilibrium at different charge levels of lithium batteries was also a topic of discussion. It was observed that the choice of metal atom significantly influenced the extent of lithium diffusion inside the olivine structure and the overall energetics of different $LiMPO_4$ compounds.

Sharma et al. [13] examined the effects of entropy production, heat transfer, and mass transfer within the context of the Jeffrey fluid model, including solar radiation as a contributing factor. The researchers used a polyvinyl alcohol–water mixture as the primary fluid medium, including copper nanoparticles and microorganisms. The ongoing research aims to create a robust numerical model that accurately represents the flow and thermal properties of a parabolic trough surface collector (PTSC) integrated into a solar plate. This endeavor is driven by the growing use of solar plates in diverse applications. It was shown that increasing the volume fraction of the nanoparticles had a positive impact on the thermal profile. However, the concept of entropy demonstrates a tendency towards increased order because of diffusion factors.

Author Contributions: This Editorial Letter was collectively authored and vetted by M.M.B., K.V., and S.I.A. All authors have read and agreed to the published version of the manuscript.

Acknowledgments: The guest editors express their gratitude to all the authors for their valuable contributions to the Special Issue and for their significant role in its successful completion. We would like to express our gratitude to all the reviewers who have contributed to the peer-review process of the submitted articles, therefore improving their overall quality and effect. We express our gratitude to the editorial assistants for their valuable contributions in ensuring a seamless and efficient process during the preparation of the complete Special Issue.

Conflicts of Interest: The authors declare no conflict of interest.

References

1. Ampah, J.D.; Yusuf, A.A.; Agyekum, E.B.; Afrane, S.; Jin, C.; Liu, H.; Fattah, I.M.R.; Show, P.L.; Shouran, M.; Habil, M.; et al. Progress and Recent Trends in the Application of Nanoparticles as Low Carbon Fuel Additives-A State of the Art Review. *Nanomaterials* **2022**, *12*, 1515. [CrossRef] [PubMed]
2. Nabwey, H.A.; Armaghani, T.; Azizimehr, B.; Rashad, A.M.; Chamkha, A.J. A Comprehensive Review of Nanofluid Heat Transfer in Porous Media. *Nanomaterials* **2023**, *13*, 937. [CrossRef] [PubMed]
3. Moghaieb, H.S.; Amendola, V.; Khalil, S.; Chakrabarti, S.; Maguire, P.; Mariotti, D. Nanofluids for Direct-Absorption Solar Collectors-DASCs: A Review on Recent Progress and Future Perspectives. *Nanomaterials* **2023**, *13*, 1232. [CrossRef] [PubMed]
4. Hu, T.; Zhang, J.; Xia, J.; Li, X.; Tao, P.; Deng, T. A Review on Recent Progress in Preparation of Medium-Temperature Solar-Thermal Nanofluids with Stable Dispersion. *Nanomaterials* **2023**, *13*, 1399. [CrossRef] [PubMed]
5. Zubair, T.; Asjad, M.I.; Usman, M.; Awrejcewicz, J. Higher-Dimensional Fractional Order Modelling for Plasma Particles with Partial Slip Boundaries: A Numerical Study. *Nanomaterials* **2021**, *11*, 2884. [CrossRef] [PubMed]
6. Qureshi, Z.A.; Bilal, S.; Shah, I.A.; Akgül, A.; Jarrar, R.; Shanak, H.; Asad, J. Computational Analysis of the Morphological Aspects of Triadic Hybridized Magnetic Nanoparticles Suspended in Liquid Streamed in Coaxially Swirled Disks. *Nanomaterials* **2022**, *12*, 671. [CrossRef] [PubMed]
7. Bhatti, M.M.; Bég, O.A.; Abdelsalam, S.I. Computational Framework of Magnetized MgO–Ni/Water-Based Stagnation Nanoflow Past an Elastic Stretching Surface: Application in Solar Energy Coatings. *Nanomaterials* **2022**, *12*, 1049. [CrossRef] [PubMed]
8. Rehman, A.; Saeed, A.; Salleh, Z.; Jan, R.; Kumam, P. Analytical Investigation of the Time-Dependent Stagnation Point Flow of a CNT Nanofluid over a Stretching Surface. *Nanomaterials* **2022**, *12*, 1108. [CrossRef] [PubMed]
9. Rizwan, M.; Hassan, M.; Makinde, O.D.; Bhatti, M.M.; Marin, M. Rheological Modeling of Metallic Oxide Nanoparticles Containing Non-Newtonian Nanofluids and Potential Investigation of Heat and Mass Flow Characteristics. *Nanomaterials* **2022**, *12*, 1237. [CrossRef] [PubMed]
10. McGlynn, R.J.; Moghaieb, H.S.; Brunet, P.; Chakrabarti, S.; Maguire, P.; Mariotti, D. Hybrid Plasma-Liquid Functionalisation for the Enhanced Stability of CNT Nanofluids for Application in Solar Energy Conversion. *Nanomaterials* **2022**, *12*, 2705. [CrossRef]
11. Perveen, S.; Nadeem, R.; Nosheen, F.; Asjad, M.I.; Awrejcewicz, J.; Anwar, T. Biochar-Mediated Zirconium Ferrite Nanocomposites for Tartrazine Dye Removal from Textile Wastewater. *Nanomaterials* **2022**, *12*, 2828. [CrossRef] [PubMed]
12. Kanungo, S.; Bhattacharjee, A.; Bahadursha, N.; Ghosh, A. Comparative Analysis of LiMPO4 (M = Fe, Co, Cr, Mn, V) as Cathode Materials for Lithium-Ion Battery Applications—A First-Principle-Based Theoretical Approach. *Nanomaterials* **2022**, *12*, 3266. [CrossRef] [PubMed]
13. Sharma, B.K.; Kumar, A.; Gandhi, R.; Bhatti, M.M.; Mishra, N.K. Entropy Generation and Thermal Radiation Analysis of EMHD Jeffrey Nanofluid Flow: Applications in Solar Energy. *Nanomaterials* **2023**, *13*, 544. [CrossRef] [PubMed]

 *nanomaterials*

Review

A Review on Recent Progress in Preparation of Medium-Temperature Solar-Thermal Nanofluids with Stable Dispersion

Ting Hu, Jingyi Zhang, Ji Xia, Xiaoxiang Li, Peng Tao * and Tao Deng *

State Key Laboratory of Metal Matrix Composites, School of Materials Science and Engineering, Shanghai Jiao Tong University, 800 Dong Chuan Road, Shanghai 200240, China
* Correspondence: taopeng@sjtu.edu.cn (P.T.); dengtao@sjtu.edu.cn (T.D.)

Abstract: Direct absorption of sunlight and conversion into heat by uniformly dispersed photothermal nanofluids has emerged as a facile way to efficiently harness abundant renewable solar-thermal energy for a variety of heating-related applications. As the key component of the direct absorption solar collectors, solar-thermal nanofluids, however, generally suffer from poor dispersion and tend to aggregate, and the aggregation and precipitation tendency becomes even stronger at elevated temperatures. In this review, we overview recent research efforts and progresses in preparing solar-thermal nanofluids that can be stably and homogeneously dispersed under medium temperatures. We provide detailed description on the dispersion challenges and the governing dispersion mechanisms, and introduce representative dispersion strategies that are applicable to ethylene glycol, oil, ionic liquid, and molten salt-based medium-temperature solar-thermal nanofluids. The applicability and advantages of four categories of stabilization strategies including hydrogen bonding, electrostatic stabilization, steric stabilization, and self-dispersion stabilization in improving the dispersion stability of different type of thermal storage fluids are discussed. Among them, recently emerged self-dispersible nanofluids hold the potential for practical medium-temperature direct absorption solar-thermal energy harvesting. In the end, the exciting research opportunities, on-going research need and possible future research directions are also discussed. It is anticipated that the overview of recent progress in improving dispersion stability of medium-temperature solar-thermal nanofluids can not only stimulate exploration of direct absorption solar-thermal energy harvesting applications, but also provide a promising means to solve the fundamental limiting issue for general nanofluid technologies.

Keywords: solar-thermal nanofluids; dispersion stability; medium-temperature nanofluid; solar collector; stabilization mechanism

 check for updates

Citation: Hu, T.; Zhang, J.; Xia, J.; Li, X.; Tao, P.; Deng, T. A Review on Recent Progress in Preparation of Medium-Temperature Solar-Thermal Nanofluids with Stable Dispersion. *Nanomaterials* **2023**, *13*, 1399. https://doi.org/10.3390/nano13081399

Academic Editors: M. M. Bhatti, Kambiz Vafai and Sara I. Abdelsalam

Received: 22 February 2023
Revised: 15 April 2023
Accepted: 16 April 2023
Published: 18 April 2023

1. Introduction

Sustainable development and modernization of human society calls for exploration of renewable energy technologies [1–3]. Among various renewable energy sources, solar energy has attracted significant research attention owing to its giant capacity, cleanness, ubiquitous availability, and diverse conversion [4]. Solar-thermal conversion is a facile and efficient way to harness the abundant solar irradiation to generate heat [5], which can be directly used for heating of buildings and water [6,7], driving industrial processes [8,9], or further converted into other forms of energy such as electricity [10] and chemical fuels [11–13]. In particular, the large-scale low-cost storage capability of solar-thermal energy, which can overcome the intermittence issue of renewable energy, would endow solar-thermal technologies with unique competitiveness [14,15]. Great efforts had been made to harvest and store the abundant solar-thermal energy as sensible heat, latent heat, or thermochemical energy to adjust the mismatch between energy supply and consumption need [16,17]. Among them, sensible heat storage within flowable liquids has evolved to be the most mature way to harvest, store, and transport solar-thermal energy.

Solar collectors, which absorb incident solar irradiation and convert it into storable heat, is one of the key components in solar-thermal systems [18]. As schemed in Figure 1, conventional surface absorption-based solar collectors utilize blackened surfaces or solar selective coatings to efficiently absorb broadband sunlight and generate heat, which is carried away by the flowing heat transfer fluids (HTFs) [19]. However, the transportation of converted heat is often limited by the low thermal conductivity of the HTFs, which causes overheating of the surface coating and sacrifices energy harvesting efficiency [19,20]. In recent years, direct absorption-based solar collectors (DASCs) [21], which make use of homogeneously dispersed photothermal particles to volumetrically harvest solar-thermal energy within the nanofluids, have been pursued as a superior way to harness solar-thermal energy [22,23]. The direct volumetric conversion and storage of solar-thermal energy as sensible heat in the thermal storage fluids avoid the slow heat transfer from the outer surfaces to the inner storage media and homogenize the temperature distribution within the storage media [24]. Such design thus can avoid the potential overheating of the outer surface, and minimize the radiation and convection heat loss from the solar collector, which in turn enables achieving high energy conversion efficiency at medium-to-high temperatures [25]. Such nanofluid-based DASCs not only improve the sustainability of solar-thermal systems [26], but also can be coupled with photovoltaic and solar steam generation systems for multifunctional applications [27].

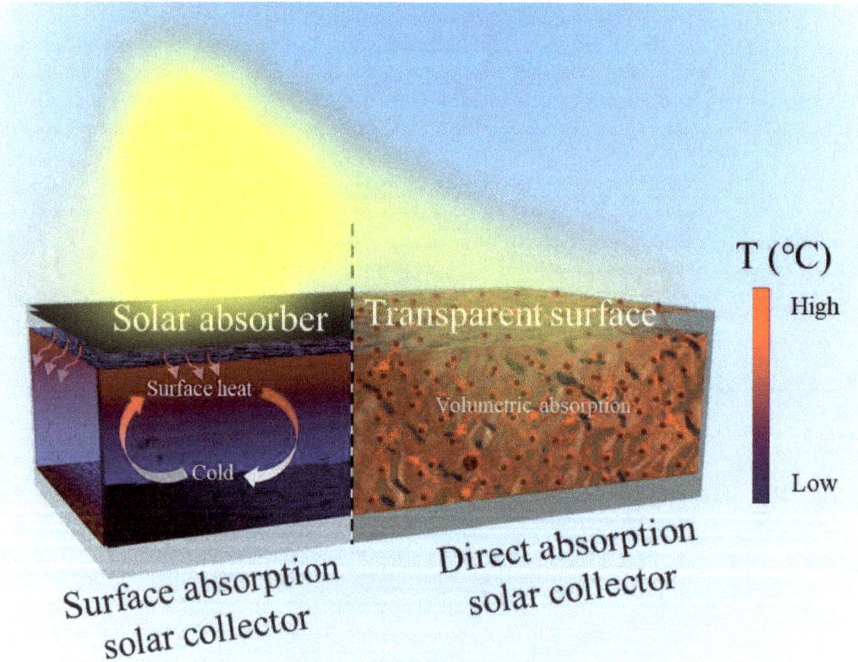

Figure 1. Schematic showing converting concentrated sunlight into medium-temperature heat by surface absorption-based solar collectors and direct absorption-based solar collectors.

The performances of DASCs are largely affected by the solar-thermal nanofluids including their dispersion behavior and key thermophysical properties such as solar absorptance, heat capacity, viscosity, and thermal conductivity [28]. In the past, intensive research efforts were directed to improving these thermophysical properties through synthesizing a variety of nanofluids loaded with plasmonic metallic particles and other type of solar absorbers [29], tailoring the loading, morphology, and size of the nanoscale solar absorbers [30], and incorporating multifunctional hybrid nanoparticles [31,32]. So far, however, these prepared solar-thermal nanofluids usually suffer from poor dispersion stability. Once forming agglomeration, not only the nanofluids lose their enhancement of thermophysical properties and volumetric solar-thermal harvesting advantages but also would cause clogging of the DASCs. While previous research efforts were concentrated on low-temperature water-based nanofluids [33,34], in recent years substantial progresses have been made in preparing stably dispersed medium-temperature solar-thermal nanofluids with an application temperature range of 100–400 °C [35,36]. Although elevated service temperatures pose even grand challenges to achieving stable dispersion, these medium-temperature nanofluids hold the promise to provide high-quality renewable heating and expand the application scope of DASCs.

Despite dispersion stability having been viewed as one of the key factors that limit the development of nanofluid technologies and their practical applications, the previous literature was mostly focused on low-temperature aqueous dispersion systems. In recent years, along with other research groups, we have explored effective strategies to improve dispersion stability and understand dispersion behavior of solar-thermal nanofluids at medium temperatures. In this review, we overview state-of-the-art developments of medium temperature solar-thermal nanofluids and specifically focus on dispersing strategies of these nanofluids. We briefly introduce the desired features of both the thermal storage fluids and solar absorbers for preparing medium-temperature solar-thermal nanofluids. We discuss the specific challenges for dispersing nanoscale solar absorbers within the nanofluids at elevated temperatures. After identifying these challenges, we describe the dispersion strategies that had been explored to improve the dispersion stability and introduce the general working principles of these strategies. We provide comprehensive overview of the dispersion approaches by introducing representative examples that apply these dispersion strategies within ethylene glycol, oil, ionic fluid, and molten salt systems. Finally, we point out the application opportunities, existing problems, and future research directions in the development and application of medium-temperature solar-thermal nanofluids.

2. Medium-Temperature Solar-Thermal Nanofluids

The desired nanofluids for direct absorption-based solar-thermal harvesting at medium temperatures should simultaneously possess the following characteristics [37]. Firstly, the nanofluids should have high solar absorptance such that they can efficiently absorb broad-spectrum sunlight and convert it into heat. Secondly, a large heat capacity is desired in order to maximize the solar-thermal energy storage density of the DASCs. Thirdly, the nanofluids should possess a high thermal conductivity to effectively exchange and release the harvested thermal energy to the application terminals. Fourthly, the nanofluids should have a low viscosity similar as that of the thermal storage fluids in order to keep the desired flowability under low mechanical pumping power. In addition, the nanofluids should have good physical and chemical stability under elevated temperatures and concentrated solar illumination conditions. Most importantly, the nanofluids should maintain their stable uniform dispersion under concentrated solar irradiation. These in turn pose strict requirements on the selection of thermal storage fluids and solar absorbers (Figure 2).

Currently, the investigated thermal storage fluids fall into four categories: ethylene glycol, oils (silicone oil, mineral oil), ionic liquids, and molten salts. As a polar working fluid, ethylene glycol can work within a temperature range of −12–197 °C and has a high specific heat capacity of 2.35 J/(g·K) due to its strong molecular hydrogen bonding [38,39]. Silicone oil and other nonpolar synthetic oils have a specific heat capacity of 1.4–1.6 J/(g·K) and can withstand a temperature up to 400 °C [40–42]. Owing to the formation of strong ionic bonds between organic cations and anions, the ionic liquids can stably operate under a high temperature up to 800 °C and possess a specific heat capacity of 1.3–2.2 J/(g·K) [43,44]. Although they have demonstrated improved thermal stability, these synthesized ionic liquids are generally expensive at laboratory scales. At industrial scale, the cost of ionic liquids produced has decreased to less than 20 US$/kg due to recent developments [45], which make them attractive for heat transfer and thermal storage. Molten salts with a specific heat capacity in the range of 1.3–1.6 J/(g·K) are the fluids for solar-thermal energy harvesting at even higher temperatures up to 900 °C [46]. To date, synthetic oils (silicone oil, Dowthermal oil) and molten salts (solar salts, Hitec salts) are the main commercial fluids used for harvesting medium-temperature and high-temperature heat in solar-thermal power plants. By comparison, molten salts have cost advantages, but the strong corrosivity and the high freezing point are the main identified issues [47,48]. Corrosion-resistant containers and pipelines made of Ni-based alloys or stainless steels are used in solar-thermal plants [49]. To avoid salt freezing at night, auxiliary heating systems are often installed, and new eutectic salts have been continuously investigated [50,51].

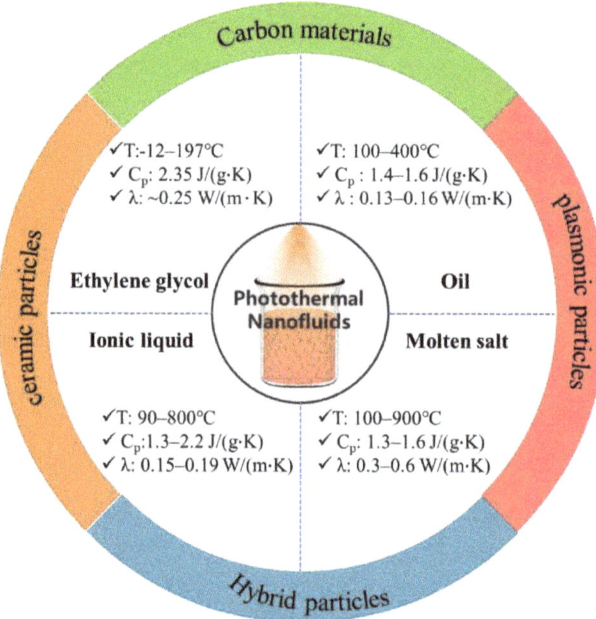

Figure 2. Representative thermal storage fluids and solar absorbers for preparing medium-temrature solar-thermal nanofluids.

A common feature of commercial thermal storage fluids is that they are transparent, most often colorless, which implies that they have very low solar absorptance. Various solar absorbers had been incorporated into the thermal storage fluids to enable direct absorption-based solar-thermal harvesting. Besides high absorptance of broadband sunlight, one key requirement is that the added solar absorbers can physically and chemically withstand the high service temperature and concentrated solar irradiation. So far, several types of absorbers including carbon nanomaterials [52], plasmonic particles [53], ceramic particles [31], and hybrid particles [54] had been explored as the promising candidates. Among them, carbon absorbers, such as graphite particle [55], reduced graphene oxide (RGO) [56], graphene [57], and carbon nanotube (CNT) [58], are the most popular choice because they are naturally black, stable against solar irradiation, and can be manufactured at large scale with abundant low-cost carbon raw materials such as graphite. In recent years, the plasmonic resonance effect of metallic nanoparticles (Au, Ag) has drawn intensive research attention for efficient photothermal conversion applications [59]. The absorption peak can be tuned through tailoring the size and the morphology of the plasmonic absorbers [60,61], which offers the opportunity to convert broad-spectrum sunlight into intensive plasmonic heat. Ceramic semiconductor particles are another category of solar absorbers, which make use of their small bandgap to absorb incident solar photons and convert them into heat. Most frequently, composition design, defect engineering and particle size are used as the means to tailor the bandgap and thereby the solar-thermal conversion performances of these ceramic solar absorber particles [62]. In comparison with metallic plasmonic particles, ceramic particles such as TiN have exhibited significantly enhanced thermal stability due to their stronger ionic bonding [63]. In addition to single-component absorbers, hybrid particles can further improve solar-thermal conversion efficiency and other thermophysical properties or combine individual functionality from many components [64,65].

3. Dispersion Challenge and Stabilization Strategy

3.1. Dispersion Challenge at Elevated Temperatures

The dispersion behavior of nanofluids is governed by the particle-particle, and particle-fluid interactions [66]. Most frequently, the added fillers do not have attractive interaction with the thermal storage fluids, i.e., the fillers like each other and prefer to stay closely together to minimize the exposed surface areas and lower the surface energy for the whole system. For the fillers that have large exposed interacting surfaces and strong interparticle attraction such as graphene, it becomes challenging to disperse them homogeneously and stably within the fluids.

If there is no attractive interaction between the fillers and the thermal storage fluid, the ubiquitous attractive van der Waals force would drive the aggregation of the added fillers. The van der Waals interaction force (F_{vdW}) between two same neighboring spherical particles with a radius of r could be described by [67,68]:

$$F_{\mathrm{vdW}} = -\frac{\left(\sqrt{A_s} - \sqrt{A_l}\right)^2 r}{12L^2} \tag{1}$$

where L is the distance between two approaching particles, A_s and A_l are the Hamaker constants of the dispersed solid particle and the thermal storage fluid, respectively. A_s is proportional to the square of the atomic density of the particle. As a type of short-range interaction, the van der Waals forces are effective only when the neighboring particles are closely contacting with each other. The high application temperature intensifies the random Brownian motion of the solar absorbers and significantly increases their colliding probability, thereby exacerbating the aggregation tendency. When the particles are approaching each other, their motion is restricted by the friction force between the particle and the thermal storage fluid (F_r). At elevated temperatures, the restriction force decreases as the viscosity of fluid drops, which also intensifies the tendency for neighboring particles to form agglomerates.

For a single particle dispersed within the thermal storage fluid, it is simultaneously subject to the gravity force (F_g), the buoyance force (F_b), and the viscous friction force (F_v). Due to the density difference between the filler and the thermal storage fluid, the heavier inorganic or metallic particle tends to gradually precipitate out of the fluids. The net gravity force (ΔF) can be described by [69]:

$$\Delta F = \frac{4\pi}{3} r^3 (\rho_s - \rho_l) g \tag{2}$$

where ρ_s and ρ_l are the density of the solid filler particle and the thermal storage fluid, and g is the acceleration of gravity (9.8 m/s^2). According to the Stokes law, the F_v can be calculated by [70]:

$$F_v = 6\pi r \eta v \tag{3}$$

where η is the viscosity of the nanofluids, and v is the sedimentation velocity. Based on force balance, the gravitational sedimentation velocity of a colloidal spherical particle (v_s) can be estimated by [71]:

$$v_s = \frac{2r^2(\rho_s - \rho_l)g}{9\eta} \tag{4}$$

A smaller particle size, a matching density and a high viscosity are favorable for reducing the sedimentation velocity. When the fluids are working under elevated temperatures, the viscosity dramatically drops, which in turn leads to rapid rise in the sedimentation velocity.

In the meanwhile, the particle is continuously bombed by the thermal storage fluid molecules undergoing random motion. The average Brownian motion velocity (v_B) of the spherical particle within a certain period (t) can be estimated by [72,73]:

$$v_B = \sqrt{\frac{k_B T}{6\pi r \eta t}} \tag{5}$$

where k_B is the Boltzmann's constant, T is the temperature of the nanofluids in the unit of Kelvin. The particles with a small size are expected to have a high Brownian motion velocity. Increasing temperature, on one hand, can decrease the viscosity of the nanofluids and accelerate the Brownian motion, which is favorable for dispersing single particles. In the meanwhile, the interparticle collision is also intensified at elevated temperatures. When the bare particles are contacting with each other, it had been previously reported that plasmonic solar absorbers such as gold nanoparticles tend to undergo coalescence at elevated temperatures and form large-sized aggregates [74,75]. Long-term heating at elevated temperatures could even cause Ostwald ripening and phase change in the nanoscale solar absorbers within the fluids [76]. For chemically stable carbon-based solar absorbers, it was also found that the nanofluids gradually lost their uniform dispersion with increasing application temperatures. In this case, increasing temperature leads to reduced viscosity of the thermal storage fluids, which cannot provide sufficient restriction forces to prevent gravitational sedimentation of the particles. For the surface-modified particles, it was found that the dynamic adsorption-desorption of surface ligands was also intensified at high temperatures [77]. After losing the protection of the surface ligands, the added particles tend to form agglomerates. The elevated temperature increases the reactivity of surface atoms of the nanofillers, the desorption kinetics of surface-capping agents, and gravitational sedimentation tendency due to reduced viscosity of the thermal storage fluids, which together add more challenges in achieving stable dispersion of medium-temperature solar-thermal nanofluids.

3.2. Stabilization Strategy

The key to achieving stable dispersion of medium-temperature solar-thermal nanofluids lies in simultaneously suppressing the van der Waals interparticle attraction and gravitational sedimentation, which can be realized through judiciously selecting the solar absorbers with proper size, density, morphology and tailoring their surface chemistry. As schemed by Figure 3, the stabilization strategies that are applicable to medium-temperature solar-thermal nanofluids can be classified into four categories: hydrogen bonding stabilization, electrostatic stabilization, steric stabilization, and self-dispersion stabilization.

3.2.1. Hydrogen Bonding Stabilization

In polar fluid systems such as ethylene glycol [78], propylene glycol [79] that have hydroxyl groups, they can form hydrogen bonds with solar absorbers that have abundant oxygen-containing surface functional groups. These hydrogen bonds provide the favorable filler-fluid interaction for the solar absorbers to overcome the van der Waals interparticle attraction. Under such circumstance, homogeneous dispersion can be achieved because the solar absorbers and the thermal storage fluid molecules prefer to mix with each other (Figure 3a). While absorbers such as RGO and MXene are naturally rich in oxygen-containing surface functional groups, the surfaces of graphene, CNTs and other solar absorbers can be modified through various kinds of surface modification techniques such as plasma treatment, washing with strong acids or grafting of polar ligands/chains.

The number density of these surface functional groups and their bonding strength with the thermal storage fluid molecules affect the dispersion state and dispersion stability of the nanofluids. The key requirements for such stabilization strategy are that the surfaces of solar absorbers should have polar functional groups to form the hydrogen bonding with the thermal storage fluids, and the particle-fluid bonding can withstand the application temperature. It is a facile way to improve the dispersion stability of solar absorbers within polar thermal storage fluids such as ethylene glycol, but it is difficult to apply such stabilization strategy in other nonpolar media such as silicone oil.

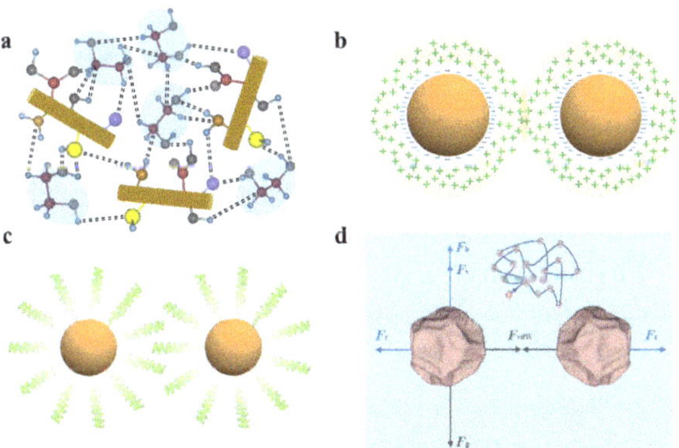

Figure 3. Schematics showing different types of stabilization strategies for preparing medium-temperature solar-thermal nanofluids: (**a**) hydrogen bonding stabilization, (**b**) electrostatic stabilization, (**c**) steric stabilization, (**d**) self-dispersion stabilization.

3.2.2. Electrostatic Stabilization

By introducing charges with the same polarity onto the surfaces of the solar absorbers through adsorption of ions or charged species, electrostatic repulsion provides the force to counterbalance the interparticle van der Waals attraction. Within the nanofluids, the charged particles are surrounded by counter-ions forming an electrical double layer, which consists of an inner tightly bound Stern layer and an outer diffusive Gouy layer [80]. When the charged particles are approaching, the electrical double layer overlaps and generates the mutual repulsion force (Figure 3b). The thickness of the electrical double layer can be characterized by the Debye length (L_D), which is expressed as [81–83]:

$$L_D = \sqrt{\frac{\varepsilon_0 \varepsilon_r k_B T}{8\pi c e^2 z^2}} \tag{6}$$

where c and z are the concentration and the valence of the counter ions, ε_0 and ε_r are the permittivity of vacuum and the dielectric constant of the dispersion fluid, and e is the unit of charge (1.6×10^{-19} C). A low concentration of counter-ions is favorable for forming a thick electrical double layer, which is desired for screening the interparticle van der Waals attraction and stabilizing the dispersion. The electrostatic stabilization is normally effective in polar fluid systems as the readily polarizable fluids have a high dielectric constant, which is favorable for increasing the thickness of the electrical double layer. Experimentally, zeta potential, which is defined as the electrical potential difference between the dispersion fluid and the bounded charge layer onto the dispersed particles, has been widely adopted as a key parameter to evaluate the dispersion stability [84]. Typically, the absolute value of the zeta potential needs to be above a threshold value in order to achieve stable dispersion [85]. Electrostatic stabilization is a popular strategy for improving dispersion of solar absorbers within polar thermal storage fluids, but the dispersion stability can be affected by the addition of other electrolytes or other fillers.

3.2.3. Steric Stabilization

Steric hindrance stabilization is another popular approach to stabilize the dispersion of colloidal particles [86,87]. It relies on surface modification of the particles with organic/inorganic ligands or polymer chains to prevent direct particle contact thereby weakening the van der Waals interparticle attraction (Figure 3c). A variety of surface modification techniques such as ligand exchange and polymer chain grafting have been developed to create the steric protection layer on the surfaces of various solar absorbers. These surface modification agents can be chemically bound or physically adsorbed onto the surface of solar absorbers. The general requirement of the surface modification agents is that they need to be compatible with the thermal storage fluids. One straightforward approach is that the surface modification agents are chemically same as the thermal storage fluids. When the modified particles are dispersed within the fluids, these decorated compatible ligands or polymer chains can be swollen by the thermal storage fluid molecules. The number density, the length distribution of the surface modification agents as well as their binding strength with solar absorbers are the key parameters affecting their dispersion stability. In comparison with hydrogen bonding stabilization and electrostatic stabilization, steric hindrance stabilization can be applied to both polar and nonpolar fluid systems, but it typically involves multi-step complex surface modification processes. When dispersing solar absorbers within polar thermal storage fluids such as ionic fluids and molten salts, steric stabilization and electrostatic stabilization strategies are often combined.

3.2.4. Self-Dispersion Stabilization

In recent years, self-dispersible nanofluids have gained intensive research attentions because the solar absorbers are designed and prepared in such a way that they can be uniformly dispersed within the thermal storage fluids without resorting to complicated post-synthesis surface modification procedures that are usually involved in other stabilization approaches [88]. As schemed by Figure 3d, this strategy makes use of the rough surface of the solar absorber to prevent large-area direction contact between neighboring particles and thus suppress the van der Waals interparticle attraction. Furthermore, the size distribution and the density of the solar absorbers are strictly controlled within a narrow range so that the Brownian motion can counterbalance the gravitational sedimentation. As a general approach, such strategy can be applied to realize homogeneous dispersion of different solar absorbers within both polar and nonpolar thermal storage fluids. The is favorite for promising scalable manufacturing of the solar-thermal nanofluids. To achieve stable dispersion at elevated temperatures, the key challenge of this dispersion strategy lies in controlled synthesis of solar absorbers that simultaneously possess appropriate surface roughness, a sufficiently small particle size, a narrow particle size distribution and a suitable density matching that of the thermal storage fluid. Evaporation-induced crumpling processes, surface etching techniques and many other approaches have been developed to prepare such self-dispersible solar absorbers and tailor their size, surface morphology and density.

4. Stably Dispersed Medium-Temperature Solar-Thermal Nanofluids

Depending on the nature of specific thermal storage fluids (ethylene glycol, oil, ionic fluid, and molten salt), various types of solar absorbers have been incorporated to prepare homogeneously dispersed medium-temperature solar-thermal nanofluids by utilizing one of the dispersion strategies or combined ones. Herein, we provide representative research examples showing how these strategies are implemented to improve the dispersion stability of the medium-temperature solar-thermal nanofluids.

4.1. Ethylene Glycol-Based Nanofluids

The hydrogen bonding between the ethylene glycol molecules and the surface functional groups on the surface of solar absorbers provides a facile way to prepare homogeneously dispersed solar-thermal nanofluids. Wang et al. [89] reported that while graphite nanoparticles were not dispersible within ethylene glycol, after acid treatment the graphite oxide nanoparticles (GONs) could retain their stable dispersion for 1 day. They further demonstrated that decorating GONs onto the surfaces of GO sheets could prolong the duration of stable dispersion to 4 weeks, but the dispersion stability of these nanofluids were tested at room temperature. At elevated temperatures, Shu et al. [90] found that while conventional water-wetted multi-layer GO sheets lost their dispersion stability within EG fluids, the ethanol-wetted single-layer GO sheets have demonstrated improved dispersion stability due to the richer oxygen-containing surface functional groups and the smaller mass of dispersed sheets. Through a one-step heating reduction, GO sheets were converted into solar-absorbing RGO, and these RGO sheets could maintain their uniform dispersion after continuous heating at 120 °C for at least 12 h. In a recent work, Lin et al. [91] synthesized hydroxyl-functionalized graphene quantum dots (OH-GQD) with a narrowly distributed size smaller than 10 nm by using a bottom-up method (Figure 4a). Compared with the large-sized GO sheets, the ultrasmall OH-GQDs simultaneously reduced the van der Waals interparticle attraction and mitigated gravitational sedimentation. The resultant EG nanofluids loaded with different concentrations of OH-GQDs could retain their stable dispersion after heating at 180 °C for 7 days and enabled direct harvesting of solar-thermal energy under concentrated solar illumination.

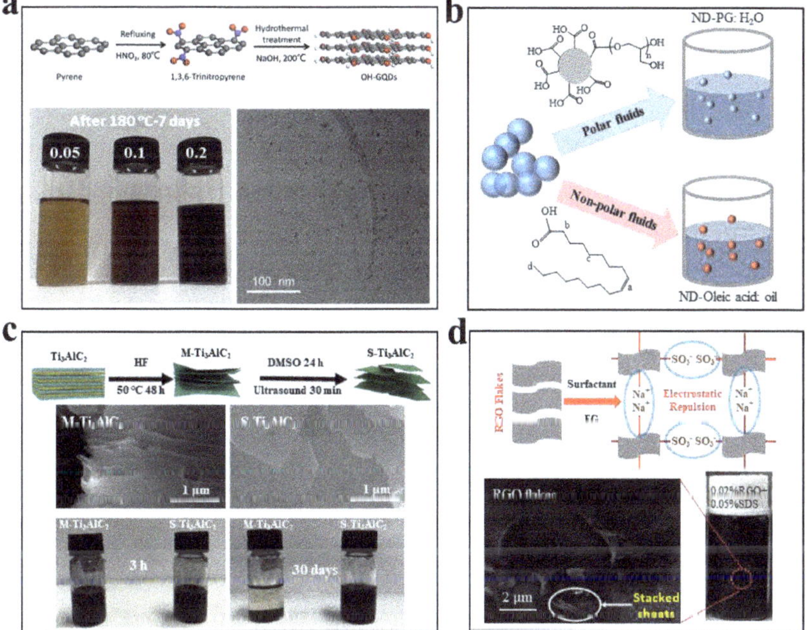

Figure 4. (**a**) The hydrogen bonding between the hydroxyl groups on the surface of GQDs and the ethylene glycol molecules enables homogenous dispersion of nanofluids with different concentrations (0.05 mg/mL, 0.1 mg/mL, 0.2 mg/mL) after heating at 180 °C for 7 days. ((**a**) reprinted/adapted with permission from Ref. [91]. 2019, Royal Society of Chemistry). (**b**) Covalent surface modification of deaggregated nanodiamond particles via formation of poly(glycidol) polymer brush chains to achieve stable dispersion within ethylene glycol. ((**b**) reprinted/adapted with permission from Ref. [92]. 2013, the American Chemical Society). (**c**) Electrostatically stabilized single-layer $Ti_3C_2T_x$ MXene homogeneously dispersed within ethylene glycol at room temperature for 30 days. ((**c**) reprinted/adapted with permission from Ref. [93]. 2021, Elsevier). (**d**) Uniform dispersion of RGO within ethylene glycol with the assistance of electrostatic stabilization by surfactants. ((**d**) reprinted/adapted with permission from Ref. [94]. 2020, Elsevier).

Surface modification of the solar absorbers with organic ligands or polymer chains that are compatible with ethylene glycol can also improve the dispersion stability. Branson et al. [92] prepared air-oxidized nanodiamond with an average particle size of ~8 nm in dimethylsulfoxide through bead-milling and ultrasonication, and then mixed the dispersion with glycidol monomer to graft poly(glycidol) polymer brushes on the surfaces of the nanodiamond. They demonstrated that the polymer brush chains were covalently bound to the surfaces of the nanodiamond. Static stable dispersion of the modified nanodiamonds within ethylene glycol fluids with a loading up to 1.9 vol% was achieved. Although detailed dispersion stability at elevated temperatures was not investigated, they reported that the compared with the pure thermal storage fluids the apparent thermal conductivity of the nanofluids could be effectively enhanced.

The polar ethylene glycol fluids also allow for dispersing solar absorbers through electrostatic stabilization. As shown by Figure 4c, Prof. Yu's team [93] compared the dispersion behavior of multilayer $Ti_3C_2T_x$ (M-$Ti_3C_2T_x$) and single-layer $Ti_3C_2T_x$ (S-$Ti_3C_2T_x$) MXene within ethylene glycol. They found that the S-$Ti_3C_2T_x$-EG nanofluids could retain their homogeneous dispersion at room temperature for 30 days and such dispersion stability was much better than that of the M-$Ti_3C_2T_x$ nanofluids. The enhanced dispersion stability was attributed to more functional groups exposed at the surfaces of the S-$Ti_3C_2T_x$ and their smaller mass. It was also found that the S-$Ti_3C_2T_x$ nanofluids had a larger zeta potential than the M-$Ti_3C_2T_x$ nanofluids under the same pH condition. When the pH value was higher than 7, the zeta potential of the S-$Ti_3C_2T_x$-EG nanofluids was larger than 30 mV, which is typically considered as a threshold value to achieve long-term electrical stabilization. Although the M-$Ti_3C_2T_x$ nanofluids had zeta potential values higher than 30 mV, the M-$Ti_3C_2T_x$ sheets were aggregated meaning that high zeta potential is not the single requirement for achieving stable dispersion. Similarly, Shah et al. [94] reported that introducing surfactants such as sodium dodecyl sulfate (SDS), sodium dodecylbenzene sulfonate (SDBS) and cetyltrimethylammonium bromide (CTAB) could effectively increase the zeta potential of the ethylene glycol nanofluids loaded with RGO particles, but the surface modified RGO still aggregated. Besides zeta potential, it was found that the initial dispersion state, the surface modification process, and the concentration of surfactants and fillers also had significance influence on the dispersion behavior and stability of the nanofluids.

4.2. Oil-Based Nanofluids

Compared with ethylene glycol fluids, silicone oil, mineral oil, and other synthetic medium-temperature solar-thermal oils, oil-based nanofluids have a wider range of working temperatures, but they are usually nonpolar. Therefore, hydrogen bonding and electrical stabilization strategies are typically not applicable, and the most frequently adopted approaches are steric stabilization and self-dispersion stabilization in these systems.

As shown in Figure 5a, Chen et al. [95] modified the surfaces of Fe_3O_4 nanoparticles through grafting polydimethylsiloxane (PDMS) chains, which are chemical compatible with the silicone oil thermal storage fluids. The control of the size of the synthesized particles and the robustness of surface binding were critical to achieve stable dispersion. To this end, they utilized benzyl alcohol as the synthetic ligand to control the size distribution of the synthesized Fe_3O_4 nanoparticles and further removed the large-sized ones through centrifugation. They synthesized phosphate-terminated PDMS chains as the surface modification agent, which can replace the synthetic ligands and the phosphate head could form robust bidentate binding with the Fe_3O_4 nanoparticles. The prepared nanofluids could retain uniform dispersion when they were exposed to direct concentrated solar illumination for harvesting of solar-thermal energy at temperatures higher than 110 °C. Dispersion stability tests indicated that the surface-modified Fe_3O_4-silicone oil nanofluids had the tendency to form agglomeration at high particle concentrations and under higher heating temperatures due to increased interparticle colliding probability and desorption of surface modification agents. Similar strategy has also been employed to modify the surface of Fe_3O_4@graphene hybrid particles for preparing homogeneously dispersed silicone oil solar-thermal nanofluids [96]. In comparison with Fe_3O_4 nanoparticles, the Fe_3O_4@graphene hybrid particles had shown improved solar absorptance and could maintain their stable dispersion within silicone oil under a heating temperature up to 150 °C.

Besides long polymer chains, densely grafted short ligands could also provide sufficient steric hindrance to suppress interparticle van der Waals attraction. Gomez-Villarejo et al. [97] adopted an in situ preparation approach and synthesized gold nanoparticles within Dowthermal oil (an eutectic mixture of biphenyl and diphenyl oxide) by using tetraoctylammonium bromide as the surfactant. They found that the surfactant could exchange with diphenyl oxide molecules to stabilize the dispersion of synthesized gold nanoparticles. Although the as-prepared nanofluids were homogeneous, aggregation and sedimentation occurred after heating the nanofluids at 300 °C and those with high concentrations had shown stronger tendency to form agglomerates. To achieve stable dispersion of GO within hydraulic oil, Li et al. [98] modified GO sheets with titanate coupling agents (Figure 5b). The isopropyl triisostearyl titanate ligands could react with hydroxyl groups on the surfaces of GO sheets and the exposed alkyl chains enabled uniform dispersion of modified sheets within hydraulic oil.

In contrast to complex surface modification, transforming planar GO sheets into crumpled RGO particles through an evaporation-induced aerosol compression process offers a facile way to prepare self-dispersible solar-thermal nanofluids [88]. However, the crumpled particles synthesized from the evaporation drying of water-wetted multi-layered GO sheets still agglomerated and precipitated out of thermal storage fluids at elevated temperatures, although they are dispersible at room temperature [88,99–101]. In a recent work, Zhang et al. [102] reported that aerosol drying of ethanol-wetted single-layer GO sheets dispersed within ethanol could generate crumpled RGO particles with more intensively deformed surfaces, a smaller particle size, a narrower particle size distribution, and a smaller mass (Figure 5c). The rough wrinkled surfaces prevent direct large-area contact between neighboring particles and thus suppressing van der Waals interparticle attraction. The small size and the small mass favor random Brownian motion of crumpled particles within the thermal storage fluids at elevated temperatures, which could counterbalance the gravitational sedimentation. These features enabled their uniform dispersion within silicone oil when the nanofluids were continuously heated at 200 °C for 14 days. The synthesized particles could be stably dispersed within both polar ethylene glycol and nonpolar oils such as silicone oil and Dowthermal A oil, and the stable operation temperature could reach 300 °C. After revealing such dispersion mechanism, this team further introduced mesopores into the crumpled RGO particles that were synthesized through evaporation drying of water-wetted GO (Figure 5d). They found that the etching of the conventional crumpled particles with hydrogen peroxide solution could lower their apparent density without affecting their original rough surface structure and narrow particle size distribution [103]. By changing the etching duration, the density of the mesoporous crumpled RGO particles could be tailored to match that of the silicone oil so that the Brownian motion could overcome the gravitational sedimentation. The introduced pores can also expose more surface area of the crumpled particles and expand their application field beyond solar-thermal conversion [104].

4.3. Ionic Liquid-Based Nanofluids

As a type of liquid salts at room temperature, ionic liquids are composed of organic or inorganic anions such as $CF_3SO_3{}^-$, $CF_3CO_2{}^-$, and halogen, and organic cations such as imidazolium, pyridinium, pyrazolium, and triazolium [105]. Although ionic liquids show liquid flowing capability and can withstand high operation temperatures, they typically have very low solar absorptance. To improve solar absorptance, various solar absorbers have been incorporated to prepare ionic liquid-based solar-thermal nanofluids with the assistance of steric hindrance stabilization and electrostatic stabilization [106].

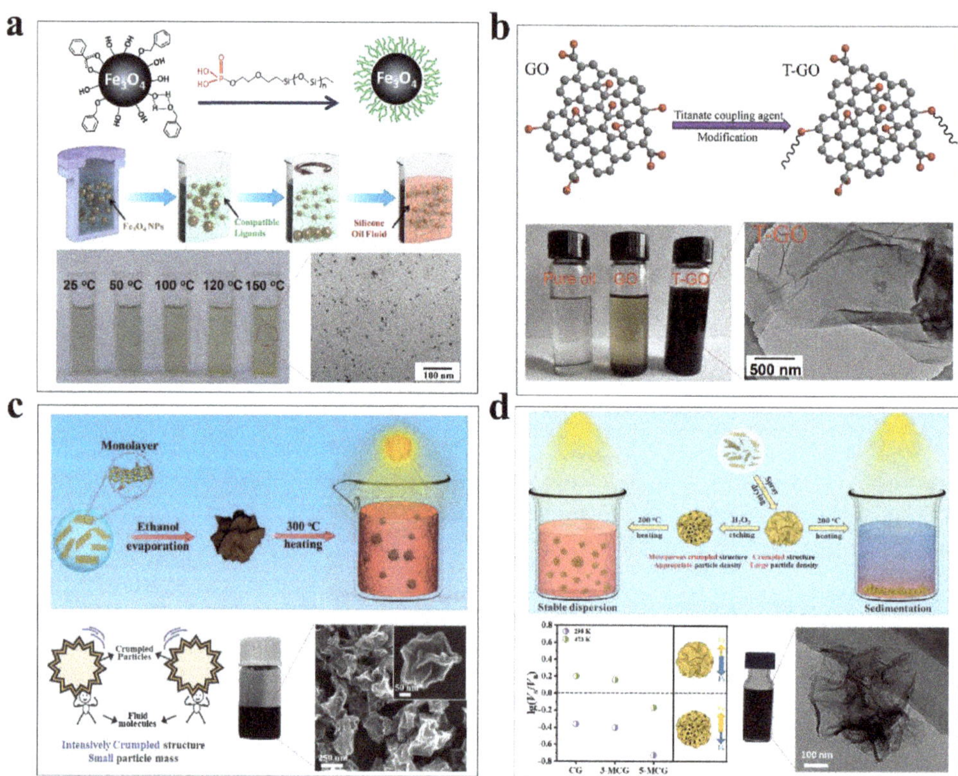

Figure 5. (**a**) Steric stabilization of Fe_3O_4 nanoparticles through surface grafting with phosphate-terminated PDMS ligands, which enable their stable uniform dispersion within silicone oil ((**a**) reprinted/adapted with permission from Ref. [95]. 2016, Royal Society of Chemistry). (**b**) Titanate coupling agent modified graphene oxide (T-GO) dispersion within hydraulic oil ((**b**) reprinted/adapted with permission from Ref. [98]. 2017, Elsevier). (**c**) Self-dispersible crumpled RGO particles stably dispersed within silicone oil due to weakened inter-particle van der Waals attraction and gravitational sedimentation ((**c**) reprinted/adapted with permission from Ref. [102]. 2022, Elsevier). (**d**) Mesoporous crumpled graphene particles as self-dispersible solar absorbers within silicone oil ((**d**) reprinted/adapted with permission from Ref. [103]. 2022, Elsevier).

Surface modification of solar absorbers to create steric hindrance is the most popular approach to achieve stable dispersion of ionic liquid nanofluids. Dash and Scott [107] reported that trace amount of 1-methylimidazole could serve as the stabilizer to homogenize the dispersion of as-synthesized gold nanoparticles when they reduced $HAuCl_4$ with $NaBH_4$ within imidazolium-based ionic liquids. Moreover, they found that the binding of 1-methylimidazole ligands helped improving the monodispersity of synthesized particles and such dispersion strategy was also applicable to bimetallic nanoparticles. As shown in Figure 6a, Gao et al. [108] covalently grafted 9-carbon chain fluorocarbon onto the surfaces of SiO_2 nanoparticles and achieved stable colloidal dispersion within 1-butyl-3-methylimidazolium tetrafluoroborate [C_4mim][BF_4]. They further analyzed that the grafted fluorocarbon layer had a thickness of ~1 nm, which could not provide sufficient steric hindrance to screen the interparticle attraction. It was revealed that it was the solvation layer surrounding the surfaces of the modified particles, which had a thickness of ~5 nm, that provided the sufficient steric stabilization.

To disperse graphene within ionic liquids, Zhang's group [109] firstly prepared carboxyl-terminated graphene (GE-COOH) and then grafted molecular chains that were similar to the 1-hexyl-3-methylimidazolium tetrafluoroborate ([HMIM]BF$_4$) by using the same synthetic reagents for synthesis of ionic liquids (Figure 6b). The modified graphene could retain their uniform dispersion when the nanofluids were subject to both continuous heating at 180 °C for one month and repeated heating/cooling from 30 °C to 180 °C for 500 cycles. By contrast, without surface grafting the GE-COOH demonstrated poor dispersion stability and precipitated out of the ionic liquids after the heating tests. The stable dispersion was attributed to the compatibility of grafted chains with ionic liquids and the high zeta potential of the modified graphene. They also reported that under the same concentration graphene had shown superior enhancement of the key thermophysical properties of the ionic liquids including thermal conductivity, solar absorptance, and photothermal conversion efficiency than graphite nanoparticles and single-wall CNTs [110].

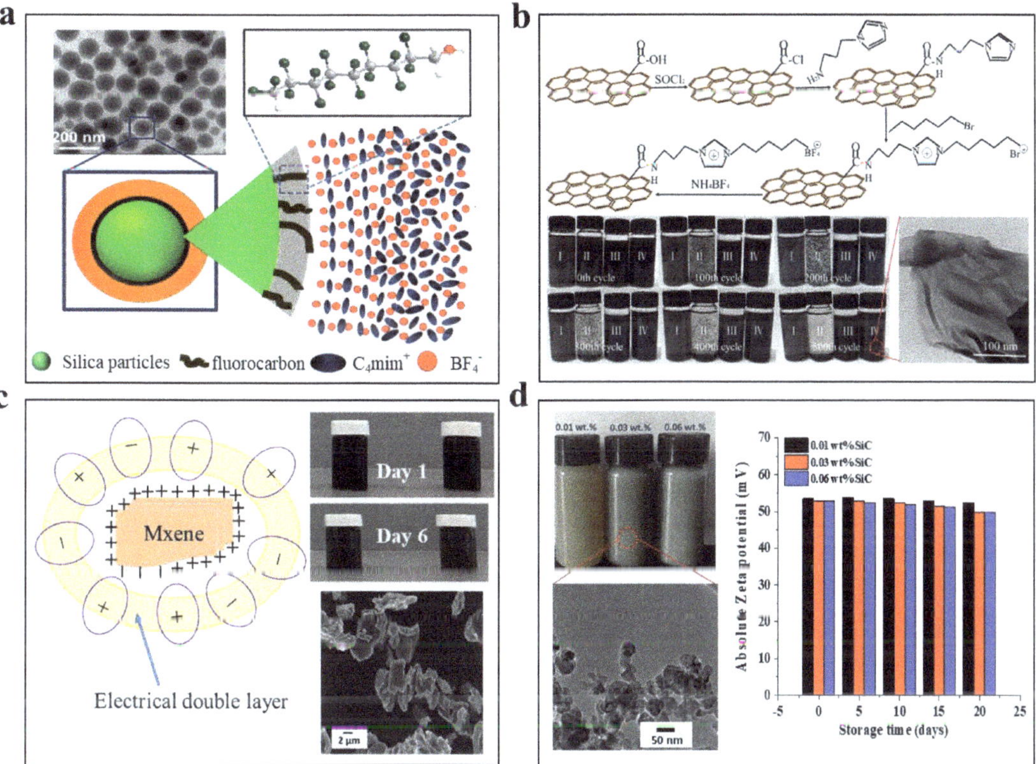

Figure 6. (**a**) Surface modification of SiO$_2$ nanoparticles with fluorocarbon brushes to create steric hinderance stabilization and enable their stable dispersion within [C$_4$mim]BF$_4$ ionic fluids ((**a**) reprinted/adapted with permission from Ref. [108]. 2015, the American Chemical Society). (**b**) Surface modification of graphene by grafting the molecular chains similar to [HMIM]BF$_4$ iconic fluids enable stable dispersion of the resultant nanofluids under medium temperatures and cycled heating/cooling conditions ((**b**) reprinted/adapted with permission from Ref. [109]. 2017, Elsevier). (**c**) Homogenous dispersion of MXene (Ti$_3$C$_2$) within [EMIM][OSO$_4$] and diethylene glycol ((**c**) reprinted/adapted with permission from Ref. [111]. 2020, Multidisciplinary Digital Publishing Institute). (**d**) Electrostatically stablized SiC nanoparticles within [HMIM]BF$_4$ ionic fluids ((**d**) reprinted/adapted with permission from Ref. [112]. 2017, Elsevier).

In parallel with surface modification with stabilizing ligands, introducing charges to the surfaces of solar absorbers can induce the formation of electrical double layer to stabilize their dispersion within ionic liquids. For example, Bakthavatchalam et al. [111] prepared multi-layer MXene through a chemical etching route and dispersed them within the mixture of diethylene glycol and 1-ethyl-3-methyl imidazolium octyl sulfate ([Emim][OSO$_4$]) ionic liquid. The resultant nanofluids had shown stable uniform dispersion during the first 4 days. In this case, the charged MXene surfaces attract [Emim] cations and [OSO$_4$] anions, which reorganized into an electrical double layer and prevented aggregation (Figure 6c). However, slight aggregation and sedimentation were observed at days 5, 6, and 7. In another work, Chen et al. [112] reported that SiC nanoparticles could be uniformly dispersed within [HMIM]BF$_4$ ionic liquids without observing obvious settlement for 20 days (Figure 6d). The absolute values of zeta potential of the prepared nanofluids were measured to be higher than 50 mV, which supported the electrostatic stabilization mechanism. It should be noted that previous theoretical and experimental studies indicated that the high charge density of ionic liquids strongly screens the electrostatic repulsion between neighboring particles and the thickness of electrical double layer is less than 1 nm, which is too thin to provide sufficient electrostatic stabilization [113,114]. Therefore, the dispersion stability mechanism of the electrostatically stabilized ionic nanofluids needs more detailed investigation. Moreover, all these reported dispersion tests were carried out at room temperature, and their dispersion behaviors at elevated temperatures need to be further investigated.

4.4. Molten Salt-Based Nanofluids

As a cousin of organic ionic liquids, inorganic molten salts not only possess high charge density but also require higher heating temperatures to keep them in the liquid state, which together pose even grand challenges in achieving stable dispersion of functional nanofillers. Within melted molten salts, the resultant Debye screening length is only ~0.1 nm, which limits the application of classical electrostatic stabilization strategy. The high operation temperature restricts the employment of steric stabilization through surface modification organic ligands that might degrade over time.

In recent years, Talapin's [115,116] group reported a strategy to disperse functional nanofillers including metals (Pt, Pd), semiconductors (CdSe/CdZnS quantum dots), rare-earth compounds (NaYF$_4$:YbEr/CaF$_2$) and magnetic particles (Fe$_3$O$_4$) in molten salts. The chemical affinity between the particle surfaces and the ions in the molten salts was identified as the key enabler to achieve stable uniform dispersion. For example, Pt and Pd nanoparticles could be homogeneously dispersed within the mixed AlCl$_3$/NaCl/KCl molten salts when there was excessive AlCl$_3$ because the electron-rich transitional metals (Pt, Pd) could form chemical bonds with AlCl$_3$ (Figure 7a). Without excessive AlCl$_3$ in the molten salts, severe agglomeration of Pt particles was observed. Similarly, CdSe nanoparticles could be dispersed within the Lewis-basic halide salts such as AlCl$_3$/NaCl/KCl and LiCl/LiI/KI, and the pseudohalide salts such as NaSCN/KSCN as well because the halide and SCN$^-$ ions could efficiently bind with Cd^{2+} at the external surfaces of CdSe particles. Molecular dynamic simulation indicated that Cl$^-$ ions formed a dense epitaxial layer and the co-ions were organized into a structured ion layer. The repulsive interaction between the ion layer is much stronger than both the van der Waals and double-layer electrostatic interaction, proving its dominant contribution to stable dispersion. Although this pioneering work provided a feasible approach to disperse various functional nanoparticles within molten salts, it is worth mentioning that the demonstrated salts typically have a relatively low melting temperature and the long-term stability of added particles at elevated temperatures has not been evaluated.

In solar-thermal storage research directions, intensive efforts were mainly devoted to incorporating SiO$_2$, Al$_2$O$_3$, MgO, and other nanoparticles into commercial solar-thermal molten salts that have been employed for thermal storage at medium-to-high temperatures [117,118]. For example, Nithiyanantham et al. [119] found that while pure Al$_2$O$_3$ and

SiO$_2$ particles precipitated or floated within solar salts (NaNO$_3$-KNO$_3$) after heating the nanofluids at 400 °C for 24 h (Figure 7b), the core-shell SiO$_2$@Al$_2$O$_3$ could create steric hindrance and obtain an intermediate density matching that of the molten salts thereby reducing agglomeration and sedimentation tendency. They utilized Al$_2$O$_3$ particles with a diameter of 12 nm as the core, and tailored the SiO$_2$ shell thickness through changing the concentration of silica precursor. It was found that the chain-like slightly aggregated SiO$_2$@Al$_2$O$_3$ could effectively enhance thermal conductivity of the nanofluids, but in the meanwhile caused higher viscosity. In a subsequent work [120], this team reported that the dispersion of SiO$_2$ particles within melted solar salts also depended on the particle size (27 nm, 450 nm, 800 nm) and larger-sized SiO$_2$ particles (800 nm) have demonstrated better dispersion (Figure 7c). They found that the dispersion stability and enhancement of thermophysical properties of the nanofluids (specific heat capacity, thermal conductivity) have opposite dependence on the size of SiO$_2$ particles.

It is worth pointing out that so far most of the reported works mainly focused on improving heat capacity and thermal conductivity of the molten salts, and the prepared nanofluids typically have low solar absorptance. Additionally, the dispersion behavior and governing mechanisms of various particles within the molten salts over a long period of operation time are awaiting further exploration. It remains a grand challenge to prepare stably dispersed molten salt nanofluids with boosted thermophysical properties for direct solar-thermal energy harvesting at medium-to-high temperatures.

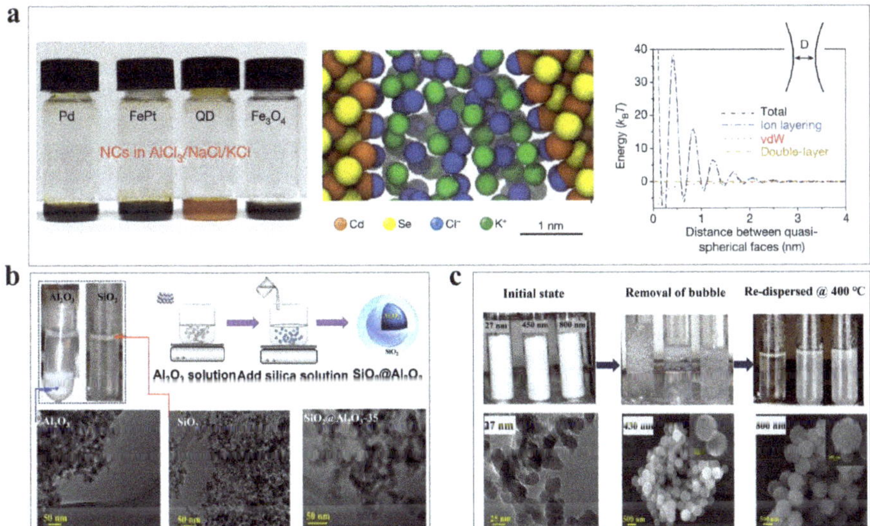

Figure 7. (**a**) Homogeneous dispersion of colloidal nanoparticles within molten salts due to chemical affinity between the nanoparticle surfaces and the molten salt ions ((**a**) reprinted/adapted with permission from Ref. [115], 2017, Springer Nature). (**b**) Core-shell SiO$_2$@Al$_2$O$_3$ nanoparticles dispersed within melted solar salts due to reduced sedimentation tendency through matching the density of particles and molten salts. ((**b**) reprinted/adapted with permission from Ref. [119], 2019, Elsevier). (**c**) Dependence of nanofluid dispersion stability on the size of SiO$_2$ particles ((**c**) reprinted/adapted with permission from Ref. [120], 2022, Elsevier).

5. Summary and Outlook

In summary, dispersion stability has been viewed as one of the long-lasting intractable issues limiting the development and practical application of nanofluid technologies. Recent years have witnessed substantial progress in understanding the dispersion behavior and developing effective strategies to prepare medium-temperature solar-thermal nanofluids homogeneously dispersed with a variety of functional nanofillers. Key thermophysical properties including solar absorptance, thermal conductivity, and specific heat capacity have shown significant enhancement. It is anticipated that these dispersed nanofluids hold the promise for expanding the usage of DASCs from low-temperature into medium-temperature solar-thermal systems with improved energy harvesting efficiency. From a broader perspective, the aggregation-resistant solar-thermal nanofluids will also enable other wide-spread thermal-related applications such as heat transfer in renewable energy systems, and thermal management of electronic devices [121].

Although the benefits for improving volumetric direct solar-thermal conversion have been widely demonstrated by the uniformly dispersed nanofluids, the influence of added solar absorbers on other important thermophysical properties such as viscosity and effective thermal conductivity has not been fully revealed [122]. Typically, the viscosity of nanofluids increases with added fillers and is affected by their loading, size, shape, dispersion state, and surface chemistry [123]. In comparison with bare fillers, the surface-modified particles could generally minimize the increasing of viscosity because these surface-capping agents improve their chemical compatibility with the thermal storage fluids. Another fact is that the viscosity of the nanofluids quickly drops with increasing application temperature. Such temperature dependence implies that the medium-temperature solar-thermal nanofluids, which have a low loading of dispersed solar absorbers, should have largely the same viscosity as the neat thermal storage fluids at elevated temperatures, and the effective viscosity can be roughly estimated by the classical Einstein model. Regarding the effective thermal conductivity of the nanofluids, most frequently the added particles have a higher thermal conductivity than the thermal storage fluids and thus can help improve the thermal conductivity and heat transfer performance. For uniform dispersion at low loading levels, the effective thermal conductivity can be estimated by the effective medium theory or many other classical models [124], which generally predict that the effective thermal conductivity is positively dependent on the volumetric fraction of the added fillers. It was often observed that increasing temperature is favorable for intensifying interparticle collision and local convection heat transfer, and thus is helpful for improving the effective heat conductivity [125]. In general, self-dispersible nanofluids should possess better heat transfer performance than the surface-modified nanofluids as the surface-capping agents typically have a low thermal conductivity and limit interparticle collision heat transfer. It should be noted that measurements of both the viscosity and the thermal conductivity of solar-thermal nanofluids currently most often were only performed at low temperatures. It is highly desired that we can take advantage of recent progresses in preparing stably dispersed medium-temperature solar-thermal nanofluids to carry out systematic characterization of the thermophysical properties at elevated temperatures and reveal the influence of added fillers on the properties of the nanofluids.

To fulfill the potential of medium-temperature nanofluid technologies, continuous research efforts from different disciplines are needed to comprehensively reveal the long-term dispersion behavior of nanofluids at elevated temperatures, establish the relationship between dispersion state and enhancement of thermophysical properties, and systematically evaluate the impact of these nanofluids on the improved performances of solar-thermal systems. To this end, different dispersion characterization techniques such as dynamic light scattering [126] and in situ transmission electronic microscopy [127,128] operating at high temperatures should be developed to facilitate the monitoring the dynamic dispersion and evolution of nanofillers within the medium-temperature nanofluids at different dimensional scales. Computational modeling tools such as molecular dynamic simulations have demonstrated unique advantages in evaluating particle-fluid and interparticle interaction at molecular and atomic levels [115]. These techniques should be combined with experimental investigation to fully elucidate the dispersion mechanisms. Another important future research direction would be developing advanced setups or tools to experimentally characterize the thermophysical properties of solar-thermal nanofluids at medium-to-high temperatures. To enable practical applications in current solar-thermal industries, on-going research efforts should be focused on the development of self-dispersible synthetic oil and molten salt-based nanofluids loaded with low concentrations of solar absorbers. Manufacturability of solar absorbers at large scale, low cost through one-step process should also be considered [129].

6. Conclusions

This work overviews recent progress in preparing medium-temperature ethylene glycol, oil, ionic liquid, and molten salt-based solar-thermal nanofluids that are uniformly dispersed with various types of solar absorber. Four categories of stabilization strategies including hydrogen bonding stabilization, electrostatic stabilization, steric stabilization, and self-dispersive stabilization have shown the capability to improve the dispersion of solar absorbers within the thermal storage fluids. By providing representative examples, the applicable dispersion strategies for different thermal storage fluidic systems have been identified. Among them, hydrogen bonding stabilization has appeared to be a facile effective way to prepare uniformly dispersed ethylene glycol-based medium-temperature nanofluids. Although the surfaces of solar absorbers could be modified with organic ligands or polymeric chains to improve their dispersion within medium-temperature oils, self-dispersion strategy has exhibited unique advantages in simple preparation processes and long-term dispersion stability. Electrostatic stabilization and self-dispersion stabilization are promising for preparing uniformly dispersed ionic liquid and molten salt-based solar-thermal nanofluids. In comparison, recently emerged self-dispersible nanofluids have demonstrated superior dispersion stability and possess unique promises to improve practical solar-thermal energy harvesting performances without sacrificing the original heat capacity and low viscosity features of the thermal storage fluids. It thus merits further development of diverse self-dispersion strategies, controllable fabrication of self-dispersible solar absorbers, and carrying out systematic investigation of their thermophysical properties and full evaluation of their direct absorption solar-thermal harvesting performances. It is hoped that the reviewed research progress in medium-temperature nanofluids and these identified research needs provide many exciting opportunities for cross-discipline collaboration to advance further development of nanofluid technologies.

Author Contributions: Conceptualization, T.H., J.Z. and P.T.; methodology, T.H., J.Z. and P.T.; validation, T.H. and P.T.; formal analysis, T.H., J.X. and P.T.; investigation, T.H. and J.Z.; resources, T.H., J.Z., J.X. and X.L.; data curation, T.H., J.Z., J.X. and X.L.; writing—original draft preparation, T.H., J.Z., J.X. and X.L; writing—review and editing, T.H., P.T. and T.D.; visualization, T.H. and P.T.; supervision, P.T. and T.D., project administration, P.T. and T.D., funding acquisition, P.T. and T.D. All authors have read and agreed to the published version of the manuscript.

Funding: This research was funded by National Natural Science Foundation of China (Grant No: 51873105), Innovation Program of Shanghai Municipal Education Commission (Grant No: 2019-01-07-00-02-E00069) and Shanghai Rising-Star Program (Grant No: 18QA1402200).

Data Availability Statement: Not applicable.

Conflicts of Interest: The authors declare no conflict of interest.

References

1. Dincer, I. Renewable energy and sustainable development: A crucial review. *Renew. Sustain. Energy Rev.* **2000**, *4*, 157–175. [CrossRef]
2. Evans, A.; Strezov, V.; Evans, T.J. Assessment of sustainability indicators for renewable energy technologies. *Renew. Sustain. Energy Rev.* **2009**, *13*, 1082–1088. [CrossRef]
3. Chu, S.; Cui, Y.; Liu, N. The path towards sustainable energy. *Nat. Mater.* **2017**, *16*, 16–22. [CrossRef] [PubMed]
4. Lewis, N.S. Research opportunities to advance solar energy utilization. *Science* **2016**, *351*, add1920. [CrossRef]
5. Thirugnanasambandam, M.; Iniyan, S.; Goic, R. A review of solar thermal technologies. *Renew. Sustain. Energy Rev.* **2010**, *14*, 312–322. [CrossRef]
6. Wang, R.Z.; Zhai, X.Q. Development of solar thermal technologies in China. *Energy* **2010**, *35*, 4407–4416. [CrossRef]

7. Kalkan, N.; Young, E.A.; Celiktas, A. Solar thermal air conditioning technology reducing the footprint of solar thermal air conditioning. *Renew. Sustain. Energy Rev.* **2012**, *16*, 6352–6383. [CrossRef]

8. Tao, P.; Ni, G.; Song, C.; Shang, W.; Wu, J.; Zhu, J.; Chen, G.; Deng, T. Solar-driven interfacial evaporation. *Nat. Energy* **2018**, *3*, 1031–1041. [CrossRef]

9. Wang, Z.; Horseman, T.; Straub, A.P.; Yip, N.Y.; Li, D.; Elimelech, M.; Lin, S. Pathways and challenges for efficient solar-thermal desalination. *Sci. Adv.* **2019**, *5*, eaax0763. [CrossRef]

10. Weinstein, L.A.; Loomis, J.; Bhatia, B.; Bierman, D.M.; Wang, E.N.; Chen, G. Concentrating solar power. *Chem. Rev.* **2015**, *115*, 12797–12838. [CrossRef]

11. Agrafiotis, C.; Roeb, M.; Sattler, C. A review on solar thermal syngas production via redox pair-based water/carbon dioxide splitting thermochemical cycles. *Renew. Sustain. Energy Rev.* **2015**, *42*, 254–285. [CrossRef]

12. Ma, J.; Jiang, B.; Li, L.; Yu, K.; Zhang, Q.; Lv, Z.; Tang, D. A high temperature tubular reactor with hybrid concentrated solar and electric heat supply for steam methane reforming. *Chem. Eng. J.* **2022**, *428*, 132073. [CrossRef]

13. Schäppi, R.; Rutz, D.; Dähler, F.; Muroyama, A.; Haueter, P.; Lilliestam, J.; Patt, A.; Furler, P.; Steinfeld, A. Drop-in fuels from sunlight and air. *Nature* **2022**, *601*, 63–68. [CrossRef] [PubMed]

14. Gur, I.; Sawyer, K.; Prasher, R. Searching for a better thermal battery. *Science* **2012**, *335*, 1454–1455. [CrossRef]

15. Tian, Y.; Zhao, C.Y. A review of solar collectors and thermal energy storage in solar thermal applications. *Appl. Energy* **2013**, *104*, 538–553. [CrossRef]

16. Suresh, C.; Saini, R.P. Review on solar thermal energy storage technologies and their geometrical configurations. *Int. J. Energy Res.* **2020**, *44*, 4163–4195. [CrossRef]

17. Tao, P.; Chang, C.; Tong, Z.; Bao, H.; Song, C.; Wu, J.; Shang, W.; Deng, T. Magnetically-accelerated large-capacity solar-thermal energy storage within high-temperature phase-change materials. *Energy Environ. Sci.* **2019**, *12*, 1613–1621. [CrossRef]

18. Kalogirou, S.A. Solar thermal collectors and applications. *Prog. Energy Combust. Sci.* **2004**, *30*, 231–295. [CrossRef]

19. Minardi, J.E.; Chuang, H.N. Performance of a "black" liquid flat-plate solar collector. *Sol. Energy* **1975**, *17*, 179–183. [CrossRef]

20. Farahat, S.; Sarhaddi, F.; Ajam, H. Exergetic optimization of flat plate solar collectors. *Renew. Energy* **2009**, *34*, 1169–1174. [CrossRef]

21. Otanicar, T.P.; Phelan, P.E.; Prasher, R.S.; Rosengarten, G.; Taylor, R.A. Nanofluid-based direct absorption solar collector. *J. Renew. Sustain. Energy* **2010**, *2*, 033102. [CrossRef]

22. Taylor, R.A.; Phelan, P.E.; Otanicar, T.P.; Walker, C.A.; Nguyen, M.; Trimble, S.; Prasher, R. Applicability of nanofluids in high flux solar collectors. *J. Renew. Sustain. Energy* **2011**, *3*, 023104. [CrossRef]

23. Lenert, A.; Wang, E.N. Optimization of nanofluid volumetric receivers for solar thermal energy conversion. *Sol. Energy* **2012**, *86*, 253–265. [CrossRef]

24. Joseph, A.; Thomas, S. Energy, exergy and corrosion analysis of direct absorption solar collector employed with ultra-high stable carbon quantum dot nanofluid. *Renew. Energy* **2022**, *181*, 725–737. [CrossRef]

25. Hussain, M.; Shah, S.K.H.; Sajjad, U.; Abbas, N.; Ali, A. Recent developments in optical and thermal performance of direct absorption solar collectors. *Energies* **2022**, *15*, 7101. [CrossRef]

26. Bretado-de los Rios, M.S.; Rivera-Solorio, C.I.; Nigam, K.D.P. An overview of sustainability of heat exchangers and solar thermal applications with nanofluids: A review. *Renew. Sustain. Energy Rev.* **2021**, *142*, 110855. [CrossRef]

27. Goel, N.; Taylor, R.A.; Otanicar, T. A review of nanofluid-based direct absorption solar collectors: Design considerations and experiments with hybrid PV/thermal and direct steam generation collectors. *Renew. Energy* **2020**, *145*, 903–913. [CrossRef]

28. Gorji, T.B.; Ranjbar, A.A. A review on optical properties and application of nanofluids in direct absorption solar collectors (DASCs). *Renew. Sustain. Energy Rev.* **2017**, *72*, 10–32. [CrossRef]

29. Kumar, S.; Chander, N.; Gupta, V.K.; Kukreja, R. Progress, challenges and future prospects of plasmonic nanofluid based direct absorption solar collectors–a state-of-the-art review. *Sol. Energy* **2021**, *227*, 365–425. [CrossRef]

30. Sun, C.; Zou, Y.; Qin, C.; Chen, M.; Li, X.; Zhang, B.; Wu, X. Solar absorption characteristics of SiO_2@ Au core-shell composite nanorods for the direct absorption solar collector. *Renew. Energ.* **2022**, *189*, 402–411. [CrossRef]

31. Li, X.; Zeng, G.; Lei, X. The stability, optical properties and solar-thermal conversion performance of SiC-MWCNTs hybrid nanofluids for the direct absorption solar collector (DASC) application. *Sol. Energy Mater. Sol. Cells* **2020**, *206*, 110323. [CrossRef]

32. Sreekumar, S.; Shah, N.; Mondol, J.D.; Hewitt, N.; Chakrabarti, S. Broadband absorbing mono, blended and hybrid nanofluids for direct absorption solar collector: A comprehensive review. *Nano Futures* **2022**, *6*, 022002. [CrossRef]

33. Tyagi, H.; Phelan, P.; Prasher, R. Predicted efficiency of a low-temperature nanofluid-based direct absorption solar collector. *J. Sol. Energy Eng.* **2009**, *131*, 041004. [CrossRef]

34. Wole-osho, I.; Okonkwo, E.C.; Abbasoglu, S.; Kavaz, D. Nanofluids in solar thermal collectors: Review and limitations. *Int. J. Thermophys.* **2020**, *41*, 157. [CrossRef]

35. Fu, B.; Zhang, J.; Chen, H.; Guo, H.; Song, C.; Shang, W.; Tao, P.; Deng, T. Optical nanofluids for direct absorption-based solar-thermal energy harvesting at medium-to-high temperatures. *Curr. Opin. Chem. Eng.* **2019**, *25*, 51–56. [CrossRef]

36. Said, Z.; Hachicha, A.A.; Aberoumand, S.; Yousef, B.A.A.; Sayed, E.T.; Bellos, E. Recent advances on nanofluids for low to medium temperature solar collectors: Energy, exergy, economic analysis and environmental impact. *Prog. Energy Combust. Sci.* **2021**, *84*, 100898. [CrossRef]

37. Taylor, R.A.; Phelan, P.E.; Otanicar, T.P.; Adrian, R.; Prasher, R. Nanofluid optical property characterization: Towards efficient direct absorption solar collectors. *Nanoscale Res. Lett.* **2011**, *6*, 225. [CrossRef]
38. Yue, H.; Zhao, Y.; Ma, X.; Gong, J. Ethylene glycol: Properties, synthesis, and applications. *Chem. Soc. Rev.* **2012**, *41*, 4218–4244. [CrossRef]
39. Ye, L.; Zhao, L.; Zhang, L.; Qi, F. Theoretical studies on the unimolecular decomposition of ethylene glycol. *J. Phys. Chem. A* **2012**, *116*, 55–63. [CrossRef]
40. Bhalla, V.; Beejawat, S.; Doshi, J.; Khullar, V.; Singh, H.; Tyagi, H. Silicone oil envelope for enhancing the performance of nanofluid-based direct absorption solar collectors. *Renew. Energy* **2020**, *145*, 2733–2740. [CrossRef]
41. Jin, H.; Andritsch, T.; Tsekmes, I.A.; Kochetov, R.; Morshuis, P.H.F.; Smit, J.J. Properties of mineral oil based silica nanofluids. *IEEE Trans. Dielectr. Electr. Insul.* **2014**, *21*, 1100–1108.
42. Gulzar, O.; Qayoum, A.; Gupta, R. Photo-thermal characteristics of hybrid nanofluids based on therminol-55 oil for concentrating solar collectors. *Appl. Nanosci.* **2019**, *9*, 1133–1143. [CrossRef]
43. Liu, J.; Wang, F.; Zhang, L.; Fang, X.; Zhang, Z. Thermodynamic properties and thermal stability of ionic liquid-based nanofluids containing graphene as advanced heat transfer fluids for medium-to-high-temperature applications. *Renew. Energy* **2014**, *63*, 519–523. [CrossRef]
44. Nieto de Castro, C.A.; Lourenço, M.J.V.; Ribeiro, A.P.C.; Langa, E.; Vieira, S.I.C.; Goodrich, P.; Hardacre, C. Thermal properties of ionic liquids and ionanofluids of imidazolium and pyrrolidinium liquids. *J. Chem. Eng. Data* **2010**, *55*, 653–661. [CrossRef]
45. Shiflett, M.B. *Commercial Applications of Ionic Liquids*; Springer: Cham, Switzerland, 2020.
46. Bauer, T.; Pfleger, N.; Laing, D.; Steinmann, W.-D.; Eck, M.; Kaesche, S. *High-Temperature Molten Salts for Solar Power Application, Molten Salts Chemistry*; Elsevier: Oxford, UK, 2013; pp. 415–438.
47. Guillot, S.; Faik, A.; Rakhmatullin, A.; Lambert, J.; Veron, E.; Echegut, P.; Bessada, C.; Calvet, N.; Py, X. Corrosion effects between molten salts and thermal storage material for concentrated solar power plants. *Appl. Energy* **2012**, *94*, 174–181. [CrossRef]
48. Bell, S.; Steinberg, T.; Will, G. Corrosion mechanisms in molten salt thermal energy storage for concentrating solar power. *Renew. Sustain. Energy Rev.* **2019**, *114*, 109328. [CrossRef]
49. Liu, B.; Wei, X.; Wang, W.; Lu, J.; Ding, J. Corrosion behavior of Ni-based alloys in molten $NaCl$-$CaCl_2$-$MgCl_2$ eutectic salt for concentrating solar power. *Sol. Energy Mater. Sol. Cells* **2017**, *170*, 77–86. [CrossRef]
50. Kearney, D.; Kelly, B.; Herrmann, U.; Cable, R.; Pacheco, J.; Mahoney, R.; Price, H.; Blake, D.; Nava, P.; Potrovitza, N. Engineering aspects of a molten salt heat transfer fluid in a trough solar field. *Energy* **2004**, *29*, 861–870. [CrossRef]
51. Peng, Q.; Ding, J.; Wei, X.; Yang, J.; Yang, X. The preparation and properties of multi-component molten salts. *Appl. Energy* **2010**, *87*, 2812–2817. [CrossRef]
52. Trong Tam, N.; Viet Phuong, N.; Hong Khoi, P.; Ngoc Minh, P.; Afrand, M.; Van Trinh, P.; Hung Thang, B.; Żyła, G.; Estellé, P. Carbon nanomaterial-based nanofluids for direct thermal solar absorption. *Nanomaterials* **2020**, *10*, 1199. [CrossRef]
53. Said, Z.; Arora, S.; Farooq, S.; Sundar, L.S.; Li, C.; Allouhi, A. Recent advances on improved optical, thermal, and radiative characteristics of plasmonic nanofluids: Academic insights and perspectives. *Sol. Energy Mater. Sol. Cells* **2022**, *236*, 111504. [CrossRef]
54. Kazemian, A.; Salari, A.; Ma, T.; Lu, H. Application of hybrid nanofluids in a novel combined photovoltaic/thermal and solar collector system. *Sol. Energy* **2022**, *239*, 102–116. [CrossRef]
55. Ladjevardi, S.M.; Asnaghi, A.; Izadkhast, P.S.; Kashani, A.H. Applicability of graphite nanofluids in direct solar energy absorption. *Sol. Energy* **2013**, *94*, 327–334. [CrossRef]
56. Chen, L.; Liu, J.; Fang, X.; Zhang, Z. Reduced graphene oxide dispersed nanofluids with improved photo-thermal conversion performance for direct absorption solar collectors. *Sol. Energy Mater. Sol. Cells* **2017**, *163*, 125–133. [CrossRef]
57. Sani, E.; Vallejo, J.P.; Cabaleiro, D.; Lugo, L. Functionalized graphene nanoplatelet-nanofluids for solar thermal collectors. *Sol. Energy Mater. Sol. Cells* **2018**, *185*, 205–209. [CrossRef]
58. Hordy, N.; Rabilloud, D.; Meunier, J.L.; Coulombe, S. High temperature and long-term stability of carbon nanotube nanofluids for direct absorption solar thermal collectors. *Sol. Energy* **2014**, *105*, 82–90. [CrossRef]
59. Mallah, A.R.; Mohd Zubir, M.N.; Alawi, O.A.; Salim Newaz, K.M.; Mohamad Badry, A.B. Plasmonic nanofluids for high photothermal conversion efficiency in direct absorption solar collectors. Fundamentals and applications. *Sol. Energy Mater. Sol. Cells* **2019**, *201*, 110084. [CrossRef]
60. Jain, P.K.; Huang, X.; El-Sayed, I.H.; El-Sayed, M.A. Noble metals on the nanoscale: Optical and photothermal properties and some applications in imaging, sensing, biology, and medicine. *Acc. Chem. Res.* **2008**, *41*, 1578–1586. [CrossRef]
61. Brongersma, M.L.; Halas, N.J.; Nordlander, P. Plasmon-induced hot carrier science and technology. *Nat. Nanotechnol.* **2015**, *10*, 25–34. [CrossRef]
62. Liu, Y.; Zhao, J.; Zhang, S.; Li, D.; Zhang, X.; Zhao, Q.; Xing, B. Advances and challenges of broadband solar absorbers for efficient solar steam generation. *Environ. Sci. Nano* **2022**, *9*, 2264–2296. [CrossRef]
63. Wang, L.; Zhu, G.; Wang, M.; Yu, W.; Zeng, J.; Yu, X.; Xie, H.; Li, Q. Dual plasmonic Au/TiN nanofluids for efficient solar photothermal conversion. *Sol. Energy* **2019**, *184*, 240–248. [CrossRef]
64. Shah, T.R.; Ali, H.M. Applications of hybrid nanofluids in solar energy, practical limitations and challenges: A critical review. *Sol. Energy* **2019**, *183*, 173–203. [CrossRef]

65. Xiong, Q.; Altnji, S.; Tayebi, T.; Izadi, M.; Hajjar, A.; Sundén, B.; Li, L.K.B. A comprehensive review on the application of hybrid nanofluids in solar energy collectors. *Sustain. Energy Technol. Assess.* **2021**, *47*, 101341. [CrossRef]

66. Yu, F.; Chen, Y.; Liang, X.; Xu, J.; Lee, C.; Liang, Q.; Tao, P.; Deng, T. Dispersion stability of thermal nanofluids. *Prog. Nat. Sci. Mater. Int.* **2017**, *27*, 531–542. [CrossRef]

67. Hamaker, H.C. The london—van der waals attraction between spherical particles. *Physica* **1937**, *4*, 1058–1072. [CrossRef]

68. Jacob, N.I.; Israelachvili, N. *Intermolecular and Surface Forces*; Academic Press: San Diego, CA, USA, 1992.

69. Cates, M.E.; Evans, M.R. *Soft and Fragile Matter: Nonequilibrium Dynamics, Metastability and Flow*; Institute of Physics Publishing: Bristol, UK, 2000.

70. Bond, W.N. LXXXII. Bubbles and drops and Stokes' law. *Lond. Edinb. Dublin Philos. Mag. J. Sci.* **1927**, *4*, 889–898. [CrossRef]

71. Loudet, J.C.; Hanusse, P.; Poulin, P. Stokes drag on a aphere in a nematic liquid crystal. *Science* **2004**, *306*, 1525. [CrossRef]

72. Rings, D.; Schachoff, R.; Selmke, M.; Cichos, F.; Kroy, K. Hot brownian motion. *Phys. Rev. Lett.* **2010**, *105*, 090604. [CrossRef]

73. Pusey, P.N. Brownian motion goes ballistic. *Science* **2011**, *332*, 802–803. [CrossRef]

74. Dutta, A.; Paul, A.; Chattopadhyay, A. The effect of temperature on the aggregation kinetics of partially bare gold nanoparticles. *RSC Adv.* **2016**, *6*, 82138–82149. [CrossRef]

75. Wang, Z.; Tao, P.; Liu, Y.; Xu, H.; Ye, Q.; Hu, H.; Song, C.; Chen, Z.; Shang, W.; Deng, T. Rapid charging of thermal energy storage materials through plasmonic heating. *Sci. Rep.* **2014**, *4*, 6246. [CrossRef] [PubMed]

76. Meli, L.; Green, P.F. Aggregation and coarsening of ligand-stabilized gold nanoparticles in poly (methyl methacrylate) thin films. *ACS Nano* **2008**, *2*, 1305–1312. [CrossRef] [PubMed]

77. Ji, X.; Copenhaver, D.; Sichmeller, C.; Peng, X. Ligand bonding and dynamics on colloidal nanocrystals at room temperature: The case of alkylamines on CdSe nanocrystals. *J. Am. Chem. Soc.* **2008**, *130*, 5726–5735. [CrossRef] [PubMed]

78. Yu, W.; Xie, H.; Li, Y.; Chen, L.; Wang, Q. Experimental investigation on the thermal transport properties of ethylene glycol based nanofluids containing low volume concentration diamond nanoparticles. *Colloids Surf. A Physicochem. Eng. Asp.* **2011**, *380*, 1–5. [CrossRef]

79. Palabiyik, I.; Musina, Z.; Witharana, S.; Ding, Y. Dispersion stability and thermal conductivity of propylene glycol-based nanofluids. *J. Nanoparticle Res.* **2011**, *13*, 5049. [CrossRef]

80. Brinker, C.J.; Scherer, G.W. *Sol-Gel Science: The Physics and Chemistry of Sol-Gel Processing*; Academic Press: San Diego, CA, USA, 2013.

81. Tadmor, R.; Hernández-Zapata, E.; Chen, N.; Pincus, P.; Israelachvili, J.N. Debye length and double-layer forces in polyelectrolyte solutions. *Macromolecules* **2002**, *35*, 2380–2388. [CrossRef]

82. Vacic, A.; Criscione, J.M.; Rajan, N.K.; Stern, E.; Fahmy, T.M.; Reed, M.A. Determination of molecular configuration by debye length modulation. *J. Am. Chem. Soc.* **2011**, *133*, 13886–13889. [CrossRef]

83. Kesler, V.; Murmann, B.; Soh, H.T. Going beyond the debye length: Overcoming charge screening limitations in next-generation bioelectronic sensors. *ACS Nano* **2020**, *14*, 16194–16201. [CrossRef]

84. Badawy, A.M.E.; Luxton, T.P.; Silva, R.G.; Scheckel, K.G.; Suidan, M.T.; Tolaymat, T.M. Impact of environmental conditions (pH, ionic strength, and electrolyte type) on the surface charge and aggregation of silver nanoparticles suspensions. *Environ. Sci. Technol.* **2010**, *44*, 1260–1266. [CrossRef]

85. Cacua, K.; Ordoñez, F.; Zapata, C.; Herrera, B.; Pabón, E.; Buitrago-Sierra, R. Surfactant concentration and pH effects on the zeta potential values of alumina nanofluids to inspect stability. *Colloids Surf. A Physicochem. Eng. Asp.* **2019**, *583*, 123960. [CrossRef]

86. Napper, D.H. *Polymeric Stabilization of Colloidal Dispersions*; Academic Press: New York, NY, USA, 1983.

87. Somasundaran, P.; Markovic, B.; Krishnakumar, S.; Yu, X. *Handbook of Surface and Colloid Chemistry*; CRC Press: Boca Raton, FL, USA, 1997.

88. Luo, J.; Jang, H.D.; Sun, T.; Xiao, L.; He, Z.; Katsoulidis, A.P.; Kanatzidis, M.G.; Gibson, J.M.; Huang, J. Compression and aggregation-resistant particles of crumpled soft sheets. *ACS Nano* **2011**, *5*, 8943–8949. [CrossRef] [PubMed]

89. Wang, B.; Hao, J.; Li, H. Remarkable improvements in the stability and thermal conductivity of graphite/ethylene glycol nanofluids caused by a graphene oxide percolation structure. *Dalton Trans.* **2013**, *42*, 5866–5873. [CrossRef] [PubMed]

90. Shu, L.; Zhang, J.; Fu, B.; Xu, J.; Tao, P.; Song, C.; Shang, W.; Wu, J.; Deng, T. Ethylene glycol-based solar-thermal fluids dispersed with reduced graphene oxide. *RSC Adv.* **2019**, *9*, 10282–10288. [CrossRef] [PubMed]

91. Lin, R.; Zhang, J.; Shu, L.; Zhu, J.; Fu, B.; Song, C.; Shang, W.; Tao, P.; Deng, T. Self-dispersible graphene quantum dots in ethylene glycol for direct absorption-based medium-temperature solar-thermal harvesting. *RSC Adv.* **2020**, *10*, 45028–45036. [CrossRef]

92. Branson, B.T.; Beauchamp, P.S.; Beam, J.C.; Lukehart, C.M.; Davidson, J.L. Nanodiamond nanofluids for enhanced thermal conductivity. *ACS Nano* **2013**, *7*, 3183–3189. [CrossRef]

93. Bao, Z.; Bing, N.; Zhu, X.; Xie, H.; Yu, W. $Ti_3C_2T_x$ MXene contained nanofluids with high thermal conductivity, super colloidal stability and low viscosity. *Chem. Eng. J.* **2021**, *406*, 126390. [CrossRef]

94. Shah, S.N.A.; Shahabuddin, S.; Sabri, M.F.M.; Salleh, M.F.M.; Ali, M.A.; Hayat, N.; Sidik, N.A.C.; Samykano, M.; Saidur, R. Experimental investigation on stability, thermal conductivity and rheological properties of rGO/ethylene glycol based nanofluids. *Int. J. Heat Mass Transf.* **2020**, *150*, 118981. [CrossRef]

95. Chen, Y.; Quan, X.; Wang, Z.; Lee, C.; Wang, Z.; Tao, P.; Song, C.; Wu, J.; Shang, W.; Deng, T. Stably dispersed high-temperature Fe_3O_4/silicone-oil nanofluids for direct solar thermal energy harvesting. *J. Mater. Chem. A* **2016**, *4*, 17503–17511. [CrossRef]

96. Tao, P.; Shu, L.; Zhang, J.; Lee, C.; Ye, Q.; Guo, H.; Deng, T. Silicone oil-based solar-thermal fluids dispersed with PDMS-modified Fe_3O_4@graphene hybrid nanoparticles. *Prog. Nat. Sci. Mater. Int.* **2018**, *28*, 554–562. [CrossRef]

97. Gómez-Villarejo, R.; Navas, J.; Martín, E.I.; Sánchez-Coronilla, A.; Aguilar, T.; Gallardo, J.J.; De los Santos, D.; Alcántara, R.; Fernández-Lorenzo, C.; Martín-Calleja, J. Preparation of Au nanoparticles in a non-polar medium: Obtaining high-efficiency nanofluids for concentrating solar power. an experimental and theoretical perspective. *J. Mater. Chem. A* **2017**, *5*, 12483–12497. [CrossRef]

98. Li, X.; Gan, C.; Han, Z.; Yan, H.; Chen, D.; Li, W.; Li, H.; Fan, X.; Li, D.; Zhu, M. High dispersivity and excellent tribological performance of titanate coupling agent modified graphene oxide in hydraulic oil. *Carbon* **2020**, *165*, 238–250. [CrossRef]

99. Luo, J.; Jang, H.D.; Huang, J. Effect of sheet morphology on the scalability of graphene-based ultracapacitors. *ACS Nano* **2013**, *7*, 1464–1471. [CrossRef] [PubMed]

100. Ma, X.; Zachariah, M.R.; Zangmeister, C.D. Crumpled nanopaper from graphene oxide. *Nano Lett.* **2012**, *12*, 486–489. [CrossRef]

101. Dou, X.; Koltonow, A.R.; He, X.; Jang, H.D.; Wang, Q.; Chung, Y.-W.; Huang, J. Self-dispersed crumpled graphene balls in oil for friction and wear reduction. *Proc. Natl. Acad. Sci. USA* **2016**, *113*, 1528–1533. [CrossRef]

102. Zhang, J.; Shu, L.; Chang, C.; Li, X.; Lin, R.; Fu, B.; Song, C.; Shang, W.; Tao, P.; Deng, T. Crumpled particles of ethanol-wetted graphene oxide for medium-temperature nanofluidic solar-thermal energy harvesting. *Carbon* **2022**, *186*, 492–500. [CrossRef]

103. Hu, T.; Zhang, J.; Whyte, J.; Fu, B.; Song, C.; Shang, W.; Tao, P.; Deng, T. Silicone oil nanofluids dispersed with mesoporous crumpled graphene for medium-temperature direct absorption solar-thermal energy harvesting. *Sol. Energy Mater. Sol. Cells* **2022**, *243*, 111794. [CrossRef]

104. Zhang, J.; Fu, B.; Song, C.; Shang, W.; Tao, P.; Deng, T. Kirigami-inspired synthesis of functional mesoporous nanospheres through spray drying of holey nanosheets. *Carbon* **2021**, *178*, 382–390. [CrossRef]

105. Austen Angell, C.; Ansari, Y.; Zhao, Z. Ionic liquids: Past, present and future. *Faraday Discuss.* **2012**, *154*, 9–27. [CrossRef]

106. Minea, A.A.; Murshed, S.M.S. A review on development of ionic liquid based nanofluids and their heat transfer behavior. *Renew. Sustain. Energy Rev.* **2018**, *91*, 584–599. [CrossRef]

107. Dash, P.; Scott, R.W.J. 1-Methylimidazole stabilization of gold nanoparticles in imidazolium ionic liquids. *Chem. Commun.* **2009**, *7*, 812–814. [CrossRef]

108. Gao, J.; Ndong, R.S.; Shiflett, M.B.; Wagner, N.J. Creating nanoparticle stability in ionic liquid $[C_4mim][BF_4]$ by inducing solvation layering. *ACS Nano* **2015**, *9*, 3243–3253. [CrossRef] [PubMed]

109. Liu, J.; Xu, C.; Chen, L.; Fang, X.; Zhang, Z. Preparation and photo-thermal conversion performance of modified graphene/ionic liquid nanofluids with excellent dispersion stability. *Sol. Energy Mater. Sol. Cells* **2017**, *170*, 219–232. [CrossRef]

110. Zhang, L.; Chen, L.; Liu, J.; Fang, X.; Zhang, Z. Effect of morphology of carbon nanomaterials on thermo-physical characteristics, optical properties and photo-thermal conversion performance of nanofluids. *Renew. Energy* **2016**, *99*, 888–897. [CrossRef]

111. Bakthavatchalam, B.; Habib, K.; Saidur, R.; Aslfattahi, N.; Rashedi, A. Investigation of electrical conductivity, optical oroperty, and stability of 2D MXene nanofluid containing ionic liquids. *Appl. Sci.* **2020**, *10*, 8943. [CrossRef]

112. Chen, W.; Zou, C.; Li, X. An investigation into the thermophysical and optical properties of SiC/ionic liquid nanofluid for direct absorption solar collector. *Sol. Energy Mater. Sol. Cells* **2017**, *163*, 157–163. [CrossRef]

113. Gebbie, M.A.; Dobbs, H.A.; Valtiner, M.; Israelachvili, J.N. Long-range electrostatic screening in ionic liquids. *Proc. Natl. Acad. Sci. USA* **2015**, *112*, 7432–7437. [CrossRef]

114. Gebbie, M.A.; Smith, A.M.; Dobbs, H.A.; Lee, A.A.; Warr, G.G.; Banquy, X.; Valtiner, M.; Rutland, M.W.; Israelachvili, J.N.; Perkin, S.; et al. Long range electrostatic forces in ionic liquids. *Chem. Commun.* **2017**, *53*, 1214–1224. [CrossRef]

115. Zhang, H.; Dasbiswas, K.; Ludwig, N.B.; Han, G.; Lee, B.; Vaikuntanathan, S.; Talapin, D.V. Stable colloids in molten inorganic salts. *Nature* **2017**, *542*, 328–331. [CrossRef]

116. Kamysbayev, V.; Srivastava, V.; Ludwig, N.B.; Borkiewicz, O.J.; Zhang, H.; Ilavsky, J.; Lee, B.; Chapman, K.W.; Vaikuntanathan, S.; Talapin, D.V. Nanocrystals in molten salts and ionic liquids: Experimental observation of ionic correlations extending beyond the debye length. *ACS Nano* **2019**, *13*, 5760–5770. [CrossRef]

117. Ma, B.; Shin, D.; Banerjee, D. Synthesis and characterization of molten salt nanofluids for thermal energy storage application in concentrated solar power plants-mechanistic understanding of specific heat capacity enhancement. *Nanomaterials* **2020**, *10*, 2266. [CrossRef]

118. Cui, L.; Yu, Q.; Wei, G.; Du, X. Mechanisms for thermal conduction in molten salt-based nanofluid. *Int. J. Heat Mass Transf.* **2022**, *188*, 122648. [CrossRef]

119. Nithiyanantham, U.; Zaki, A.; Grosu, Y.; González-Fernández, L.; Igartua, J.M.; Faik, A. SiO_2@Al_2O_3 core-shell nanoparticles based molten salts nanofluids for thermal energy storage applications. *J. Energy Storage* **2019**, *26*, 101033. [CrossRef]

120. Nithiyanantham, U.; Zaki, A.; Grosu, Y.; González-Fernández, L.; Anagnostopoulos, A.; Navarro, M.E.; Ding, Y.; Igartua, J.M.; Faik, A. Effect of silica nanoparticle size on the stability and thermophysical properties of molten salts based nanofluids for thermal energy storage applications at concentrated solar power plants. *J. Energy Storage* **2022**, *51*, 104276. [CrossRef]

121. Moore, A.L.; Shi, L. Emerging challenges and materials for thermal management of electronics. *Mater. Today* **2014**, *17*, 163–174. [CrossRef]

122. Apmann, K.; Fulmer, R.; Soto, A.; Vafaei, S. Thermal conductivity and viscosity: Review and optimization of effects of nanoparticles. *Materials* **2021**, *14*, 1291. [CrossRef] [PubMed]

123. Bashirnezhad, K.; Bazri, S.; Safaei, M.R.; Goodarzi, M.; Dahari, M.; Mahian, O.; Dalkılıça, A.S.; Wongwises, S. Viscosity of nanofluids: A review of recent experimental studies. *Int. Commun. Heat Mass Transf.* **2016**, *73*, 114–123. [CrossRef]

124. Younes, H.; Mao, M.; Murshed, S.S.; Lou, D.; Hong, H.; Peterson, G.P. Nanofluids: Key parameters to enhance thermal conductivity and its applications. *Appl. Therm. Eng.* **2022**, *207*, 118202. [CrossRef]

125. Yu Hua, L.; Wei, Q.; Jian Chao, F. Temperature dependence of thermal conductivity of nanofluids. *Chinese Phys. Lett.* **2008**, *25*, 3319. [CrossRef]

126. Navarrete, N.; Gimeno-Furió, A.; Forner-Escrig, J.; Juliá, J.E.; Mondragón, R. Colloidal stability of molten salt-based nanofluids: Dynamic light scattering tests at high temperature conditions. *Powder Technol.* **2019**, *352*, 1–10. [CrossRef]

127. Uemura, N.; Egoshi, T.; Murakami, K.; Kizuka, T. High-power laser irradiation for high-temperature in situ transmission electron microscopy. *Micron* **2022**, *157*, 103244. [CrossRef]

128. Kim, J.S.; LaGrange, T.; Reed, B.W.; Taheri, M.L.; Armstrong, M.R.; King, W.E.; Browning, N.D.; Campbell, G.H. Imaging of transient structures using nanosecond in situ TEM. *Science* **2008**, *321*, 1472–1475. [CrossRef] [PubMed]

129. Zhu, J.; Zhang, J.; Lin, R.; Fu, B.; Song, C.; Shang, W.; Tao, P.; Deng, T. Rapid one-step scalable microwave synthesis of $Ti_3C_2T_x$ MXene. *Chem. Commun.* **2021**, *57*, 12611–12614. [CrossRef] [PubMed]

 *nanomaterials*

Review

Nanofluids for Direct-Absorption Solar Collectors—DASCs: A Review on Recent Progress and Future Perspectives

Hussein Sayed Moghaieb [1], Vincenzo Amendola [2], Sameh Khalil [1], Supriya Chakrabarti [1], Paul Maguire [1] and Davide Mariotti [1,*]

1 School of Engineering, Ulster University, 2-24 York Street, Belfast BT15 1AP, UK
2 Dipartimento di Scienze Chimiche, Universita' degli Studi di Padova, Via Marzolo 1, 35131 Padova, Italy
* Correspondence: d.mariotti@ulster.ac.uk

Abstract: Owing to their superior optical and thermal properties over conventional fluids, nanofluids represent an innovative approach for use as working fluids in direct-absorption solar collectors for efficient solar-to-thermal energy conversion. The application of nanofluids in direct-absorption solar collectors demands high-performance solar thermal nanofluids that exhibit exceptional physical and chemical stability over long periods and under a variety of operating, fluid dynamics, and temperature conditions. In this review, we discuss recent developments in the field of nanofluids utilized in direct-absorption solar collectors in terms of their preparation techniques, optical behaviours, solar thermal energy conversion performance, as well as their physical and thermal stability, along with the experimental setups and calculation approaches used. We also highlight the challenges associated with the practical implementation of nanofluid-based direct-absorption solar collectors and offer suggestions and an outlook for the future.

Keywords: nanofluids; solar thermal energy conversion; direct-absorption solar collectors

 check for updates

Citation: Moghaieb, H.S.; Amendola, V.; Khalil, S.; Chakrabarti, S.; Maguire, P.; Mariotti, D. Nanofluids for Direct-Absorption Solar Collectors—DASCs: A Review on Recent Progress and Future Perspectives. *Nanomaterials* 2023, 13, 1232. https://doi.org/10.3390/nano13071232

Academic Editors: M. M. Bhatti, Kambiz Vafai, Sara I. Abdelsalam, Antonio Di Bartolomeo and Henrich Frielinghaus

Received: 28 February 2023
Revised: 22 March 2023
Accepted: 27 March 2023
Published: 30 March 2023

1. Introduction

Researchers from a wide range of fields are engaged in research to find solutions to meet the expected rise in renewable energy demands as a result of the growing impact of climate change and the rapid depletion of primary energy resources [1–3]. Since solar energy is free, clean, abundant, sustainable, renewable, nonlocalized, and generally has no net negative environmental impacts, it is being adopted at a rapid rate, increasing by 8.3% annually [3–5]. Governments, therefore, support technologies that can harvest solar energy efficiently and it is anticipated that by 2100, solar energy will account for nearly 70% of all global energy consumption [6,7]. Solar energy can be utilized in different forms, most notably solar-to-electrical (photovoltaic), solar-to-chemical (photosynthesis) and solar-to-thermal (photothermal) energy conversion [8,9]. For solar-to-thermal energy conversion (STC) systems, solar thermal collectors are widely used to deliver the generated thermal energy to a variety of domestic and industrial applications, such as solar refrigeration, space heating or cooling and electricity generation [6,10–13]. The traditional types of solar thermal collectors rely on solid absorbers for harvesting and converting sunlight into heat, which is then thermally transferred to working fluids, typically water, glycols, or thermal oils (Figure 1A). At the elevated surface temperature of the absorber, heat is lost at significant rates via conduction to other components of the collector, as well as convection and radiation to the environment, reducing the overall efficiency of the collector (Figure 1B). This is further complicated by issues with corrosion and chemical material degradation, which shorten the lifetime of collectors, and the absorber weight impacts installation and maintenance [14,15]. The idea of direct-absorption solar collectors (DASCs), Figure 1C, was put forth to completely remove the solid absorber and directly expose the working fluid to solar radiation for it to be absorbed and converted into heat in a volumetric STC process, overcoming the issues associated with surface-based collectors [16–18]. DASCs also enable the development of novel and adaptable configurational arrangements without being constrained by the incorporation of multiple components in conventional collectors, such as piping systems and surface absorbers.

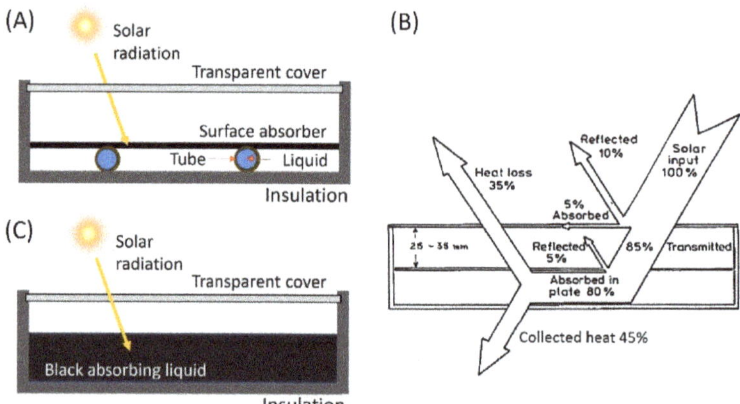

Figure 1. (**A**) Surface-based absorption solar collector with (**B**) the flow of both heat and solar light in a typical liquid flat plate collector. (**C**) Direct (volume-based) absorption solar collector (DASC). Reproduced with permission from [19].

Whilst solar energy is distributed over three main spectral regions (5% from the ultraviolet 300–400 nm, 43% from the visible 400–700 nm, and 52% from the near-infrared 700–2500 nm), common working fluids in solar collectors such as water, glycols, and thermal oils are good absorbers in the ultraviolet (UV) and infrared (IR) regions, both making up 57% of the total solar energy. However, they are transparent in the visible (Vis) region, meaning that ~43% of the solar energy is not efficiently captured and is instead lost as transmitted light. For instance, Otanicar et al. [20] measured the transmittance of four of the most common heat transfer liquids (water, ethylene glycol, propylene glycol, and Therminol VP-1) over a wavelength range of 200–1500 nm for a sample thickness of 10 mm. They found that water exhibited the highest absorption compared to the other fluids but could only absorb up to 13% of the incoming solar energy. This is made worse by their poor thermal conductivities [21]. In earlier attempts, black dyes were added to these conventional working fluids to improve STC performance; however, this led to the light-induced and thermal degradation of the dyes at high operational temperatures [17,22]. Suspending small solid particles with dimensions in the order of the millimetre and micrometre ranges in conventional fluids was also investigated owing to the high thermophysical and optical properties of solids compared with fluids [23]. However, DASCs suffered several operational issues such as accumulation of the suspended particles that resulted in low stability and surface erosion of pipes as well as the need for more pumping power to overcome the pressure drop brought on by the increase in friction [21].

Thanks to the rapid advances in nanotechnology, a new generation of smaller solid particles with dimensions << 1 μm (referred to as nanoparticles, NPs) was offered for their dispersion in conventional heat transfer liquids, forming nanofluids (NFs) [24]. In 1993, Masuda et al. [25] measured the thermal conductivity of three different types of metal oxide NPs suspended in water as the base fluid and they reported significant improvements in the thermal conductivity. Later, in 1995, the term nanofluid was first introduced by Choi and Eastman [26] after they worked on a copper–water NF and reported a remarkable increase in thermal conductivity over the base fluid. Since then, research on NFs has seen rapid growth in different aspects that are related to the synthesis of NPs and the preparation techniques of the corresponding NFs. Initially, NFs were investigated with respect to their thermal properties and their heat transfer characteristics. The study of their solar absorption characteristics and stability for their use in DASCs is far more recent [27–29]. NPs made of metals, metal oxides, and carbons (e.g., carbon black, graphene, carbon nanotubes, and carbon nanofibers) can be synthesized in different morphologies including fibres, tubes, spheres, and other shapes offering a wide variety of different types of nanoscale materials [30–32]. The synthesis of NPs to be used for STC systems can rely on several techniques that have found wide applicability. In this case, NP synthesis and NF preparation can be viewed as two distinct consecutive steps [33,34]. However, NPs can be also synthesized directly in the base fluid using methods such as chemical reduction in liquids [35] or physical vapour condensation [36], which require a high level of control and post-synthesis processing [37]. Generally, the overall performance of NF-based DASCs is mainly determined by the design and materials selection of the collector, and more importantly, by the characteristics of the utilized NFs.

This review is focused on NPs and corresponding NFs, their properties with relevance for DASCs, including their optical properties, resulting in STC performance and stability, with emphasis on experimental work. We have included relevant discussions on the characterization of both optical properties and stability. Important figures of merit are also discussed, which require increased attention to achieve effective and efficient STC and long-term stability under various operating conditions. The review covers the fundamentals of these parameters and various methods for measuring them before analysing the most recent progress made in research over the past few years. The review, therefore, provides a new materials science and physics perspective, that departs from relevant reviews available in the literature, and where material properties are directly linked to the NF performance. It offers an assessment of energy conversion paths, also providing the experimental methodologies to assess light–NF interactions and discuss quantitative fundamental measurements that can be used for full-scale and DASC system simulations.

2. Nanoparticles Used for DASC Nanofluids

2.1. A Classification of Nanoparticles Used for Nanofluids

A diverse range of nanomaterials has been investigated for their potential use in NFs for DASCs. It is useful to classify these with respect to their material composition and, for instance, consider metal, metal oxide, and carbon-based NPs [38,39]. Metal and carbon-based NPs generally exhibit strong absorption across the solar spectrum and specifically in the visible range; however, they often present issues in terms of stability (e.g., agglomeration and chemical degradation, and especially oxidation in high-temperature DASCs) [40,41]. Metal NPs can, in some cases, exhibit plasmonic effects, which enhance absorption in narrow spectral ranges [42]. Metal oxides are materials that offer exceptional chemical stability in a wide range of conditions and an overall reduced environmental impact; however, due to their wide energy bandgap, the optical properties can be unsuitable for solar energy absorption [43,44]. Various doping and defect engineering strategies can be used to enhance metal oxide optical properties; these can be used to lower the bandgap or to Increase the carrier density and introduce plasmonic behaviour [45,46]. Semiconducting NPs with lower bandgap values than metal oxides could, in principle, offer more favourable optical properties compared to metal oxides. However, semiconducting NPs have attracted very limited interest in DASC applications, possibly due to costs or toxicity of the raw materials (e.g., Ga, Pb, and Ge), complex and expensive synthesis methodologies (e.g., for nitrides) as well as limited stability and their propensity for chemical degradation (e.g., Si oxidation).

The synthesis of NPs and dispersion methods in NFs can vary depending on the material and the chosen technique, which significantly impacts the resulting NF properties [47]. Methods such as chemical reduction, seed-mediated growth, sol–gel, hydrothermal, and precipitation can be used for synthesizing plasmonic metals and metal oxides [48,49]. Carbon-based nanomaterials can be synthesized through chemical vapour deposition (CVD), the Hummers method, and hydrothermal synthesis [50,51]. After synthesis, NPs can be dispersed into the base fluid using ultrasonication, magnetic stirring, or high-speed stirring and surfactants, and stabilizing agents or gum Arabic can be added to ensure stability and prevent agglomeration or sedimentation [41,52]. While NPs dispersed in the base fluids are often made from the same type of NP, i.e., single-component NF, multi-component NFs have also been widely investigated, where the NPs in the NF are made of different materials and/or have different sizes, and/or shapes. While there are initial claims of broader absorption features, multicomponent NFs have shown to add nonnegligible complexity to the NF preparation with limited benefits [53,54].

In this section, we highlight examples of metal, metal oxide, and carbon-based NPs used in single-component as well as multicomponent NFs. A list of relevant work is also summarized in Table 1.

2.2. Metallic Nanoparticles

Plasmonic metals, including gold, silver, and copper, hold great promise for application in DASCs due to their inherent plasmonic absorption properties, which facilitate more efficient light absorption and scattering compared to bulk materials [8]. These metals are characterized by the presence of free electrons in their conduction bands. The collective oscillation of these free electrons in response to an electromagnetic field is known as a plasmon, which confined within an NP results in a localized surface plasmon. After absorbing light energy, plasmons can transfer energy to the lattice structure of the metal via electron–phonon coupling. resulting in the generation of heat [55]. Factors such as size, shape, and material composition of plasmonic nanostructures considerably influence their efficiency in light-to-heat conversion [56]. Additionally, the narrow bandwidth of the plasmonic effect can negatively impact the application of plasmonic NPs in DASCs, and the simultaneous excitation of transversal and longitudinal modes should be preferred in DASC NFs (e.g., using asymmetric nanostructures) [57].

Numerous recent studies have explored the synthesis of metal NPs for DASCs, emphasizing their potential to boost the efficiency of solar thermal collectors. In a study that employed surface modification, Chen et al. [58] generated Au NPs with sizes of 25 nm, 33 nm, and 40 nm using a seed-mediated synthesis method, as depicted in Figure 2B. Through a series of wet chemical reactions, ultrasonication, and heating, an aqueous solution of $hAuCl_4$ was reduced to Au NPs using sodium borohydride ($NaBH_4$) as a reducing agent and cetrimonium chloride (CTAC) as a stabilizer. The particles from each sample were collected via centrifugation at 8000 revolutions per minute (rpm) for 10 min and were re-dispersed in water. In a recent study by Gupta et al. [59], spherical gold Au NPs with a size of ~30 nm were synthesized using the Turkevich method. Chloroauric acid was boiled and a trisodium citrate dihydrate solution was gradually added, resulting in a colour change from transparent to red, indicating Au NPs formation. After allowing the solution to cool, polyvinylpyrrolidone (PVP) was added and the mixture was stirred for 10 h to create more stable PVP-capped Au NPs. Concurrently, an Azadirachta indica leaf extract was prepared by boiling leaves in ethanol. Three heat transfer fluids were then formulated: one containing Au NPs in deionized water, one combining Au NPs with the natural extract, and the last one incorporating only the natural extract in deionized water. These NF samples were subsequently compared in a DASC for effective temperature gain and photo-thermal conversion efficiency. Sharaf et al. [60] examined the impact of four distinct surfactants on Au NPs prepared through a chemical reduction method, as shown in Figure 2C. They first synthesized citrate-capped Au NPs (CIT–Au NPs) with an average size of 13 nm by reducing $hAuCl_4$ with trisodium citrate. Subsequently, they coated the CIT–Au NPs with three different polymers: bovine serum albumin (BSA), polyvinylpyrrolidone (PVP) and polyethylene glycol (PEG). After filtering the samples using a stirred cell to remove excess polymer molecules, they diluted them to achieve a concentration of 0.16 mg/mL for each NF. As surfactants can add complexity to the synthesis method and complicate heat transfer at the NP–fluid interface, McGlynn et al. [61] produced Au NPs by plasma-induced nonequilibrium electrochemistry from an aqueous gold salt precursor (hAuCl$_4$.3H$_2$O) without utilizing reducing or stabilizing agents, as illustrated in Figure 2A. The Au NPs were then formed into nanofluids (NFs) by drying and re-dispersing them in ethylene glycol (EG) using sonication at a concentration of 0.0055 vol%. A variety of shapes, including spherical, hexagonal, and nanorods, were observed among the Au NPs, with a mean size of 27 nm. Kimpton et al. [62] synthesized Ag NPs through a reducing–oxidizing method using citrate and PVP to achieve electrostatic and steric stabilization. Subsequently, they coated the Ag NPs with silica (SiO$_2$) using tetraethyl orthosilicate (TEOS) in a basic ethanolic solution, resulting in SiO$_2$@Ag NPs. The final solution was diluted with water to yield 6 mL of SiO$_2$@Ag NPs.

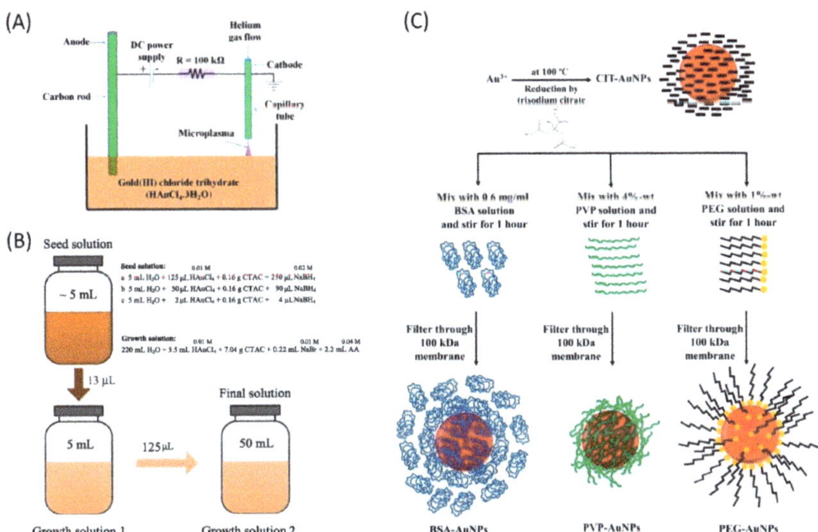

Figure 2. Three different methods of Au NP synthesis with and without surface modifications. (**A**) Atmospheric-pressure microplasma setup, adapted from [61]. (**B**) Seed-mediated synthesis method, adapted from [58]. (**C**) Polymer coating procedure of Au NPs prepared by chemical reduction, adapted from [60].

2.3. Carbon-Based Nanomaterials

Carbon-based nanomaterials, such as carbon nanotubes, graphene, and graphite, exhibit strong light absorption due to their electronic structure and high aspect ratio. These materials possess a conjugated π-electron system that facilitates the free movement of electrons. Light absorption results in the excitation of electrons to higher energy states and the subsequent relaxation of these electrons release energy as heat. In addition, the strong electron–electron interactions and the unique band structure lead to the generation of nonequilibrium hot carriers. The energy of these hot carriers can be transferred to the lattice, generating heat and contributing to solar thermal conversion. The efficiency of light-to-heat conversion in carbon-based nanomaterials is influenced by factors such as material structure, size, and surface properties [24,29,63–66].

Recently, McGlynn et al. [67] produced carbon nanotubes (CNTs) using a floating-catalyst chemical vapour deposition system, which generated ribbon-like macroscopic assemblies of continuous length. The synthesis process entailed flowing ferrocene, thiophene, and methane through a furnace, with hydrogen serving as a carrier gas. Following cooling, a dense carbon aerogel formed, which was drawn into a long black "sock". Specific gas flows and furnace temperatures were maintained to obtain ribbon-like CNT assemblies, which were then pressed between glass slides to create flat ribbon-like materials. Prior to plasma functionalization, the CNT ribbon samples were annealed in a standard air atmosphere. A plasma-induced nonequilibrium electrochemistry system was employed for the functionalization of the CNT ribbons, utilizing direct-current microplasma generated in helium between a nickel capillary tube and the electrolyte surface. The CNT ribbons underwent treatment at a constant current for 15 min and two distinct plasma–liquid treatments were investigated: one consisting of a 1:9 ethanol:water ratio and another containing 10% ethylenediamine in the same stock solution. Following plasma treatments, the ribbons were rinsed in distilled water and disentangled by sonication to form NFs. Yu et al. [66] prepared carbon-based nanofluids using a modified vortex trap method, which utilizes water mist to collect carbon NPs produced by oxygen–acetylene combustion, equipped with an external circulation device that improved carbon NP collection efficiency and production rate. In a separate study, Chen et al. [68] prepared carbon quantum dots and dispersed them in PEG-200 using a microwave heating method. They employed a 700 W home-used microwave oven to heat 5 mL of PEG-200. This heating process led to the breakdown of PEG chains, which subsequently oxidized into various fragments in the presence of oxygen, ultimately forming a surface passivation layer. The degree of carbonization was controlled by adjusting the microwave heating duration. Guo et al. [69] prepared three NFs of graphene oxide (GO), multiwalled CNTs, and Ti_3C_2 in 1-Butyl-3-methylimidazolium tetrafluoroborate. Commercial samples of GO and CNTs were used, while Ti_3C_2 was synthesized using a solvent etching method. Briefly, they dissolved 8 g of lithium fluoride in 100 mL of HCl solution (10 M) and added 5 g of Ti_3AlC_2 MAX to polyethylene while stirring in a water bath. The final suspension was stirred for two days and washed with water, and a thin layer of $Ti_3C_2T_x$ was collected after multiple centrifugations and ultrasonication steps. In an extensive study on CNTs, Mesgari et al. [70] chemically functionalized single-, double-, and multiwalled CNTs and produced three different NFs with water, propylene glycol (PG), and Therminol. The NFs were prepared using probe-type sonication in a water–ice bath.

2.4. Metal Oxide Nanoparticles

Metal oxides, including copper oxide, zinc oxide, titanium dioxide, and iron oxide, can absorb solar radiation with limitations imposed by their bandgap. Upon light absorption, electrons are excited from the valence band to the conduction band, generating electron-hole pairs [71]. The relaxation from excited states releases energy in the form of heat [72]. The efficiency of light-to-heat conversion in metal oxides is highly dependent on factors such as material composition, crystal structure, and surface properties.

Copper oxide NPs have emerged as a prominent material in solar thermal conversion research due to their unique properties. In a recent study conducted by Moghaieb et al. [29], the authors synthesized surfactant-free CuO_x NPs by plasma-induced nonequilibrium electrochemistry, which featured a sacrificial copper foil anode and a capillary tube cathode. The researchers prepared a 6 mL electrolyte solution by combining equal parts of sodium chloride, sodium nitrate, and sodium hydroxide, without the addition of reducing, capping, or stabilizing agents. They maintained atmospheric-pressure microplasma for 30 min at the electrolyte surface, leading to a visible colour change from transparent to orange. Following centrifugation, heat drying, and annealing at 400 °C for 6 h, a completely black powder was obtained, suggesting a phase transformation due to thermal oxidation. Both annealed and nonannealed CuO_x NP powders were then separately re-dispersed in ethylene glycol (EG) at different volume fractions and subjected to sonication to achieve homogenous dispersions. In an extensive study by Milanese et al. [73], the researchers investigated a broad array of metal oxide water-based NFs, encompassing commercially available Al_2O_3, CuO, and TiO_2, as well as CeO_2, ZnO, and Fe_2O_3, prepared via the hydrolytic synthesis method. For instance, the CeO_2 NPs were synthesized by dissolving 2.6 mmol $Ce(NO_3)_3 \cdot 6H_2O$ in methanol, followed by the addition of acetylacetone. Subsequently, a 30 wt% ammonia solution in water was introduced to the mixture, which was then vigorously stirred for 24 h. The final suspensions of CeO_2, ZnO, and Fe_2O_3 NPs were acquired after centrifugation, ensuring optimal particle collection. Moreover, magnetic metal oxides have gained considerable interest due to their unique magnetic properties.

In a study by Balakin et al. [74], the authors prepared a magnetic Fe_2O_3/water NF by mechanically stirring Fe_2O_3 NPs in distilled water and subjecting the suspension to bath-type sonication at 130 W for 30 min. They refrained from using surfactants to avoid potential electromagnetic and rheological impacts on the solar thermal conversion process. In a related study, Hosseini and Dehaj [75] prepared Fe_2O_3/water and Fe_3O_4/water NFs by employing more powerful sonication (probe ultrasonicator at 300 W) and incorporating the surfactant sodium dodecylbenzene sulfonate (SDBS) with a neutral pH to improve colloidal behaviour. Afterwards, the mixture was re-sonicated using a bath-type sonicator. They ultimately obtained two NFs at volume fractions of 0.005%, 0.01%, and 0.02%, further expanding the range of potential applications in solar thermal conversion processes.

Considering the plasmonic absorption properties of Ag NPs and antimony-doped tin oxide (ATO) NPs in the visible and near-infrared spectra, Sreekumar et al. [76] developed a multicomponent ATO–Ag/water NF to enable broader absorption across these regions and potentially enhance STC efficiency. They prepared the mixture through a reduction reaction involving Sb_2O_5, $SnCl_2$, HCl, $AgNO_3$, and ATO NPs. Following the chemical reduction process, the ATO–Ag NP solution was washed and dried in a hot air oven at 60 °C. The ATO–Ag NPs were finally re-dispersed in distilled water and sodium dodecyl sulfate (SDS) surfactant was added to improve the stability of the NFs. As an example of a semiconducting nitride, titanium nitride (TiN), which exhibits strong plasmonic absorption, can be highlighted. Wang et al. [77] prepared TiN NPs/EG by dispersing 0.1135 g of TiN NPs in 1000 mL of EG, followed by simultaneous stirring and ultrasonication for 30 min, resulting in a concentration of 0.01 wt.%. Further dilutions using ultrasonication yielded concentrations of 0.001 wt.%, 0.003 wt.%, 0.005 wt.%, and 0.007 wt.%.

Table 1. The synthesis and morphology of various types of nanomaterials and the preparation, optical properties, and stability of their nanofluids.

Nanofluid	NP Synthesis Method	NP Morphology	NP Loading	Stabilizer	NF Preparation	Measured Optical Property	Stability Tests	[Ref] Year
- Au NPs/EG - CuO QDs/EG - Au NPs–CuO - QDs/EG	Plasma-induced nonequilibrium electrochemistry (PiNE) synthesis at atmospheric pressure	- Different shapes of 27 nm mean diameter - Nearly spherical of 3.3 nm mean diameter	- 0.0055 vol% - 0.0017 vol%	Electrostatic stabilization	Heat drying of the as-prepared samples followed by sonication in EG	Absorption and scattering	11 weeks of ambient storage	[61] 2020
Au NPs/water	Seed-mediated synthesis method	Average sizes of 25 nm, 33 nm, and 40 nm	0.07 mg/L, 0.18 mg/L, and 0.39 mg/L	CTAC surfactant and ultrasonication (30 min at 20 °C)	Ultrasonication for 30 min at 20 °C	Absorbance	16 h of ambient storage and heating at 90 °C	[58] 2016
Au NPs/water	Citrate, PEG, PVP, and BSA coatings	-	0.1611 mg/mL	PEG, PVP, and BSA dispersants	The excessive polymer was filtered, and samples were diluted	-	3 years of ambient storage, continuous irradiation, and cyclic irradiation	[60] 2021
TiN/EG	Commercial	30 nm diameter	0.01, 0.001, 0.003, 0.005, and 0.007 wt.%	-	30 min sonication and dilution	-	-	[43] 2021
ATO–Ag/water	Chemical reduction	Nonspherical of an average size <40 nm	0.01–0.2 wt%	-	Dispersion in distilled water by adding SDS surfactant	Extinction	-	[76] 2020
- Ag NPs/water - SiO$_2$ NPs/water - Ag NPs@SiO$_2$/water	A reducing–oxidizing method with adding citrate followed by Coating Ag NPs with SiO$_2$ using TEOS	- 38–44 nm diameter - 141 nm diameter	6 mL	- TEOS shell - PVP dispersant	- Dilution	-	3 weeks of ambient storage and solar radiation (simulated and real)	[62] 2020
- Al$_2$O$_3$/water - CuO/water - TiO$_2$/water - ZnO/water - CeO$_2$/water - Fe$_2$O$_3$/water	- Commercial (Al$_2$O$_3$, CuO and TiO$_2$) - Hydrolytic synthesis (ZnO, CeO$_2$, and Fe$_2$O$_3$)	-	0.05–1 vol%.	-	Direct dispersion and centrifuge	Extinction	-	[73] 2016
- Fe$_2$O$_3$/water	Commercial	60 nm (110 nm after heating)	0.5–2 wt%	-	Mechanical stirring and sonication for 30 min at 130 W	-	-	[74] 2022
- Fe$_2$O$_3$/water - Fe$_3$O$_4$/water	Commercial	20–50 nm diameter	0.001, 0.01, and 0.02 vol%	SDBS dispersant	Ultrasonication for 20–60 min at 300 W	-	-	[75] 2022

Table 1. *Cont.*

Nanofluid	NP Synthesis Method	NP Morphology	NP Loading	Stabilizer	NF Preparation	Measured Optical Property	Stability Tests	[Ref] Year
- CQDs/PEG-200	- Microwave heating method - 5 mL	Less than 10 nm dimeter	-	PEG-200	Stirring of 22 mL (~500 rpm)	-	One month of ambient storage	[68] 2022
- GO/[BMIM]BF$_4$ - MWCN/[BMIM]BF$_4$ - Ti$_3$C$_2$/[BMIM]BF$_4$	- Commercial (GO and MWCN) - Solvent etching (Ti$_3$C$_2$)	- 0.5–5 μm diameter - 3–5 nm and 8–15 nm inner and outer diameters	- 0.001, 0.002, 0.005, 0.02, and 0.04 wt% - 0.02 wt% - 0.02 wt%		- Ultrasonication for 7 min (400 W, 20 kHz)	-	-	[69] 2022
- SWCNTs/water, /PG, or /therminol-55 - MWCNTs/ water, /PG, or /therminol-55	Each material was chemically functionalized by acid and potassium persulfate (KPS)	-	-	- Oxygenated, carboxyl, and hydroxyl groups - SDBS dispersant	- Ultrasonication at 20 W in a water–ice bath for 20 min	-	Working temperature change (80–250 °C)	[70] 2016
- CIT–Au NPs/water - PEG–Au NPs/water	- Citrate reduction by the Turkevich method	Near-spherical particles of 10 nm diameter	- 0.0358–0.358 mg/mL	- CIT coating - PEG coating	The as-made samples were diluted after post-synthesis purification	- Extinction	16 months of ambient storage and heating for 12 h	[78] 2019
- MWCNT/water - MWCNT–TiN/water	Commercial	5–10 nm and 20–30 nm inner and outer diameters and 10–30 μm length - TiN of 20 nm diameter	10, 20, and 30 ppm	CTAB surfactant	Magnetic stirring at 800 r/min for 30 min - Sonication for 60 min at 750 W and 20 kHz	-	One week of ambient storage	[79] 2022
- Ag/water - Fe$_3$O$_4$/water - Graphite/water	Commercial	- 50 nm - 15 nm - 20 nm	-	TPABr surfactant and sulfuric acid treatment	-	-	-	[80] 2016
SiC–MWCNTs/EG	-	- SiC: 40 nm in diameter - MWCNT: 20 nm diameter and 30 μm length	0.01–1 wt%	PVP–K30 and hexane dispersant and ultrasonication for 1 h	-	-	One month of ambient storage, mass fraction, temperature, and thermal cycling	[81] 2020
- Graphene/water	Commercial	Nanoplates of 2 nm thickness and 2 μm diameter	0.00025–0.005 wt%	-	-	Extinction	-	[82] 2016

Table 1. *Cont.*

Nanofluid	NP Synthesis Method	NP Morphology	NP Loading	Stabilizer	NF Preparation	Measured Optical Property	Stability Tests	[Ref] Year
- CeO_2/Chloroform - Fe_2O_3/Chloroform	-	- Spherical $(3.0 \pm 0.4$ nm) - Spherical $(4.3 \pm 1.9$ nm)	- 3 mg/mL - 5 mg/mL	-	-	Absorbance	-	[73] 2016
- Ag/water - Fe_3O_4/water - Graphite/water	-	- Commercial - Mean diameter is 50 (Ag), 15 (Fe_3O_4), and 20 (graphite)	40 ppm	-	-	Extinction	-	[80] 2016
- Ag NPs/water - Ag NPs@SiO2/water	-	- 41 nm diameter - 50–100 nm diameter	-	TEOS shell, Na_2SiO_3 shell, and citrate coating	-	-	Cyclic heating and cyclic UV radiation	[83] 2018
- Al_2O_3/water - TiO_2/water	-	- 13 nm - ~21 nm	0.1 and 0.3 vol%	-	-	Absorption and scattering	-	[84] 2014
- CuO/water–EG - Al_2O_3/water–EG - CuO–Al_2O_3/water–EG	-	- Spherical of 40 nm average size - Spherical of ~100 nm average size	- 0.0001–0.0003 vol% - 0.004–0.008 vol%	SHMP surfactant, pH control, and ultrasonication	-	Extinction	Two weeks of ambient storage	[85] 2016
ZrC/water	-	40 nm diameter	-	Ultrasonication for 10 min)	-	-	-	[86] 2019
- GE/water - GO/water - RGO/water	-	GO: nanosheets of 0.55–1.2 nm thickness and 0.5–3 μm size	-	PVP	-	-	Two months of ambient storage	[87] 2017
- GO/EG–water - rGO/EG–water - rGO/water	-	-	-	- UV exposure - PVP dispersant	-	-	Two months of ambient storage	[88] 2019
- MWCNTs/water	-	20 nm diameter and 1–25 μm length	-	SDBS, CTAB, SDS, and Triton X-100	-	-	- One month of ambient storage	[89] 2018

Table 1. *Cont.*

Nanofluid	NP Synthesis Method	NP Morphology	NP Loading	Stabilizer	NF Preparation	Measured Optical Property	Stability Tests	[Ref] Year
- BN/EG - BN–CB–/EG	-	-	- 30–90 - 4–15 (ppm)	-	-	-	-	[90] 2021
- CuO/EG	- Plasma-induced nonequilibrium electrochemistry (PiNE)	- flower-like morphology of sizes up to 200 nm	- 0.001–0.01 vol%	- No surfactant used	- Ultrasonication in EG	- Absorption and scattering	- 30 days of ambient storage - Cyclic heating	[29] 2023
- Au NPs/Water - Au NPs/ Azadirachta Indica–water	- Turkevich method	- 30 nm	- 4 mg/L	- PVP	-	- Absorbance	- 30 days of ambient storage	[59] 2023
- C NPs/Water-PSS	- Modified vortex trap method (VTM)	- amorphous carbon, graphene oxide, and reduced graphene oxide with a mean particle size of 30 nm	- 0.025–0.5 wt%	-	-	-	- Ambient storage	[66] 2023

3. Energy Flow in Solar Thermal Nanofluids Used in DASCs

Light interaction with an NF can give rise to several physical and chemical phenomena (Figure 3), establishing a flow pattern for the energy originating from solar photons. Here we consider phenomena taking place within the NF, as interface interactions such as reflection, etc., at the front- and back-end of the NF can be discussed under textbook optics. The extent of energy exchange and the mechanisms involved strongly depends on the material characteristics of the NP in the NFs [91–96] as well as the size/shape of the NPs and their concentration in the NF. With reference to Figure 3, heating of the working fluid, hence its temperature increase, is ultimately the desired 'finish-line' for the energy flow. Note that the temperature of the NPs may not be at equilibrium with the fluid temperature under solar irradiation, hence the NPs and the working fluid should be considered separately.

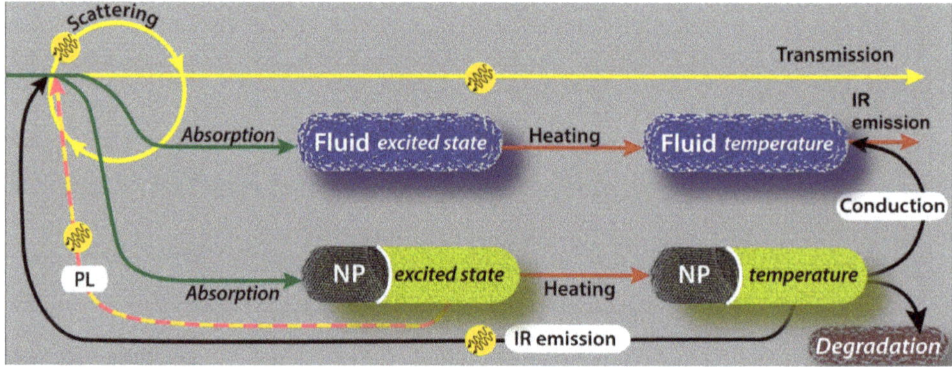

Figure 3. Energy flow in solar thermal nanofluids (nanoparticles and base fluid).

Transmission, scattering, and absorption are the three main light–matter interaction events to be considered. As illustrated in Figure 3, the two components of NFs, i.e., the NPs and the base fluid, interact differently with light. While both can absorb light to varying degrees with different patterns over the solar spectrum, only NPs can scatter light. Transmitted light is lost energy with no possibility of being recovered. Light that is scattered when travelling through the NF [96] can be in part 'recycled' with chances of being completely lost through transmission, scattered again, or eventually absorbed. Hence, scattering can only lead to partial light re-absorption and should be minimized in favour of direct absorption. Absorbed light, either by the working fluid or the NPs, is energy that becomes available for conversion into heat; however, several energy-loss paths are still possible, especially after absorption by the NPs.

Direct light absorption by the working fluid is a very efficient energy conversion process as the only loss mechanism after absorption is represented by the fluid IR emission if the NF is adequately insulated. NP light absorption initially produces an excited state for electrons and the subsequent decay is highly dependent upon the material characteristics and composition. For instance, metallic NPs generally exhibit fast phonon emissions [97,98], hence increasing the NP temperature very quickly before any other loss mechanism can take place. Some metals or materials with high carrier concentrations can exhibit plasmonic absorption, i.e., enhanced absorption at specific wavelength ranges. This can be beneficial if the absorption range is within the visible and still relatively broad. However, plasmonic electronic excitation is often abrupt and generates high electric fields at the NP surface, which could result in cavitation (in extreme cases) or sudden temperature changes at the NP–fluid interface that could be detrimental to the chemical stability of the NFs.

In nonmetallic NPs that feature an energy bandgap, such as metal oxide or semiconducting NPs, electrons decay through phonon emissions down to the bandgap edge; there, phonon emission is either prohibited or sufficiently slow for other mechanisms to take place. A similar process can take place due to surface states, where electron relaxation through phonon emission can be prevented. For instance, photoluminescence (PL), i.e., the re-emission of a photon of a lower wavelength than the one that was absorbed, can occur in these cases. Similar to scattering, PL can recycle the light, and minimizing this phenomenon in the UV–Vis range in favour of direct absorption is preferred. Slow electron decay, especially due to surface states can also enhance NP–fluid chemical reactions and possibly pose a challenge to NF stability through its degradation. Hence fast phonon emission is the preferred channel also for metal oxide and semiconducting NPs, where a higher bandgap generally reduces heating and increases energy losses.

Upon heating the NP and a temperature rise, heat conduction from the NPs to the working fluid takes place. Stabilizers or surfactants that are often used in chemical synthesis introduce an additional thermal resistance which could locally increase the temperature at the interface between the NPs and the working fluid. Localized heating due to surfactants or via plasmonic absorption may have an impact on the chemical stability of the NFs and can represent a form of energy loss that has received little attention so far. Localized heating can result in the degradation of NPs and agglomeration over time under several operating conditions such as high solar and thermal loadings as well as solar and thermal cycling. This type of energy loss to chemical energy (transformation of matter) is extremely important not only because it reduces the STC efficiency but also because it can impact the chemical/physical stability of the NFs, hence their long-term applicability.

As NPs exhibit temperature values that are higher than the working fluid, heat conduction occurs only in one direction, i.e., from the NPs to the working fluid. While NP IR emissions can be significant in some cases, this can be often re-absorbed directly by the working fluid and should not represent a major loss mechanism.

4. Fundamental Optical Characterisation of Nanofluids for DASCs

Conventional base fluids used in DASCs are strong absorbers in the UV and IR regions, which represents 57% of the total solar energy; however, they are transparent in the Vis part, meaning that 43% is mostly lost as transmitted light [20]. Figure 4a reports the solar irradiance together with the transmittance spectra for four relevant working fluids used in DASCs. The optical characteristics of the working fluids result in power absorbed (Figure 4b), which clearly shows that very little solar irradiance is used up to ~900 nm. The main functionality in adding NPs to these base fluids is, therefore, to compensate for this by absorbing the solar energy present in the Vis region. In this view, studying the absorption and scattering behaviours of NFs is essential in evaluating their potential as working media in DASCs since this determines to what extent the NF can effectively utilize solar radiation energy [99,100]. For the performance improvement of DASCs, it is ultimately desirable to increase absorption, which can be impacted by the shape, size, material type, and concentration of the NPs, as well as the type of base fluid [101].

Experimentally, when taking measurements in the lab, the light interaction with an NF sample is characterized by the transmittance ($T(\lambda) = I_{tr}(\lambda)/I_{inc}(\lambda)$), absorptance ($Abs(\lambda) = I_{abs}(\lambda)/I_{inc}(\lambda)$), and scattering ($S(\lambda) = I_{sct}(\lambda)/I_{inc}(\lambda)$), which correspond to fractions, with values from 0 to 1, of the incident light intensity ($I_{inc}(\lambda)$), hence

$$T(\lambda) + Abs(\lambda) + S(\lambda) = 1 \tag{1}$$

All the optical quantities are wavelength-dependent (λ); however, here below we have omitted this dependence for simplicity. Furthermore, the interaction of light with an NF sample is three-dimensional in nature; however, a one-dimensional analysis is applicable for the measurements of the fundamental physical NF properties where the sample size can be minimized in two of the three dimensions and ensuring re-absorption is negligible. The light that is not transmitted (i.e., $Abs + S$ or $1-T$) is also referred to as attenuated light. From a practical point of view, both scattered and transmitted light represents the light that has escaped the sample, the latter without any change in the direction of the propagation. However, experimental measurements of S and T include more than purely scattering and transmission phenomena and their values result from the sum of all of the light exiting the sample. If, for example, samples exhibit strong PL or IR emissions (Figure 3), the corresponding light emission is measured as either scattering or transmission. As previously discussed, PL and IR emissions should be generally considered losses to be minimized and should be expected to be negligible. Optical measurements of NF samples then allow the extraction of fundamental physical NF properties such as extinction (μ_{ext}), absorption (μ_{abs}), and scattering (μ_{sct}) coefficients measured in m^{-1}, which can then be used, for instance, to simulate full DASC system performance in three dimensions [29].

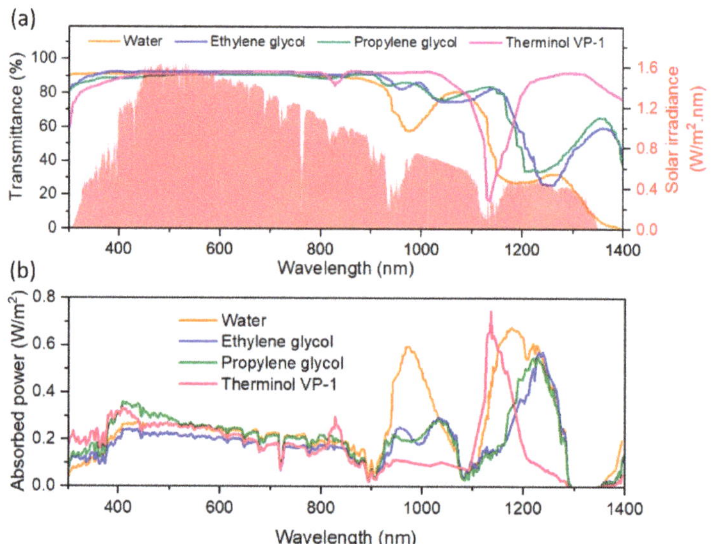

Figure 4. (**a**) Spectral solar irradiance at Earth's level (air mass of 1.5, AM1.5) [102] and (**b**) transmittance of common base fluids used in DASCs for a one cm sample thickness (transmittance for Terminol VP-1 is taken from [29]).

The total attenuation of light while travelling through an NF sample with a given depth (or thickness, *L*) can be evaluated by measuring the transmittance using a spectrophotometer. Most conventional spectrophotometers can cover electromagnetic spectral ranges between 300 nm and 900 nm, including the ultraviolet and visible regions, as well as part of the near-infrared, and this can extend into higher wavelengths depending on the type of the instrument used, e.g., up to 3000 nm. The transmittance is determined by the ratio of the light intensity transmitted through the NF sample (I_{tr}), and collected by the detector to the intensity of the incident light, i.e., the intensity of the light source incident on the sample and prior to interacting with the sample (I_{inc}), see Figure 5A. The latter needs to be measured with an appropriate reference sample, which can take into account instrumental losses including those due to the cuvette absorption and reflection at the interfaces. As most working fluids have negligible attenuation in relevant UV and Vis ranges, in order to evaluate the impact of NFs, it is convenient to use a reference with just the working fluid (i.e., without NPs). This will, however, limit the validity of the NF optical properties to these ranges where the working fluid has little interaction with the light (e.g., <900 nm). In order to further extend the validity of the measurements to the IR, more complex calculations and verifications would need to be introduced.

The extinction coefficient of the NF can be calculated from the Beer–Lambert law [96]:

$$T(L) = \frac{I_{tr}}{I_{inc}} = e^{-\mu_{ext}L} \tag{2}$$

$$\mu_{ext} = -\frac{lnT}{L} \tag{3}$$

Attenuation is due to both absorption as well as scattering, hence the extinction coefficient is the sum of the corresponding coefficients:

$$\mu_{ext} = \mu_{abs} + \mu_{sct} \tag{4}$$

The contribution of scattering can be neglected in some conditions, e.g., for NPs with sizes much smaller than the incident wavelength [103–105] and at low particle concentrations with no agglomeration (<0.01% volume fraction) [104]. In this case, the extinction coefficient can be representative of the absorption coefficient ($\mu_{ext} \approx \mu_{abs}$) [96]. Larger particles, high particle concentrations and/or the presence of agglomeration can lead to significant scattering, which requires the spectrophotometer instrument to be equipped with an integrating sphere system (Figure 5B) to yield transmittance and scattering separately. Depending on the spectrometer and integrating sphere, separate measurements of T and S may not be possible, so scattering is calculated from the measurements of T and $(T + S)$, by difference. In general, it is difficult to have a priori knowledge of the light–NF interactions and the general case where both absorption and scattering coefficients are not negligible should be assumed.

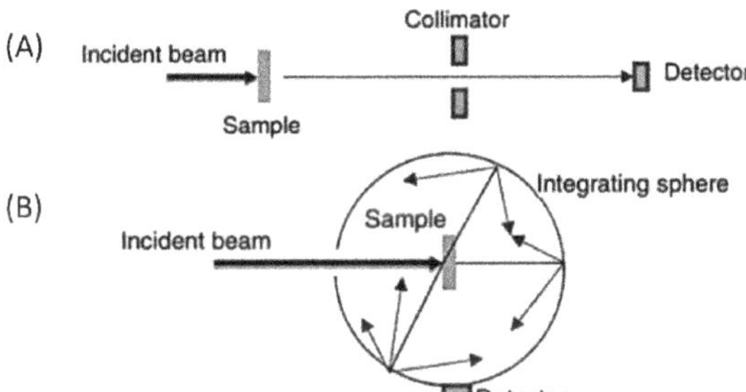

Figure 5. Schematic shows methods of direct measurements of (**A**) transmittance using the transmission port and (**B**) transmittance and scattering combined in the integrating sphere compartment [106].

The light absorbed (I_{abs}) by an NF sample of depth L can be described in terms of the transmitted light intensity I_{tr} at depth x, multiplied by the absorption coefficient, and where Equation (2) has been used:

$$I_{abs} = \int_0^L \mu_{abs} I_{tr}(x) \cdot dx = \mu_{abs} I_{inc} \int_0^L e^{-\mu_{ext} x} \cdot dx = \mu_{abs} I_{inc} \left(\frac{1 - e^{-\mu_{ext} L}}{\mu_{ext}} \right) = \mu_{abs} \frac{I_{inc}(1 - T)}{\mu_{ext}} \tag{5}$$

The absorption coefficient can be then determined by rearranging Equation (5) and substituting the values produced by the measurements of T and $(T + S)$ as follows:

$$\mu_{abs} = I_{abs} \frac{\mu_{ext}}{I_{inc}(1 - T)} = \frac{I_{abs}}{I_{inc}} \frac{\mu_{ext}}{(1 - T)} = [1 - (S + T)] \frac{\mu_{ext}}{(1 - T)} = -\frac{[1 - (S + T)]}{(1 - T)} \frac{lnT}{L} \tag{6}$$

Consequently, the scattering coefficient can be simply determined by subtracting the absorption coefficient in Equation (6) from the extinction coefficient in Equation (4).

$$\mu_{sct} = -\frac{lnT}{L} - \left\{ -\frac{[1 - (S + T)]}{(1 - T)} \frac{lnT}{L} \right\} = \frac{lnT}{L} \left\{ \frac{[1 - (S + T)]}{(1 - T)} - 1 \right\} \tag{7}$$

This approach was used, for instance, to evaluate the properties of surfactant-free Au NPs prepared by McGlynn et al. [61], which exhibited a variety of shapes and sizes, with most particles consisting of small spherical NPs with an average diameter of 27 nm.

They evaluated the absorption and scattering behaviours of their NF (Au NPs in ethylene glycol) across the wavelength range of 300–900 nm, and the results revealed that both absorption and scattering coefficients increased with the NP concentrations, as expected. In addition, the absorption coefficient is dominated by the plasmonic absorption between 540 nm and 560 nm (red in Figure 6b), which is much higher than the corresponding scattering coefficient (red in Figure 6c) for most of the spectral range.

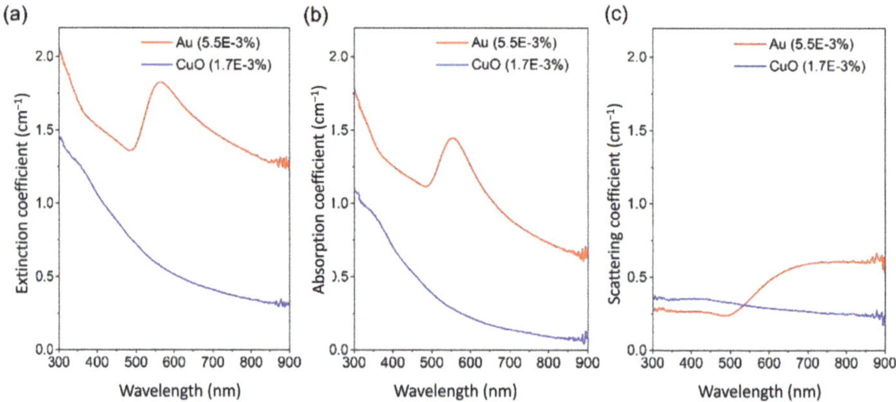

Figure 6. (a) Extinction, (b) absorption, and (c) scattering coefficients for gold (red) and copper oxide (blue) NPs in an ethylene glycol base fluid. Adapted with permission from [61].

Many reports use absorbance to characterize NFs; however, this parameter does not allow for a direct comparison with other NFs and cannot be used directly in simulations [58,78]. The transmittance of the water-based metal oxide NFs of Al_2O_3, CuO, TiO_2, ZnO, CeO_2, and Fe_2O_3 prepared by Milanese et al. [73] at 0.05–1 vol% were measured over 200–1300 nm using a 3 mm path-length quartz cuvette. As depicted in Figure 7a, at 0.05 vol%, the six materials exhibited three transmittance patterns where CuO and Fe_2O_3 transmitted light only in the near-IR region at wavelengths above 800 nm and 600 nm, respectively, while TiO_2, ZnO, and CeO_2 transmitted more in the Vis and near-IR region at wavelengths above ~380 nm; the last material, Al_2O_3, had the lowest performance with most of the light transmitted. In Figure 7b, the distance at which all incident light was attenuated was calculated at different concentrations for the six materials and showed that Al_2O_3 is the worst material while the other five materials have much higher attenuation rates within short distances of less than 0.02 m at a concentration as low as 0.05 vol%. The extinction coefficient of ATO–Ag NPs [76] with an average particle size of 40 nm and dispersed in water at 0.01–0.2 wt% was calculated over 300–900 nm, and its peak was 8.2 cm^{-1} at 300 nm at 0.2 wt% and dropped to 3.7 cm^{-1} at 900 nm. It was also observed that the extinction coefficient exhibited approximately a linear function with the mass fraction. By dispersing graphene nanoplates with a thickness of 2 nm and diameter of less than 2 μm in deionized water at 0.00025–0.005 wt% [82], the NFs demonstrated a strong absorption band in the range of 250–300 nm, with complete absorption across the range of 1400–1550 and above 1850 nm. In addition, the NF with the highest concentration of 0.005 wt% exhibited the highest extinction coefficient with a peak of 6.5 cm^{-1} at 300 nm, which decreased gradually to below 4.5 cm^{-1} at wavelengths above 700 nm. Milanese et al. [73] studied the effect of temperature change within a range of 100–500 °C in 100 °C steps on the optical properties of ZnO, CeO_2 (~4.3 nm in diameter), and Fe_2O_3 (~3.0 nm in diameter) in chloroform solution at 3 mg/mL and 5 mg/mL for Fe_2O_3 and CeO_2 NPs, respectively.

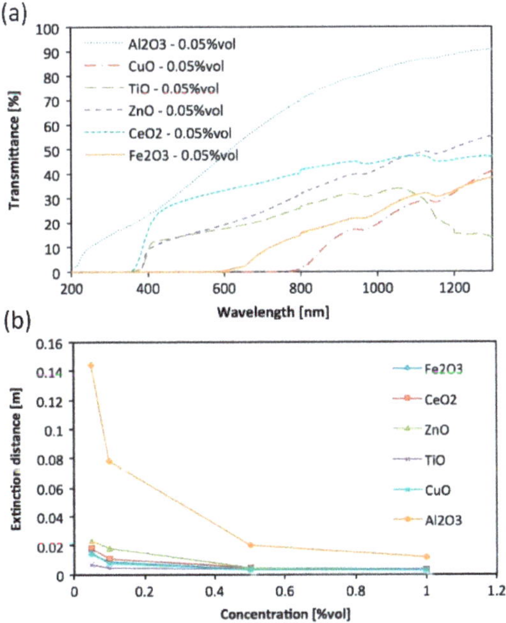

Figure 7. (**a**) Comparison between the transmittance of several NFs as a function of wavelength. (**b**) The extinction distance for investigated NFs as a function of NP concentration, in the case of wavelength equal to 1300 nm. Adapted from [73].

Similar to absorbance, the use of attenuation or extinction coefficients, or transmission only, cannot differentiate between absorbed and scattered light, hence the results can be misinterpreted. In general, assessing the optical properties of NFs using transmittance or absorbance or extinction coefficient only can be misleading. For instance, we have produced NFs from three different types of CuO particles. We have used quantum dots (QDs) [107], commercially available NPs and microparticles (MPs) (Sigma Aldrich with CAS numbers of 1317-38-0 and 1317-38-0, respectively), and each sample of the three CuO particles was dispersed in 10 mL of ethylene glycol (Sigma, CAS number 107-21-1) at a volume fraction of 0.005%. The re-dispersion process was performed for 1 h in a bath-type sonicator (VWR USC300TH, 230 V, 50 Hz, 370 VA) and then for 3 min in a probe-type sonicator (VCX-130PB ultrasonic processor, Sonics Materials) at 80% (104 W) for high dispersion homogeneity. The impact of the different sizes on the scattering is immediately clear in Figure 8c.

In this case, using the extinction coefficient to characterize the optical behaviour of QDs (red in Figure 8a) is a suitable approach for the part of the wavelength range under analysis as scattering (red in Figure 8c) is very low in comparison to the absorption coefficient (red in Figure 8b). However, the scattering coefficient of NPs (blue in Figure 8c) is comparable to the absorption coefficient (blue in Figure 8b), hence the first cannot be neglected. This is even more obvious for MPs where the extinction coefficient (green in Figure 8a) is very high throughout the full wavelength range. However, the analysis of absorption and scattering separately (green in Figure 8b,c, respectively) shows that much of the extinction is due to scattering in the range 800–900 nm. In this specific case, the higher scattering is most likely all due to the relationship between incident wavelength and particle size as per Mie's theory. However, we should note that scattering can take over the absorption also due to other light–NP interaction phenomena, such as is the case for NPs with plasmonic properties (e.g., red in Figure 6c) as plasmon resonance enhances scattering above the plasmonic peak wavelength. It is therefore important to evaluate NF properties appropriately for these to be used in assessing NF performance and in more complex analytical and simulation efforts.

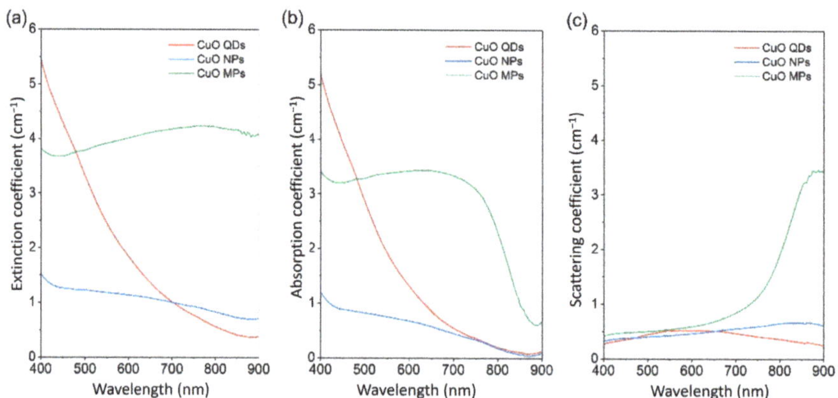

Figure 8. Coefficients of (**a**) extinction, (**b**) absorption, and (**c**) scattering of three different types of CuO at different sizes dispersed in ethylene glycol (mean diameter: ~3.5 nm QDs, ~50 nm NPs, ~3–10 μm MPs).

5. Characterization of the Stability of Nanofluids for DASC Applications

Nanofluids are colloids of nanoparticles within base fluids and it is critical for the performance of NF-based DASCs to address the long-term stability under various thermal and solar radiation operating conditions. Dispersed particles are prone to destabilization under the effect of forces, such as Van der Waal and gravity, which can induce agglomeration leading to the formation of larger particles/agglomerates and enhancing sedimentation. Other forces, such as buoyancy, steric, and electrostatic forces, can play opposite roles in improving particle stability, and nanofluid stability is dependent on which group of forces dominates. The formation of large agglomerates leads to a sequence of operational issues and contributes to the deterioration of the thermal and optical properties of NFs, affecting the overall performance of DASCs. Such destabilization/stabilization forces are highly affected by the operating conditions (temperatures, particle loading, and incident solar radiation). In addition, the chemical stability of NFs and the NPs or working fluid separately is important to maintain NF properties across operational conditions.

High operating temperatures increase Brownian motion leading to higher particle collision frequencies, also enhanced by a reduction in viscosity of the base fluid at higher temperatures. Temperature change also has an impact on the surface charge and zeta potential of the particles [108]. In addition, an increase in temperature can lead to the degradation of surfactants in the case of steric-stabilized NPs [109]. Direct exposure of the NFs to concentrated radiation (intensity higher than one sun) can cause the chemical deterioration of surfactants and functional groups and even of the chemical composition or phase of the NPs, particularly from UV radiation [110]. Furthermore, medium- and high-temperature DASCs operate at high solar radiation intensity, which requires relatively high particle loadings, i.e., more particles per unit volume and consequently shorter interparticle distance, which induce agglomeration of particles at higher rates compared to low particle loadings. Moreover, due to heating during exposure to sunlight in the daytime and cooling during nighttime, this daily cyclic heating/cooling impacts several parameters, including changes in the morphology of the NPs that could promote agglomeration and chemical reactions such as oxidation [83,111] as well as a reduction in thermophysical properties [112]. For these reasons, the chemical stability of NPs needs to be assessed in the specific working fluid as several chemical reactions could be taking place at the NP–fluid solid–liquid interface that would not normally take place in other environmental conditions.

The stability of NFs can be assessed following various methodologies, which can produce operational stability parameters and an understanding of relevant underlying mechanisms. The observation of discolouration and/or particle sedimentation of the NFs is perhaps one of the simplest and most direct assessment methods. However, seldom can it provide quantifiable operational measures and scientific insights. On the other hand, measuring the changes in the optical properties, morphology, surface chemical composition, etc., of the NPs can be more helpful in analysing and understanding stability.

Sharaf et al. [78] extensively investigated the stability of polymer-coated Au NPs/water against several destabilization factors, including ambient storage for 16 months, elevated temperature, continuous and cyclic modes of solar radiation, and intensified UV radiation. In their work, they tested four samples: CIT–Au NPs and CIT–Au NPs coated using three different polymers to give BSA–AuNPs, PVP–AuNPs, and PEG–AuNPs. The results revealed that the ambient dark storage for three years did not affect the stability and PEG–AuNPs demonstrated the best long-term stability. The absorbance measurements showed that no plasmonic wavelength shifts or shoulder evolutions took place in all four NFs. The absorbance peak intensity, however, increased for PVP– and BSA–Au NPs by 9.0% and 9.5%, and this was attributed to changes in the thickness of the adsorbed polymer coating over time, while the peak intensity of CIT–Au NPs dropped by 9.6% due to sedimentation. In addition, after the samples were exposed to broadband solar radiation at different intensities for 12 h, CIT–Au NPs resulted to be the most stable with no changes in absorbance intensity, while both PEG– and PVP–Au NPs showed a drop in absorbance and the hydrodynamic diameter under radiation exposure. However, no shifting or broadening in the absorbance peaks of PEG– and PVP–Au NPs were observed, indicating that they experienced no agglomeration after being exposed to radiation. Furthermore, cyclic radiation for 20 cycles at one-sun intensity for 8 h followed by 16 h in the dark produced more significant changes in absorbance and hydrodynamic diameter compared to exposure to continuous radiation, with the CIT–Au NPs less impacted than the polymer-coated samples. Moreover, they tested the NFs at elevated temperatures of 55 °C, 70 °C, and 85 °C for 12 h, and all NFs showed excellent thermal stability confirmed by their absorbance measurements. The stability of the same NFs (aqueous NFs with Au NPs coated by CIT, PEG, PVP, and BSA) was then investigated for a much longer period [60]. Over three years of ambient storage, continuous irradiation at varying intensity for 12 h and 20 consecutive cycles of one-sun illumination (8 h under light, followed by 16 h in the dark), the NFs exhibited no shifts in their plasmonic absorption peak wavelength and insignificant changes in the hydrodynamic diameter. They also reported that CIT-coated AuNPs were the most stable under continuous solar radiation. The effect of ambient storage on the stability of CTAC surfactant-based Au NPs/water was investigated by Chen et al. [58] and they found that the NFs were stable after days but the increase in temperature up to 90 °C worsened the stability and rapid sedimentation appeared several hours after the experiments. They attributed the post-experiment instability to agglomeration induced by Brownian motion and surfactant degradation at high temperatures.

Gorji and Ranjbar [80] measured the zeta potential of three different types of materials (Ag NPs, Fe_3O_4 NPs, and graphite) in water and found that the graphite/water exhibited the highest value of 45.1 mV compared to 41.3 mV and 38.7 mV for Fe_3O_4/water and Ag NPs/water, respectively, all at a pH of 9.5–10. Coating Ag NPs with SiO_2 and then dispersing them in water to form Ag NPs@SiO_2/water NFs resulted in better stability when compared with uncoated Ag NPs in the same base fluid, which exhibited poor stability under both heating and UV radiation cycles [83]. Kimpton et al. [62] investigated the stability of both Ag NPs/water with and without coating with SiO_2. They monitored the stability of both NFs over three weeks of ambient storage, as well as under exposure for a period of two weeks to simulated and natural solar radiation, the latter during the first half of June 2018 in the UK. As shown in Figure 9, they assessed the impacts of destabilization factors by studying the UV–Vis spectra and the morphology of the particles via transmission electron microscopy (TEM). Their results revealed a lot of interesting points when comparing bare Ag NPs and SiO_2-coated Ag NPs, as well as differences due to the use of simulated and natural light. For instance, SiO_2-coated Ag NPs showed less stability than the uncoated Ag NPs, as the former showed a higher reduction and position shifting in the absorption peak after ambient dark storage and upon exposure to simulated light. This lack of stability was explained by the changes in the morphology of the SiO2@Ag NPs samples verified by TEM. They found a reduction in the size of Ag NPs inside SiO_2 shells by about 10 nm suggesting that the silver was being etched inside the coating. In addition, they found degradation of the SiO_2 shell, since it had reduced in size, and some Ag NPs were found no longer coated. Based on these results, they carried out further modifications by using thicker SiO_2 coatings to improve stability. However, they found that increasing the coating thickness accelerated the change in morphology of Ag NPs under exposure to light. On the other side, exposure to natural sunlight revealed similar results to that obtained under exposure to simulated light in terms of the shift of the absorbance peak position, but different in the reduction in the peak values. They attributed this to a faster change in the morphology of the particles.

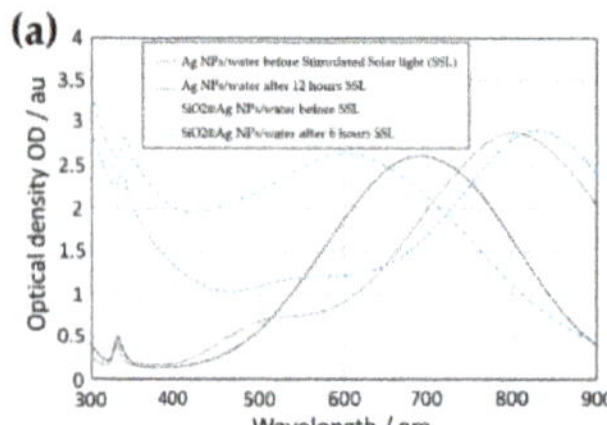
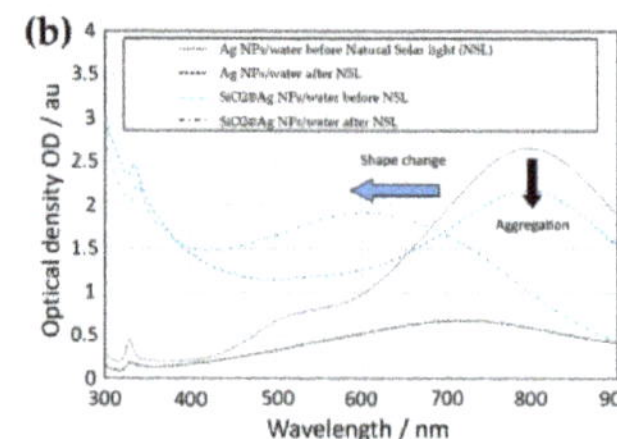
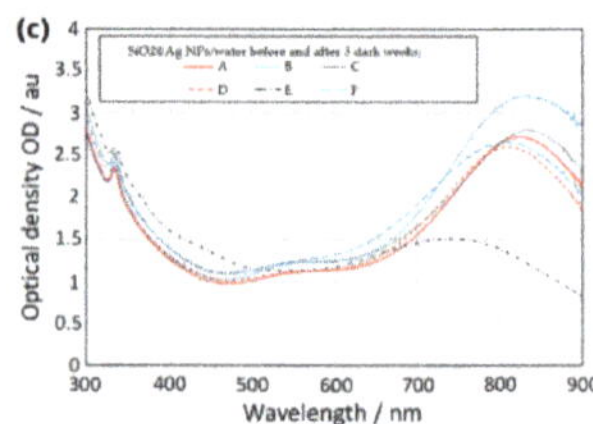

Figure 9. UV–Vis spectra of Ag NPs and SiO2@Ag NPs in water: (**a**) after exposure to stimulated solar light, (**b**) after exposure to natural solar light for two weeks, and (**c**) after storage in the dark for three weeks (adapted from [62]).

Zheng et al. [79] studied the dispersion stability of CTAB surfactant-based NFs of CNT/water and CNT–TiN/water over ambient storage of one week. At different mass fractions of up to 30 ppm, they found no precipitation after the one-week test, and to confirm this, they recorded the time-dependent absorbance spectra of the single- and multicomponent NFs. At low concentrations, they found overlapping in the spectra between the as-prepared samples and the same samples after seven days, while samples with higher concentrations exhibited some changes in absorbance.

Chen et al. [68] found that CQDs/PEG-200 prepared by microwave heating are stable over 30 days of ambient storage. They observed light sedimentation from optical images and very small changes in transmittance spectra. GO/water–EG NFs with GO of 0.06 wt% were irradiated with an ultraviolet lamp for 240–615 s to obtain a series of GO//water–EG NFs [88]. For comparison purposes, a reduced GO/water NF was also prepared at 0.02 wt% using PVP surfactant followed by UV irradiation for 340 s. After two months of ambient storage, GO/water–EG NFs obtained under irradiation durations of ≤465 s exhibited no sedimentation, while some sediments were observed at longer duration. When the temperature changed from 30 °C to 70 °C, GO/water–EG NFs at 0.06 wt% showed an increase in zeta potential, while the water-based GO NFs at 0.02 wt% showed the opposite, highlighting the effect of the base fluid used on the stability of NFs.

In a study on NF samples of two metal oxides of spherical Al_2O_3 NPs of 40 nm diameter and spherical CuO NPs of 100 nm diameter, as well as the combination of both (Al_2O_3-CuO NPs) in EG [85], the stability of the samples was tested over two weeks of ambient storage. NFs based on metal oxides offer interesting features for DASC applications due to their chemical inertness and thermal stability. Whilst many stoichiometric oxides exhibit very large bandgap values and low or no absorption in the solar spectral range, it is possible to produce metal oxide NPs with strong absorption also in the visible [29]. Pure CuO, for instance, deviating from the characteristics of most oxides, has a relatively low bandgap (~1.2 eV [113]), which is suitable for absorbing solar energy. However, the bandgap is impacted by quantum confinement and given the relatively large CuO Bohr radius (~28.27 nm [114,115]), the size of CuO NPs should be sufficiently large not to experience significant bandgap widening. Adversely, larger NPs will tend to affect the NF stability. This can be shown again by comparing the stability of the same CuO QDs, NPs, and MPs that we reported in Figure 8. In order to assess the long-term static stability of NFs, we have monitored the transmittance over days (Figure 10a–c), with the NF stored in a dark space at room temperature without any form of mixing or shaking ('static stability'). The results show that while QDs (Figure 10c) exhibit superior stability compared to MPs (Figure 10a) or NPs (Figure 10b), their transmittance is relatively high, a consequence of quantum confinement and widening of the bandgap (to 2.2 eV [116,117]). As the cumulative transmittance increases with the size of the CuO particles, the static stability worsens and the best compromise between optical performance and stability should be found. However, static stability tests can be viewed as too severe for DASC applications, considering that NFs are generally in some form of motion when used in DASCs. Here, we show that after a simple manual shaking ("after shaking" in Figure 10a–c), the transmittance of all different types of CuO particles return to the values measured at the start (D0)

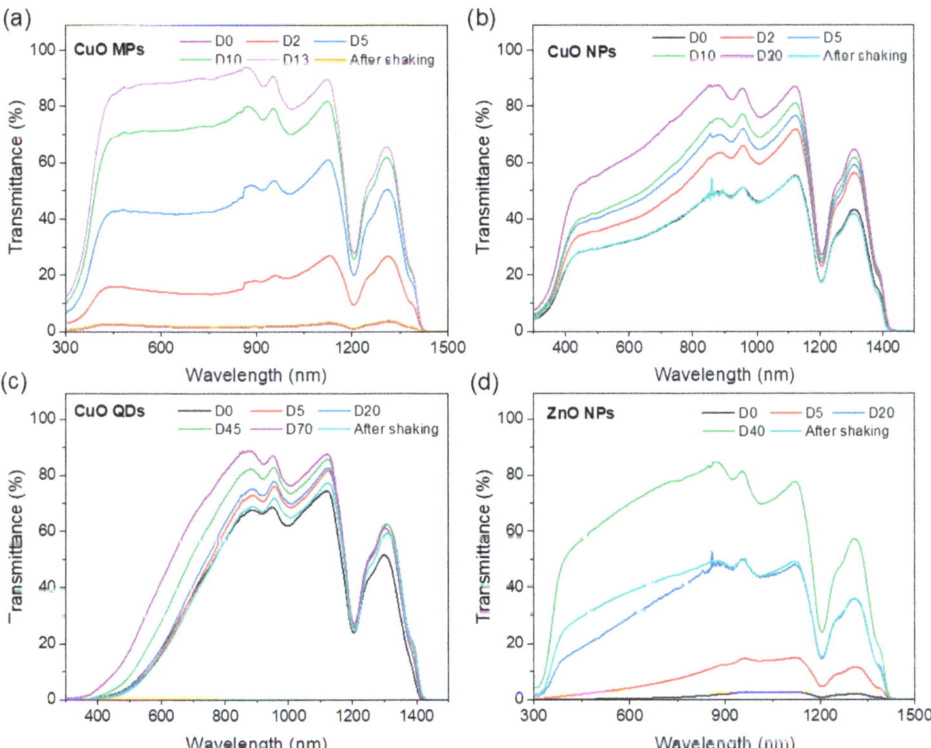

Figure 10. Transmittance variation in (**a**) CuO MPs, (**b**) CuO NPs, and (**c**) CuO QDs in addition to (**d**) ZnO NPs in EG across a wavelength range of 300–1500 nm. The transmittance measurements were performed in a static mode, i.e., without shaking, and were then manually shaken before the last measurement "After shaking".

The manipulation of the morphology or chemical composition of metal oxide NPs also offers other opportunities to enhance the combined optical-stability performance of NFs. In Figure 10d, we report the stability of NFs produced with oxygen-deficient ZnO NPs. The results show substantial improvements in the optical properties with very low cumulative transmittance at day 0 (D0) compared to the expected high transparency of ZnO NPs, due to the very high ZnO bandgap. While the stability of these NFs with ZnO NPs requires some optimization, these still outperform other CuO-based NFs after 20 days.

Performance of Nanofluid-Based DASCs

DASCs are used in solar thermal systems, as illustrated in Figure 11a. The overall performance of DASC systems is impacted by various losses. Before reaching the nanofluid surface, the incident solar radiation intensity (I_0) is reduced by reflection at the surface of the transparent glass layer ($I_{ref_{glass}}$) as well as absorption through it ($I_{abs\,glass}$). It is worth mentioning that despite this loss in incident radiation intensity, the glass layer plays an important role in improving the overall performance of the DASCs by reducing the thermal energy loss to the surrounding, which is relatively high compared to the loss in radiation intensity. The amount of solar radiation reaching the nanofluid (I_{nf}) undergoes an STC process through which the solar radiation is absorbed and scattered while travelling through the NF, as shown in Figure 11c (see also Figure 3). Here, we refer to the total radiation absorbed as $I_{abs\,nf}$, while we use $I_{sc\,nf}$ to refer to the amount of light that is scattered or that escaped the NF without re-absorption. Furthermore, thermal energy is then generated (Q_{gen}) from the absorbed radiation via an STC process. With the increase in the NF temperature above the surrounding temperature, a portion of the generated thermal energy is lost to the surrounding, as shown in Figure 11b, by convection ($Q_{loss\,conv}$) and radiation ($Q_{loss\,rad}$). Thus, the thermal energy gain within the NF, i.e., the net thermal energy ($Q_{net\,nf}$), can be obtained using the energy balance as follows:

$$Q_{net\,nf} = I_0 - \left[\left(I_{ref_{glass}} + I_{abs\,glass}\right) + I_{sc\,nf} + \left(Q_{loss\,conv} + Q_{loss\,rad}\right)\right] \quad (8)$$

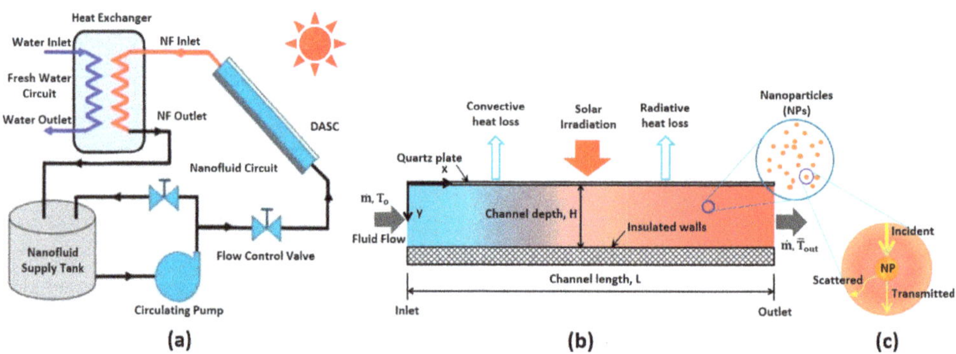

Figure 11. DASC schematics; (**a**) simplified closed-loop system for heat transfer. Adapted from [24], (**b**) irradiation and heat loss sources. Adapted from [118], (**c**) nanoparticle and solar irradiation interaction.

The net thermal energy can be also determined at a given mass flow rate ($\dot{m}$), specific heat capacity (c_p), and temperature rise (ΔT) of the NF:

$$Q_{net\,nf} = \dot{m} \times c_p \times \Delta T \quad (9)$$

The overall efficiency of a DASC system, (η_{DASC}), can be determined from the ratio of the net thermal energy stored by the nanofluid and the incident solar radiation, as follows:

$$\eta_{DASC} = \frac{Q_{net\,nf}}{I_0} \quad (10)$$

Other aspects to be considered when evaluating the overall efficiency of a DASC system include the diffraction and reflection of light at the outer and inner surfaces of the glass layer, as well as at the interfaces between the NF and the DASC walls that contain the NF, and the variation in the ambient temperature.

DASCs can take a variety of geometries/shapes to achieve the desired performance within a defined range of operating temperatures. For instance, Chen et al. [58] described the effect of DASC design on the plasmonic absorption-induced STC of Au NPs/water NFs in a static mode under simulated solar radiation of 450 W/m². They tested flat and cubic DASCs made of polymethyl methacrylate (PMMA). The flat- and cubic-shaped DASCs were designed with inner dimensions of 100 mm × 100 mm × 6 mm and 50 mm × 50 mm × 48 mm, respectively, and both were thermally insulated. As depicted in Figure 12, they found that at a concentration of Au NPs of as low as 0.000008 wt%, the flat-shaped DASC system improved the STC efficiency by 21.3%, which was higher than that of the cube-shaped DASC (19.9%) at the same concentration. They also included the impact of variation in the size of Au NPs (25 nm, 33 nm, and 40 nm) to see if this can alter the performance results of both shapes. The STC efficiency in the flat DASC increased by reducing the NP size, while this size reduction did not influence the cube-shaped DASC. However, the cube-shaped DASC exhibited higher overall efficiency since the flat-shaped DASC exhibited higher heat loss. From this, we can see that the performance of DASC systems can be tuned by their size and shape.

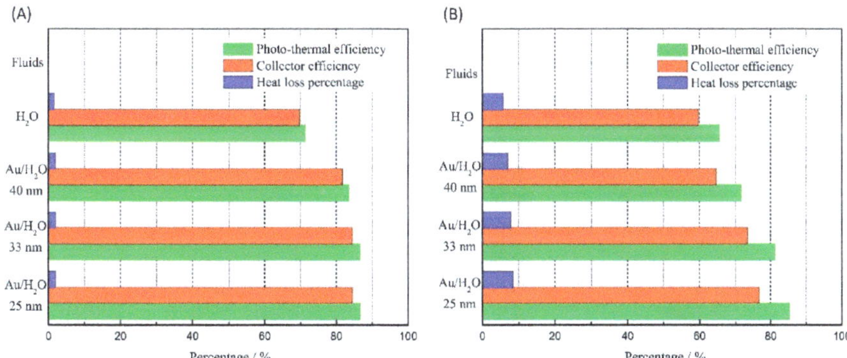

Figure 12. Photo-thermal and collector efficiencies as well as the heat loss of different NFs in (**A**) cube-shaped and (**B**) flat DASCs (adapted from [58]).

The effect of NP concentration and solar radiation intensity were studied by Wang et al. [77]. They investigated the effect of these parameters on the absorption and STC performance of EG-based TiN NFs. They used TiN NPs at various concentrations (0.001–0.01 wt.%) in static mode in a thermally insulated (with aerogel) DASC made of quartz with a radius and height of 1.75 cm and 4 cm, respectively. They irradiated the samples with solar radiation of intensity up to two suns (2000 W/m²) and studied the effect of two different directions of the radiation on the DASC performance: side and bottom. The absorption performance of the NFs increased by ~50% with the increase in temperature from 0 °C to 60 °C. At a concentration of 0.003 wt.%, the bottom irradiation mode achieved a conversion efficiency of 45% higher than that of side irradiation. However, the latter saved nearly 40% of the time required to reach steady-state temperature when compared with the side radiation mode. In addition, both radiation direction modes exhibited uniform temperatures over the NF depth of 1 cm. Ag/water, Fe₃O₄/water and graphite/water were tested in a dynamic mode (fluid circulation) [80] in a rectangular channel made of stainless steel (12 mm × 5 mm × 2 mm), with the top glazing made of low-reflectance borosilicate glass of 4 mm and the inner surface of the bottom wall covered with reflective steel. The DASC was thermally insulated by a 3 cm thick Styrofoam with a thermal conductivity of 0.03 W/m K. At varying simulated solar radiation of 600 W/m^2, 800 W/m^2 and 1000 W/m^2, Fe$_3$O$_4$/water showed the highest outlet temperature and collector efficiency at all radiation flux values followed by graphite/water, with Ag/water being the lowest.

Fe$_2$O$_3$/water NFs at two concentrations of 0.5 wt% and 2 wt% were investigated by Balakin et al. [74] in a dynamic mode with Reynolds numbers up to 1000 in a DASC of 400 mm long transparent glass tube with an internal diameter of 4 mm. The NFs reached a maximum thermal efficiency of 65% at the highest concentration of 2 wt%, while the lowest concentration of 0.5 wt% achieved 8%. In another similar work [75] Fe$_2$O$_3$/water NFs were used in an evacuated tubular DASC made of a borosilicate glass tube of 140 cm in length and containing a smaller glass tube of 2.5 cm and 4 cm inner and outer diameters, respectively. The flow rate was 400 mL/min (Reynold number of 340) and a surface area of 0.055 cm^2 was exposed to solar radiation. At a 4 cm sample depth, Fe$_2$O$_3$/water at 0.02 vol% absorbed all incident radiation while the other NF absorbed 17% less at the same volume fraction. In terms of STC efficiency, Fe$_2$O$_3$/water confirmed its better performance over Fe$_3$O$_4$/water by a maximum efficiency of 73%, which is higher by 22% than the maximum efficiency reached by Fe$_2$O$_3$ NF at 0.02 vol%. CQDs of less than 10 nm in diameter prepared by microwave heating [68] were dispersed in PEG-200 and investigated in a static mode under simulated solar radiation of 1–4 sun for 60 min under continuous magnetic stirring in a DASC made from a quartz vessel of 3 cm × 3 cm × 3 cm with a sponge plate on the top to reduce sample evaporation. The results revealed that extending the microwave heating duration has an advantageous impact on STC efficiency. In addition, the efficiency reached its maximum of 81% when exposed to 1 sun and microwave heating was 24 min, which is triple the efficiency of the base fluid PEG-200 (27%).

Guo et al. [50] produced NFs with GO (0.5–5 μm diameter), CNTs (−5 nm and 8–15 nm inner and outer diameters), and Ti$_3$C$_2$ at concentrations from 0.001 wt% to 0.04 wt% for investigating their STC performance in a static mode in a quartz DASC of 2.5–4 cm height. The collector was covered with a sponge, except for the top part, and it was exposed to varying solar concentrations from 1 sun to 2.5 suns. Ti$_3$C$_2$ outperformed both GO and CNTs by achieving an STC efficiency of 20% and 29% higher. The STC efficiency of the 0.04 wt% Ti$_3$C$_2$ NF was 2.7 times that of the pure working fluid when irradiated for 6500 s at a solar concentration ratio of 2.5 with a DASC height of 2.5 cm. In terms of concentration effect, they found that 0.02 wt% was the optimal particle concentration since no significant enhancement in performance was found at higher concentrations. Zheng et al. [79] studied the solar absorption and STC efficiency of a single-component NF of CNT/water (5–10 nm and 20–30 nm inner and outer diameters and 10–30 μm length) and a muti-component NF of CNT–TiN/water (TiN NPs are 20 nm in diameter) at mass fractions of 10 ppm, 20 ppm, and 30 ppm. The DASC that they used for this work was a cylinder made of acrylic with a 15 mm height, 26.4 mm inner diameter and 2 mm wall thickness. To reduce heat losses, they covered the DASC top with a 1 mm quartz glass of over 90% light transmittance and they also used an aluminium mirror at the bottom to reflect transmitted light back to the NFs. In a static mode and under one sun radiation intensity, the STC efficiency of the single- and multicomponent NF reached 25% and 26%, respectively, which is higher by 8.7% and 13.0%, respectively, compared to that of the base fluid. The increase in STC efficiency by the CNT–TiN/water over the single-component NF was attributed to the plasmonic absorption of TiN NPs. The increase in mass fraction above 10 ppm increased scattering by TiN NPs, which negatively impacted the efficiency.

The physical shape of NPs plays a crucial role in determining the performance of NFs. Nanoparticles with different shapes have distinct physicochemical properties, such as surface area, surface energy, and surface charge. These properties affect the absorption, stability, and thermal conductivity of NFs. Maheshwary et al. [119] explored the impact of NP shapes on the thermal conductivity and stability enhancement of NFs. The study investigated the thermal conductivity and stability of five different water-based NFs, CuO, MgO, TiO$_2$, ZrO$_2$, and Al$_2$O$_3$, with different shapes of spherical, cubic, and rod-shaped NPs.

All NFs were prepared with the same concentration of 2.5 wt%. It was revealed that the NF of cubic Al_2O_3 showed the highest thermal conductivity over the base fluid by 3.13 times. However, the cubic-shaped NPs exhibited lower stability compared to spherical and rod shapes. Chen et al. [120] reported that cubic Au NPs produce higher solar heating and higher absorption than spherical and cylindrical NPs, as shown in Figure 13. Additionally, Wang et al. [121] indicated that the absorption, in terms of both magnitude and bandwidth, of the cubic core–shell carbon Au NPs was better than spherical NPs.

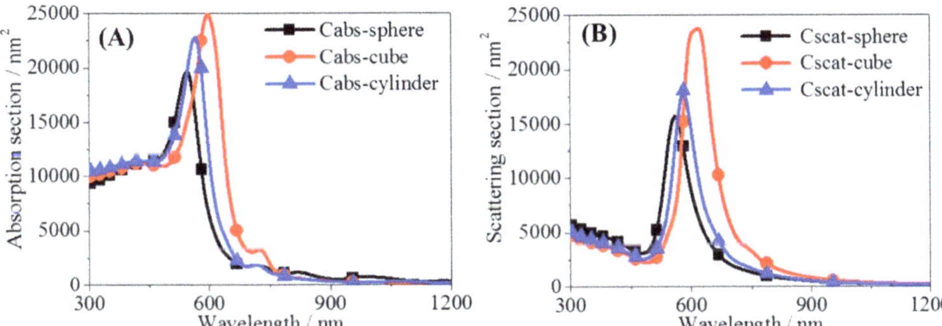

Figure 13. Optical properties of different Au nanostructures; (**A**) absorption section, and (**B**) scattering section. Adapted from [120].

Using mixed NP morphologies can enhance the performance of NFs in terms of absorption and thermal conductivity. Mixed Au NPs with different shapes were used by Duan et al. [122] to prepare a plasmonic water-based NF and study this for DASC applications. A plasmonic-blended NF was created by mixing Au NPs of three distinct shapes (spherical, rod, and star-shaped) in water. Both experimental and numerical studies showed that the optical and thermal properties of blended NFs indicate that the extinction spectra of Au blended NFs were wider in comparison to single-component nanofluids. Moreover, the photothermal properties of these three mixed shapes of NPs showed a higher temperature rise and higher heat collection ability, and the experimental results aligned with the numerical analysis findings.

NFs have shown promising potential in the performance enhancement of DASCs by increasing the heat transfer rate and increasing the absorption of the solar spectrum. However, there are some limitations and disadvantages that need to be considered before implementing NFs in DASCs. One of the primary concerns is the stability of NFs since NPs tend to agglomerate and settle down over time and negatively affect the homogeneity of the fluid properties. Consequently, the usability of NFs will be limited for some applications, especially those that demand a higher concentration of NPs [63,123]. Corrosion is also another challenge, as some NPs can corrode the components of the solar collector, leading to material degradation and reduced efficiency [92]. The viscosity of NFs is also a problem, as NFs have higher viscosities compared to traditional heat transfer fluids, which means that they encounter greater resistance when flowing through a system. As a result, the pressure drop in the system increases and more power is needed to maintain the flow rate [124]. Overall, while the deployment of NFs for DASCs still presents some challenges, scientific and technological progress has shown that these can be overcome in the future.

6. Conclusions: Challenges, Research Gaps and Future Perspectives

From the literature reviewed above, researchers have clearly made commendable progress in the development of efficient and stable NFs for DASCs. However, research work is still required to bring this new technology from lab experiments to large-scale production and attract investments from the industrial sector. The commercialization of

NF-based DASCs will depend on completely resolving issues such as the instability of NFs, high costs and the synthesis complexity of nanomaterials [125], increased pumping power and the appearance of mechanical erosion and chemical corrosion due to deposition of NPs on the inner surfaces of flow channels [126], in addition to possible environmental concerns [127,128]. An in-depth understanding is required for sedimentation, agglomeration, morphological transformations, and the chemical deterioration of NPs in base fluids over long periods of operation, at elevated temperatures, and under thermal and solar cycling processes, as these factors are all critical to the optical, thermal, and rheological properties of NFs. Many optimization avenues are available and, for instance, NP loading, flow conditions, DASC system geometry, and NP chemical/structural properties can all be studied and optimized to overcome the current challenges.

High concentrations of NPs increase scattering at the surface, increasing radiative losses, which further increase with particles of large sizes, as well as higher heat losses due to the overheating of the NF surface [128]. At very low NP loadings, e.g., less than 0.1 vol%, such limitations may not affect the applicability of NFs in DASCs. Another benefit of using very low particle loadings in DASCs is that it alleviates the problem of nanomaterials synthesis, which can be costly and sophisticated in some cases [129]. However, because low particle loading may compromise the ability of NFs to maintain high solar absorption and thermal conduction of the NFs, further research work is still needed to develop innovative solutions for simple and cost-effective techniques for both the synthesis of nanomaterials and the preparation of stable NFs. Since DASCs are a green technology that depends on solar energy, it is highly important to use NFs with no environmental impacts. In some cases, NFs may have environmental and toxicological effects [130–132], hence the selection of appropriate size, surface properties, and chemical composition is also important.

Solar thermal energy conversion technologies can greatly benefit from ongoing and future NF research. NF fundamental and physical properties can be reliably assessed in a laboratory setting and can provide valuable information in the development of full DASC systems aided by computer simulations. The availability of sophisticated synthesis methodologies in combination with fundamental materials science investigations can produce very high specification NFs at sufficiently low costs for the full deployment of NF-based DASCs.

Author Contributions: Conceptualization, H.S.M. and D.M.; methodology, H.S.M., V.A. and D.M.; validation, H.S.M. and V.A.; formal analysis, H.S.M., S.K., S.C., P.M. and D.M.; investigation, H.S.M. and V.A.; resources, V.A., S.C., P.M. and D.M.; data curation, H.S.M., S.K., S.C. and D.M.; writing—original draft preparation, H.S.M., S.K., S.C. and D.M.; writing—review and editing, H.S.M., V.A., S.K., S.C., P.M. and D.M.; visualization, H.S.M., S.K., S.C., P.M. and D.M.; supervision, S.C. and D.M.; project administration, V.A., S.C. and D.M.; funding acquisition, V.A., S.C., P.M. and D.M. All authors have read and agreed to the published version of the manuscript.

Funding: This research was funded by EPSRC grant number EP/V055232/1 and EP/M024938/1 and the Department for Economy-Northern Ireland grant number USI160.

Data Availability Statement: The data presented in this study are available on request from the corresponding author.

Conflicts of Interest: The authors declare no conflict of interest.

References

1. Alper, A.; Oguz, O. The role of renewable energy consumption in economic growth: Evidence from asymmetric causality. *Renew. Sustain. Energy Rev.* **2016**, *60*, 953–959. [CrossRef]
2. Ohler, A.; Fetters, I. The causal relationship between renewable electricity generation and GDP growth: A study of energy sources. *Energy Econ.* **2014**, *43*, 125–139. [CrossRef]
3. IEA. *World Energy Outlook 2017*; IEA: Paris, France, 2017.
4. Kabir, E.; Kumar, P.; Kumar, S.; Adelodun, A.A.; Kim, K.-H. Solar energy: Potential and future prospects. *Renew. Sustain. Energy Rev.* **2018**, *82*, 894–900. [CrossRef]

5. Conti, J.; Holtberg, P.; Diefenderfer, J.; LaRose, A.; Turnure, J.T.; Westfall, L. *International Energy Outlook 2016 with Projections to 2040*; USDOE Energy Information Administration (EIA): Washington, DC, USA, 2016.
6. Kalogirou, S.A. Solar thermal collectors and applications. *Prog. Energy Combust. Sci.* **2004**, *30*, 231–295. [CrossRef]
7. Birol, F. *World Energy Outlook*; International Energy Agency: Paris, France, 2010; Volume 1.
8. Mallah, A.R.; Zubir, M.N.M.; Alawi, O.A.; Newaz, K.M.S.; Badry, A.B.M. Plasmonic nanofluids for high photothermal conversion efficiency in direct absorption solar collectors: Fundamentals and applications. *Sol. Energy Mater. Sol. Cells* **2019**, *201*, 110084. [CrossRef]
9. Huide, F.; Xuxin, Z.; Lei, M.; Tao, Z.; Qixing, W.; Hongyuan, S. A comparative study on three types of solar utilization technologies for buildings: Photovoltaic, solar thermal and hybrid photovoltaic/thermal systems. *Energy Convers. Manag.* **2017**, *140*, 1–13. [CrossRef]
10. Nwaji, G.N.; Okoronkwo, C.A.; Ogueke, N.V.; Anyanwu, E.E. Hybrid solar water heating/nocturnal radiation cooling system I: A review of the progress, prospects and challenges. *Energy Build.* **2019**, *198*, 412–430. [CrossRef]
11. Pérez-Aparicio, E.; Lillo-Bravo, I.; Moreno-Tejera, S.; Silva-Pérez, M. Economical and environmental analysis of thermal and photovoltaic solar energy as source of heat for industrial processes. In *AIP Conference Proceedings*; AIP Publishing: Long Island, NY, USA, 2017; Volume 1850, p. 180005.
12. Goel, N.; Taylor, R.A.; Otanicar, T.J.R.E. A review of nanofluid-based direct absorption solar collectors: Design considerations and experiments with hybrid PV/Thermal and direct steam generation collectors. *Renew. Energy* **2020**, *145*, 903–913. [CrossRef]
13. Bhalla, V.; Tyagi, H. Parameters influencing the performance of nanoparticles-laden fluid-based solar thermal collectors: A review on optical properties. *Renew. Sustain. Energy Rev.* **2018**, *84*, 12–42. [CrossRef]
14. Gorji, T.B.; Ranjbar, A.A.J.R.; Reviews, S.E. A review on optical properties and application of nanofluids in direct absorption solar collectors (DASCs). *Renew. Sustain. Energy Rev.* **2017**, *72*, 10–32. [CrossRef]
15. Phelan, P.; Otanicar, T.; Taylor, R.; Tyagi, H. Trends and opportunities in direct-absorption solar thermal collectors. *J. Therm. Sci. Eng. Appl.* **2013**, *5*, 021003. [CrossRef]
16. Minardi, J.E.; Chuang, H.N. Performance of a "black" liquid flat-plate solar collector. *Sol. Energy* **1975**, *17*, 179–183. [CrossRef]
17. Huang, B.-J.J.; Wung, T.-Y.Y.; Nieh, S. Thermal analysis of black liquid cylindrical parabolic collector. *Sol. Energy* **1979**, *22*, 221–224. [CrossRef]
18. Arai, N.; Itaya, Y.; Hasatani, M. Development of a "volume heat-trap" type solar collector using a fine-particle semitransparent liquid suspension (FPSS) as a heat vehicle and heat storage medium Unsteady, one-dimensional heat transfer in a horizontal FPSS layer heated by thermal radiatio. *Sol. Energy* **1984**, *32*, 49–56. [CrossRef]
19. DeWinter, F. *Solar Collectors, Energy Storage, and Materials*; MIT Press: Cambridge, MA, USA, 1990; Volume 5.
20. Otanicar, T.P.; Phelan, P.E.; Golden, J.S. Optical properties of liquids for direct absorption solar thermal energy systems. *Sol. Energy* **2009**, *83*, 969–977. [CrossRef]
21. Das, S.K.; Choi, S.U.S.; Patel, H.E.J.H.t.e. Heat transfer in nanofluids—A review. *Heat Transf. Eng.* **2006**, *27*, 3–19. [CrossRef]
22. Mercatelli, L.; Sani, E.; Fontani, D.; Zaccanti, G.; Martelli, F.; Di Ninni, P. Scattering and absorption properties of carbon nanohorn-based nanofluids for solar energy applications. *J. Eur. Opt. Soc.-Rapid Publ.* **2011**, *6*, 11025. [CrossRef]
23. Duncan, A.B.; Peterson, G.P. Review of microscale heat transfer. *Appl. Mech. Rev.* **1994**, *47*, 397–428. [CrossRef]
24. Chamsa-Ard, W.; Brundavanam, S.; Fung, C.C.; Fawcett, D.; Poinern, G. Nanofluid types, their synthesis, properties and incorporation in direct solar thermal collectors: A review. *Nanomaterials* **2017**, *7*, 131. [CrossRef]
25. Masuda, H.; Ebata, A.; Teramae, K. Alteration of thermal conductivity and viscosity of liquid by dispersing ultra-fine particles. Dispersion of Al_2O_3, SiO_2 and TiO_2 ultra-fine particles. *Netsu Bussei* **1993**, *7*, 227–233. [CrossRef]
26. Choi, S.U.S.; Eastman, J.A. *Enhancing Thermal Conductivity of Fluids with Nanoparticles*; Argonne National Lab: Lemont, IL, USA, 1995.
27. Mahian, O.; Kianifar, A.; Sahin, A.Z.; Wongwises, S. Performance analysis of a minichannel-based solar collector using different nanofluids. *Energy Convers. Manag.* **2014**, *88*, 129–138. [CrossRef]
28. Raj, P.; Subudhi, S. A review of studies using nanofluids in flat-plate and direct absorption solar collectors. *Renew. Sustain. Energy Rev.* **2018**, *84*, 54–74. [CrossRef]
29. Moghaieb, H.S.; Padmanaban, D.B.; Kumar, P.; Haq, A.U.; Maddi, C.; McGlynn, R.; Arredondo, M.; Singh, H.; Maguire, P.; Mariotti, D. Efficient solar-thermal energy conversion with surfactant-free Cu-oxide nanofluids. *Nano Energy* **2023**, *108*, 108112. [CrossRef]
30. Khan, M.; Tahir, M.N.; Adil, S.F.; Khan, H.U.; Siddiqui, M.R.H.; Al-warthan, A.A.; Tremel, W. Graphene based metal and metal oxide nanocomposites: Synthesis, properties and their applications. *J. Mater. Chem. A* **2015**, *3*, 18753–18808. [CrossRef]
31. Bordiwala, R.V. Green Synthesis and Applications of Metal Nanoparticles—A Review article. *Results Chem.* **2023**, *5*, 100832. [CrossRef]
32. Joudeh, N.; Linke, D. Nanoparticle classification, physicochemical properties, characterization, and applications: A comprehensive review for biologists. *J. Nanobiotechnol.* **2022**, *20*, 262. [CrossRef] [PubMed]
33. Wang, H.; Li, X.; Luo, B.; Wei, K.; Zeng, G. The MXene/water nanofluids with high stability and photo-thermal conversion for direct absorption solar collectors: A comparative study. *Energy* **2021**, *227*, 120483. [CrossRef]
34. Zhang, Y.; Chen, Y.; Westerhoff, P.; Hristovski, K.; Crittenden, J.C. Stability of commercial metal oxide nanoparticles in water. *Water Res.* **2008**, *42*, 2204–2212. [CrossRef]

35. Liu, M.-S.; Lin, M.C.-C.; Tsai, C.Y.; Wang, C.-C. Enhancement of thermal conductivity with Cu for nanofluids using chemical reduction method. *Int. J. Heat Mass Transf.* **2006**, *49*, 3028–3033. [CrossRef]

36. Eastman, J.A.; Choi, S.U.S.; Li, S.; Yu, W.; Thompson, L.J. Anomalously increased effective thermal conductivities of ethylene glycol-based nanofluids containing copper nanoparticles. *Appl. Phys. Lett.* **2001**, *78*, 718–720. [CrossRef]

37. Yu, W.; Xie, H. A review on nanofluids: Preparation, stability mechanisms, and applications. *J. Nanomater.* **2012**, *2012*, 435873. [CrossRef]

38. Alaghmandfard, A.; Ghandi, K. A comprehensive review of graphitic carbon nitride (g-C3N4)–metal oxide-based nanocomposites: Potential for photocatalysis and sensing. *Nanomaterials* **2022**, *12*, 294. [CrossRef] [PubMed]

39. Kakavelakis, G.; Petridis, K.; Kymakis, E. Recent advances in plasmonic metal and rare-earth-element upconversion nanoparticle doped perovskite solar cells. *J. Mater. Chem. A* **2017**, *5*, 21604–21624. [CrossRef]

40. Fu, Y.; Zeng, G.; Lai, C.; Huang, D.; Qin, L.; Yi, H.; Liu, X.; Zhang, M.; Li, B.; Liu, S. Hybrid architectures based on noble metals and carbon-based dots nanomaterials: A review of recent progress in synthesis and applications. *Chem. Eng. J.* **2020**, *399*, 125743. [CrossRef]

41. Li, Z.; Wang, L.; Li, Y.; Feng, Y.; Feng, W. Carbon-based functional nanomaterials: Preparation, properties and applications. *Compos. Sci. Technol.* **2019**, *179*, 10–40. [CrossRef]

42. Wang, L.; Hasanzadeh Kafshgari, M.; Meunier, M. Optical properties and applications of plasmonic-metal nanoparticles. *Adv. Funct. Mater.* **2020**, *30*, 2005400. [CrossRef]

43. Parashar, M.; Shukla, V.K.; Singh, R. Metal oxides nanoparticles via sol–gel method: A review on synthesis, characterization and applications. *J. Mater. Sci. Mater. Electron.* **2020**, *31*, 3729–3749. [CrossRef]

44. Concina, I.; Vomiero, A. Metal oxide semiconductors for dye-and quantum-dot-sensitized solar cells. *Small* **2015**, *11*, 1744–1774. [CrossRef] [PubMed]

45. Liu, X.; Swihart, M.T. Heavily-doped colloidal semiconductor and metal oxide nanocrystals: An emerging new class of plasmonic nanomaterials. *Chem. Soc. Rev.* **2014**, *43*, 3908–3920. [CrossRef]

46. Runnerstrom, E.L.; Bergerud, A.; Agrawal, A.; Johns, R.W.; Dahlman, C.J.; Singh, A.; Selbach, S.M.; Milliron, D.J. Defect engineering in plasmonic metal oxide nanocrystals. *Nano Lett.* **2016**, *16*, 3390–3398. [CrossRef] [PubMed]

47. Urmi, W.T.; Rahman, M.; Kadirgama, K.; Ramasamy, D.; Maleque, M. An overview on synthesis, stability, opportunities and challenges of nanofluids. *Mater. Today Proc.* **2021**, *41*, 30–37. [CrossRef]

48. Xia, Y.; Xiong, Y.; Lim, B.; Skrabalak, S.E. Shape-controlled synthesis of metal nanocrystals: Simple chemistry meets complex physics? *Angew. Chem. Int. Ed.* **2009**, *48*, 60–103. [CrossRef]

49. Sugimoto, T. *Fine Particles: Synthesis, Characterization, and Mechanisms of Growth*; CRC Press: Boca Raton, FL, USA, 2000; Volume 92.

50. Zhang, L.; Webster, T.J. Nanotechnology and nanomaterials: Promises for improved tissue regeneration. *Nano Today* **2009**, *4*, 66–80. [CrossRef]

51. Speranza, G. Carbon nanomaterials: Synthesis, functionalization and sensing applications. *Nanomaterials* **2021**, *11*, 967. [CrossRef] [PubMed]

52. Wang, L.; Fan, J. Nanofluids research: Key issues. *Nanoscale Res. Lett.* **2010**, *5*, 1241–1252. [CrossRef]

53. Mittal, A.; Brajpuriya, R.; Gupta, R. Solar Steam Generation Using Hybrid Nanomaterials to Address Global Environmental Water Pollution and Shortage Crisis. *Mater. Today Sustain.* **2023**, *21*, 100319. [CrossRef]

54. Sreekumar, S.; Shah, N.; Mondol, J.D.; Hewitt, N.; Chakrabarti, S. Broadband absorbing mono, blended and hybrid nanofluids for direct absorption solar collector: A comprehensive review. *Nano Futur.* **2022**, *6*, 022002. [CrossRef]

55. Wang, M.; Ye, M.; Iocozzia, J.; Lin, C.; Lin, Z. Plasmon—mediated solar energy conversion via photocatalysis in noble metal/semiconductor composites. *Adv. Sci.* **2016**, *3*, 1600024. [CrossRef]

56. Paściak, A.; Pilch-Wróbel, A.; Marciniak, Ł.; Schuck, P.J.; Bednarkiewicz, A. Standardization of methodology of light-to-heat conversion efficiency determination for colloidal nanoheaters. *ACS Appl. Mater. Interfaces* **2021**, *13*, 44556–44567. [CrossRef] [PubMed]

57. Zheng, J.; Cheng, X.; Zhang, H.; Bai, X.; Ai, R.; Shao, L.; Wang, J. Gold nanorods: The most versatile plasmonic nanoparticles. *Chem. Rev.* **2021**, *121*, 13342–13453. [CrossRef]

58. Chen, M.; He, Y.; Zhu, J.; Kim, D.R.J.E.C.; Management. Enhancement of photo-thermal conversion using gold nanofluids with different particle sizes. *Energy Convers. Manag.* **2016**, *112*, 21–30. [CrossRef]

59. Gupta, V.K.; Kumar, S.; Kukreja, R.; Chander, N. Experimental thermal performance investigation of a direct absorption solar collector using hybrid nanofluid of gold nanoparticles with natural extract of Azadirachta Indica leaves. *Renew. Energy* **2023**, *202*, 1021–1031. [CrossRef]

60. Sharaf, O.Z.; Rizk, N.; Munro, C.J.; Joshi, C.P.; Anjum, D.H.; Abu-Nada, E.; Martin, M.N.; Alazzam, A. Radiation stability and photothermal performance of surface-functionalized plasmonic nanofluids for direct-absorption solar applications. *Sol. Energy Mater. Sol. Cells* **2021**, *227*, 111115. [CrossRef]

61. McGlynn, R.; Chakrabarti, S.; Alessi, B.; Moghaieb, H.S.; Maguire, P.; Singh, H.; Mariotti, D. Plasma-induced non-equilibrium electrochemistry synthesis of nanoparticles for solar thermal energy harvesting. *Sol. Energy* **2020**, *203*, 37–45. [CrossRef]

62. Kimpton, H.; Cristaldi, D.A.; Stulz, E.; Zhang, X. Thermal performance and physicochemical stability of silver nanoprism-based nanofluids for direct solar absorption. *Sol. Energy* **2020**, *199*, 366–376. [CrossRef]

63. Kumar, S.; Chander, N.; Gupta, V.K.; Kukreja, R. Progress, challenges and future prospects of plasmonic nanofluid based direct absorption solar collectors—A state-of-the-art review. *Sol. Energy* **2021**, *227*, 365–425. [CrossRef]

64. Borode, A.; Ahmed, N.; Olubambi, P. A review of solar collectors using carbon-based nanofluids. *J. Clean. Prod.* **2019**, *241*, 118311. [CrossRef]

65. Trong Tam, N.; Viet Phuong, N.; Hong Khoi, P.; Ngoc Minh, P.; Afrand, M.; Van Trinh, P.; Hung Thang, B.; Żyła, G.; Estellé, P. Carbon nanomaterial-based nanofluids for direct thermal solar absorption. *Nanomaterials* **2020**, *10*, 1199. [CrossRef]

66. Yu, S.-P.; Teng, T.-P.; Huang, C.-C.; Hsieh, H.-K.; Wei, Y.-J. Performance Evaluation of Carbon-Based Nanofluids for Direct Absorption Solar Collector. *Energies* **2023**, *16*, 1157. [CrossRef]

67. McGlynn, R.J.; Moghaieb, H.S.; Brunet, P.; Chakrabarti, S.; Maguire, P.; Mariotti, D. Hybrid Plasma–Liquid Functionalisation for the Enhanced Stability of CNT Nanofluids for Application in Solar Energy Conversion. *Nanomaterials* **2022**, *12*, 2705. [CrossRef]

68. Chen, X.; Xiong, Z.; Chen, M.; Zhou, P. Ultra-stable carbon quantum dot nanofluids for direct absorption solar collectors. *Sol. Energy Mater. Sol. Cells* **2022**, *240*, 111720. [CrossRef]

69. Guo, J.; Wang, F.; Li, S.; Wang, Y.; Hu, X.; Zu, D.; Shen, Y.; Li, C. Enhanced optical properties and light-to-heat conversion performance of Ti3C2/[BMIM]BF4 nanofluids based direct absorption solar collector. *Sol. Energy Mater. Sol. Cells* **2022**, *237*, 111558. [CrossRef]

70. Mesgari, S.; Taylor, R.A.; Hjerrild, N.E.; Crisostomo, F.; Li, Q.; Scott, J. An investigation of thermal stability of carbon nanofluids for solar thermal applications. *Sol. Energy Mater. Sol. Cells* **2016**, *157*, 652–659. [CrossRef]

71. Torokcr, M.C.; Carter, E.A. Transition metal oxide alloys as potential solar energy conversion materials. *J. Mater. Chem. A* **2013**, *1*, 2474–2484. [CrossRef]

72. Zhang, Y.; Wang, J.; Qiu, J.; Jin, X.; Umair, M.M.; Lu, R.; Zhang, S.; Tang, B. Ag-graphene/PEG composite phase change materials for enhancing solar-thermal energy conversion and storage capacity. *Appl. Energy* **2019**, *237*, 83–90. [CrossRef]

73. Milanese, M.; Colangelo, G.; Cretì, A.; Lomascolo, M.; Iacobazzi, F.; De Risi, A.J.S.E.M.; Cells, S. Optical absorption measurements of oxide nanoparticles for application as nanofluid in direct absorption solar power systems–Part I: Water-based nanofluids behavior. *Sol. Energy Mater. Sol. Cells* **2016**, *147*, 315–320. [CrossRef]

74. Balakin, B.V.; Stava, M.; Kosinska, A. Photothermal convection of a magnetic nanofluid in a direct absorption solar collector. *Sol. Energy* **2022**, *239*, 33–39. [CrossRef]

75. Hosseini, S.M.S.; Dehaj, M.S. The comparison of colloidal, optical, and solar collection characteristics between Fe2O3 and Fe3O4 nanofluids operated in an evacuated tubular volumetric absorption solar collector. *J. Taiwan Inst. Chem. Eng.* **2022**, *135*, 104381. [CrossRef]

76. Sreekumar, S.; Joseph, A.; Sujith Kumar, C.S.; Thomas, S. Investigation on influence of antimony tin oxide/silver nanofluid on direct absorption parabolic solar collector. *J. Clean. Prod.* **2020**, *249*, 119378. [CrossRef]

77. Wang, K.; He, Y.; Zheng, Z.; Gao, J.; Kan, A.; Xie, H.; Yu, W. Experimental optimization of nanofluids based direct absorption solar collector by optical boundary conditions. *Appl. Therm. Eng.* **2021**, *182*, 116076. [CrossRef]

78. Sharaf, O.Z.; Rizk, N.; Joshi, C.P.; Abi Jaoudé, M.; Al-Khateeb, A.N.; Kyritsis, D.C.; Abu-Nada, E.; Martin, M.N. Ultrastable plasmonic nanofluids in optimized direct absorption solar collectors. *Energy Convers. Manag.* **2019**, *199*, 112010. [CrossRef]

79. Zheng, N.; Yan, F.; Wang, L.; Sun, Z. Photo-thermal conversion performance of mono MWCNT and hybrid MWCNT-TiN nanofluids in direct absorption solar collectors. *Int. J. Energy Res.* **2022**, *46*, 8313–8327. [CrossRef]

80. Gorji, T.B.; Ranjbar, A.A.J.S.E. A numerical and experimental investigation on the performance of a low-flux direct absorption solar collector (DASC) using graphite, magnetite and silver nanofluids. *Sol. Energy* **2016**, *135*, 493–505. [CrossRef]

81. Li, X.; Zeng, G.; Lei, X. The stability, optical properties and solar-thermal conversion performance of SiC-MWCNTs hybrid nanofluids for the direct absorption solar collector (DASC) application. *Sol. Energy Mater. Sol. Cells* **2020**, *206*, 110323. [CrossRef]

82. Vakili, M.; Hosseinalipour, S.M.; Delfani, S.; Khosrojerdi, S.J.S.E.M.; Cells, S. Photothermal properties of graphene nanoplatelets nanofluid for low-temperature direct absorption solar collectors. *Sol. Energy Mater. Sol. Cells* **2016**, *152*, 187–191. [CrossRef]

83. Taylor, R.A.; Hjerrild, N.; Duhaini, N.; Pickford, M.; Mesgari, S. Stability testing of silver nanodisc suspensions for solar applications. *Appl. Surf. Sci.* **2018**, *455*, 465–475. [CrossRef]

84. Said, Z.; Saidur, R.; Rahim, N.A. Optical properties of metal oxides based nanofluids. *Int. Commun. Heat Mass Transf.* **2014**, *59*, 46–54. [CrossRef]

85. Menbari, A.; Alemrajabi, A.A.; Ghayeb, Y.J.E.C.; Management. Experimental investigation of stability and extinction coefficient of Al2O3–CuO binary nanoparticles dispersed in ethylene glycol–water mixture for low-temperature direct absorption solar collectors. *Energy Convers. Manag.* **2016**, *108*, 501–510. [CrossRef]

86. Wang, K.; He, Y.; Kan, A.; Yu, W.; Wang, D.; Zhang, L.; Zhu, G.; Xie, H.; She, X. Significant photothermal conversion enhancement of nanofluids induced by Rayleigh-Bénard convection for direct absorption solar collectors. *Appl. Energy* **2019**, *254*, 113706. [CrossRef]

87. Chen, L.; Liu, J.; Fang, X.; Zhang, Z. Reduced graphene oxide dispersed nanofluids with improved photo-thermal conversion performance for direct absorption solar collectors. *Sol. Energy Mater. Sol. Cells* **2017**, *163*, 125–133. [CrossRef]

88. Xu, X.; Xu, C.; Liu, J.; Fang, X.; Zhang, Z. A direct absorption solar collector based on a water-ethylene glycol based nanofluid with anti-freeze property and excellent dispersion stability. *Renew. Energy* **2019**, *133*, 760–769. [CrossRef]

89. Choi, T.J.; Jang, S.P.; Kedzierski, M.A. Effect of surfactants on the stability and solar thermal absorption characteristics of water-based nanofluids with multi-walled carbon nanotubes. *Int. J. Heat Mass Transf.* **2018**, *122*, 483–490. [CrossRef]

90. Hazra, S.K.; Michael, M.; Nandi, T.K. Investigations on optical and photo-thermal conversion characteristics of BN-EG and BN/CB-EG hybrid nanofluids for applications in direct absorption solar collectors. *Sol. Energy Mater. Sol. Cells* **2021**, *230*, 111245. [CrossRef]

91. Cao, S.; Jiang, Q.; Wu, X.; Ghim, D.; Derami, H.G.; Chou, P.-I.; Jun, Y.-S.; Singamaneni, S. Advances in solar evaporator materials for freshwater generation. *J. Mater. Chem. A* **2019**, *7*, 24092–24123. [CrossRef]

92. Joseph, A.; Thomas, S. Energy, exergy and corrosion analysis of direct absorption solar collector employed with ultra-high stable carbon quantum dot nanofluid. *Renew. Energy* **2022**, *181*, 725–737. [CrossRef]

93. Heberlein, J.; Mentel, J.; Pfender, E. The anode region of electric arcs: A survey. *J. Phys. D Appl. Phys.* **2009**, *43*, 23001. [CrossRef]

94. Krishna, K.S.; Biswas, S.; Navin, C.V.; Yamane, D.G.; Miller, J.T.; Kumar, C.S.S.R.J.J. Millifluidics for chemical synthesis and time-resolved mechanistic studies. *JoVE* **2013**, *81*, e50711.

95. Martinsson, E.; Shahjamali, M.M.; Large, N.; Zaraee, N.; Zhou, Y.; Schatz, G.C.; Mirkin, C.A.; Aili, D.J.S. Influence of surfactant bilayers on the refractive index sensitivity and catalytic properties of anisotropic gold nanoparticles. *Small* **2016**, *12*, 330–342. [CrossRef]

96. Bohren, C.F.; Huffman, D.R. *Absorption and Scattering of Light by Small Particles*; John Wiley & Sons: Hoboken, NJ, USA, 2008.

97. Saavedra, J.; Asenjo-Garcia, A.; García de Abajo, F.J. Hot-electron dynamics and thermalization in small metallic nanoparticles. *Acs Photonics* **2016**, *3*, 1637–1646. [CrossRef]

98. Jain, P.K.; Qian, W.; El-Sayed, M.A. Ultrafast electron relaxation dynamics in coupled metal nanoparticles in aggregates. *J. Phy. Chem. B* **2006**, *110*, 136–142. [CrossRef]

99. Mehrali, M.; Ghatkesar, M.K.; Pecnik, R. Full-spectrum volumetric solar thermal conversion via graphene/silver hybrid plasmonic nanofluids. *Appl. Energy* **2018**, *224*, 103–115. [CrossRef]

100. Bhanvase, B.; Barai, D. 5-Electrical, optical, and tribological properties of the nanofluids. In *Nanofluids for Heat and Mass Transfer*; Bharat Bhanvase, D.B., Ed.; Academic Press: Cambridge, MA, USA, 2021; pp. 167–190. [CrossRef]

101. Khlebtsov, N.G.; Trachuk, L.A.; Mel'nikov, A.G. The effect of the size, shape, and structure of metal nanoparticles on the dependence of their optical properties on the refractive index of a disperse medium. *Opt. Spectrosc.* **2005**, *98*, 77–83. [CrossRef]

102. *SATM-G173*; Standard Tables for Reference Solar Spectral Irradiances: Direct Normal and Hemispherical on 37 Tilted Surface. American Society for Testing Materials: West Conshocken PA, USA, 2007; Volume 14.

103. Fan, X.; Zheng, W.; Singh, D.J. Light scattering and surface plasmons on small spherical particles. *Light Sci. Appl.* **2014**, *3*, e179. [CrossRef]

104. Said, Z.; Sajid, M.; Saidur, R.; Mahdiraji, G.; Rahim, N. Evaluating the optical properties of TiO_2 nanofluid for a direct absorption solar collector. *Numer. Heat Transf. Part A Appl.* **2015**, *67*, 1010–1027. [CrossRef]

105. Wriedt, T. Mie theory: A review. In *The Mie Theory*; Springer: Berlin/Heidelberg, Germany, 2012; pp. 53–71.

106. Madsen, S.J.; Wilson, B.C. Optical properties of brain tissue. In *Optical Methods and Instrumentation in Brain Imaging and Therapy*; Springer: Berlin/Heidelberg, Germany, 2013; pp. 1–22.

107. Amendola, V.; Meneghetti, M. Laser ablation synthesis in solution and size manipulation of noble metal nanoparticles. *Phys. Chem. Chem. Phys.* **2009**, *11*, 3805–3821. [CrossRef]

108. Chen, Y.; Huang, Y.; Li, K. Temperature effect on the aggregation kinetics of CeO_2 nanoparticles in monovalent and divalent electrolytes. *J. Environ. Anal. Toxicol.* **2012**, *2*, 158–162.

109. Ali, N.; Teixeira, J.A.; Addali, A. A review on nanofluids: Fabrication, stability, and thermophysical properties. *J. Nanomater.* **2018**, *2018*, 6978130. [CrossRef]

110. Hjerrild, N.E.; Scott, J.A.; Amal, R.; Taylor, R.A. Exploring the effects of heat and UV exposure on glycerol-based Ag-SiO2 nanofluids for PV/T applications. *Renew. Energy* **2018**, *120*, 266–274. [CrossRef]

111. Nine, M.J.; Chung, H.; Tanshen, M.R.; Osman, N.A.B.A.; Jeong, H. Is metal nanofluid reliable as heat carrier? *J. Hazard. Mater.* **2014**, *273*, 183–191. [CrossRef]

112. Nguyen, C.T.; Desgranges, F.; Roy, G.; Galanis, N.; Maré, T.; Boucher, e.; Mintsa, H.A. Temperature and particle-size dependent viscosity data for water-based nanofluids–hysteresis phenomenon. *Int. J. Heat Fluid Flow* **2007**, *28*, 1492–1506. [CrossRef]

113. Balamurugan, B.; Mehta, B. Optical and structural properties of nanocrystalline copper oxide thin films prepared by activated reactive evaporation. *Thin Solid Film.* **2001**, *396*, 90–96. [CrossRef]

114. Addala, S.; Bouhdjer, L.; Chala, A.; Bouhdjar, A.; Halimi, O.; Boudine, B.; Sebais, M. Structural and optical properties of a NaCl single crystal doped with CuO nanocrystals. *Chin. Phys. B* **2013**, *22*, 098103. [CrossRef]

115. Piri, F.; Afarani, M.S.; Arabi, A.M. Synthesis of copper oxide quantum dots: Effect of surface modifiers. *Mater. Res. Express* **2019**, *6*, 125006. [CrossRef]

116. Padmanaban, D.B.; McGlynn, R.; Byrne, E.; Velusamy, T.; Swadźba-Kwaśny, M.; Maguire, P.; Mariotti, D. Understanding plasma–ethanol non-equilibrium electrochemistry during the synthesis of metal oxide quantum dots. *Green Chem.* **2021**, *23*, 3983–3995. [CrossRef]

117. Velusamy, T.; Liguori, A.; Macias-Montero, M.; Padmanaban, D.B.; Carolan, D.; Gherardi, M.; Colombo, V.; Maguire, P.; Svrcek, V.; Mariotti, D. Ultra-small CuO nanoparticles with tailored energy-band diagram synthesized by a hybrid plasma-liquid process. *Plasma Proc. Poly.* **2017**, *14*, 1600224. [CrossRef]

118. Qin, C.; Kang, K.; Lee, I.; Lee, B.J. Optimization of a direct absorption solar collector with blended plasmonic nanofluids. *Sol. Energy* **2017**, *150*, 512–520. [CrossRef]

119. Maheshwary, P.B.; Handa, C.C.; Nemade, K.R.; Chaudhary, S.R. Role of nanoparticle shape in enhancing the thermal conductivity of nanofluids. *Mater. Today Proc.* **2020**, *28*, 873–878. [CrossRef]
120. Chen, M.; He, Y.; Wang, X.; Hu, Y. Numerically investigating the optical properties of plasmonic metallic nanoparticles for effective solar absorption and heating. *Sol. Energy* **2018**, *161*, 17–24. [CrossRef]
121. Wang, Z.; Quan, X.; Zhang, Z.; Cheng, P. Optical absorption of carbon-gold core-shell nanoparticles. *J. Quant. Spectrosc. Radiat. Transf.* **2018**, *205*, 291–298. [CrossRef]
122. Duan, H.; Zheng, Y.; Xu, C.; Shang, Y.; Ding, F. Experimental investigation on the plasmonic blended nanofluid for efficient solar absorption. *Appl. Therm. Eng.* **2019**, *161*, 114192. [CrossRef]
123. Elsheikh, A.H.; Sharshir, S.W.; Mostafa, M.E.; Essa, F.A.; Ahmed Ali, M.K. Applications of nanofluids in solar energy: A review of recent advances. *Renew. Sustain. Energy Rev.* **2018**, *82*, 3483–3502. [CrossRef]
124. Razi, P.; Akhavan-Behabadi, M.A.; Saeedinia, M. Pressure drop and thermal characteristics of CuO–base oil nanofluid laminar flow in flattened tubes under constant heat flux. *Int. Commun. Heat Mass Transf.* **2011**, *38*, 964–971. [CrossRef]
125. Ahmadlouydarab, M.; Ebadolahzadeh, M.; Ali, H.M. Effects of utilizing nanofluid as working fluid in a lab-scale designed FPSC to improve thermal absorption and efficiency. *Phys. A Stat. Mech. Its Appl.* **2020**, *540*, 123109. [CrossRef]
126. Celata, G.P.; D'Annibale, F.; Mariani, A.; Sau, S.; Serra, E.; Bubbico, R.; Menale, C.; Poth, H. Experimental results of nanofluids flow effects on metal surfaces. *Chem. Eng. Res. Des.* **2014**, *92*, 1616–1628. [CrossRef]
127. Radwan, A.; Ahmed, M.; Ookawara, S. Performance enhancement of concentrated photovoltaic systems using a microchannel heat sink with nanofluids. *Energy Convers. Manag.* **2016**, *119*, 289–303. [CrossRef]
128. Hussein, A.K. Applications of nanotechnology to improve the performance of solar collectors–Recent advances and overview. *Renew. Sustain. Energy Rev.* **2016**, *62*, 767–792. [CrossRef]
129. Sabiha, M.A.; Saidur, R.; Hassani, S.; Said, Z.; Mekhilef, S. Energy performance of an evacuated tube solar collector using single walled carbon nanotubes nanofluids. *Energy Convers. Manag.* **2015**, *105*, 1377–1388. [CrossRef]
130. Elsaid, K.; Olabi, A.G.; Wilberforce, T.; Abdelkareem, M.A.; Sayed, E.T. Environmental impacts of nanofluids: A review. *Sci. Total Environ.* **2021**, *763*, 144202. [CrossRef]
131. Mo, Y.; Gu, A.; Tollerud, D.J.; Zhang, Q. Nanoparticle toxicity and environmental impact. In *Oxidative Stress and Biomaterials*; Elsevier: Amsterdam, The Netherlands, 2016; pp. 117–143.
132. Bakand, S.; Hayes, A.; Dechsakulthorn, F. Nanoparticles: A review of particle toxicology following inhalation exposure. *Inhal. Toxicol.* **2012**, *24*, 125–135. [CrossRef]

 *nanomaterials*

Review

A Comprehensive Review of Nanofluid Heat Transfer in Porous Media

Hossam A. Nabwey [1,2,*], Taher Armaghani [3], Behzad Azizimehr [3], Ahmed M. Rashad [4] and Ali J. Chamkha [5]

[1] Department of Mathematics, College of Science and Humanities in Al-Kharj, Prince Sattam Bin Abdulaziz University, Al-Kharj 11942, Saudi Arabia

[2] Department of Basic Engineering Science, Faculty of Engineering, Menoufia University, Shebin El-Kom 32511, Egypt

[3] Department of Engineering, West Tehran Branch, Islamic Azad University, Tehran 1477893855, Iran

[4] Department of Mathematics, Faculty of Science, Aswan University, Aswan 81528, Egypt

[5] Faculty of Engineering, Kuwait College of Science and Technology, Doha District, Kuwait City 35004, Kuwait

* Correspondence: h.mohamed@psau.edu.sa or eng_hossam21@yahoo.com

Abstract: In the present paper, recent advances in the application of nanofluids in heat transfer in porous materials are reviewed. Efforts have been made to take a positive step in this field by scrutinizing the top papers published between 2018 and 2020. For that purpose, the various analytical methods used to describe the flow and heat transfer in different types of porous media are first thoroughly reviewed. In addition, the various models used to model nanofluids are described in detail. After reviewing these analysis methods, papers concerned with the natural convection heat transfer of nanofluids in porous media are evaluated first, followed by papers on the subject of forced convection heat transfer. Finally, we discuss articles related to mixed convection. Statistical results from the reviewed research regarding the representation of various parameters, such as the nanofluid type and the flow domain geometry, are analyzed, and directions for future research are finally suggested. The results reveal some precious facts. For instance, a change in the height of the solid and porous medium results in a change in the flow regime within the chamber; as a dimensionless permeability, the effect of Darcy's number on heat transfer is direct; and the effect of the porosity coefficient has a direct relationship with heat transfer: when the porosity coefficient is increased or decreased, the heat transfer will also increase or decrease. Additionally, a comprehensive review of nanofluid heat transfer in porous media and the relevant statical analysis are presented for the first time. The results show that Al2O3 nanoparticles in a base fluid of water with a proportion of 33.9% have the highest representation in the papers. Regarding the geometries studied, a square geometry accounted for 54% of the studies.

Keywords: porous media; heat transfer; Darcy–Brinkman–Forchheimer; MHD

Citation: Nabwey, H.A.; Armaghani, T.; Azizimehr, B.; Rashad, A.M.; Chamkha, A.J. A Comprehensive Review of Nanofluid Heat Transfer in Porous Media. *Nanomaterials* **2023**, *13*, 937. https://doi.org/10.3390/nano13050937

Academic Editors: M. M. Bhatti, Kambiz Vafai and Sara I. Abdelsalam

Received: 17 January 2023
Revised: 26 February 2023
Accepted: 2 March 2023
Published: 4 March 2023

1. Introduction

A porous medium or a porous material is a solid material that contains pores. Depending on the ability of the porous medium to allow fluids to pass through it under the influence of external forces, it is classified as a permeable or a nonpermeable porous medium. Porous media may be non-dispersed or post-dispersed, homogeneous or heterogeneous, and multi-structure or the result of a combination of different structures [1]. Porosity, the main characteristic property of a porous medium, is a measure of the empty spaces in a material and is a fraction of the volume of the empty spaces over the total volume between 0 and 1. Most natural porous materials have a porosity of 0.6 (excluding hair). However, manufactured materials, such as metal foams, can have a porosity of up to 0.99. Table 1 shows the porosity value for different materials [2].

Table 1. The porosity of various materials [2].

Material	Porosity $\varnothing$
Agar–agar	0.57–0.66
Black slate powder	0.12–0.34
Brick	0.45
Catalyst (Fischer–Tropsch, granules only)	
Cigarette	0.17–0.49
Cigarette filters	0.02–0.12
Coal Concrete (ordinary mixes)	~0.10
Concrete (bituminous)	
Copper powder (hot-compacted)	0.09–0.34
Corkboard	
Fiberglass	0.88–0.93
Granular crushed rock	0.45
Hair (on mammals)	0.95–0.99
Hair felt	
Leather	0.56–0.59
Hair felt limestone (dolomite)	0.04–0.10
Leather	0.37–0.50
Sand sandstone ("oil sand")	0.08–0.38
Silica grains	0.65
Silica powder	0.37–0.49
Soil	0.43–0.54
Spherical packings (shaken well)	0.36–0.43
Wire crimps	0.68–0.76

Investigating heat transfer characteristics in porous media is of great interest in various industries and engineering domains such as heat exchangers, heat storage, geothermal systems, and drying techniques. Including a porous material in a mechanical system can be considered a passive method for heat transfer enhancement. Indeed, the presence of a porous material alters the flow patterns and improves the overall thermal conductivity of the system [3–7].

A further improvement in heat transfer can be achieved by adding conductive nanoparticles to the base fluid. The working fluid is then considered a nanofluid. The added nanoparticles can be metallic and made of metals such as aluminum and copper, among others, or metal oxides, or non-metallic materials such as carbon. Hybrid nanofluids refer to the fluids in which two or more types of nanoparticles are dispersed. The main objective of dispersing nanoparticles in the fluid is to increase its thermal conductivity [8]. Nonetheless, adding nanoparticles to a fluid presents a drawback that should be avoided. Increasing the volume fraction of the nanoparticles above a certain level increases the fluid viscosity and may, as a result, hinder heat transfer.

Due to the double importance of nanofluids and porous materials in heat transfer enhancement, significant research has been performed to study the thermal characteristics of nanofluid flow in porous media. The present paper summarizes the recent papers dealing with this topic. Considering the large number of articles published since the last review article (2017 to the present) and their analytical complexities, the authors divide the articles into two parts: 2018–2020 and 2020 until the present. These parts are presented in two articles. The paper is organized as follows: in Section 2, the various models of fluid flow and heat transfer in porous media are recalled and the classifications of the nanofluids are presented. In Section 3, the works dealing with heat transfer in porous media are presented and are classified into free convection and forced convection. Finally, Section 4 presents a general conclusion.

2. Methods and Materials

2.1. Methods of Analysis of Porous Material

In this section, the various analytical methods used to describe porous media are presented based on the scale length of the medium.

2.1.1. Microporous Medium

Microporous media are at the size scale of the nanometer. Examples of such media include activated carbon, silica gels, carbon molecular sieves, and some crystalline structures such as zeolites.

2.1.2. Mesoporous Media

The pore size of mesoporous media is between 2 and 50 nm. This size applies to non-organic jellies such as alumina, silica powders, porous glass, and columnar or non-columnar bricks [2]. Since this review considers the heat transfer of nanofluids in porous media and nanofluids have particles with a diameter of more than 10 nm, the two methods described above cannot be used for this purpose. Therefore, we describe the macroscopic method in the following sections.

2.1.3. Macroporous Media

Macroporous media have a size scale of greater than 50 nanometers. Macroporous media are widely found in nature, such in soil, broken rocks, sandstones, wood materials, and various types of food. Foods are generally macroporous but have different porosity size properties in some cases. This feature is indicated by the distribution of pore size, which indicates two or more sizes. Manufactured materials such as thermal insulation materials, silicate, ceramics, cement, synthetic resins, and many other artificial materials are also macroporous materials. They have many applications, such as in fluid filters, electronic components, complex fiber structures, bioceramics, fluid chromatography, and biotechnology [2].

2.1.4. Macroscopic Governing Equations

The macroscopic equation is written as follows [2]:

$$\hat{\nabla}.v = 0 \tag{1}$$

Considering the Boussinesq assumption, the macroscopic momentum is as follows:

$$\rho_f\left[\frac{\partial v}{\partial t} + \hat{\nabla}.\left(\frac{vv}{\varnothing}\right)\right] = -\hat{\nabla}_p + \mu_f\hat{\nabla}^2 v + B - \rho_f\varnothing B\left(\hat{T}_f - \hat{T}_\infty\right)g \tag{2}$$

where

$$B = -\frac{1}{v}\int_{A_{fs}} p_f dS + \frac{\mu_f}{v}\int_{A_{fs}}\left(\nabla v_f\right).dS \tag{3}$$

and

$$p = \varnothing \hat{p}_f \tag{4}$$

In these formulas $p, \varnothing, \rho,$ and μ are the pressure, porosity, density, and viscosity, respectively. The subscripts s and f are related to the solid and fluid.

The macroscopic energy equations for the fluid and solid are as follows:

$$\varnothing(\rho C_p)_f\left[\frac{\partial \hat{T}_f}{\partial t} + \hat{\nabla}.\left(\hat{v}_f\hat{T}_f\right)\right] = \varnothing\hat{\nabla}.\left[\left(k_f + k'\right)\hat{\nabla}T_f\right] + q_{sf} \tag{5}$$

$$(1-\varnothing)(\rho C_p)_s\frac{\partial \hat{T}_s}{\partial t} = (1-\varnothing)\left[\hat{\nabla}.(k_s\hat{\nabla}\hat{T}_s)\right] - q_{sf} \tag{6}$$

In these formulas, q_{sf} is the heat transfer between the solid matrix and fluid flow, and k represents the conductivity.

2.2. Nanofluid

Generally, a nanofluid is obtained by dispersing a certain volume fraction of nanoparticles in a base fluid. Nanoparticles are produced at different sizes from 10 nm to 100 nm, depending on their application. Since most of the fluids used for heat transfer have a low conductivity, heat transfer can be significantly improved with a uniform distribution of nanoparticles within the base fluid. Therefore, the typical size of nanofluid particles is greater than the typical pore size of microporous media and of the same order of magnitude as the typical pore size of mesoporous media. Therefore, nanofluids can only be used in macroporous media. Figure 1 shows a picture of the Titania nanoparticles [4].

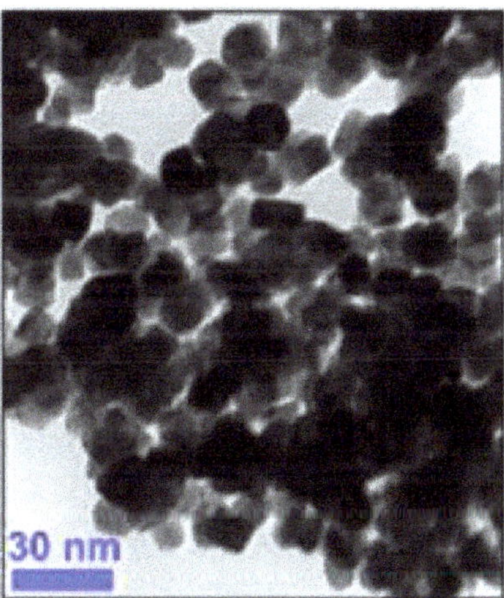

Figure 1. Titania nanoparticles (30 nm).

Some of the exceptional properties of nanoparticles include their non linear relationship between the conductivity and the concentration of solids, the low momentum of the particles, a higher mobility than microparticles, the strong dependence of conductivity on temperature, a strong increase in heat flux in the boiling region, and acceptable viscosity. These properties, provided to the fluids, are considered some of the most suitable and strongest choices.

- Thermal conductivity: the coefficient of the thermal conductivity of nanofluids depends on parameters such as the composition of the chemical percentage of nanoparticles, the volume percentage of nanoparticles, the surface-active substances, and the temperature. The coefficient of thermal conductivity is also influenced by mechanisms such as the Brownian motion of nanoparticles in the fluid, which increases mixing in the fluid, facilitates heat transfer and increases the coefficient of thermal conductivity;

- Size reduction: the small size of the nanoparticles reduces their motion and contact with the solid wall, reduces momentum, and ultimately reduces the possibility of erosion of parts such as heat exchangers, pipelines, and pumps;
- Stability: nanoparticles are less likely to be precipitated due to their low weight and small size, which prevents the problem of nanoparticle suspension caused by sedimentation. Presently, the instability of nanofluids hinders the application of nanofluids. The stability of the nanofluid means that the nanoparticles do not accumulate and precipitate at a significant rate and, as a result, the concentration of the floating nanoparticles is constant. Stokes' law can be used to calculate the settling velocity of spherical particles in a quiescent fluid. This equation is obtained from the balance of gravity, buoyancy, and drag forces that act on particles. The stability of the nanofluid is a necessary condition for optimizing the properties of the nanofluid. Three general methods to increase nanofluid stability are:
- Adding a surfactant;
- Controlling the pH of the nanofluid;
- Ultrasonic vibration.

2.2.1. Nanofluid Evaluation Methods

Researchers use different perspectives to analyze nanoscale behavior. Therefore, the modeling and formulation of nanoscale behavior also differs based on these perspectives. From one point of view, a certain volume fraction of nanoparticles is combined with the base fluid, but the nanofluid concentration does not change with respect to the initial concentration. The concentration of the nanofluid remains constant in different areas, but the newly formed fluid (nanofluid) has improved thermophysical properties compared to the base fluid. On the other hand, the movement of nanoparticles relative to the base fluid is not considered. Hence, this method or perspective is called a homogeneous or single-phase method. In the homogeneous method, three equations of mass conservation, momentum conservation, and energy conservation form the main structure of the model, and the thermophysical properties of the nano-fluid are substituted in conservation equations. The second method of analyzing nanoscale behavior is somewhat different from the homogeneous method. In this method, the movement of the nanoparticles changes the volume fraction of the nanoparticles (relative to the initial concentration) in different areas of the flow domain. Therefore, this method is called a non-homogeneous or pseudo-two-phase method. Several factors, such as gravitational force, Brownian forces, and thermophoresis forces, affect the heat transfer in this model. However, the Brownian and thermophoresis forces are dominant forces, according to ref. [8]. Because nanofluids consist of solid particles and a base fluid, one of the modeling approaches is to solve the conservation equations for the base fluid and the nanoparticles separately and is called the two-phase model.

2.2.2. Hybrid Nanofluid

Another type of nanofluid in which two or more nanoparticles exist within the base fluid is called a hybrid nanofluid [6]. When one nanoparticle type is dispersed in a combination of two or more fluids it is also called a hybrid nanofluid. The heat conductivity is improved significantly when using this type of nanofluid. For example, although the addition of aluminum oxide nanoparticles to water increases the conductivity, the addition of copper nanoparticles to the aluminum–water nanofluid can further increase the conductivity relative to the base fluid. Figure 2 shows the aluminum and copper hybrid nanofluid presented in ref. [9].

Figure 2. Aluminum and copper hybrid nanofluid.

2.3. Fluid Flow Models

2.3.1. Darcy's Equation

Based on the experimental data, Darcy's equation provides a linear proportion between the fluid's volumetric average velocity (v) and its pressure difference (Δp) in addition to a porous media.

$$v = k\frac{\Delta p}{\Delta x} \tag{7}$$

In this formula, $\frac{\Delta p}{\Delta x}$ is the pressure gradient. Hydraulic conductivity is an expression of how easily a fluid can flow throughout the hollows of a medium. Darcy's equation is valid for creeping (viscous fluids having slow motion), isothermal incompressible flows [2].

2.3.2. Hazen–Darcy Equation

Darcy's experiments were performed with a single-type fluid at a constant temperature; therefore, Darcy's equation did not include the fluid's viscosity, μ. By providing a specific permeability relation, $K = k\mu$, Darcy's equation is redefined in a viscosity-dependent form as follows:

$$v = \left(\frac{K}{\mu}\right)\frac{\Delta p}{\Delta x} \tag{8}$$

The specific permeability K is considered a hydraulic parameter and is independent of the fluid's characteristics. The above-mentioned formula is the Hazen–Darcy equation, which is known as Darcy's law [2].

2.3.3. Hazen–Dupuit–Darcy Equation

The quadratic Hazen–Dupuit–Darcy equation is derived from the analysis of steady-state, open-channel flows. The equation is defined based on an equilibrium state between the gravitational force and the shearing resistance as follows:

$$0 = \frac{\partial p}{\partial x} + \frac{\mu}{K}v - C\rho v^2 \tag{9}$$

The second and third terms on the right-hand side express the viscous drag and geometrical drag, respectively. It is worthwhile to mention that the equation is only valid for steady-state and one-dimensional flows [2].

2.3.4. Brinkman–Hazen–Dupuit–Darcy Equation

Brinkman realized that when the permeability is low, the shearing tension of the fluid's viscosity can be negligible in comparison to the viscous drag. Therefore, by adding the shearing tension term (a Laplacian one), the Brinkman–Hazen–Dupuit–Darcy equation is derived.

$$0 = -\nabla p + \mu \nabla^2 v - \frac{\mu}{K}\phi v + \frac{C_F}{K^{1/2}}\rho\phi^2 |v|v \tag{10}$$

where $|v|$ is the magnitude of the velocity. The fluid is capable of transforming the viscous shearing tension independent of the viscous drag. Both the K and C_F parameters have a wide variety in porous media in the engineering paradigms; for example, as packed beds, windowing environments, metal foams, and aerodynamic gels [2].

2.3.5. Brinkman–Forchheimer Equation

The modified equation is as follows.

$$\rho\left[\frac{1}{\varphi}\frac{\partial v}{\partial t} + \frac{1}{\varphi^2}(v.\nabla v)\right] = -\nabla p + \mu_e \nabla^2 v - \frac{\mu}{K}v - \frac{c_F\rho}{K^{1/2}}vv \tag{11}$$

The equation is applied for an incompressible fluid. The inertia term on the left-hand side is calculated by conventional averaging. The first viscosity term in the Brinkman equation and the last term in Forchheimer's equation, the C_F, is the Forchheimercoefficient. For more information about the mentioned formula, please see ref. [2].

2.4. Heat Transfer Models

Generally, there are two heat transfer models based on the volumetric averaging method: the single-equation model, which is based on the LTE (local thermal equilibrium) assumption, and the two-equation model, which is based on the LNTE (the local, non-thermal equilibrium) assumption. The two models will be discussed separately in terms of their assumptions and their application limitations [2].

2.4.1. LTE

By distinguishing the gradient operator in the microscopic and macroscopic coordinates, a simple form of the volumetric average is provided. By averaging the microscopic equations on a representative elementary volume (REV), the macroscopic equations for the mandatory relocation of an incompressible flow in a variant porous medium are derived.

$$(\rho C_p)_m \frac{\partial \hat{T}}{\partial t} + \overline{\nabla}.[v\hat{T}] = \alpha_f\left(\frac{k_m}{k_f} + \frac{k'}{k_f}\right)\hat{\nabla}^2\hat{T} \tag{12}$$

According to the LTE assumptions, the single-equation model regarding the energy equation is highly capable of saving calculation time. The LTE assumption is that the temperature difference between the solid object and the fluid is negligible or is very small in comparison to the whole system's temperature difference. This hypothesis can describe the inaccuracies in the energy transfer model.

2.4.2. LNTE

If the heat transfer between the solid object and the liquid is permissible, it leads to:

$$(1 - \varphi)(\rho c)_s \frac{\partial T_s}{\partial t} = (1 - \varphi)\nabla.(k_s \nabla T_s) + (1 - \varphi)q_s''' + h\left(T_f - T_s\right) \tag{13}$$

$$\varphi(\rho c_P)_f \frac{\partial T_f}{\partial t} + (\rho c_P)v.\nabla T_f = \varphi\nabla\left(k_f \nabla T_f\right) + \varphi q_f''' + h\left(T_s - T_f\right) \tag{14}$$

In these formulas, q''' represents the volumetric heat generation, and h represents the convection heat transfer coefficient between the solid matrix and fluid flow.

2.4.3. Buongiorno's Heterogeneous Model

Heat transfer among nanoparticles due to the effects of two phenomena, the thermophoresis sliding velocity and Brownian motion, results in a kind of heterogeneity in the nanofluid which is known for transfer or as Buongiorno's model. In Buongiorno's heterogeneous model, the thermophoresis and Brownian forces are the dominant forces exerted on the nanoparticles.

The thermophoresis force acts against the temperature gradient and tends to carry the nanoparticles from warm regions to cold ones. In contrast, the Brownian motion tends to transfer the nanoparticles from high-concentration regions to low-concentration regions. Due to the movement of nanoparticles in the base fluid, two important effects appear. The first is that the nanofluid in the low-concentration regions is lightweight and tends to move upward, while the nanofluid in the high-concentration regions is heavy and tends to move downward. The second is that the movements of the nanoparticles lead to energy transfer due to mass transfer [10].

The continuity equation, or the concentration of the particles using Buongiorno's model in dimensional form, is as follows:

$$\frac{1}{\varepsilon} V.\nabla \varphi = \nabla.[D_B \nabla \varphi + \frac{D_T}{T} \nabla \varphi] \tag{15}$$

D_B and D_T are the Brownian motion and thermophoresis coefficient, respectively:

$$D_B = \frac{k_B T}{3\pi \mu d_{np}} \tag{16}$$

$$D_T = (\frac{0.26k}{2k + k_{np}})(\frac{\mu}{\rho})\varphi \tag{17}$$

where k_B and d_{np} are the Boltzmann constant and the particle diameter, respectively.
For LTE, the energy equation in porous media is:

$$\rho c_p V.\nabla T = \nabla.k_m \nabla T + \varepsilon \rho_{np} c_{p,np} [D_B \nabla \varphi.\nabla T + D_T \frac{\nabla T.\nabla T}{T}] \tag{18}$$

where k_m is the effective thermal conductivity of the porous medium.

3. Results and Discussion

3.1. Free Convection Heat Transfer

When there is a temperature difference between a hot source and its surrounding fluid, the fluid's density varies in terms of the temperature variation and the buoyancy forces come into play. Therefore, the fluid begins to flow on the solid's surface in accordance with the temperature difference. In this case, heat transfer that occurs without any external stimulus and is completely autonomous is called free or natural convectional heat transfer.

Considering the importance of heat transfer, all its aspects and solutions should be taken into account to maximize the heat transfer speed. Natural heat transfer happens through the random motions of the molecules and the bulk motion of the fluid. Therefore, utilizing compounds with high heat conduction ratios in comparison to the fluids is useful. In this regard, using tiny particles with high heat conduction is considered. Studies showed that the smaller the size and dimension of the particles, the greater the effect on heat transfer rate [9–13].

3.2. Integrated Free Convectional Heat Transfer

Integrated free convectional heat transfer makes sense when two phenomena are combined: first, heat transfer due to the fluid's displacement, and conduction heat transfer due to the contact between the fluid and the solid object. The type of medium that can host the integrated free convectional heat transfer phenomenon is a porous medium containing a fluid and nanoparticles. The integrated free convectional heat transfer phenomenon has several industrial applications, such as for particle storage, filtering, gas drying, underground pollution, maintaining cooling radioactive waste containers, soil cleaning by steam injection, heat insulation for buildings, solar collector technologies, electronic cooling, and many more [14].

Yekani-Motlagh et al. [15] investigated the free convection of a two-phase nanofluid in an inclined, porous, semi-annulus enclosure. They used Fe_3O_4-water as their magnetic nanofluid and used Darcy and Boungiorno's models. They concluded that the Nusselt number increases when the nanoparticle volume fraction is increased.

The characteristics of the research papers related to free convection heat transfer and their results are summarized in Table 2.

Table 2. Research paper characteristics related to the free convection heat transfer.

Ref	Geometry Description	Nanofluid	Methodology	Results	Decision Variables
[15]	Inclined, porous, semi-annulus enclosure	Magnetic Fe_3O_4 -water	Free convection, Buongiorno and Darcy models, FVM, SIMPLE	- Adding nanoparticle volume fraction → Nu increases - Increase in porosity number → Nu increases	$10 \leq Ra \leq 1000$ Porosity number = 0.4, 0.7 $0 \leq \varphi \leq 0.04$ $0 \leq$ inclination angle of cavity ≤ 90
[16]	Square enclosure and convection around a circular cylinder, different geometries of cylinders	Ag-water	Free convection, Darcy–Brinkman model	- Porous layer thickness increases (20% to 80%) → free convection performance decreases (up to 50%)	$10^3 < Ra < 10^6$ $10^{-5} < Da < 10^{-1}$ 0% < thickness of porous layer < 100% 1 < thermal conductivity ratio $0 < \varphi < 0.1$
[17]	Square enclosure	MWCNT–Fe_3O_4/water	Free convective MHD, MRT, Lattice–Boltzmann	- Increase in Ra → increase in heat transfer rate - Increase in Ha → decrease in Ra - Increase in Nu (+4.9%)	$10^{-2} < Da < 10^{-1}$; $10^3 < Ra < 10^5$; 0.4 < porosity < 0.9; $0 < \varphi < 0.003$; $0 < Ha < 50$;
[18]	Inclined square enclosure and exothermic chemical reaction administered by Arrhenius kinetics	Tilted nanofluid	Free convective Buongiorno nanofluid model, FEM	- Re increases → Nu decreases	Dissemination of streamlines; isotherms; iso-concentrations; and average Nusselt number
[19]	Square cavity and linearly heated left wall with composite nanofluid–porous layers	Cu-water	Free convection, Galerkin finite element method, Darcy–Brinkmann model	- Increase in Ra → intense streamlines	$\varphi = 0.1$; $10^{-7} \leq Da \leq 1$; $10^3 \leq Ra \leq 10^7$
[20]	Inverse T-shaped cavity	MWCNT–Fe_3O_4/water	Free convection MHD, extended Darcy–Brinkman–Forchheimer model	- Lower inclination angle → higher Nu - Lower values of ratio of dimensionless convection coefficient and the magnetic field viscosity parameter → significant heat transfer enhancement	$0 \leq$ magnetic field viscosity parameter ≤ 1; $0.7 \leq$ porosity ratio ≤ 1.4; $0 \leq$ magnetic field inclination angle $\leq \pi$; $0 \leq$ ratio of dimensionless convection coefficient ≤ 10; $Ha = 20$; $Ra = 10^5$

Table 2. *Cont.*

Ref	Geometry Description	Nanofluid	Methodology	Results	Decision Variables
[21]	Square cavity and two semicircular heat sources in the wall	MWCNT–Fe_3O_4/water	Free convection, FEM	- $Ra = 1 \times 10^4 \rightarrow Nu$ increases with magnetic number	$100 <$ Magnetic number < 5000; $0.2 <$ Strength ratio of magnetic sources < 5; $0 < Ha < 50$; $0.1 <$ porosity coefficient < 9
[22]	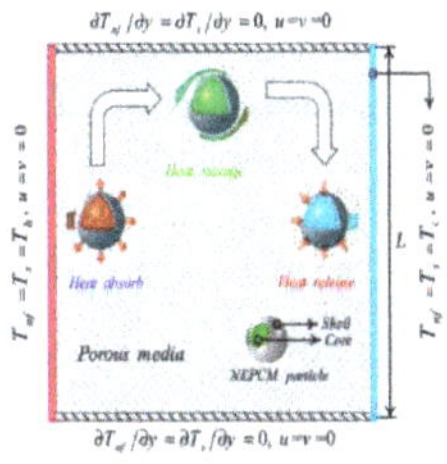	Nano-Encapsulated Phase Change Materials (NEPCM)	Free convection. local thermal non-equilibrium (LTNE)	- Increase in thermal conductivity of porous medium $\rightarrow$ and increase in heat transfer	$0 \leq \varphi \leq 0.05$
[23]	Transient natural convection and a square cavity, considering nanoparticle sedimentation	Al_2O_3/water	Free convection	- Nu decreased - Reduction in convection heat transfer	$10^4 < Ra < 10^7$; $10^{-5} < Da < 10^{-2}$
[24]	Square cavity	Ag–MgO/water	Free convection, LTNE, Darcy model, Galerkin FEM	- Increase in $Ra \rightarrow$ increase in the vortex's strength - Increase in heat transfer (5.85 times)	$10 \leq Ra \leq 1000$; $0.1 \leq \varepsilon \leq 0.9$; $0 \leq \varphi \leq 0.02$; $1 \leq H \leq 1000$
[25]	Inclined enclosure with wavy walls and partially layered porous medium	Cu-Al_2O_3 water	Free convection, Galerkin FEM, Darcy–Brinkman model	- Increase in heat transfer	$0 <$ inclination angle < 90; $10^4 \leq Ra \leq 10^7$; $10^{-2} \leq Da \leq 10^{-5}$; $0.2 \leq$ porous layer width ≤ 0.8; $1 \leq$ number of undulations ≤ 4; $0 \leq \varphi \leq 0.2$
[26]	Eccentricity heat source and porous annulus	Cu-water	Free convection	- Increase in heat transfer	$0 \leq \psi \leq 0.04$, $10^3 \leq Ra \leq 10^6$; $10^{-4} \leq Da \leq 10^{-1}$;
[27]	Transient natural convection and non-Darcy porous cavity with an inner solid body	Al_2O_3 -Water	Free convection, Buongiorno model, Brinkman–Forchheimer extended Darcy formulation. FDM	- Higher Da$\rightarrow$ uniform nanoparticle distribution - Increasing porosity $\rightarrow$ uniform nanoparticle distribution - Maximum Nu enhancement is approximately 30%	The porosity of the porous medium; Darcy number; The nanoparticles' average volume fraction
[28]	Inner corrugated cylinders inside wavy enclosure and porous–nanofluid layers	Ag nanofluid	Free convection	- Increase in Ra and Da $\rightarrow$ increase in fluid flow strength and shear layer thickness - Increase in porous layer thickness$\rightarrow$ decrease in heat transfer	$10^6 \geq Ra \geq 10^3$; $0.1 \geq Da \geq 0.00001$; $0.2 \geq$ vertical location (H) ≥ -0.2; $6 \geq$ number of sinusoidal inners; cylinders (N) ≥ 3

Table 2. *Cont.*

Ref	Geometry Description	Nanofluid	Methodology	Results	Decision Variables
[29]	Inverse T-shaped cavity and trapezoidal heat source in the wall	Fe_3O_4-water	Free convection, magnetic field dependent (MFD), FEM	- Local and average Nu increased	Darcy, Hartmann, and Rayleigh numbers; inclination angle; cavity aspect ratio
[30]	Spherical electronic device	Cu-water	Free convection, SIMPLE algorithm	- Heat transfer increases - Average Nu increases	$6.5 \times 10^6 < Ra < 1.32 \times 10^9$; $0 < \varphi < 10\%$; $0 <$ thermal conductivity of the porous material's matrix < 40
[31]	Tilted hemispherical enclosure	Water-ZnO	Free convection experiment	Increase in heat transfer	$0 <$ inclination angle < 90; $0 < \varphi < 8.22\%$
[32]	Wavy-walled porous cavity and inner solid cylinder	Al_2O_3/water	Free convection, FEM, Forchheimer–Brinkman extended Darcy model, Boussinesq approximation	- Higher values of Da → heat transfer enhancement	$0 \leq \varphi \leq 0.04$; $10^{-6} < Da < 10^{-2}$; $0.2 \leq \varepsilon \leq 0.8$
[33]	Partitioned porous cavity for application in solar power plants	MWCNT–Fe_3O_4/water	Free convection, CFD method, volume averaging the microscopic equations	- Increase in Da, Ra → Nu_{ave} increases	$103 < Ra < 106$; $0.5<$ porosity coefficient ratio < 1.8; $0 < \varphi < 0.003$; $0.1 < Ri < 20$; $0.01 < Da < 100$; Thermal conductivity ratio $= 0.2, 0.4, 1, 5$
[34]	Square cavity and inner sinusoidal vertical interface	Ag/water	Free convection, Galerkin FEM	- Increase in Da, Pr → Nu_{ave} increases	$0.6 <$ power law index < 1.4; $10^{-5} < Da < 10^{-1}$; $0 < \varphi < 0.2$; $1 <$ undulation number (N) < 4; $0.015 < Pr < 13.4$; $Ra = 10^5$
[35]	Hot rectangular cylinder and cold circular cylinder	copper–water	Free convection, Brinkman-extended Darcy model, Brinkman correlation	- Heat transfer enhanced	Rayleigh number; Hartmann number; Darcy number; magnetic field inclination angle; nanoparticles volume fraction; nanoparticles shape factor; nanoparticles material; nanofluid thermal conductivity; dynamic viscosity models; nanofluid electrical conductivity correlation on streamlines; isotherms; local and average Nusselt numbers

Table 2. *Cont.*

Ref	Geometry Description	Nanofluid	Methodology	Results	Decision Variables
[36]	Partially heated enclosure	Al_2O_3/water	Free convection, FEM, Brinkman equation	- Heat transfer rate augmented - Ra, Da increases → average velocity	$10^3 < Ra < 10^6$; $0 < \varphi < 5\%$; $0 < Ha < 100$; $0.001 < Da < 1$
[37]	I -shaped cavity	Cu–water	Free convection, MHD, FDM	- Ha increases → Nu decreases - Ra increases → Nu increases - Maximum Nu occurs at B = 0.2 - Minimum Nu occurs at B = 0.8	Ha; nanofluid volume fraction; heat source size; location and angle of magnetic field on heat transfer; entropy generation; thermal performance
[38]	Porous enclosure	Cu, Al_2O_3 and TiO_2/water	Free convection, MHD	- Increase in magnetic field intensity→ heat transfer deterioration - Enlarging nanoparticles, denser nanoparticles→ heat transfer deterioration	$0 \leq Ha \leq 50$; Nanoparticle volume fraction; Nanoparticle diameter
[39]	Inclined cavity	Al_2O_3-water	Free convection Entropy generation	- Increase in chamber angle → increase in heat transfer - Adding nanoparticle volume fraction → increase in heat transfer	Rayleigh number Hartmann number; magnetic field angle changes; chamber angle changes; entropy parameter; radiation parameter; volume percent of nanoparticles
[40]	Cubical electronic component and hemispherical cavity	Water-ZnO	Free convection, control volume method	- Inclination increases → Nu_{ava} decreases - Nanofluid concentration increases → heat transfer increases	$0 <$ volume fraction $< 10\%$; Nu_{ave}
[41]	Inverted T-shape	MWCNT–Fe_3O_4/water	Free convection, thermal transmission	- Ha increases → Nu_{ave} decreases	Heat transfer performance; flow structures
[42]	Inverse T-shaped cavity and trapezoidal heat source in wall with wavy Wall	Magnetic Al_2O_3/water	Free convection, FEM, Koo–Kleinstreuer–Li (KKL) correlations	- Increase in Ra, decrease in Ha → increase in flow intensity -	Heat generation parameter, the shape factor of nanoparticles; Hartmann number; nanoparticle concentration, displacement of the trapezoidal heater wall; Rayleigh number; the amplitude of wavy wall

3.3. Forced Convection Heat Transfer

Forced convection occurs when the fluid motion is generated by an external source and not only by the density difference inside the fluid.

Sheikholeslami et al. [43] performed analyses on the effects of Reynolds number, the volumetric quotient of the water–calcium oxide nanofluid, and the Hartmann and Darcy numbers on the forced convectional heat transfer in a container of a hot liquid. They realized that the heat profile decreases with the increase of the voltage, but heat conduction is improved by an increase in the Darcy and Reynolds numbers. In addition, the studies indicate that there is an inverse relationship between the temperature gradient and the Hartmann number.

Furthermore, Sheikholeslami et al. [44] evaluated the forced convectional heat transfer of water–aluminum oxide in the presence of a magnetic field and realized that the nanofluid velocity profile is in direct relation to the Reynolds number of the volumetric quotient of aluminum oxide but in an inverse relation to the Hartmann number. In addition, by the increase in the Lorentz force, the convectional heat transfer decreases. It was also revealed that the temperature gradient on the moving surface increases as the hot surface velocity and the volumetric aluminum oxide quotient increase. They reported that increasing the Reynold number resulted in the Nusselt number increasing. In addition, increasing the Hartmann number augmented heat transfer.

Ferdows and Alzahrani [45] numerically investigated the possibility of similar solutions as well as dual-branch solutions to evaluate the performance of nanoparticles associated with water: the base water. A steady-state condition was considered through a moving, flat, porous plate in the presence of magnetic fields. They concluded that the largest velocity profile and the smallest temperature distribution refer to the Cu-water nanofluid.

Table 3 summarizes several studies dealing with the forced convection of nanofluids.

Table 3. Research paper characteristics related to the forced convection heat transfer.

Ref	Geometry Description	Nanofluid	Methodology	Results	Decision Variables
[45]	Moving surface	Water- based nanoparticles: copper (Cu), alumina (Al2O3), and titania (TiO2)	Forced convection, MHD	- Smallest temperature distribution: Cu - Largest velocity profile: Cu	Skin friction coefficient; Local Nu number
[46]	Channel, staggered, and in-line arrangements of square pillars	Al_2O_3-water	Forced convection, First and Second laws of thermodynamics, FVM	- Nu increases and decreases for the Re and nanofluid volume fraction. - The Al_2O_3 nanoparticles participation in the base fluid decreases the entropy generation. - The entropy generation and the Be decrease and increase with the nanofluid particle volume fraction.	Porosity: 0.84, 0.75, 0.91; Re = 10, 200, 300; Nanofluid V.F. = 4%
[47]	Multi-layered, U-shaped vented cavity and wall corrugation effects	CNT-water	Forced convection, FEM	- Heat transfer enhancement	$100 < Re < 1000$; $0 < Ha < 50$; $10^{-4} < Da < 5 \times 10^{-2}$
[48]	Lid-driven cavity and hot sphere obstacle	Al_2O_3-water	Forced convection, Lattice–Boltzmann method	- Rate of heat transfer enhances with the rise of permeability of porous media and velocity of lid wall. This is due to an enhanced temperature gradient with the increase of Da and Re.	$0.001 < Da < 100$; $0 < Ha < 40$; $30 < Re < 180$
[49]	U-bend pipe	Al_2O_3-CuO-water	Forced convection, FEM, Darcy–Brinkman–Forchheimer equation	- Decrease in Da $\rightarrow$ increase in Nu_{ave} $\rightarrow$ Increase in pressure drop	$10^{-4} < Da < 10^{-1}$
[50]	Cylinder	Al_2O_3–CuO–water	Forced convection, MHD, FVM	- Decrease in Da $\rightarrow$ increase in Nu $\rightarrow$ increase in Ha - Increase in Da and Ha $\rightarrow$ decrease in pressure drop - Adding metal nanoparticles $\rightarrow$ increase in Da and Ha	Nu number; $0.0001 < Da < 0.1$; $0 < Ha < 40$; Magnetic field orientation

Table 3. *Cont.*

Ref	Geometry Description	Nanofluid	Methodology	Results	Decision Variables
[51]	Annulus with porous ribs	Al_2O_3-water	Forced convective	- Increase in porous ribs $\rightarrow$ increase in pressure drop	Nu
[52]	Horizontal plate	Water-based Cu/Alumina/Titania	Forced convection	- Heat transfer rate is higher for Cu than others - Increase in nanoparticles V.F. $\rightarrow$ increase in heat transfer rate	Nanoparticle volume fraction; porosity of porous media; thermal conductivity of porous media; effect of nanoparticle type on its heat transfer

3.4. Mixed Convection

Khademi et al. studied the mixed convection of nanofluids on a sloped, flat surface in a porous medium with the presence of a magnetic field. The boundary layer equations were solved numerically using the DQM method. This research identified a decrease in the Nusselt number [53].

Bondarenko et al. evaluated the mixed convection heat transfer of a nanofluid in a square container with insulated lateral walls and cold top and bottom walls. The boundary layer equations were solved using the FDM method [54]. Various research papers that considered the mixed convection of nanofluids are presented in Table 4.

Table 4. Research paper characteristics related to the mixed convection heat transfer.

Ref	Geometry Description	Nanofluid	Methodology	Results	Decision Variables
[53]	Inclined flat plate	Water-Cu	Mixed convection, MHD, DQM	Nu reduced	$Ra = 10^5$; $Ha = 25$
[54]	Lid-driven enclosure and two adherent porous blocks	Alumina/water	Mixed convection	-Ri < 1 $\rightarrow$ heat transfer enhancement - Ri $\geq$ 1 $\rightarrow$ reduction in heat transfer	$0.01 \leq Ri \leq 10$; $0 \leq \varphi \leq 0.04$
[55]	Rotating circular cylinder and trapezoidal enclosure	Cu-water	Mixed convection, MHD	Decrease in stream function values $\rightarrow$ vertical magnetic field - Increase in Ha $\rightarrow$ increase in Nu_{ave} - Increase in Ha, thermal conductivity rate, cylinder radius, Da $\rightarrow$ increase in Nu_{ave} - Decrease in Ri $\rightarrow$ increase in Nu_{ave}	0 < Ha < 100; 1 < Thermal conductivity ratio < 10; −5 < angular rotational velocity < 5; 0.01 < Ri < 100; 0 < Inclination angle < 90; 0.2 < Cylinder radius < 0.4; 10^{-5} < Da < 10^{-1}; 0 < nanofluid concentration < 0.1
[56]	Square cavity with inlet and outlet ports	Water-based nanofluid	Mixed convection, Brownian diffusion, thermophoresis, FDM	- Increase in Re $\rightarrow$ cooling improvement - Ra = 10 $\rightarrow$ Nu = 1.071 - Ra = 100 $\rightarrow$ Nu = 3.104 - Ra = 1000 $\rightarrow$ Nu = 13.839 - Ra = 10000 $\rightarrow$ Nu = 49.253 $- \varnothing = 0.01 \rightarrow Nu = 31.6043$ $- \varnothing = 0.02 \rightarrow Nu = 31.2538$ $- \varnothing = 0.03 \rightarrow Nu = 30.829$	$10^4 < Ra < 10^6$; $Pr = 6.82$; $10^{-5} < Da < 10^{-6}$; $50 < Re < 300$; $\varepsilon = 0.5$; $Le = 1000$

Table 4. *Cont.*

Ref	Geometry Description	Nanofluid	Methodology	Results	Decision Variables
[57]	H-shaped cavity with cooler and heater cylinders	Cu-water	Mixed convection, Boussinesq approximation	- Increase in AR $\rightarrow$ decrease in heat transfer rate increase in Da, decrease in Ri $\rightarrow$ increase in heat transfer rate	$10^{-4} \leq$ Da $\leq 10^{-2}$; $1 \leq$ Ri ≤ 100; $1.4 \leq$ AR ≤ 1.6;
[58]	Trapezoidal chamber	Cu-Al$_2$O$_3$/water	Mixed convection, FDM	- Increase in Re $\rightarrow$ increase in energy transport and convective circulation - Increase in Da $\rightarrow$ heat transfer enhancement	Reynolds number; Darcy number; nanoparticle volume fraction
[59]	Inclined cavity	Cu-water	Mixed convection, Darcy–Brinkman–Forchheimer model, SIMPLE algorithm	- Heat transfer rate increases with increasing Da.	
[60]	Lid-driven square cavity	Al$_2$O$_3$/water	Mixed convection	- Decrease in Ri $\rightarrow$ increase in momentum - Ri = 100 $\rightarrow$ decrease in Darcy effects - Changing nanoparticles volume fraction and Da $\rightarrow$ significant changes in streamlined pattern - Higher Ri $\rightarrow$ more buoyancy effects - Increase in Da and Ri $\rightarrow$ less fluid resistance and more momentum penetration - Increase in Da $\rightarrow$ decrease in temperature, more uniformity in heat transfer	Ri = 0.01, 10 and 100; $10^{-4} \leq$ Da $\leq 10^{-2}$; $0 \leq \varphi \leq 0.04$
[61]	Stretching surface	—	Mixed convection, MHD	- For m < 1 $\rightarrow$ increase in velocity results in an increase in thermophoresis - For m > 1 $\rightarrow$ increase in velocity results in a decrease in thermophoresis.	Effects of buoyancy parameter; magnetic parameter; Brownian motion; thermophoresis parameter, etc., on velocity, temperature, and nanoparticle volume fraction
[62]	Square cavity and two rotating cylinders	Al$_2$O$_3$/water	Mixed convection	- Heat transfer enhancement (+ 20.4%)	
[63]	Triangular shape, partitioned, lid-driven square cavity involving a porous compound	Ag–MgO/water	Mixed convection MHD	- Nu enhancement (14.7%)	< Ri < 100; 0 < Ha < 60; 10^{-4} < Da < 5 $\times 10^{-2}$; $0 < \varphi < 0.01$
[64]	Vertical surface	Cu-water	Mixed convection, Laplace transform technique Crank Nicolson method	- Increase in magnetic field strength $\rightarrow$ and decrease in fluid velocity - Porosity increases $\rightarrow$ fluid velocity decreases	Magnetic parameter; porosity parameter; thermal and solute Grashof number; nanoparticle volume fraction parameter; time; Schmidt number; chemical reaction parameter; Prandtl number

Table 4. *Cont.*

Ref	Geometry Description	Nanofluid	Methodology	Results	Decision Variables
[65]	Inclined cavity and porous layer	Cu-water	Mixed convection, incompressible smoothed particle hydrodynamics (ISPH)	- Ri increases → Nu$_{ave}$ decreases - φ increases → overall heat transfer increases	$0.001 < Ri < 100$; $10^{-5} < Da < 10^{-2}$; $0 < \varphi < 0.05$
[66]		CuO–Water	Mixed convection, entropy generation, Buongiorno's two-phase model	- Increase in volume concentration → increase in Nu$_{ave}$ - Maximum enhancement In cooling performance was 17.75%	- Volume concentration; development of a new predictive correlation
[67]	Gamma-shaped cavity	CuO–Water	Mixed convection, Entropy generation, FVM	- Increase in the Nusselt number with the volume fraction is more pronounced for the smallest heat source, a heat source placed at the lowest height from the bottom side, the lowest volumetric heat generation, the lowest imposed magnetic field, the lowest Darcy number, and for a porous media with the lowest solid to fluid thermal conductivity ratio. Increasing the nanoparticle volume fraction has a higher impact on the production of entropy than the enhancement in the heat transfer rate.	- Hartmann number; nanoparticle volume fraction; the length and location of a heat source
[68]	Rotating triangle chamber	Graphene Oxide generalized hybrid	Mixed convection utilizing bvp4c solver	The velocity upsurges due to the dimensionless radius of the slender body parameter in case of the assisting flow.	$0.025 \leq \varphi \leq 0.035$

3.5. Overall Review of Papers

Considering that the most important purpose of adding nanoparticles to a base fluid is the possibility of increasing the heat transfer, in the following section and in Table 5, the selected articles presented in Tables 2–4 in the previous section were examined and studied based on the increase or decrease of the Nusselt number, and a report of the increase or decrease in the Nusselt number is presented as a percentage. In addition, due to the importance of the size of nanoparticles, their sizes are presented in a separate column.

In most articles about the natural and mixed convection of a nanofluid in porous media, the Nusselt number is increased by adding the volume fraction of nanoparticles; however, some articles also reported a decrease. When nanoparticles are added to the base fluid, the conductivity definitely increases. Viscosity also increases; therefore, in low-velocity flows, such as free displacement heat transfer, the decrease in speed due to the increase in viscosity is quite evident and effective and it can dominate the increase in conductivity, causing a decrease in displacement heat transfer. The same applies to a low-speed combined heat transfer. In forced heat transfer, the increase in nanoparticles increases the heat transfer.

Additionally, in some cases in which an increase in the nanoparticle diameter is observed, the natural convection heat transfer rate is increased by more than 5%. By increasing the nanoparticle diameter in the articles, the forced convection heat transfer rate was decreased.

Table 5. Nu changes with nanofluid concentration and size.

Ref	Nu	Volume Fraction and Size
[15]	Nu increases 0.35% Heat transfer increases 0.48%	$0 \leq \varphi \leq 0.04$ $0.384 \leq d_p \leq 2.5$
[17]	Nu increases 4.9% Heat transfer increases 6.72%	$0 < \varphi < 0.003$ $1.46 \leq d_p \leq 2.8$
[20]	Nu increases 3.6% Heat transfer increases 5.94%	$0 < \varphi < 0.3$ $0.65 \leq d_p \leq 1.41$
[21]	Nu increases 6.01% Heat transfer increases 9.24%	$\varphi < 0.3$ $4.5 \leq d_p \leq 7.6$
[23]	Nu decreased 6.64% Heat transfer decreased 8.21%	$\varphi < 0.05$ $0.51 \leq d_p \leq 2.3$
[26]	Nu increases 44.44%	$0 \leq \phi \leq 0.04$ $0.5 \leq d_p \leq 20$
[29]	Nu decreased 0.69% Heat transfer decreased 0.8%	$0 \leq \phi \leq 0.1$ $5 \leq d_p \leq 15$
[32]	Nu increases 5.69% Heat transfer increases 9.28%	$\varphi \leq 0.04$ $0.385 \leq d_p \leq 33$
[35]	Nu increases 24.98%	$\varphi < 0.003$ $8 \leq d_p \leq 29$
[50]	Nu increases 0.5% Heat transfer increases 2.3%	$\varphi < 0.005$ $0.52 \leq d_p \leq 7.5$
[53]	Nu decreased 30%	$\varphi < 0.15$ $2 \leq d_p \leq 24$
[56]	Nu increases 66.6%	$0 \leq \varphi \leq 0.04$ $4 \leq d_p \leq 14$
[57]	Nu decreased 12.28%	$0 \leq \varphi \leq 0.01$ $0.89 \leq d_p \leq 1.3$
[60]	Nu decreased 26.08%	$\varphi \leq 0.04$ $0.384 \leq d_p \leq 47$
[62]	Nu increases 5.97% Heat transfer increases 20.4%	$\varphi \leq 2$ $3 \leq d_p \leq 25$
[63]	Nu increases 14.7%	$\varphi \leq 0.01$ $7 \leq d_p \leq 44$
[65]	Nu decreased 68.75%	$\varphi \leq 0.05$ $1.94 \leq d_p \leq 6.29$

3.6. Statistical Results

In this section, the statistical distribution of different parameters in the published papers on nanofluids in porous materials is presented. By reviewing the published research, it was revealed that Al_2O_3 nanoparticles in a base fluid of water with a proportion of 33.9% have the highest representation in the papers. Copper nanomaterials and Fe_3O_4-water, with a representation of 32.14% and 12.5%, respectively, occupy the second and third ranks. The share of each nanofluid is depicted in Table 6.

Table 6. The share of each nanoparticle in published studies.

Nanoparticle	Share (%)
Al_2O_3-Water	33.9
Cu-Water	32.14
Fe_3O_4-Water	12.5
Ag-Water	5.35
Ag-MgO	3.75
Other nanoparticles	12.5

In general, it can be said that the most popular nanoparticle is the alumina nanoparticle due to its very good dispersion in the base fluid. Of course, the role of the stability of the nanofluid in the porous media is more important. Metal oxide nanoparticles have shown a dual behavior in free convection heat transfer which has been reported to increase in some cases and decrease in some cases. However, in forced convection heat transfer, the increase of any type of nanoparticle increases the heat transfer.

Regarding the studied geometries, a square geometry accounted for 54% of the studies. Hole and circular geometries were represented with a share of 17% and 13%, respectively, and occupy the next ranks. The details are depicted in Table 7.

Table 7. The share of each of the studied geometries in published papers.

Geometry	Share (%)
Square	54
Hole	18
Circular	12
Other shapes	16

Figure 3 illustrates the proportion of the models that were used for evaluation. Based on the figure, Darcy–Brinkman–Forchheimer has the most share at 16%, followed by the Darcy–Brinkman and Darcy models, with shares of 6% and 2%, respectively.

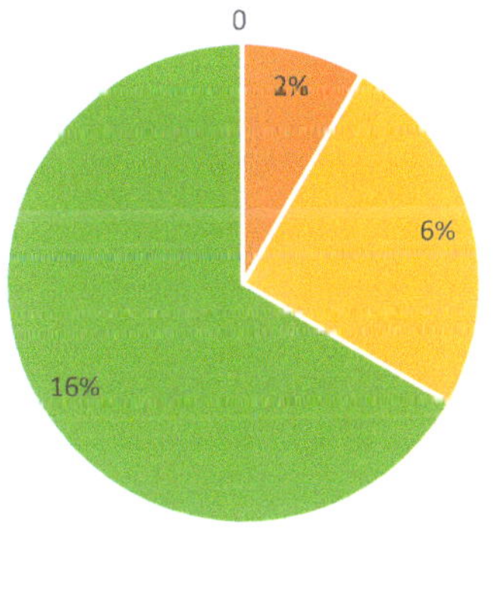

Figure 3. The contribution of each of the models used to model the system in the published papers.

Figure 4 illustrates the share of the models used in the presence and absence of a magnetic field.

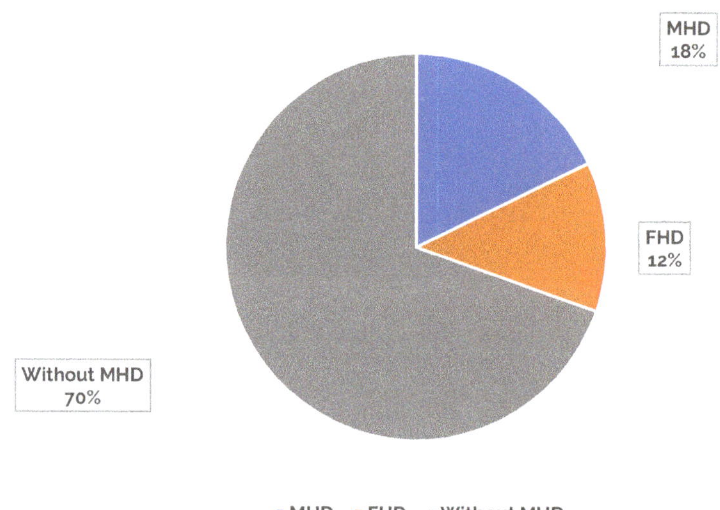

Figure 4. The presence or absence of magnetic fields to model the system in published papers.

Figure 5 depicts the contribution of scientific publishers Science Direct, Springer, Tandf Online, John Wiley, and ASME in published papers related to this specific topic. According to this figure, the highest share of the studies was indexed in Science Direct at 56%.

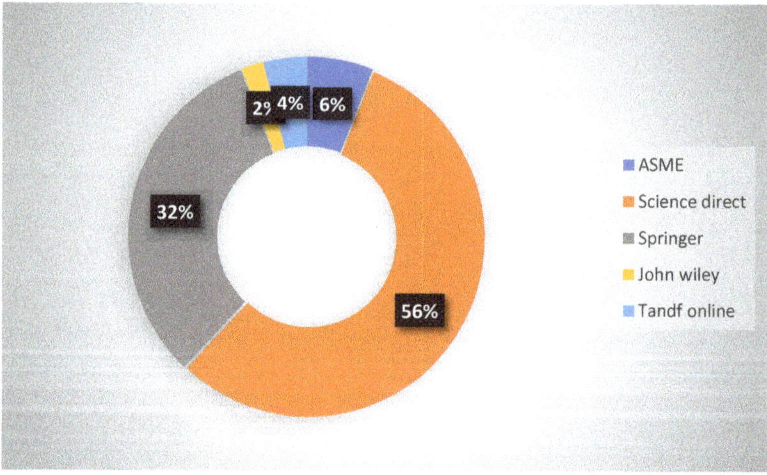

Figure 5. The contribution of each publisher of the published papers related to nanofluids in a porous material.

In the statistical analysis section, the items that are important for heat transfer, according to the authors, were examined:

- The type of nanoparticle;
- The geometry under consideration.

Additionally, considering that the major contribution of the articles was related to numerical modeling, the model selected for investigating the flow in the porous medium was analyzed. It may be said that the last two forms are not directly related to the subject;

however, as we know, external forces, including the magnetic force, have an effect on heat transfer, and approximately 30% of the subject articles dealt with it.

4. Conclusions

In this paper, works published between 2018 and 2020 that concern nanofluid heat transfer within porous media are reviewed. As the latest review in this field was published at 2017, it can be found at ref. [69]. Additionally, a good book was published in this field [70]. However, in this article, a statistical analysis of the published articles was performed. This has not been done before.

The results showed that heat transfer is improved by adding nanoparticles to the base fluid. On the contrary, by increasing the volume concentration of nanoparticles, the viscosity effect overcomes the thermal conductivity and hinders the heat transfer. The heat transfer is also reduced with an increase in the nanoscale conductivity relative to the porous matrix, and heat transfer is enhanced with the rise of the Darcy number and the porosity coefficient. Some important results emerged from this study. They are presented as follows:

1. As a dimensionless permeability, hte effect of the Darcy number on heat transfer is direct. Therefore, as the Darcy number increases, the heat transfer increases. As the Darcy number decreases, the heat transfer decreases;
2. The effect of the porosity coefficient has a direct relationship with the heat transfer, and the heat transfer will increase or decrease by increasing or decreasing the porosity coefficient;
3. A change in the width of the solid and porous medium results in a change in the flow regime inside the chamber and has an increasing or decreasing effect on heat transfer;
4. A change in the height of the solid and porous medium results in a change in the flow regime within the chamber and has an increasing or decreasing effect on heat transfer;
5. Increasing the Ra number increases the heat transfer.

A detailed analysis of the share of each nanoparticle type and geometrical configuration revealed a disparity among the published papers. Future research should be directed towards the nanofluids that were less frequently investigated and should address new types of nanoparticles and hybrid nanofluids. Similarly, novel geometrical configurations should be subjected to more examination, such as circular and triangular enclosures. Finally, the economic aspect has not been widely considered for the various engineering applications. Studies addressing such aspects and evaluating the energy and long-term costs and gains of implementing nanofluids instead of conventional fluids in applications involving porous media should be emphasized in the future.

Finally, the main shortcomings inherent to the use of nanofluids in porous media are the increasing pressure drop, especially in forced convection, and the stability of the nanofluids (reduction due to the impact of nanoparticles on the solid matrix). It is proposed that research is performed on these problems so that they may be solved.

Author Contributions: Conceptualization, H.A.N., B.A., T.A., A.M.R., A.J.C.; Supervision, T.A. and H.A.N.; Investigation, H.A.N., B.A., T.A., A.M.R., A.J.C.; Methodology, H.A.N., B.A., T.A.; Writing—original draft, H.A.N., B.A., T.A.; Writing—review & editing, H.A.N., T.A., A.M.R., A.J.C.; Software, B.A., T.A.; Formal Analysis, H.A.N., B.A., T.A., A.M.R., A.J.C.; Project Adminstration H.A.N. All authors have read and agreed to the published version of the manuscript.

Funding: This study was supported by the Prince Sattam bin Abdulaziz University through project number (2022/RV/14).

Data Availability Statement: Data are available upon request.

Acknowledgments: This study was supported by the Prince Sattam bin Abdulaziz University through project number (2022/RV/14).

Conflicts of Interest: The authors declare no conflict of interest.

Nomenclature

Boltzmann constant	k_B
Bejan Number	Be
Brownian motion	D_B
Conduction factor for base fluid	k_{bf}
Conduction factor for nanoparticles	k_p
Darcy velocity	v
Diameter of the particle	d_p
Effective viscosity	μ_e
Fluid's viscosity	μ
Forchheimer's dimensionless factor	CF
Hydraulic conduction	k
Magnetic field	M
Mass quotient of each phase	C_k
Nanofluid volume fraction	φ
Nanoparticle	n_p
Porosity	φ
Pressure	P
Pressure in saturated state	Ps
Specific permeability	K
Thermophoresis Coefficient	D_T
Volumetric average velocity	v

Abbreviations

AR	Aspect Ratio
CFD	Computational Fluid Dynamics
CNT	Carbon nanotube
Da	Darcy
DQM	Differential Quadrature Method
FDM	Finite Difference Method
FEM	Finite Element Method
FVM	Finite Volume Method
Ha	Hartman Number
ISPH	Incompressible Smoothed Particle Hydrodynamics
KKL	Koo–Kleinstreuer–Li correlations
Le	Lewis Number
LNTE	Local Non-Thermal Equilibrium
LTE	Local Thermal Equilibrium
MFD	magnetic field dependent
MHD	Magnetohydrodynamics
NEPCM	Nano-Encapsulated Phase Change Materials
Nu	Nusselt Number
Pr	Prandtle Number
Ra	Raighly Number
Re	Reynolds Number
Ri	Richardson Number
SIMPLE	Semi-Implicit Method for Pressure Linked Equations
V. F.	Volume Fraction

References

1. Lapwood, E.R. Convection of a fluid in a porous medium. In *Mathematical Proceedings of the Cambridge Philosophical Society*; Cambridge University Press: Cambridge, UK, 1948; Volume 44, pp. 508–521. [CrossRef]
2. Bejan, A.; Dincer, I.; Lorente, S.; Miguel, A.F.; Reis, A.H. *Porous and Complex Flow Structures in Modern Technologies*; Springer: New York, NY, USA, 2004.
3. Whitaker, S. Flow in porous media I: A theoretical derivation of Darcy's law. *Transp. Porous Media* **1986**, *1*, 3–25. [CrossRef]

4. Murshed, S.M.S.; Leong, K.C.; Yang, C. Thermophysical and electrokinetic properties of nanofluids—A critical review. *Appl. Therm. Eng.* **2008**, *28*, 2109–2125. [CrossRef]

5. Meng, X.; Yang, D. Critical Review of Stabilized Nanoparticle Transport in Porous Media. *J. Energy Resour. Technol.* **2019**, *141*. [CrossRef]

6. Boccardo, G.; Tosco, T.; Fujisaki, A.; Messina, F.; Raoof, A.; Aguilera, D.R.; Crevacore, E.; Marchisio, D.L.; Sethi, R. A review of transport of nanoparticles in porous media: From pore-to macroscale using computational methods. *Nanomater. Detect. Remov. Wastewater Pollut.* **2020**, 351–381.–381. [CrossRef]

7. Ling, X.; Yan, Z.; Liu, Y.; Lu, G. Transport of nanoparticles in porous media and its effects on the co-existing pollutants. *Environ. Pollut.* **2021**, *283*, 117098. [CrossRef]

8. Hyun, J.M.; Choi, B.S. Transient natural convection in a parallelogram-shaped enclosure. *Int. J. Heat Fluid Flow* **1990**, *11*, 129–134. [CrossRef]

9. Takabi, B.; Shokouhmand, H. Effects of Al_2O_3-Cu/water hybrid nanofluid on heat transfer and flow characteristics in turbulent regime. *Int. J. Mod. Phys. C* **2015**, *26*, 1550047. [CrossRef]

10. Nield, D.A.; Bejan, A. *Convection in Porous Media*; Springer: New York, NY, USA, 2013.

11. Buongiorno, J. Convective transport in nanofluids. *J. Heat Transfer.* **2006**, *128*, 240–250. [CrossRef]

12. Jou, R.Y.; Tzeng, S.C. Numerical research of natural convective heat transfer enhancement filled with nanofluids in rectangular enclosures. *Int. Commun. Heat Mass Transf.* **2006**, *33*, 727–736. [CrossRef]

13. Agarwal, D.K.; Vaidyanathan, A.; Sunil Kumar, S. Synthesis and characterization of kerosene-alumina nanofluids. *Appl. Therm. Eng.* **2013**, *60*, 275–284. [CrossRef]

14. Motlagh, S.Y.; Golab, E.; Sadr, A.N. Two-phase modeling of the free convection of nanofluid inside the inclined porous semi-annulus enclosure. *Int. J. Mech. Sci.* **2019**, *164*, 105183. [CrossRef]

15. Hussain, S.H.; Rahomey, M.S. Comparison of Natural Convection Around a Circular Cylinder with Different Geometries of Cylinders Inside a Square Enclosure Filled with Ag Nanofluid Superposed Porous Nanofluid Layers. *J. Heat Transfer.* **2019**, *141*, 022501. [CrossRef]

16. Sajjadi, H.; Amiri Delouei, A.; Izadi, M.; Mohebbi, R. Investigation of MHD natural convection in a porous media by double MRT lattice Boltzmann method utilizing MWCNT–Fe_3O_4/water hybrid nanofluid. *Int. J. Heat Mass Transf.* **2019**, *132*, 1087–1104. [CrossRef]

17. Rahman, M.M.; Pop, I.; Saghir, M.Z. Steady free convection flow within a titled nanofluid saturated porous cavity in the presence of a sloping magnetic field energized by an exothermic chemical reaction administered by Arrhenius kinetics. *Int. J. Heat Mass Transf.* **2019**, *129*, 198–211. [CrossRef]

18. Al-Srayyih, B.M.; Gao, S.; Hussain, S.H. Effects of linearly heated left wall on natural convection within a superposed cavity filled with composite nanofluid-porous layers. *Adv. Powder Technol.* **2019**, *30*, 55–72. [CrossRef]

19. Mehryan, A.M.; Sheremet, M.A.; Soltani, M.; Izadi, M. Natural convection of magnetic hybrid nanofluid inside a double-porous medium using two-equation energy model. *J. Mol. Liq.* **2019**, *277*, 959–970. [CrossRef]

20. Izadi, M.; Mohebbi, R.; Delouei, A.A.; Sajjadi, H. Natural convection of a magnetizable hybrid nanofluid inside a porous enclosure subjected to two variable magnetic fields. *Int. J. Mech. Sci.* **2019**, *151*, 154–169. [CrossRef]

21. Ghalambaz, M.; Hashem Zadeh, S.M.; Mehryan, S.A.M.; Haghparast, A.; Zargartalebi, H. Free convection of a suspension containing nano-encapsulated phase change material in a porous cavity; local thermal non-equilibrium model. *Heliyon* **2020**, *6*, e03823. [CrossRef]

22. Baghsaz, S.; Rezanejad, S.; Moghimi, M. Numerical investigation of transient natural convection and entropy generation analysis in a porous cavity filled with nanofluid considering nanoparticles sedimentation. *J. Mol. Liq.* **2019**, *279*, 327–341. [CrossRef]

23. Mehryan, S.A.M.; Ghalambaz, M.; Chamkha, A.J.; Izadi, M. Numerical study on natural convection of Ag–MgO hybrid/water nanofluid inside a porous enclosure: A local thermal non-equilibrium model. *Powder Technol.* **2020**, *367*, 443–455. [CrossRef]

24. Kadhim, H.T.; Jabbar, F.A.; Rona, A. Cu-Al_2O_3 hybrid nanofluid natural convection in an inclined enclosure with wavy walls partially layered by porous medium. *Int. J. Mech. Sci.* **2020**, *186*, 105889. [CrossRef]

25. Gholamalipour, P.; Siavashi, M.; Doranehgard, M.H. Eccentricity effects of heat source inside a porous annulus on the natural convection heat transfer and entropy generation of Cu-water nanofluid. *Int. Commun. Heat Mass Transf.* **2019**, *109*, 104367. [CrossRef]

26. Alsabery, A.I.; Ismael, M.A.; Chamkha, A.J.; Hashim, I. Effect of nonhomogeneous nanofluid model on transient natural convection in a non-Darcy porous cavity containing an inner solid body. *Int. Commun. Heat Mass Transf.* **2020**, *110*, 104442. [CrossRef]

27. Abdulkadhim, A.; Hamzah, H.K.; Ali, F.H.; Abed, A.M.; Abed, I.M. Natural convection among inner corrugated cylinders inside a wavy enclosure filled with nanofluid superposed in porous–nanofluid layers. *Int. Commun. Heat Mass Transf.* **2019**, *109*, 104350. [CrossRef]

28. Molana, M.; Dogonchi, A.S.; Armaghani, T.; Chamkha, A.J.; Ganji, D.D.; Tlili, I. Investigation of Hydrothermal Behavior of Fe_3O_4-H_2O Nanofluid Natural Convection in a Novel Shape of Porous Cavity Subjected to Magnetic Field Dependent (MFD) Viscosity. *J. Energy Storage* **2020**, *30*, 101395. [CrossRef]

29. Baïri, A. Using nanofluid saturated porous media to enhance free convective heat transfer around a spherical electronic device. *Chin. J. Phys.* **2020**, *70*, 106–116. [CrossRef]

30. Baïri, A. Experimental study on enhancement of free convective heat transfer in a tilted hemispherical enclosure through Water-ZnO nanofluid saturated porous materials. *Appl. Therm. Eng.* **2019**, *148*, 992–998. [CrossRef]
31. Alsabery, A.I.; Mohebbi, R.; Chamkha, A.J.; Hashim, I. Effect of the local thermal non-equilibrium model on natural convection in a nanofluid-filled wavy-walled porous cavity containing an inner solid cylinder. *Chem. Eng. Sci.* **2019**, *201*, 247–263. [CrossRef]
32. Mohebbi, R.; Mehryan, S.A.M.; Izadi, M.; Mahian, O. Natural convection of hybrid nanofluids inside a partitioned porous cavity for application in solar power plants. *J. Therm. Anal. Calorim.* **2019**, *137*, 1719–1733. [CrossRef]
33. Al-Amir, Q.R.; Ahmed, S.Y.; Hamzah, H.K.; Ali, F.H. Effects of Prandtl Number on Natural Convection in a Cavity Filled with Silver/Water Nanofluid-Saturated Porous Medium and Non-Newtonian Fluid Layers Separated by Sinusoidal Vertical Interface. *Arab. J. Sci. Eng.* **2019**, *44*, 10339–10354. [CrossRef]
34. Dogonchi, S.; Sheremet, M.A.; Ganji, D.D.; Pop, I. Free convection of copper–water nanofluid in a porous gap between the hot rectangular cylinder and cold circular cylinder under the effect ofthe inclined magnetic field. *J. Therm. Anal. Calorim.* **2019**, *135*, 1171–1184. [CrossRef]
35. Akhter, R.; Ali, M.M.; Alim, M.A. Hydromagnetic natural convection heat transfer in a partially heated enclosure filled with porous medium saturated by a nanofluid. *Int. J. Appl. Comput. Math.* **2019**, *5*, 1–27. [CrossRef]
36. Armaghani, T.; Chamkha, A.; Rashad, A.M.; Mansour, M.A. Inclined magneto: Convection, internal heat, and entropy generation of nanofluid in an I-shaped cavity saturated with porous media. *J. Therm. Anal. Calorim.* **2020**, *142*, 2273–2285. [CrossRef]
37. Benos, L.T.; Polychronopoulos, N.D.; Mahabaleshwar, U.S.; Lorenzini, G.; Sarris, I.E. Thermal and flow investigation of MHD natural convection in a nanofluid-saturated porous enclosure: An asymptotic analysis. *J. Therm. Anal. Calorim.* **2019**, *143*, 751–765. [CrossRef]
38. Hajatzadeh Pordanjani, A.; Aghakhani, S.; Karimipour, A.; Afrand, M.; Goodarzi, M. Investigation of free convection heat transfer and entropy generation of nanofluid flow inside a cavity affected by the magnetic field and thermal radiation. *J. Therm. Anal. Calorim.* **2019**, *137*, 997–1019. [CrossRef]
39. Baïri, A. New correlations for free convection with water-ZnO nanofluid saturated porous medium around a cubical electronic component in the hemispherical cavity. *Heat Transf. Eng.* **2020**, *41*, 1275–1287. [CrossRef]
40. Izadi, M.; Oztop, H.F.; Sheremet, M.A.; Mehryan, S.A.M.; Abu-Hamdeh, N. Coupled FHD–MHD free convection of a hybrid nanoliquid in an inversed T-shaped enclosure occupied by partitioned porous media. *Numer. Heat Transf. Part A Appl.* **2019**, *76*, 479–498. [CrossRef]
41. Sadeghi, M.S.; Tayebi, T.; Dogonchi, A.S.; Armaghani, T.; Talebizadehsardari, P. Analysis of hydrothermal characteristics of magnetic Al_2O_3-H_2O nanofluid within a novel wavy enclosure during natural convection process considering internal heat generation. *Math. Methods Appl. Sci.* **2020**, mma.6520. [CrossRef]
42. Sheikholeslami, M.; Hayat, T.; Muhammad, T.; Alsaedi, A. MHD forced convection flow of a nanofluid in a porous cavity with the hot elliptic obstaclethrough Lattice Boltzmann method. *Int. J. Mech. Sci.* **2018**, *135*, 532–540. [CrossRef]
43. Sheikholeslami, M.; Hayat, T.; Alsaedi, A. Numerical simulation of nanofluid forced convection heat transfer improvement in the existence of magnetic field using lattice Boltzmann method. *Int. J. Heat Mass Transf.* **2017**, *108*, 1870–1883. [CrossRef]
44. Ferdows, M.; Alzahrani, F. Dual solutions of nanofluid forced convective flow with heat transfer and porous media past a moving surface. *Phys. A Stat. Mech. Its Appl.* **2020**, *551*, 124075. [CrossRef]
45. Torabi, M.; Torabi, M.; Ghiaasiaan, S.M.; Peterson, G.P. The effect of Al_2O_3-water nanofluid on the heat transfer and entropy generation of laminar forced convection through isotropic porous media. *Int. J. Heat Mass Transf.* **2017**, *111*, 804–816. [CrossRef]
46. Selimefendigil, F.; Öztop, H.F. Magnetohydrodynamics forced convection of nanofluid in multi-layered U-shaped vented cavity with a porous region considering wall corrugation effects. *Int. Commun. Heat Mass Transf.* **2020**, *113*, 104551. [CrossRef]
47. Sheikholeslami, M. Influence of magnetic field on Al_2O_3-H_2O nanofluid forced convection heat transfer in a porous lid-driven cavity with hot sphere obstacle through LBM. *J. Mol. Liq.* **2018**, *263*, 472–488. [CrossRef]
48. Moghadasi, H.; Aminian, E.; Saffari, H.; Mahjoorghani, M.; Emamifar, A. Numerical analysis on laminar forced convection improvement of hybrid nanofluid within a U-bend pipe in porous media. *Int. J. Mech. Sci.* **2020**, *179*, 105659. [CrossRef]
49. Aminian, E.; Moghadasi, H.; Saffari, H. Magnetic field effects on forced convection flow of a hybrid nanofluid in a cylinder filled with porous media: A numerical study. *J. Therm. Anal. Calorim.* **2020**, *141*, 2019–2031. [CrossRef]
50. Siavashi, M.; Talesh Bahrami, H.R.; Aminian, E.; Saffari, H. Numerical analysis on forced convection enhancement in an annulus using porous ribs and nanoparticle addition to base fluid. *J. Cent. South Univ.* **2019**, *26*, 1089–1098. [CrossRef]
51. Tlili, I.; Rabeti, M.; Safdari Shadloo, M.; Abdelmalek, Z. Forced convection heat transfer of nanofluids from a horizontal plate with convective boundary condition and a line heat source embedded in porous media. *J. Therm. Anal. Calorim.* **2020**, *141*, 2081–2094. [CrossRef]
52. Khademi, R.; Razminia, A.; Shiryaev, V.I. Conjugate-mixed convection of nanofluid flow over an inclined flat plate in porous media. *Appl. Math. Comput.* **2020**, *366*, 124761. [CrossRef]
53. Bondarenko, D.S.; Sheremet, M.A.; Oztop, H.F.; Abu-Hamdeh, N. Mixed convection heat transfer of a nanofluid in a lid-driven enclosure with two adherent porous blocks. *J. Therm. Anal. Calorim.* **2019**, *135*, 1095–1105. [CrossRef]
54. Ali, F.H.; Hamzah, H.K.; Hussein, A.K.; Jabbar, M.Y.; Talebizadehsardari, P. MHD mixed convection due to a rotating circular cylinder in a trapezoidal enclosure filled with a nanofluid saturated with a porous media. *Int. J. Mech. Sci.* **2020**, *181*, 105688. [CrossRef]

55. Sheremet, M.A.; Roşca, N.C.; Roşca, A.V.; Pop, I. Mixed convection heat transfer in a square porous cavity filled with a nanofluid with suction/injection effect. *Comput. Math. Appl.* **2018**, *76*, 2665–2677. [CrossRef]
56. Li, Z.; Barnoon, P.; Toghraie, D.; Balali Dehkordi, R.; Afrand, M. Mixed convection of non-Newtonian nanofluid in an H-shaped cavity with cooler and heater cylinders filled by a porous material: Two-phase approach. *Adv. Powder Technol.* **2019**, *30*, 2666–2685. [CrossRef]
57. Cimpean, D.S.; Sheremet, M.A.; Pop, I. Mixed convection of hybrid nanofluid in a porous trapezoidal chamber. *Int. Commun. Heat Mass Transf.* **2020**, *116*, 104627. [CrossRef]
58. Rajarathinam, M.; Nithyadevi, N.; Chamkha, A.J. Heat transfer enhancement of mixed convection in an inclined porous cavity using Cu-water nanofluid. *Adv. Powder Technol.* **2018**, *29*, 590–605. [CrossRef]
59. Nazari, S.; Ellahi, R.; Sarafraz, M.M.; Safaei, M.R.; Asgari, A.; Akbari, O.A. Numerical study on mixed convection of a non-Newtonian nanofluid with porous media in a two lid-driven square cavity. *J. Therm. Anal. Calorim.* **2020**, *140*, 1121–1145. [CrossRef]
60. Ghosh, S.; Mukhopadhyay, S. MHD mixed convection flow of a nanofluid past a stretching surface of variable thickness and vanishing nanoparticle flux. *Pramana J. Phys.* **2020**, *94*, 1–12. [CrossRef]
61. Tahmasbi, M.; Siavashi, M.; Abbasi, H.R.; Akhlaghi, M. Mixed convection enhancement by using optimized porous media and nanofluid in a cavity with two rotating cylinders. *J. Therm. Anal. Calorim.* **2020**, *141*, 1829–1846. [CrossRef]
62. Selimefendigil, F.; Chamkha, A.J. MHD mixed convection of Ag–MgO/water nanofluid in a triangular shape partitioned lid-driven square cavity involving a porous compound. *J. Therm. Anal. Calorim.* **2020**, *143*, 1467–1484. [CrossRef]
63. Madhura, K.R.; Iyengar, S.S. Impact of Heat and Mass Transfer on Mixed Convective Flow of Nanofluid Through Porous Medium. *Int. J. Appl. Comput. Math.* **2017**, *3*, 1361–1384. [CrossRef]
64. Aly, M.; Raizah, Z.A.S. Mixed Convection in an Inclined Nanofluid Filled-Cavity Saturated with a Partially Layered Porous Medium. *J. Therm. Sci. Eng. Appl.* **2019**, *11*, 041002. [CrossRef]
65. Asadi, A.; Molana, M.; Ghasemiasl, R.; Armaghani, T.; Pop, M.I.; Saffari Pour, M. A New Thermal Conductivity Model and Two-Phase Mixed Convection of CuO–Water Nanofluids in a Novel I-Shaped Porous Cavity Heated by Oriented Triangular Hot Block. *Nanomaterials* **2020**, *10*, 2219. [CrossRef] [PubMed]
66. Chamkha, J.; Mansour, M.A.; Rashad, A.M.; Kargarsharifabad, H.; Armaghani, T. Magnetohydrodynamic Mixed Convection and Entropy Analysis of Nanofluid in Gamma-Shaped Porous Cavity. *J. Thermophys. Heat Transf.* **2020**, *34*, 836–847. [CrossRef]
67. Khan, U.; Zaib, A.; Sheikholeslami, M.; Wakif, A.; Baleanu, D. Mixed convective radiative flow through a slender revolution bodies containing molybdenum-disulfide graphene oxide along with generalized hybrid nanoparticles in porous media. *Crystals* **2020**, *10*, 771. [CrossRef]
68. Choi, S.U.S. *Enhancing Thermal Conductivity of Fluids with Nanoparticles*; No. ANL/MSD/CP-84938; CONF-951135-29; Argonne National Lab.(ANL): Argonne, IL, USA, 1995.
69. Kasaeian, A.; Daneshazarian, R.; Mahian, O.; Kolsi, L.; Chamkha, A.J.; Wongwises, S.; Pop, I. Nanofluid flow and heat transfer in porous media: A review of the latest developments. *Int. J. Heat Mass Transf.* **2017**, *107*, 778–791.
70. Kandelousi, M.; Ameen, S.; Akhtar, M.S.; Shin, H.S. *Nanofluid Flow in Porous Media*; IntechOpen: London, UK, 2020.

nanomaterials

Entropy Generation and Thermal Radiation Analysis of EMHD Jeffrey Nanofluid Flow: Applications in Solar Energy

Bhupendra Kumar Sharma [1], Anup Kumar [1], Rishu Gandhi [1], Muhammad Mubashir Bhatti [2,3,*] and Nidhish Kumar Mishra [4]

1. Department of Mathematics, Birla Institute of Technology and Science, Pilani 333031, India
2. College of Mathematics and Systems Science, Shandong University of Science and Technology, Qingdao 266590, China
3. Material Science, Innovation and Modelling (MaSIM) Research Focus Area, North-West University, Mafikeng Campus, Private Bag X2046, Mmabatho 2735, South Africa
4. Department of Basic Science, College of Science and Theoretical Studies, Saudi Electronic University, Riyadh 11673, Saudi Arabia
* Correspondence: mmbhatti@sdust.edu.cn or mubashirme@hotmail.com

Abstract: This article examines the effects of entropy generation, heat transmission, and mass transfer on the flow of Jeffrey fluid under the influence of solar radiation in the presence of copper nanoparticles and gyrotactic microorganisms, with polyvinyl alcohol–water serving as the base fluid. The impact of source terms such as Joule heating, viscous dissipation, and the exponential heat source is analyzed via a nonlinear elongating surface of nonuniform thickness. The development of an efficient numerical model describing the flow and thermal characteristics of a parabolic trough solar collector (PTSC) installed on a solar plate is underway as the use of solar plates in various devices continues to increase. Governing PDEs are first converted into ODEs using a suitable similarity transformation. The resulting higher-order coupled ODEs are converted into a system of first-order ODEs and then solved using the RK 4th-order method with shooting technique. The remarkable impacts of pertinent parameters such as Deborah number, magnetic field parameter, electric field parameter, Grashof number, solutal Grashof number, Prandtl number, Eckert number, exponential heat source parameter, Lewis number, chemical reaction parameter, bioconvection Lewis number, and Peclet number associated with the flow properties are discussed graphically. The increase in the radiation parameter and volume fraction of the nanoparticles enhances the temperature profile. The Bejan number and entropy generation rate increase with the rise in diffusion parameter and bioconvection diffusion parameter. The novelty of the present work is analyzing the entropy generation and solar radiation effects in the presence of motile gyrotactic microorganisms and copper nanoparticles with polyvinyl alcohol–water as the base fluid under the influence of the source terms, such as viscous dissipation, Ohmic heating, exponential heat source, and chemical reaction of the electromagnetohydrodynamic (EMHD) Jeffrey fluid flow. The non-Newtonian nanofluids have proven their great potential for heat transfer processes, which have various applications in cooling microchips, solar energy systems, and thermal energy technologies.

Keywords: solar radiations; exponential heat source; copper nanoparticles; gyrotactic motile microorganisms; EMHD; Joule heating

Citation: Sharma, B.K.; Kumar, A.; Gandhi, R.; Bhatti, M.M.; Mishra, N.K. Entropy Generation and Thermal Radiation Analysis of EMHD Jeffrey Nanofluid Flow: Applications in Solar Energy. *Nanomaterials* **2023**, *13*, 544. https://doi.org/10.3390/nano13030544

Academic Editor: Henrich Frielinghaus

Received: 17 December 2022
Revised: 26 January 2023
Accepted: 26 January 2023
Published: 29 January 2023

1. Introduction

Solar energy is one of the most significant renewable energy sources with negligible adverse environmental effects. This technology provides clean and inexhaustible energy without burning any fuel. The use of solar energy is directly connected to the natural resources of heat, water, and electricity. It is presently a difficult issue in the field of solar energy storage to improve the thermal efficiency of solar collectors to meet the demands and needs of energy in industrial and technical applications. Solar collector performance and operation are also impacted by inadequate heat transmission and thermophysical properties of the base fluid. Numerous efforts have been made in this direction to improve the thermophysical properties of the base fluids. Nanofluids are the next-generation fluids that not only discover surprising thermal features but also result in high thermal properties. Adding nanoparticles to fluids can enhance heat transmission and solar energy storage. Mushtaq et al. [1] investigated the impact of solar radiation on the 2D stagnation-point flow of a nanofluid. The MHD 3D Jeffrey fluid flow characteristics for heat transfer in solar energy applications were analyzed by Shehzad et al. [2]. The 3D nanofluid flow of the boundary layer for solar energy applications was investigated by Khan et al. in [3,4]. The effect of silver and copper nanoparticles on the unsteady MHD Jeffrey fluid flow was described by Zin et al. [5]. Reddy et al. [6] reviewed the applications of nanofluids in solar conversion systems and performance enhancement. Daniel et al. [7] described the impact of thermal radiation on the flow of EMHD nanofluid over a nonuniformly thick stretching sheet. The effects of thermal radiation on the EMHD nanofluid flow over a stretching sheet with nonuniform thickness was described by Daniel et al. [7]. Wahab et al. [8] discussed the potential of nanofluids in the efficiencies of solar collectors. Numerical simulation for the effects of solar energy on the turbulence of MHD flow of cross nanofluid was explored by Azam et al. [9]. The radiation effects on the peristaltic flow of gold nanoparticles with double-diffusive convection across an asymmetric channel were discussed by Sunitha et al. [10]. The thermal behavior of a hybrid nanofluid flow in the presence of radiative solar energy inside a microchannel was studied by Acharya [11]. The significance of copper and alumina nanoparticles migrating across the water with a convectively heated surface was analyzed by Song et al. [12]. Jamshed et al. examined the heat transmission in Maxwell nanofluid flow across an exponentially uniform stretchable plate within the parabolic trough solar collector. The optimization of solar energy using Sutterby hybrid nanofluid in solar HVAC subjected to expanding sheets was also addressed by Jamshed et al. [13]. Recently, some research ([14–17]) has examined the effect of nanoparticles in different types of fluids. A comparison of thermal characteristics of Cu and Ag nanoparticle suspensions in sodium alginate for Sutterby nanofluid flow in solar collectors was conducted by Bouslimi et al. [18]. Shahzad et al. [19] investigated the thermal cooling efficiency of a solar water pump by using Oldroyd-B (aluminum alloy–titanium alloy/engine oil) hybrid nanofluid.

Thermophoresis is the force exerted by particles suspended in a fluid in response to the heat gradient, also known as thermophoresis or thermal diffusion. The random motion of particles suspended in the liquid results in collisions when it collides with another particle. Furthermore, these collisions of particles cause a random or zigzag motion. It involves the transfer of energy between the particles, and this phenomenon is known as Brownian diffusion. Pakravan and Yaghoubi [20] explored the heat transfer with Dufour and the thermophoresis effect of nanoparticles. Anbuchezhian et al. [21] studied heat transfer corresponding to solar radiations under the effects of the thermophoresis and Brownian motion. Kandasamy et al. [22] presented the effect of thermophoresis and Brownian motion under temperature stratification owing to sun radiation on nanofluid flow. Hayat et al. [23] examined the heat transfer with thermophoretic diffusion and Brownian motion through a stretching cylinder of the Jeffrey nanofluid flow. The effects of magnetohydrodynamics, thermophoresis, and Brownian motion on the radiative nanofluid flow across a rotating sheet were examined by Mabood et al. [24]. Sulochana et al. [25] reviewed the impact of transpiration on a Carreau nanofluid flow with Brownian motion and thermophoresis diffusion via a stretching sheet. Applications of solar radiation under the cumulative effects of thermophoresis and Brownian motion for CuO–Water nanofluid by natural convection were investigated by Astanina et al. [26]. Awan et al. [27] analyzed the radiative heat transfer in MHD nanofluid flow with the effects of thermophoresis, Brownian motion, and solar energy. Rekha et al. [28] investigated the influence of thermophoretic particle deposition under solar radiation impacts on heat transfer for nanofluid flow in various geometries. Recently, some research has been conducted to study the thermophoresis and

Brownian motion impacts under different types of considerations (in the articles [29–31]) past vertical and horizontal surfaces.

Entropy is a system's thermal energy which is unavailable for any valuable work. Entropy generation generally measures the performance of engineering systems. Entropy production improves a system's performance and destroys significant amounts of energy. Studying the entropy generation in solar collectors is essential to optimize the amount of energy accessible. Lowering the entropy generation rate in constructing practical systems is preferable. The effect of direct absorption solar collectors for heat transfer and entropy generation was studied by Parvin et al. [32]. The efficiency of entropy formation in the viscous fluid flow past a porous rotating disk was explored by Khan et al. [33] by using the theory for the partial slip under nonlinear radiation effect. Analysis of entropy generation of a two-phase thermosyphon for solar collectors was explored by Wang et al. [34]. Several types of research ([35–37]) have been conducted to examine the influence of entropy generation in the existence of moving gyrotactic microorganisms with different considerations. The dynamics of mixed Marangoni convective flow for entropy generation using aluminum oxide and copper nanofluids in solar energy storage was analyzed by Li et al. [38]. Recently, Sharma et al. [39] discussed the entropy optimization in MHD thermally radiating flow with hybrid nanoparticles through a tapered multistenosed artery. Khanduri and Sharma [40] analyzed the entropy generation assuming the viscosity and thermal conductivity of the fluid to be temperature-dependent. Entropy generation for nonlinear radiation and thermodynamic behavior of a hybridized Prandtl–Eyring magneto-nanofluid for a solar aircraft were studied by Salawu et al. [41].

The gyrotactic motile microorganism is one of the particular microorganisms which moves due to the torque caused by viscous and gravity forces present in the system. The upswing suspension of gyrotactic microorganisms in the bioconvection mechanism will enhance nanoparticle stability. The system's stability to support the suspension of nanoparticles with moving gyrotactic microorganisms was discussed by Avramenko and Kuznetsov [42]. Later, Avramenko and Kuznetsov [43] also investigated the thermal stability in the bioconvection of gyrotactic microorganisms with a primary vertical temperature gradient. Mutuku and Makinde [44] examined the bioconvection driven by a new type of water-based nanofluid that is hydromagnetically flowing across a porous vertical moving surface suspended with nanoparticles. The impact of solar radiation on gyrotactic microorganisms' bioconvection nano-fluid flow was explored by Acharya et al. [45]. The behavior of the nanoparticles and microorganisms on the MHD Jeffrey nanofluid flow due to a rotating vertical cone was described by Saleem et al. [46]. The entropy of Maxwell nanofluid considering the existence of motile microorganisms with heterogeneous–homogeneous reactions was explored by Sohail et al. [47]. Song et al. [48] studied the mixed bioconvection flow of nanofluid with varying Prandtl numbers across a vertically moving thin needle for the aspects of solar energy. MHD Jeffrey nanofluid flow past a vertical stretching sheet under the effects of thermal partial and radiation slip with motile microorganisms was investigated by Naidu et al. [49]. Gyrotactic microorganisms in MHD flow through porous material via an inclined elongating sheet to examine heat transfer was carried out by Sharma et al. [50]. Bhatti et al. [51] explored the gyrotactic microorganisms swimming between rotating circular plates embedded in porous medium for thermal energy storage. Gyrotactic microorganisms swimming under the solar biomimetic system above the esophagus for hyperbolic tangent blood nanomaterial were examined by Hussain and Farooq [52].

Many industrial operations, including electricity production, heating or cooling processes, chemical processes, and microelectronics, depend on common fluids, including water, ethylene glycol, and heat transfer oil. In thermal engineering devices, these fluids cannot attain effective heat transfer rates due to their relatively low thermal conductivity. Utilizing ultrafine solid particles suspended in ordinary fluids to boost their heat conductivity is one method for overcoming this barrier. From the literature [53], it is noted that polyvinyl alcohol–water-based fluid with copper nanoparticles has higher temperature profiles and heat transfer rates. The previous literature is limited to the entropy generation of the Jeffrey nanofluid flows in heat and mass transfer processes over the stretching surfaces of uniform thickness. No attempts have been made to analyze the entropy generation of the EMHD Jeffrey fluid flow with motile gyrotactic microorganisms via vertical surfaces stretching nonlinearly with nonuniform thickness under the influence of solar radiations. Therefore, this study investigates the entropy generation and solar radiation effects in the presence of motile gyrotactic microorganisms and copper nanoparticles with polyvinyl alcohol–water as the base fluid. The influence of the source terms such as viscous dissipation, Ohmic heating, exponential heat source, and chemical reaction of the EMHD Jeffrey fluid flow past a vertical nonlinearly elongating surface of nonuniform thickness was studied.

2. Formulation of Model

2.1. Physical Assumptions

Consider time-independent, laminar, 2D, incompressible, non-Newtonian Jeffrey nanofluid flow with motile gyrotactic microorganisms incorporating Copper nanoparticles in the polyvinyl alcohol–water base fluid. The flow is subjected to a nonlinear elongating surface with nonuniform thickness under Joule heating, exponential heat source, viscous dissipation, and chemical reaction. It is also assumed that the motile microorganisms swimming direction and speed are unaffected by the suspended nanoparticles. The x_1^* and y_1^* are the axes along the vertical and horizontal direction, having corresponding components of velocities as u_1^* and v_1^*, respectively. The influence of the induced magnetic field is minimal and is therefore neglected by assuming a very small magnetic Reynolds number ($Re << 1$).

The current density formulated by Ohm's law is $J = \sigma(E + V \times B)$, where E is the strength of the applied electric field, V refers the fluid velocity, and B is the strength of the applied magnetic field. In fluid flows of high electrical conductivity, flow control can be made in the absence of any external electric field. For that case, the volume density induced due to the Lorentz force is formulated as $F = J \times B$. Therefore, in this case, the moderate applied magnetic field will induce a sufficiently strong electrical current density without any external electric field. Hence, the expression for the electrical current density and Lorentz force becomes $J = \sigma(V \times B)$ and $F = \sigma(V \times B) \times B$. However, in the case of low electrical conductivity of fluids, the current density in the absence of an electric field is minimal to achieve the flow control, even for the magnetic field of several Teslas. Therefore, an external electric field is needed for efficient flow control. The electrical current density in this case is formulated as $J = \sigma(E + V \times B)$. Figure 1 illustrates the mathematical model. The stress tensor components for the Jeffrey fluid model are given by

$$T = -PI + S, \tag{1}$$

where S in mathematical form is expressed as

$$S = \frac{\mu_{nf}}{1 + \lambda_1}\left[A + \lambda_2 \frac{dA}{dt}\right], \tag{2}$$

A is the first-order Rivlin Erection tensor given by

$$A = \nabla V + (\nabla V)',$$

where $'$ denotes the transpose.

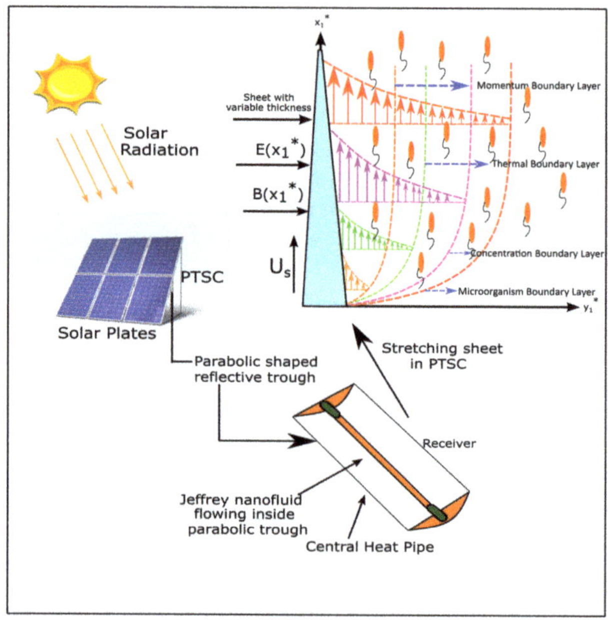

Figure 1. An illustration of the mathematical model.

2.2. Governing Equations of the Physical Model

The governing equations for Jeffrey nanofluid flow derived using the prior assumptions, the order of magnitude approach, and the typical Boussinesq approximation for the boundary layer are ([54])

$$\frac{\partial u_1^*}{\partial x_1^*} + \frac{\partial v_1^*}{\partial y_1^*} = 0, \tag{3}$$

$$u_1^* \frac{\partial u_1^*}{\partial x_1^*} + v_1^* \frac{\partial u_1^*}{\partial x_1^*} = \frac{\nu_{nf}}{(1+\lambda_1)} \left[\frac{\partial^2 u_1^*}{\partial y_1^{*2}} + \lambda_2 \left(u_1^* \frac{\partial^3 u_1^*}{\partial x_1^* \partial y_1^{*2}} + \frac{\partial u_1^*}{\partial y_1^*} \frac{\partial^2 u_1^*}{\partial x_1^* \partial y_1^*} + v_1^* \frac{\partial^3 u_1^*}{\partial y_1^{*3}} \right. \right.$$
$$\left. \left. + \frac{\partial v_1^*}{\partial y_1^*} \frac{\partial^2 u_1^*}{\partial y_1^{*2}} \right) \right] - \frac{\sigma_{nf}}{\rho_{nf}} (B^2 u_1^* - EB) + g(\beta_T)_{nf}(T_1^* - T_\infty^*) + g(\beta_C)_{nf}(C_1^* - C_\infty^*)$$
$$- g(\beta_N)_{nf}(N_1^* - N_\infty^*), \tag{4}$$

$$u_1^* \frac{\partial T_1^*}{\partial x_1^*} + v_1^* \frac{\partial T_1^*}{\partial y_1^*} = \frac{\kappa_{nf}}{(\rho Cp)_{nf}} \frac{\partial^2 T_1^*}{\partial y_1^{*2}} + \frac{16\sigma^* T_\infty^3}{3\kappa^* (\rho c_p)_{nf}} \frac{\partial^2 T_1^*}{\partial y_1^{*2}} + \tau \left(D_B \frac{\partial C_1^*}{\partial y_1^*} \frac{\partial T_1^*}{\partial y_1^*} \right.$$
$$+ \frac{D_T}{T_\infty^*} \left(\frac{\partial T_1^*}{\partial y_1^*} \right)^2 \right) + \frac{\mu_{nf}}{(\rho Cp)_{nf}(1+\lambda_1)} \left[\left(\frac{\partial u_1^*}{\partial y_1^*} \right)^2 + \lambda_2 \left(u_1^* \frac{\partial u_1^*}{\partial y_1^*} \frac{\partial^2 u_1^*}{\partial x_1^* \partial y_1^*} + v_1^* \frac{\partial u_1^*}{\partial y_1^*} \frac{\partial^2 u_1^*}{\partial y_1^{*2}} \right) \right]$$
$$+ \frac{\sigma_{nf}}{(\rho Cp)_{nf}} (u_1^* B(x_1^*) - E(x_1^*))^2 + \frac{Q_e^* (T_s^* - T_\infty^*)}{(\rho Cp)_{nf}} \exp\left(-n_1 y_1^* \sqrt{\frac{(n+1)a}{2\nu}} (x_1^* + b)^{\frac{n-1}{2}} \right), \tag{5}$$

$$u_1^* \frac{\partial C_1^*}{\partial x_1^*} + v_1^* \frac{\partial C_1^*}{\partial y_1^*} = \frac{D_T}{T_\infty^*} \frac{\partial^2 T_1^*}{\partial y_1^{*2}} + D_B \frac{\partial^2 C_1^*}{\partial y_1^{*2}} - K_r(C_1^* - C_\infty^*), \tag{6}$$

$$u_1^* \frac{\partial N_1^*}{\partial x_1^*} + v_1^* \frac{\partial N_1^*}{\partial y_1^*} + \left[\frac{\partial}{\partial y} \left(N \frac{\partial C_1^*}{\partial y_1^*} \right) \right] \frac{bW_c}{C_w - C_\infty} = D_m \frac{\partial^2 N_1^*}{\partial y_1^{*2}}. \tag{7}$$

where $B(x_1^*) = B_0(x_1^* + b)^{\frac{n-1}{2}}$ is the applied magnetic field, $E(x_1^*) = E_0(x_1^* + b)^{\frac{n-1}{2}}$ is the applied electric field in a direction perpendicular to the elongating surface, $U_s(x_1^*) = a(x_1^* + b)^n$ is the velocity of the elongating surface at $y_1^* = A(x_1^* + b)^{\frac{(1-n)}{2}}$, and n is the power index.

2.3. Similarity Transformations

In fluid mechanics, the majority of known exact solutions are similarity solutions in which the number of independent variables is reduced by one or more. Similarity solutions are typically asymptotic problem solutions used to gain physical insight into the characteristics of complex fluid movements. These solutions represent the physical, as well as dynamic, and thermal parameters of the actual situation and their influence. Similarity transformations introduced to obtain the dimensionless form of governing equation for this Jeffrey fluid model are

$$u_1^* = \frac{\partial \psi}{\partial y_1^*} = U_s(x_1^*)G'(\zeta), v_1^* = -\frac{\partial \psi}{\partial x_1^*} = -\sqrt{\frac{(n+1)va}{2}}(x_1^* + b)^{\frac{n-1}{2}}\left(G(\zeta) + \eta\frac{n-1}{n+1}G'(\zeta)\right),$$

$$\theta = \frac{T_1^* - T_\infty^*}{T_s^* - T_\infty^*}, \phi = \frac{C_1^* - C_m^*}{C_s^* - C_\infty^*}, \varsigma = \frac{N_1^* - N_m^*}{N_s^* - N_\infty^*}, \zeta = y_1^*\sqrt{\frac{n+1}{2}\frac{a}{v}}(x_1^* + b)^{\frac{n-1}{2}},$$

$$\psi = G(\zeta)\sqrt{\frac{2}{n+1}va}(x_1^* + b)^{\frac{n+1}{2}}. \tag{8}$$

After employing the similarity transformations, we obtain the following governing equations:

$$G'''(\zeta) + (1 + \beta_1)\frac{\varepsilon_2}{\varepsilon_1}[G''(\zeta)G(\zeta) - \frac{2n}{n+1}G'^2(\zeta)] + (1 + \beta_1)\frac{\varepsilon_5}{\varepsilon_1}[M(E_1 - G'(\zeta))]$$
$$+ (1 + \beta_1)\frac{\varepsilon_2\varepsilon_4}{\varepsilon_1}[Gr\theta(\zeta) + Gc\phi(\zeta) - Nc\xi(\zeta)]$$
$$+ \beta_2\left[\frac{3n-1}{2}(G''(\zeta))^2 + (n-1)G'(\zeta)G'''(\zeta) - \frac{n+1}{2}G(\zeta)G^{iv}(\zeta)\right] = 0, \tag{9}$$

$$\theta''(\zeta) + Pr[Nt\theta'^2(\zeta) + Nb\theta'(\zeta)\phi'(\zeta)] + \frac{\varepsilon_3}{(\varepsilon_6 + Nr)}Pr[G(\zeta)\theta'(\zeta)]$$
$$+ \frac{\varepsilon_5}{(\varepsilon_6 + Nr)}[PrEcM(G'(\zeta) - E_1)^2] + \frac{1}{(\varepsilon_6 + Nr)}Pr[Q_e\exp(-n_1\zeta)] + \frac{PrEc\varepsilon_1}{(1 + \beta_1)(\varepsilon_6 + Nr)}$$
$$\times\left[G''^2(\zeta) + \beta_2\left(\frac{3n-1}{2}G'(\zeta)(G''(\zeta))^2 - \left(\frac{n+1}{2}\right)G(\zeta)G''(\zeta)G'''(\zeta)\right)\right] = 0, \tag{10}$$

$$\phi''(\zeta) + LeG(\zeta)\phi'(\zeta) + \frac{Nt}{Nb}\theta''(\zeta) - K\phi = 0, \tag{11}$$

$$\xi''(\zeta) + LbC(\zeta) \quad Pc[\phi''(\zeta)(\xi + \delta_N) + \xi'(\zeta)\phi'(\zeta)] - 0. \tag{12}$$

where $\varepsilon_1 = \frac{1}{(1-\varphi)^{2.5}}$, $\varepsilon_2 = 1 - \varphi + \varphi(\frac{\rho_p}{\rho_f})$, $\varepsilon_3 = 1 - \varphi + \varphi(\frac{\rho C_p}{\rho C_f})$, $\varepsilon_4 = 1 - \varphi + \varphi(\frac{\rho\beta_p}{\rho\beta_f})$, $\varepsilon_5 = 1 + \frac{3(\sigma_p - \sigma_f)\varphi}{(\sigma_p + 2\sigma_f) - (\sigma_p - \sigma_f)\varphi}$, and $\varepsilon_6 = \frac{(\kappa_p + 2\kappa_f + 2\varphi(\kappa_p - \kappa_f))}{(\kappa_p + 2\kappa_f - 2\varphi(\kappa_p - \kappa_f))}$. The dimensionless parameters obtained in the present fluid model are mentioned in the Table 1.

Table 1. Dimensionless parameter of the fluid model ([29,54]).

Magnetic field parameter, $M = \frac{2\sigma B_0^2}{\rho_f a(n+1)}$	Electric field parameter, $E_1 = \frac{E_0}{B_0 a(x_1^*+b)^n}$
Grashof number, $Gr = \frac{2g\beta_T(T_s^*-T_\infty^*)}{a^2(n+1)(x_1^*+b)^{2n-1}}$	Solutal Grashof number, $Gc = \frac{2g\beta_C(C_s^*-C_\infty^*)}{a^2(n+1)(x_1^*+b)^{2n-1}}$
Bioconvection Rayleigh parameter, $Nc = \frac{2g\beta_N(N_s^*-N_\infty^*)}{a^2(n+1)(x_1^*+b)^{2n-1}}$	Radiation parameter, $Nr = \frac{16\sigma^* T_\infty^{*3}}{3\kappa_f \kappa^*}$
Eckert number, $Ec = \frac{U_s^2}{C_p(T_s^*-T_\infty^*)}$,	Prandtl number, $Pr = \frac{\mu C_p}{\kappa}$,
Temperature difference parameter, $\delta = \frac{T_s^*-T_\infty^*}{T_\infty^*}$	Exponential heat source parameter, $Q_e = \frac{2Q_e^* \nu \theta(\eta)}{(n+1)\kappa a(x_1^*+b)^{\frac{(n+1)}{2}}}$
Brownian diffusion parameter, $Nb = \frac{\rho c_p D_B(C_s^*-C_\infty^*)}{\rho_f C_p \nu}$	Thermophoresis diffusion parameter, $Nt = \frac{\rho C_p D_T(T_s^*-T_\infty^*)}{\rho_f C_p \nu T_\infty^*}$
Chemical reaction parameter, $K = \frac{2K_r}{(n+1)a(x_1^*+b)^{n-1}}$	Lewis number, $Le = \frac{\nu}{D_B}$
Bioconvection Lewis number, $Lb = \frac{\nu}{D_m}$	Peclet number, $Pe = \frac{bW_c}{D_m}$
Diffusion parameter, $L = \frac{RD_B(C_s^*-C_\infty^*)}{\kappa}$	Bioconvection diffusion parameter, $L^* = \frac{RD_B(N_s^*-N_\infty^*)}{\kappa}$
Concentration difference parameter, $\delta_1 = \frac{C_s^*-C_\infty^*}{C_\infty^*}$	Microorganism concentration difference parameter, $\delta_N = \frac{N_\infty^*}{N_s^*-N_\infty^*}$

2.4. Boundary Conditions

The boundary conditions of the flow system are ([54])

$$u_1^* = U_s(x_1^*), \quad v_1^* = 0, \quad T_1^* = T_s^*, \quad C_1^* = C_s^*, \quad N_1^* = N_s^*, \quad at \quad y_1^* = A(x_1^*+b)^{\frac{1-n}{2}},$$

$$u_1^* \to 0, \quad T_1^* \to T_\infty^*, \quad C_1^* \to C_\infty^*, \quad N_1^* \to N_\infty^*, \quad as \quad y_1^* \to \infty. \tag{13}$$

On employing the similarity transformation mentioned in Equation (8), the nondimensionalized form of the boundary conditions is

$$G(\zeta) = \alpha\frac{1-n}{1+n}, \quad G'(\zeta) = 1, \quad \theta(\zeta) = 1, \quad \phi(\zeta) = 1, \quad \xi(\zeta) = 1 \quad at\ \zeta = 0, \tag{14}$$

$$G'(\zeta) \to 0, \quad \theta(\zeta) \to 0, \quad \phi(\zeta) \to 0, \quad \xi(\zeta) \to 0 \quad as\ \zeta \to \infty. \tag{15}$$

2.5. Numerical Methodology

This section uses a numerical methodology to obtain the solution of the nondimensional higher-order coupled ODEs. In this methodology, higher-order ODEs are converted into a system of first-order ODEs with corresponding boundary conditions and then solved using the RK 4th-order method with shooting technique. To implement the RK 4th-order method, we need four initial conditions for the momentum equation, two initial conditions for the energy equation, two initial conditions for the concentration equation, and two for the equation of concentration of microorganisms. However, we only have two initial conditions given for the momentum equation and one initial condition given for the energy, concentration, and microorganism distribution equation. Consequently, the remainder of them are derived by updating the original predictions using Newton's approach. The flow chart illustrating the numerical methodology used in the present analysis is shown in Figure 2. Let us assume

$$\begin{cases} G = F, \quad G' = F_1, \quad G'' = F_2, \quad G''' = F_3, \\ \theta = F_4, \quad \theta' = F_5, \quad \theta'' = F_6, \\ \phi = F_7, \quad \phi' = F_8, \quad \phi'' = F_9, \\ \xi = F_{10}, \quad \xi' = F_{11}, \end{cases} \tag{16}$$

The system of first-order differential equation is given by

$$\begin{cases} F' = F_1, \\ F_1' = F_2, \\ F_2' = F_3, \\ F_3' = \frac{2F_3}{(n+1)\beta_2 F} + \frac{2(1+\beta_1)\varepsilon_2}{\beta_2(n+1)F\varepsilon_1}[FF_2 - \frac{2n}{n+1}F_1^2] + \frac{2\varepsilon_5(1+\beta_1)}{\varepsilon_1\beta_2(n+1)F}M(E_1 - F_1) + \frac{2\varepsilon_2\varepsilon_4(1+\beta_1)}{\varepsilon_1\beta_2(n+1)F}[GrF_4 + GcF_7 \\ \qquad\qquad - NcF_{10}] + \frac{2}{(n+1)F}\left[\left(\frac{3n-1}{2}\right)(F_2)^2 + (n-1)F_1F_3\right] \\ F_4' = F_5, \\ F_5' = F_6 = -\frac{\varepsilon_3}{\varepsilon_6}\frac{Pr}{(1+Nr)}[FF_5] - Pr[NbF_5F_8 + NtF_5^2] + \frac{\varepsilon_5}{\varepsilon_6}PrMEc(\Gamma_1 - E_1)^2 + \frac{1}{\varepsilon_6}Pr[Q_e exp(-n_1\eta)] \\ \qquad\qquad + \frac{PrEc\varepsilon_1}{(1+\beta_1)\varepsilon_6}\left[F_2^2 + \beta_2\left(\frac{3n-1}{2}FF_2^2 - \frac{n+1}{2}FF_2F_3\right)\right] \\ F_7' = F_8, \\ F_8' = F_9 = -\frac{Nt}{Nb}F_6 - LeFF_8 + LeKF_7, \\ F_{10}' = F_{11}, \\ F_{11}' = -LbFF_{11} + Pe[F_9(F_{10} + \delta_N) + F_8F_{11}] \end{cases} \tag{17}$$

and the corresponding boundary conditions become

$$\begin{cases} F(\zeta) = \alpha\frac{1-n}{1+n}, \quad F_1(\zeta) = 1, \quad F_4(\zeta) = 1, \quad F_7(\zeta) = 1, \quad F_{10}(\zeta) = 1 \quad \text{at } \zeta = 0, \\ F_1(\zeta) \to 0, \quad F_4(\zeta) \to 0, \quad F_7(\zeta) \to 0, \quad F_{10}(\zeta) \to 0 \quad \text{as } \zeta \to \infty. \end{cases} \tag{18}$$

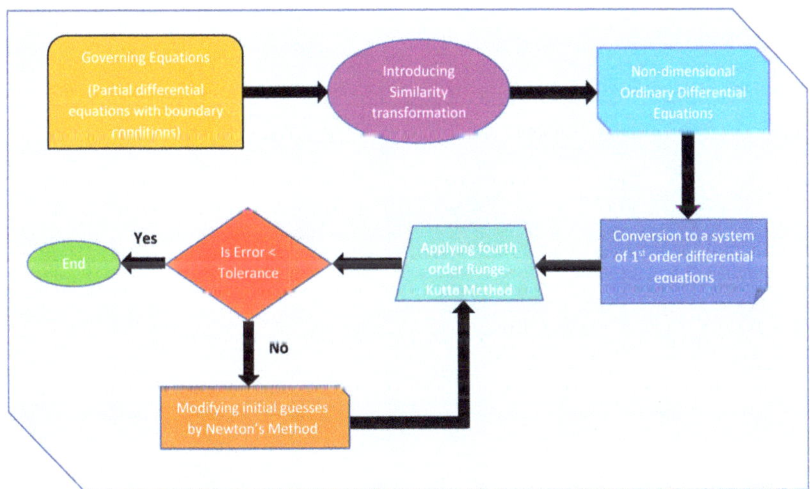

Figure 2. Flowchart presenting numerical methodology.

3. Result and Discussion

The numerical findings computed with the help of the above methodology are discussed in this section for various emergent parameter values. The remarkable impacts of pertinent parameters are shown in graphical forms in terms of fluid velocity $G'(\zeta)$, dimensionless temperature $\theta(\zeta)$, dimensionless concentration of the fluid $\phi(\zeta)$, microorganism density $\xi(\zeta)$, entropy generation number N_s, Bejan number Be, drag coefficient C_f, local Nusselt number Nu_x, Sherwood number Sh_x, and the local gyrotactic microorganism density Nn_x. The thermophysical parameters of nanofluid are mentioned using Table 2. Table 3 depicts the thermophysical properties of Copper nanoparticles, polyvinyl alcohol (PVA), and water. The default values of the parameters considered in the present analysis are listed in Table 4.

Table 2. Thermophysical features of nanofluid ([29,54]).

Properties	Mathematical Expression for Nanofluid
Viscosity	$\mu_{nf} = \dfrac{\mu_f}{(1-\phi_1)^{2.5}}$
Density	$\rho_{nf} = (1-\phi_1)\rho_f + \phi_1 \rho_{s_1}$
Heat Capacity	$(\rho C_p)_{nf} = (1-\phi_1)(\rho C_p)_f + \phi_1 (\rho C_p)_{s_1}$
Thermal Conductivity	$\dfrac{k_{nf}}{k_f} = \dfrac{k_{s_1}+(m-1)k_f-(m-1)\phi_1(k_f-k_{s_1})}{k_{s_1}+(m-1)k_f+\phi_1(k_f-k_{s_1})}$
Electrical Conductivity	$\dfrac{\sigma_{nf}}{\sigma_f} = \dfrac{\sigma_{s_1}+(m-1)\sigma_f-(m-1)\phi_1(\sigma_f-\sigma_{s_1})}{\sigma_{s_1}+(m-1)\sigma_f+\phi_1(\sigma_f-\sigma_{s_1})}$
Thermal Expansion Coefficient	$(\beta_T)_{nf} = (1-\phi_1)(\beta_T)_f + \phi_1 (\beta_T)_{s_1}$
Concentration Thermal Expansion Coefficient	$(\beta_C)_{nf} = (1-\phi_1)(\beta_C)_f + \phi_1 (\beta_C)_{s_1}$
Microorganism Thermal Expansion Coefficient	$(\beta_N)_{nf} = (1-\phi_1)(\beta_N)_f + \phi_1 (\beta_N)_{s_1}$

Table 3. Thermophysical properties of nanoparticles and basefluids ([12,53]).

Physical Properties	Copper	PVA	Water
Density [ρ (kg/m^3)]	8933	1020	997
Thermal Conductivity [κ (W/mK)]	400	0.2	0.613
Electrical Conductivity [σ (S/m)]	5.96×10^7	11.7×10^{-6}	0.05
Thermal Expansion Coefficient [$\beta_T \times 10^{-5}$ (K^{-1})]	1.67	2.5	21
Specific Heat Capacity [C_p (J/kgK)]	385	2000	4179

Table 4. Default Values of emerging parameters.

Parameters	Values	Parameters	Values	Parameters	Values	Parameters	Values
α	0.7	Pr	7.743	β_1	1	β_2	1
Ec	0.5	$Nb = Nt$	1	Le	0.3	M	0.1
$Gr = Gc$	0.3	Nc	0.1	n	0.5	E_1	0.1
m	0.9	E_t	1	K	0.5	φ	0.01
δ	1	δ_1	0.5	δ_N	1	Qe	0.3
$Pe = Lb$	0.3	L	1	L^*	1	n_1	3

The verification of the numerical scheme employed is mandatory, and therefore, the study by Sharma et al. [29] is used to validate the results obtained from the above methodology. In the absence of a few physical factors, the current model reduces to that of [29]. The influence of microorganisms, solar radiations, and nanoparticle volume fraction in the present analysis and the effect of thermal heat source and activation energy in [29] was disregarded to verify the results with those of Sharma et al. [29]. Figure 3a,b displays validation plots, with Figure 3a displaying the validation graph of velocity profiles and Figure 3b showing the validation graph of temperature profiles. Based on these Figures, the current analysis's findings are in excellent accord with the work of Sharma et al. [29].

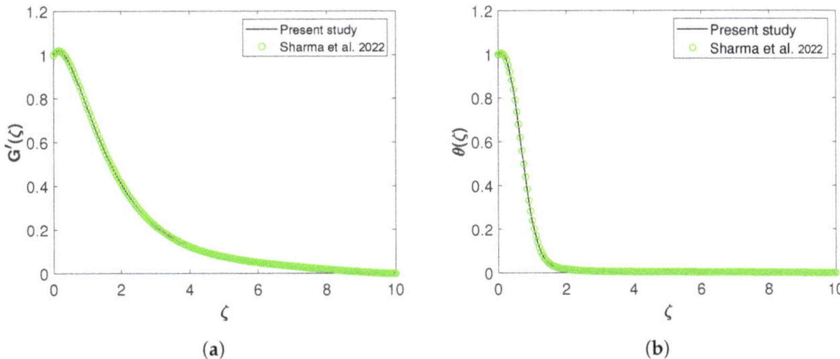

Figure 3. Comparative analysis of (**a**) velocity profile $G'(\zeta)$ for $Gc=1$ and (**b**) temperature profile $\theta(\zeta)$ for $Pr = 7$. Sharma et al. [27].

The plots of fluid velocity against the different physical parameters are depicted in Figure 4a–h. The physical impact of β_1 on fluid velocity is highlighted in Figure 4a. This Figure illustrates that fluid velocity diminishes with an increment in β_1. Higher values of retardation time enhance the momentum boundary layer of the flow, since retardation time varies inversely with the Deborah number β_1. So, the escalating values of β_1 reduce the fluid velocity. Figure 4b represents the fluid's velocity behavior corresponding to Deborah number β_2. The velocity of the fluid rises with an increment in β_2. The boundary layer of the flow field improves as the retardation time parameter is increased. Therefore, the velocity profile of the flow field enhances with the augmenting values of β_2. Figure 4c demonstrates the magnetic field's effect on fluid velocity. Intriguingly, this Figure indicates that the velocity of the fluid initially drops as the M increases and then exhibits the opposite behavior as the similarity variable reaches a specific value. This outcome is consistent with the findings of [7]. The increased magnetic field's Lorentz force serves as a frictional force. This frictional force resists the fluid flow, but when the magnetic field strengthens in the presence of an electric field, the electric field produces an accelerating force. Thus, the velocity profile changes its behavior close to the stretching surface and increases after a certain distance from the wall. Figure 4d represents the influence of E_1 on the fluid velocity, demonstrating that fluid velocity increases as E_1 increases. Increasing values of E_1 accelerate the nanofluid flow, enhancing the fluid's velocity. The fluid's velocity profiles for the thickness parameter α are displayed in Figure 4e. The increasing wall thickness parameter results in a thicker momentum boundary layer, retarding the fluid's velocity. Figure 4f represents the fluid velocity variation corresponding to the Grashof number Gr, which reveals that the velocity of the fluid boosts with Gr. Physically, the growing values of Gr increase the temperature gradient and improve the buoyancy force, which accelerates the fluid flow. Figure 4g depicts the effect of Gc and the flow field velocity. This figure reveals that augmenting values of Gc enhance the fluid velocity because of a stronger buoyancy force. This force produces a pressure gradient in the flow field, which causes fluid acceleration. Hence, fluid velocity raises with escalating values of Gc. The effect of the velocity profile against bioconvection Rayleigh number Nc is portrayed in Figure 4h. This figure shows that higher bioconvection Rayleigh number Nc values resist the upward nanofluid flow.

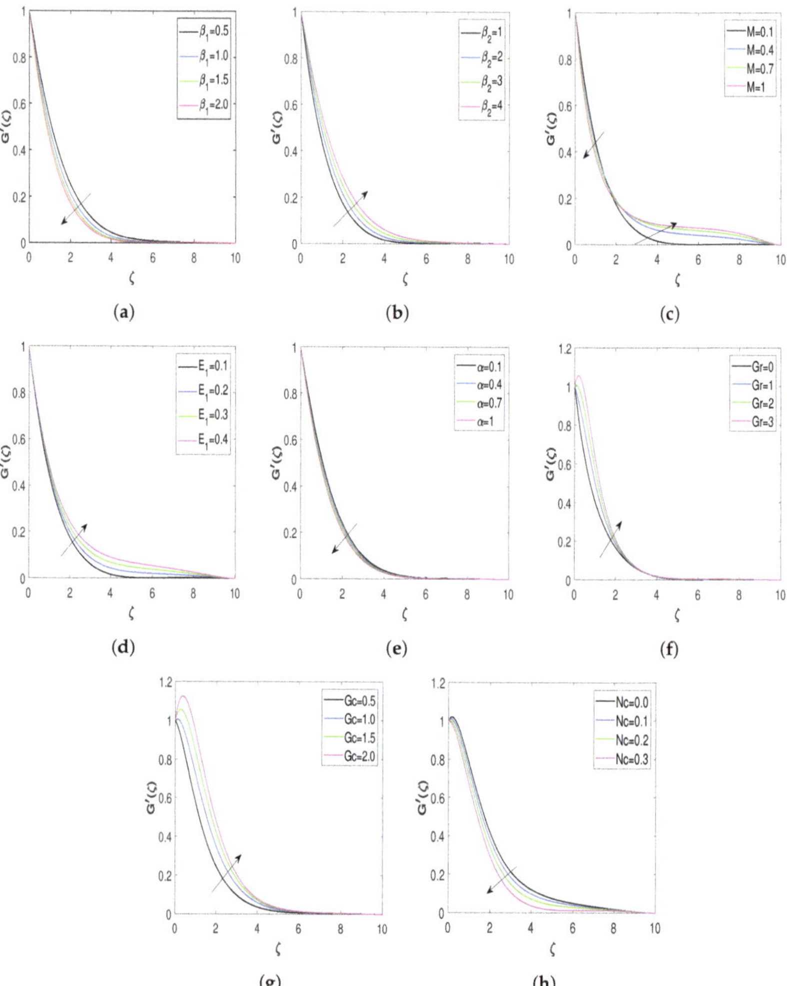

Figure 4. Nondimensional velocity profiles for different values of influential parameters. (**a**) Velocity profile against β_1; (**b**) velocity profile against β_2; (**c**) velocity profile against M; (**d**) velocity profile against E_1; (**e**) velocity profile against α; (**f**) velocity profile against Gr; (**g**) velocity profile against Gc; and (**h**) velocity profile against Nc.

The remarkable impact of numerous physical parameters on the nondimensional temperature of the fluid is presented in Figures 5a–h. Fluid temperature profiles for parameter M are shown in Figure 5a. A rise in the values of M improves the fluid's temperature, since an increase in the magnetic field produces a stronger Lorentz force. Figure 5b portrays the influence of E_1 on the nondimensional temperature of the fluid. The temperature profiles show declination up to a particular value of the similarity variable ζ but then modify its behavior and depict an inclining effect as it approaches farther from the wall of the stretching surface. The heat transfer features via Prandtl number Pr are illustrated in Figure 5c. The temperature of the flow field diminishes with the augmenting values of Pr. The flow becomes viscous-dominant for higher Prandtl number values, and the thermal boundary layer becomes thinner due to viscous dominance. Figure 5d represents the variation of fluid temperature versus Ec. The characterization of viscous heat dissipation is defined by Ec. A higher Eckert number causes more friction on the adjacent fluid layers. An increase in frictional forces magnifies the flow field's internal heat energy. Therefore, the temperature rises with Ec. Figure 5e highlights the behavior of Nt on the fluid temperature. The growth in the values of Nt modifies the fluid temperature. Thermophoresis is the force responded to by suspended particles of the fluid owing to its thermal gradient. An increase in thermal gradient enhances the flow field's temperature. Therefore, fluid temperature enhances with growing values of Nt. Figure 5f displays the impact of Nb on the fluid temperature. This graph illustrates that a rise in Nb raises the temperature. The random motion of particles under suspension in liquid is called Brownian motion. A particle changes its path when it collides with another particle. Furthermore, these collisions cause a random or zigzag motion, and the collision involves the transfer of energy between the particles. Therefore, the augmenting values of Nb improve the temperature profile. Variation of the exponential heat source Qe with the temperature of the flow field is shown in Figure 5g. The escalating values of Qe raise the fluid temperature. Generally, an exponential heat source energizes the flow field, improving the fluid temperature. Figure 5h reveals the impact of Nr on the temperature, which clarifies that an increasing radiation parameter increases the temperature of the fluid. The thermal radiation provides thermal energy to the stretching surface, and then the temperature profile of the fluid enhances due to the conduction between the surface and fluid. Figure 5i represents the influence of φ on the fluid temperature. As per this figure, an increase in the volume fraction improves heat transfer. An increment in φ improves the Jeffrey nanofluid's thermal conductivity, which improves the flow's thermal boundary layer. Hence, thermal profiles escalate with the volume fraction of the nanoparticle.

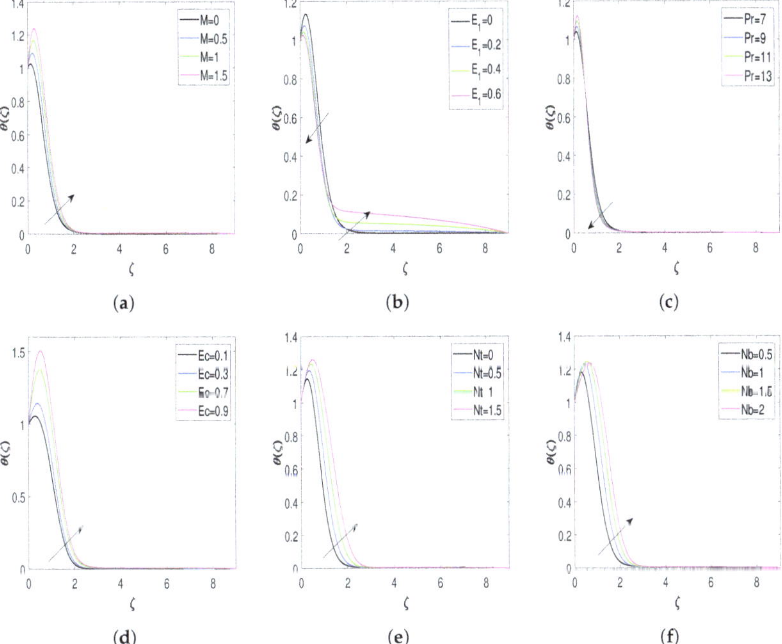

Figure 5. *Cont.*

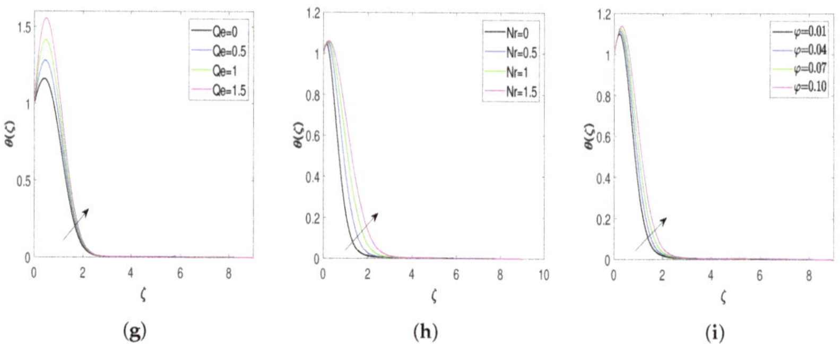

(g) **(h)** **(i)**

Figure 5. Nondimensional temperature profiles for different values of influential parameters. (**a**) Temperature profile against M; (**b**) temperature profile against E_1; (**c**) temperature profile against Pr; (**d**) temperature profile against Ec; (**e**) temperature profile against Nt; (**f**) temperature profile against Nb; (**g**) temperature profile against Q_e; (**h**) temperature profile against Nr; and (**i**) temperature profile against φ.

The influence of Lewis number Le and chemical reaction parameter K on the concentration of the fluid is shown in Figure 6a,b. The behavior of Le concerning the fluid's concentration is displayed in Figure 6a, which declares that the concentration of the flow field diminishes with the growing values of Le. Higher values of Le reduce the solute diffusivity, which drops the concentration of the nanofluid, and the mass transfer rate accelerates at the stretching surface. Therefore, the concentration of the nanofluid reduces with Le. The effect of K on the fluid's concentration is highlighted in Figure 6b. Increasing values of K decelerate mass diffusivity and diminish the concentration boundary layer in the flow field. Hence, the fluid's concentration decreases with K, which is in good agreement with [29].

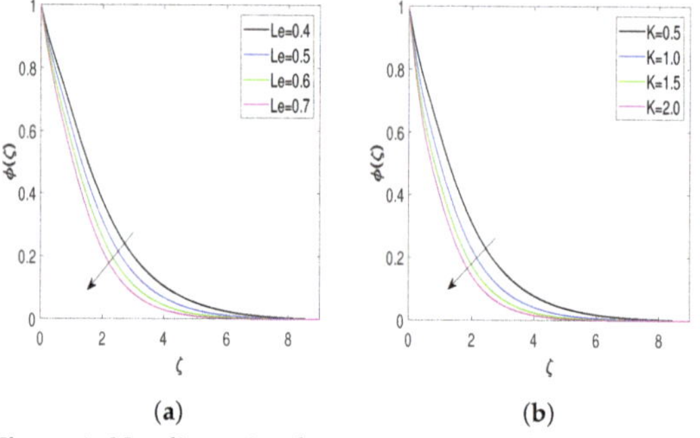

(a) **(b)**

Figure 6. Nondimensional concentration profiles for flow parameters. (**a**) Concentration profile against Le and (**b**) concentration profile against K.

The behavior of the density of microorganism distribution ξ against various physical parameters are displayed in Figure 7a–c. The Lb features on the density of microorganisms are depicted in Figure 7a. The microorganisms' concentration distribution reduces with Lb. Due to the fact that a greater Lb indicates a weaker Brownian motion diffusion coefficient, swimming microorganisms have a relatively shallow penetration depth. Hence, the concentration of microorganism distribution diminishes with Lb, and this result is in validation with the result in the previously published literature [55]. The variation of concentration of motile microorganism distribution versus Peclet number Pe is plotted in Figure 7b. There is a decline in the gyrotactic microorganisms' concentration with a rise in the values of Pe. The impact of δ_N on the motile microorganisms' density is described in Figure 7c. The motile microorganism density decreases with the enhancement in the microorganism difference parameter. This behavior of δ_N is precisely similar to the result in the literature [49].

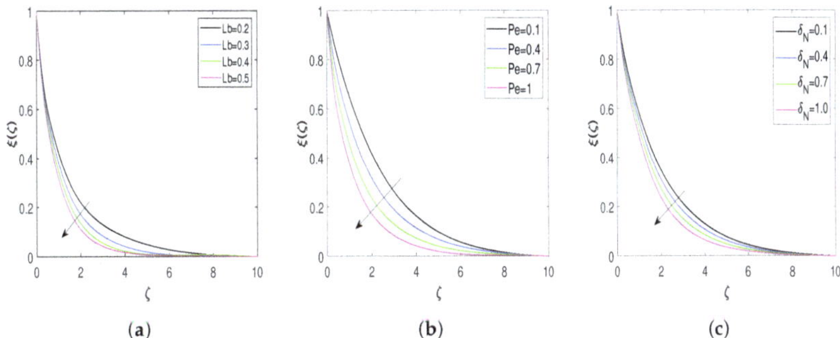

Figure 7. Distribution of microorganisms for different flow parameters. (**a**) Microorganism distribution for Lb; (**b**) Microorganism distribution for Pe; (**c**) Microorganism distribution for δ_N.

3.1. Entropy Generation Model

Entropy generation of this Jeffrey fluid model due to momentum, heat transfer, mass transfer, and density of microorganism distribution is given by

$$
E_g = \left(\kappa_{nf} + \frac{16\sigma^* T_\infty^3}{3\kappa^* (\rho c_p)_{nf}} \right) \frac{(\nabla T_f)^2}{(T_\infty^*)^2} + \mu_{nf} \frac{F}{T_\infty^*} + \frac{\sigma_{nf}}{(T_\infty^*)^2} (u_1^* B - E)^2 + RD \frac{(\nabla C_f)^2}{C_\infty^*}
$$
$$
+ RD \frac{\nabla C_f \nabla T_f}{T_\infty^*} + RD \frac{(\nabla N_f)^2}{N_\infty^*} + RD \frac{\nabla N_f \nabla T_f}{T_\infty^*}. \tag{19}
$$

The mathematical expression of entropy generation of this fluid model is

$$
E_g = \frac{1}{(T_\infty^*)^2} \left(\kappa_{nf} + \frac{16\sigma^* T_\infty^3}{3\kappa^* (\rho c_p)_{nf}} \right) \left(\frac{\partial T_1^*}{\partial y_1^*} \right)^2
$$
$$
+ \frac{\mu_{nf}}{T_\infty^* (1 + \lambda_2)} \left[\left(\frac{\partial u_1^*}{\partial y_1^*} \right)^2 + \lambda_1 \left(u_1^* \frac{\partial u_1^*}{\partial y_1^*} \frac{\partial^2 u_1^*}{\partial x_1^* \partial y_1^*} + v_1^* \frac{\partial u_1^*}{\partial y_1^*} \frac{\partial^2 u_1^*}{\partial y_1^{*2}} \right) \right] + \frac{\sigma_{nf}}{(T_\infty^*)^2} (u_1^* B - E)^2
$$
$$
+ \frac{RD_h}{C_\infty^*} \left(\frac{\partial C_1^*}{\partial y_1^*} \right)^2 + \frac{RD_h}{T_\infty^*} \left(\frac{\partial C_1^*}{\partial y_1^*} \right) \left(\frac{\partial T_1^*}{\partial y_1^*} \right) + \frac{RD_h}{N_\infty^*} \left(\frac{\partial N_1^*}{\partial y_1^*} \right)^2 + \frac{RD_h}{T_\infty^*} \left(\frac{\partial N_1^*}{\partial y_1^*} \right) \left(\frac{\partial T_1^*}{\partial y_1^*} \right). \tag{20}
$$

Mathematical expression of the entropy generation into nondimensionalized form:

$$
N_s = (\varepsilon_6 + Nr)\delta \theta'^2 + \varepsilon_1 \frac{PrEc}{(1 + \beta_1)} \left[G''^2 + \beta_2 \left(\frac{3n-1}{2} G'G''^2 - \frac{n+1}{2} GG''G''' \right) \right] +
$$
$$
\varepsilon_5 M Pr Ec (G' - E_1)^2 + L\theta'\psi' + L^* \frac{1}{\delta \delta_N} \zeta'^2 + L \frac{\delta_1}{\delta} \phi'^2 + L^* \theta' \zeta'. \tag{21}
$$

The mathematical expression of the Bejan number is

$$
Be = \frac{(\varepsilon_6 + Nr)\delta \theta'^2 + L\theta'\phi' + L^* \frac{1}{\delta \delta_N} \zeta'^2 + L \frac{\delta_1}{\delta} \phi'^2 + L^* \theta' \zeta'}{N_s}. \tag{22}
$$

The impacts of the Prandtl number, Eckert number, diffusion parameter, and microorganism diffusion parameter on the entropy generation number (N_s) and Bejan number (Be) are presented in Figures 8 and 9. The behavior of entropy N_s for the Pr is displayed in Figure 8a, demonstrating that enhancing the Prandtl number escalates the entropy formation in the flow field. Figure 8b represents the impact of the Ec on the entropy generation, which declares that enhancement in the Ec improves the rate of entropy generation. A rise in Eckert's number amplifies the difference between the enthalpy and kinetic energy of the flow boundary layer, hence increasing the flow field's irreversibility. Hence, the rate of entropy generation enhances for the larger Eckert number. The behavior of entropy corresponding to the diffusion parameter is highlighted in Figure 8c. From this Figure, it is clear that the entropy rate enhances due to a rise in the diffusion parameter. The flow field becomes more disorderly as the diffusion parameter rises, which causes irreversibility to rise and, as a result, accelerates entropy formation. Figure 8d illustrates the entropy variation for the bioconvection diffusion parameter. Diffusion of bioconvection increases the system's irreversibility. Hence, the higher values of L^* increase the entropy generation rate. Figure 9a,b represents the Bejan number profiles for Eckert and Prandtl numbers. There is a decline in Be profiles with an increment in Pr and Ec. Figure 9c,d reveals the influence of the diffusion and bioconvection diffusion parameters on the Bejan number. According to these figures, increasing the values of L and L^* enhances the Bejan number.

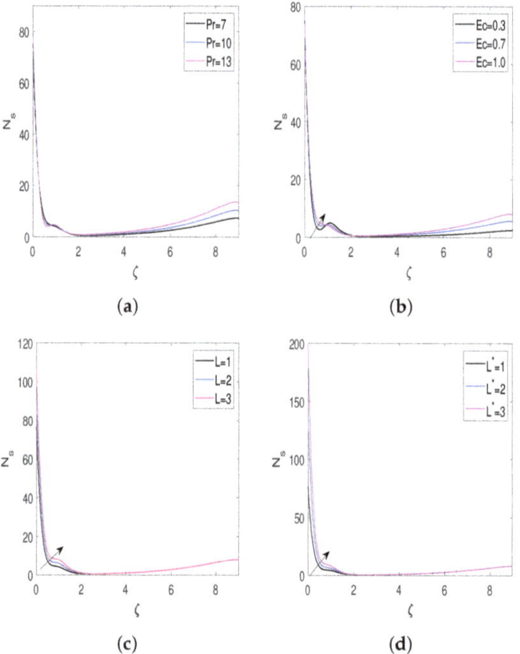

Figure 8. Variation in entropy for different flow parameters. (**a**) Entropy versus Pr; (**b**) entropy versus Ec; (**c**) entropy versus L; and (**d**) entropy versus L^*.

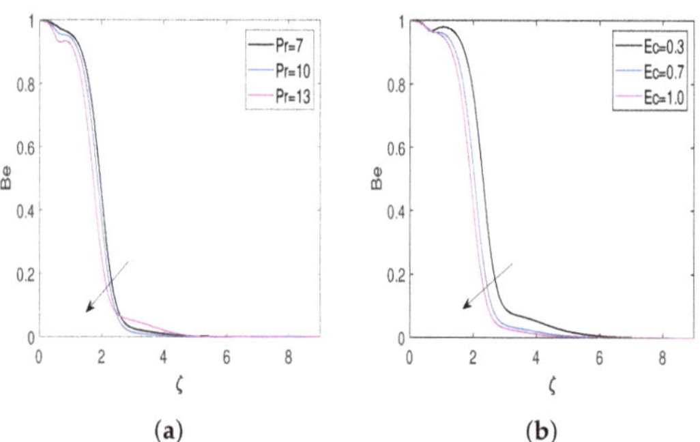

Figure 9. *Cont.*

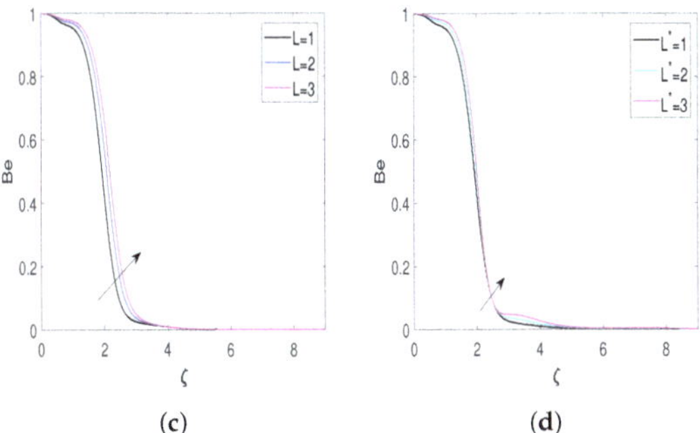

(c) (d)

Figure 9. Bejan number profiles for different influential parameters. (**a**) Bejan number versus *Pr*; (**b**) Bejan number versus *Ec*; (**c**) Bejan number versus *L*; and (**d**) Bejan number versus *L**.

Contour plots illustrate the impact of different influential flow parameters on entropy generation parameter N_s and Bejan number Be in the Figure 10a–l. Figure 10a shows the variation of β_1 and β_2 on the entropy N_s. An increment in β_1 and β_2 reduces the entropy values. The behavior of β_1 and β_2 versus Be is displayed in Figure 10f. The Be rises with β_1 and β_2, but the growth rate of the Bejan number due to β_1 is more compared with β_2. The contour plot illustrating the effect of M and E_1 on entropy is shown in Figure 10b. An opposite effect on entropy is observed with M and E_1, i.e., an increment in M declines the entropy, while the increasing values of E_1 enhance the entropy. Figure 10g represents the behavior of M and E_1 on Be. The Bejan number diminishes for the higher values of E_1 and improves with the growth in M. The influence on entropy via Nt and Nb are depicted in Figure 10c. This figure declares that entropy increases with the escalating values of Nb, while it shows declination with Nt. The contour for the Bejan number showing the influence of Nt and Nb is highlighted in Figure 10h. According to this figure, larger values of Nt cause the Bejan number to decrease, but higher values of Nb cause the Bejan number to increase. The behavior of K and Le versus entropy is shown in Figure 10d. This figure declares that entropy declines with an increase in K and grows with an increase in Le. Figure 10i displays the variation in Bejan number with K and Le. Be values get reduced with increasing K and decreasing Le. Figure 10e shows the influence of δ_N and Lb on entropy formation. The entropy formation rate boosts due to an increase in both δ_N and Lb. Bejan number variations for δ_N and Lb are displayed in Figure 10j. The Bejan number also shows enhancement for augmenting values of δ_N and Lb. Figure 10f,l displays the impact of radiation parameter Nr and exponential heat source Qe on N_s and Be, which shows that an increase in Nr and Qe increases both the entropy and Bejan number of the flow field.

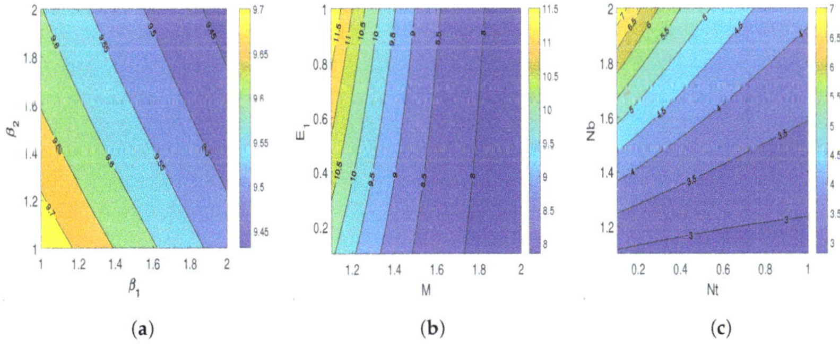

(a) (b) (c)

Figure 10. *Cont.*

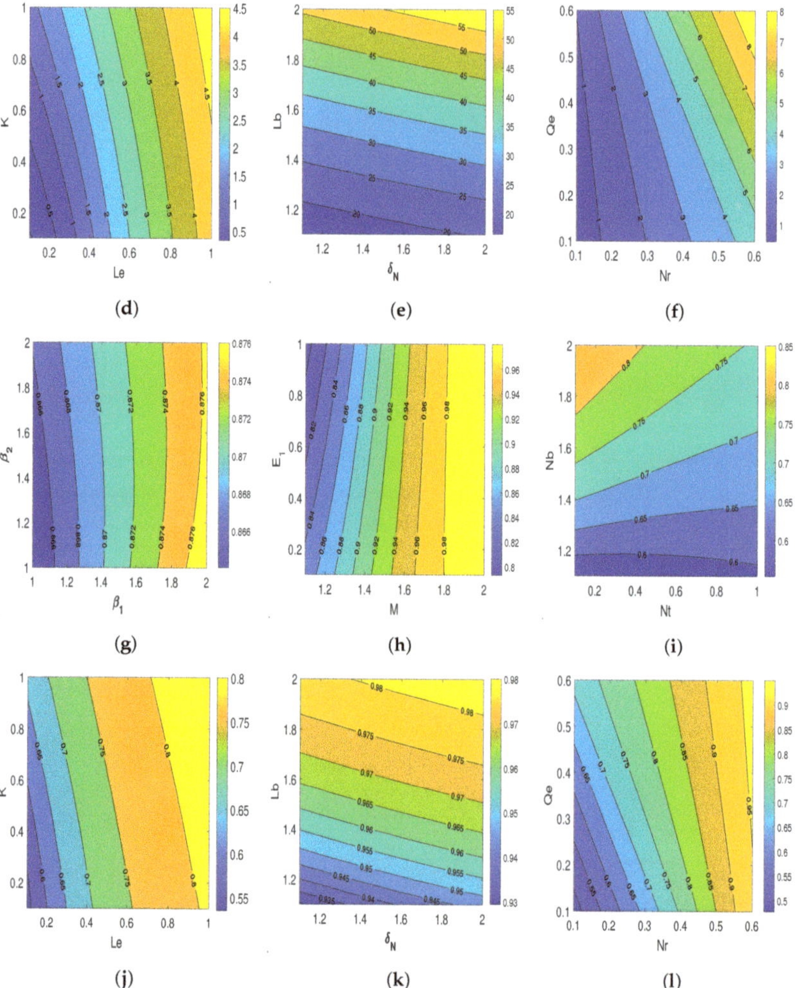

Figure 10. Contour plots illustrating the effect of various influential parameters on entropy generation and Bejan number. (**a**) Entropy via β_1 and β_2; (**b**) entropy via M and E_1; (**c**) entropy via Nt and Nb; (**d**) entropy via K and Le; (**e**) entropy via δ_N and Lb; (**f**) entropy via Nr and Qe; (**g**) Bejan number via β_1 and β_2; (**h**) Bejan number via M and E_1; (**i**) Bejan number via Nt and Nb; (**j**) Bejan number via K and Le; (**k**) Bejan number via δ_N and Lb; and (**l**) Bejan number via Nr and Qe.

3.2. Physical Quantities of Engineering Interest

Drag coefficient (C_f),

$$C_f = \frac{\tau_s^*}{\rho(U_s)^2}, \quad \tau_s^* = \frac{\mu_{nf}}{1+\lambda_1}\left[\frac{\partial u_1^*}{\partial y_1^*} + \lambda_2\left(u_1^*\frac{\partial^2 u_1^*}{\partial x_1^*\partial y_1^*} + v_1^*\frac{\partial^2 u_1^*}{\partial y_1^{*2}}\right)\right]\Bigg|_{y_1^*=A(x_1^*+b)^{\frac{1-n}{2}}}, \tag{23}$$

Local Nusselt number microorganisms (Nu_x),

$$Nu_x = \frac{(x_1^*+b)q_s}{\kappa_{nf}(T_s^*-T_\infty^*)}, \quad q_s = -\kappa_{nf}\frac{\partial T_1^*}{\partial y_1^*}\Bigg|_{y_1^*=A(x_1^*+b)^{\frac{1-n}{2}}}, \tag{24}$$

Sherwood number (Sh_x)

$$Sh_x = \frac{(x_1^* + b)q_m}{D_B(C_s^* - C_\infty^*)}, \quad q_m = -D_B\frac{\partial C_1^*}{\partial y}\bigg|_{y_1^* = A(x_1^* + b)^{\frac{1-n}{2}}}, \tag{25}$$

Local density of microorganisms (Nn_x),

$$Nn_x = \frac{(x_1^* + b)q_m}{D_B(C_s^* - C_\infty^*)}, \quad q_n = -D_n\frac{\partial N_1^*}{\partial y}\bigg|_{y_1^* = A(x_1^* + b)^{\frac{1-n}{2}}}. \tag{26}$$

Using similarity transformation variables, the nondimensionlized forms of the above quantities are given by

$$Re_x^{1/2}C_f = \sqrt{\frac{n+1}{2}}\frac{1}{1+\beta_1}\left[G''(\eta) + \beta_2\left(\frac{3n-1}{2}G'(\eta)G''(\eta) - \frac{n+1}{2}G(\eta)G'''(\eta)\right)\right]\bigg|_{\eta=0},$$

$$Re_x^{-1/2}Nu_x = -\sqrt{\frac{n+1}{2}}\theta'(\eta)\bigg|_{\eta=0}, \quad Re_Y^{-1/2}Sh_Y = -\sqrt{\frac{n+1}{2}}\phi'(\eta)\bigg|_{\eta=0},$$

$$Re_x^{-1/2}Nn_x = -\sqrt{\frac{n+1}{2}}\zeta'(\eta)\bigg|_{\eta=0}. \tag{27}$$

The drag coefficient (C_f), dimensionless parameter of heat transfer coefficient in terms of Nusselt number (Nu_x), dimensionless parameter of mass transfer coefficient in terms of Sherwood number (Sh_x), and the local density of the microorganisms (Nn_x) results are illustrated using surface plots. The effects of the magnetic field parameter M and Deborah number β_1 versus drag coefficient are plotted in Figure 11a. It is observed that the larger magnetic field parameter increases the drag coefficient, while the larger β_1 diminishes the drag coefficient. Figure 11b illustrates the surface plot of the electric field parameter E_1 and Deborah number β_2 versus drag coefficient. This plot declares that the drag coefficient improves with the escalation in E_1, but the effect of the E_1 is much less than that of Deborah number β_2. Figure 11c represents the Nusselt number results of Deborah numbers β_1 and β_2. It is noticed that the Nusselt number at the stretching surface reduces with growing values of β_1, while the heat transfer rate escalates with Deborah number β_2. Figure 11d reveals the behavior of the Nu_x corresponding to Nt and Nb. This Figure clarifies that the Nusselt number diminishes with enhancement in both Nt and Nb. The nature of the M and Qe on the Nusselt number is presented in Figure 11e, which declares that an increment in Qe enhances Nu_x. In contrast, the growth in M reduces the Nusselt number. Figure 11f declares the effect of the Sh_x for Le and δ. In this Figure, the Sherwood number escalates with the augmenting values of δ and Le. The variation of the Sherwood number versus Nb and K is displayed in Figure 11g. This figure reveals that higher values of Nb enlarge the Sherwood number, while the Sherwood number reduces with enhancing K. Figure 11h represents the effects of δ and δ_N on Nn_x. It is examined that the local density of the microorganisms rises with δ and δ_N. The temperature difference parameter is more dominant than the microorganism difference parameter. Figure 11i illustrates the effect of Lb and Pe on Nn_x. As per this figure, it is seen that augmenting the values of bioconvection Lb and Pe enlarges the local density of the motile gyrotactic microorganisms. Moreover, the Lb is more dominant than the Peclet number Pe.

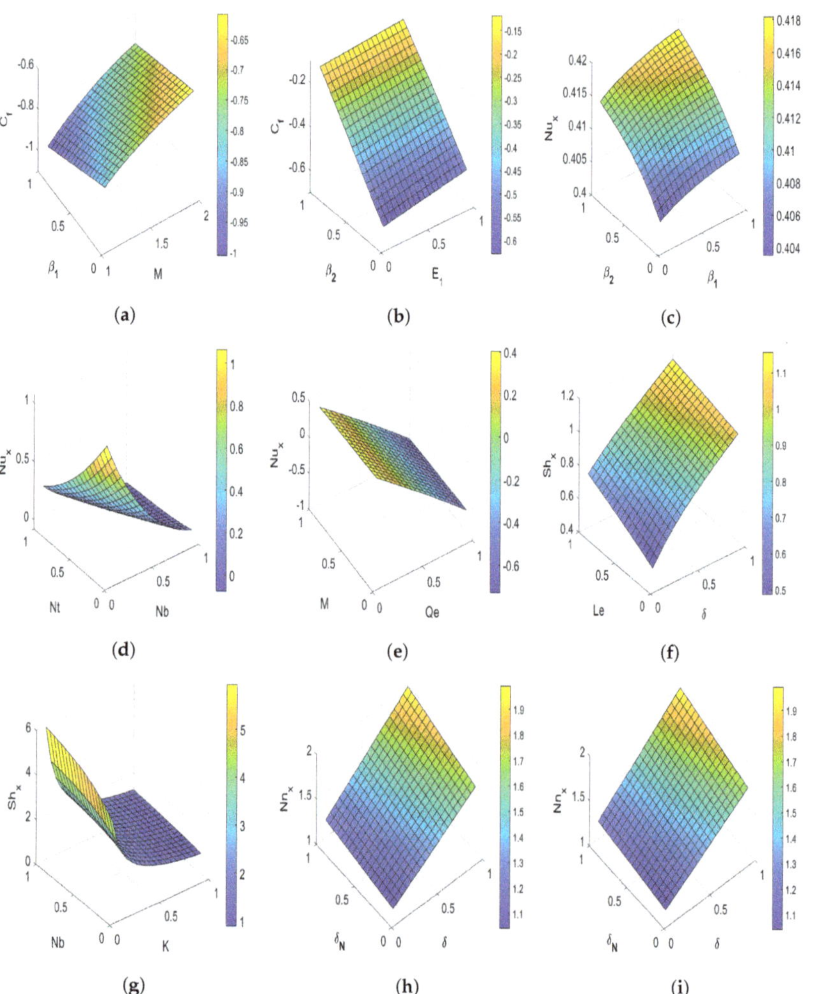

Figure 11. Surface plots displaying the effects of various influence parameters on C_f, Nu_x, Sh_x, and Nn_x. (**a**) C_f versus M and β_1; (**b**) C_f versus E_1 and β_2; (**c**) Nu_x versus β_1 and β_2; (**d**) Nu_x versus Nt and Nb; (**e**) Nu_x versus M and Qe; (**f**) Sh_x versus δ and Le; (**g**) Sh_x versus K and Nb; (**h**) Nn_x versus δ and δ_N; and (**i**) Nn_x versus Pe and Lb.

4. Conclusions

This study discusses the EMHD Jeffrey nanofluid flow through a nonlinear vertically stretching surface of nonuniform thickness, considering the effects of solar radiation and the existence of gyrotactic microorganisms and copper nanoparticles with polyvinyl alcohol–water base fluid. The influence of the source terms such as Joule heating, exponential heat source, viscous dissipation, and chemical reaction is analyzed. The equations governing the fluid flow and boundary conditions are nondimensionalized using suitable similarity transformations. Higher-order nondimensional ODEs are reduced to a system of first-order ODEs and then solved using the RK 4th-order method and the shooting technique. The following are the conclusions of this study:

- Deborah number β_1 diminishes the velocity profile, while Deborah number β_2 enhances the velocity profile.
- Fluid velocity shows enhancement for rising values of E_1, Gr, and Gc, while fluid velocity decays for augmenting values of M, α, and Nc.
- The increment in M, E_1, Ec, Nt, Nb, Qe, and φ enhances the temperature profile, whereas the temperature profile decays for the magnifying values of Pr.
- The concentration profile decays for increasing values of Le and K.
- The rate of entropy increases with an increment in Pr and Ec, while the Bejan number shows declination.
- Both the entropy formation rate and Bejan number enhance with the increment diffusion parameter L and bioconvection diffusion parameter L^*.
- Drag coefficient improves with the growing values of M, E_1, and β_2, while the drag coefficient reduces with an increase in β_1.
- Nusselt number enlarges with the enhancement in β_2, Ec, and Qe. Moreover, it diminishes with a higher β_1, Pr, Nt, Nb, and M.
- Sherwood number escalates with the augmenting values of δ and Le. Moreover, it will reduce with augmenting values of K.
- Nn_x improves with the growth in δ, Lb, and Pe, while it decreases with δ_N.

The previous literature is limited to the entropy generation of the Jeffrey nanofluid flows in heat and mass transfer processes over the stretching surfaces of uniform thickness. No attempts have been made to analyze the entropy generation of the EMHD Jeffrey fluid flow with motile gyrotactic microorganisms via vertical surfaces stretching nonlinearly with nonuniform thickness under the influence of solar radiations. Therefore, this study investigates entropy generation and solar radiation effects in the presence of motile gyrotactic microorganisms and copper nanoparticles with polyvinyl alcohol–water as the base fluid, as well as the influence of the source terms, such as viscous dissipation, Ohmic heating, exponential heat source, and chemical reaction of the EMHD Jeffrey fluid flow past a vertical nonlinearly elongating surface of nonuniform thickness. This research could aid in constructing solar heat engines, solar thermochemical heat pumps, solar ponds, and household solar water heaters, among others. The mechanism of bioconvection will improve the stability of nanoparticles. The motile gyrotactic microorganisms are crucial, because a better liquid mixture is required for a better biological process. Entropy production improves system performance, updates systematic presentations, and destroys energy. So, this study can be helpful for such types of applications.

Optimizing the heat transfer in solar energy systems is a peculiar feature of storing thermal energy in the maximum possible amount from solar radiation. In this direction, nanofluids are proved to be very useful in improving the heat transfer properties of base fluids. In the future, the present work can be extended by using hybrid and ternary nanofluids, using other base fluids as heat transfer fluids, which may result in better heat transfer characteristics. A sensitivity analysis to determine the effective parameters affecting the heat transfer of hybrid nanofluid flow in the parabolic trough collectors in different base fluids can be performed.

Author Contributions: Conceptualization, B.K.S. and A.K.; methodology, R.G.; formal analysis, M.M.B.; investigation, N.K.M.; writing—original draft preparation, R.G.; writing—review and editing, M.M.B; visualization, A.K.; supervision, B.K.S. All authors have read and agreed to the published version of the manuscript.

Funding: This research received no external funding.

Institutional Review Board Statement: Not applicable.

Informed Consent Statement: Not applicable.

Data Availability Statement: Not applicable.

Conflicts of Interest: The authors declare no conflict of interest.

Nomenclature

Br	Brinkmann number	Q_e	Exponential heat source parameter
T^*_∞	Ambient temperature (unit: K)	Sh_x	Sherwood number
C^*_∞	Ambient concentration (unit: mol/m^3)	T^*_s	Temperature at the surface of the wall (unit: K)
C_f	Skin friction coefficient	U_s	Stretching sheet velocity (unit: m/ s)
C^*_s	Concentration at the surface of the wall (unit: mol/m^3)	u^*_s	Velocity component in x-direction (unit: m/s)
C_p	Specific heat capacity (unit: mol Jkg^{-1} K^{-1})	v^*_s	Velocity component in y-direction (unit: m/s)
D_B	Brownian diffusion coefficient (unit: kg m^{-1} s^{-1})	***Greek Letters***	
D_T	Thermophoresis diffusion coefficient (unit: kg m^{-1} s^{-1} K^{-1})	α	Thickness parameter
E_1	Electric field parameter	α_1	Thermal diffusivity (unit: s/m^2)
E_a	Dimensional activation energy (unit: KJ/mol)	β_C	Buoyancy force due to concentration (unit: K^{-1})
E_t	Nondimensional activation energy	β_T	Buoyancy force due to temperature (unit: K^{-1})
Ec	Eckert number	β_1	Relaxation and retardation time ratio parameter
G'	Dimensionless velocity	β_2	Retardation time parameter
Gr	Grashof number	ζ	Similarity variable
Gc	Solutal Grashof number	κ	Thermal conductivity (unit: W/(m.K))
K_r	Chemical reaction rate (unit: mol/s)	μ	Dynamic viscosity (unit: Pa.s)
K	Chemical reaction parameter	v	Fluids' kinematic viscosity (unit: m$^2 \cdot$ s^{-1})
Le	Lewis number	λ_1	Relaxation and retardation time ratio
Lb	Bioconvection Lewis number	L^*	Microorganism diffusion parameter
m	Fitted rate constant	λ_2	Retardation time (unit: s)
M	Magnetic field parameter	ϕ	Dimensionless concentration (unit: mol/m^3)
Nt	Thermophoresis diffusion parameter	ρ	Density of fluid (unit: Kg/m^3)
Nu_x	Nusselt number	ψ	Stream function (unit: Kg/(m.s))
Nb	Brownian diffusion parameter	σ	Electrical conductivity (unit: S/m)

References

1. Mushtaq, A.; Mustafa, M.; Hayat, T.; Alsaedi, A. Nonlinear radiative heat transfer in the flow of nanofluid due to solar energy: A numerical study. *J. Taiwan Inst. Chem. Eng.* **2014**, *45*, 1176–1183. [CrossRef]
2. Shehzad, S.A.; Hayat, T.; Alsaedi, A.; Obid, M.A. Nonlinear thermal radiation in three-dimensional flow of Jeffrey nanofluid: A model for solar energy. *Appl. Math. Comput.* **2014**, *248*, 273–286. [CrossRef]
3. Khan, J.A.; Mustafa, M.; Hayat, T.; Farooq, M.A.; Alsaedi, A.; Liao, S. On model for three-dimensional flow of nanofluid: An application to solar energy. *J. Mol. Liq.* **2014**, *194*, 41–47. [CrossRef]
4. Khan, J.A.; Mustafa, M.; Hayat, T.; Alsaedi, A. Three-dimensional flow of nanofluid over a non-linearly stretching sheet: An application to solar energy. *Int. J. Heat Mass Transf.* **2015**, *86*, 158–164. [CrossRef]
5. Zin, N.A.M.; Khan, I.; Shafie, S. The impact silver nanoparticles on MHD free convection flow of Jeffrey fluid over an oscillating vertical plate embedded in a porous medium. *J. Mol. Liq.* **2016**, *222*, 138–150.
6. Reddy, K.; Kamnapure, N.R.; Srivastava, S. Nanofluid and nanocomposite applications in solar energy conversion systems for performance enhancement: A review. *Int. J. Low-Carbon Technol.* **2017**, *12*, 1–23. [CrossRef]
7. Daniel, Y.S.; Aziz, Z.A.; Ismail, Z.; Salah, F. Impact of thermal radiation on electrical MHD flow of nanofluid over nonlinear stretching sheet with variable thickness. *Alex. Eng. J.* **2018**, *57*, 2187–2197. [CrossRef]
8. Wahab, A.; Hassan, A.; Qasim, M.A.; Ali, H.M.; Babar, H.; Sajid, M.U. Solar energy systems–potential of nanofluids. *J. Mol. Liq.* **2019**, *289*, 111049. [CrossRef]
9. Azam, M.; Shakoor, A.; Rasool, H.; Khan, M. Numerical simulation for solar energy aspects on unsteady convective flow of MHD Cross nanofluid: A revised approach. *Int. J. Heat Mass Transf.* **2019**, *131*, 495–505. [CrossRef]
10. Sunitha, G. Influence of thermal radiation on peristaltic blood flow of a Jeffrey fluid with double diffusion in the presence of gold nanoparticles. *Informatics Med. Unlocked* **2019**, *17*, 100272.
11. Acharya, N. On the flow patterns and thermal behaviour of hybrid nanofluid flow inside a microchannel in presence of radiative solar energy. *J. Therm. Anal. Calorim.* **2020**, *141*, 1425–1442. [CrossRef]
12. Song, Y.Q.; Obideyi, B.; Shah, N.A.; Animasaun, I.; Mahrous, Y.; Chung, J.D. Significance of haphazard motion and thermal migration of alumina and copper nanoparticles across the dynamics of water and ethylene glycol on a convectively heated surface. *Case Stud. Therm. Eng.* **2021**, *26*, 101050. [CrossRef]
13. Jamshed, W.; Eid, M.R.; Safdar, R.; Pasha, A.A.; Mohamed Isa, S.S.P.; Adil, M.; Rehman, Z.; Weera, W. Solar energy optimization in solar-HVAC using Sutterby hybrid nanofluid with Smoluchowski temperature conditions: A solar thermal application. *Sci. Rep.* **2022**, *12*, 11484. [CrossRef] [PubMed]
14. Sharma, B.K.; Poonam.; Chamkha, A.J. Effects of heat transfer, body acceleration and hybrid nanoparticles (Au-Al$_2$O$_3$) on MHD blood flow through a curved artery with stenosis and aneurysm using hematocrit-dependent viscosity. *Waves Random Complex Media* **2022**, 1–31. [CrossRef]

15. Gandhi, R.; Sharma, B.; Kumawat, C.; Bég, O.A. Modeling and analysis of magnetic hybrid nanoparticle (Au-Al$_2$O$_3$/blood) based drug delivery through a bell-shaped occluded artery with Joule heating, viscous dissipation and variable viscosity effects. *Proc. Inst. Mech. Eng. Part E J. Process. Mech. Eng.* **2022**, *236*, 09544089221080273. [CrossRef]
16. Sharma, B.; Kumawat, C.; Vafai, K. Computational biomedical simulations of hybrid nanoparticles (Au-Al$_2$O$_3$/blood-mediated) transport in a stenosed and aneurysmal curved artery with heat and mass transfer: Hematocrit dependent viscosity approach. *Chem. Phys. Lett.* **2022**, *800*, 139666.
17. Bhatti, M.; Ellahi, R.; Doranehgard, M.H. Numerical study on the hybrid nanofluid (Co3O4-Go/H2O) flow over a circular elastic surface with non-Darcy medium: Application in solar energy. *J. Mol. Liq.* **2022**, *361*, 119655. [CrossRef]
18. Bouslimi, J.; Alkathiri, A.A.; Althagafi, T.M.; Jamshed, W.; Eid, M.R. Thermal properties, flow and comparison between Cu and Ag nanoparticles suspended in sodium alginate as Sutterby nanofluids in solar collector. *Case Stud. Therm. Eng.* **2022**, *39*, 102358. [CrossRef]
19. Shahzad, F.; Jamshed, W.; Eid, M.R.; Safdar, R.; Putri Mohamed Isa, S.S.; El Din, S.M.; Mohd Nasir, N.A.A.; Iqbal, A. Thermal cooling efficacy of a solar water pump using Oldroyd-B (aluminum alloy-titanium alloy/engine oil) hybrid nanofluid by applying new version for the model of Buongiorno. *Sci. Rep.* **2022**, *12*, 19817. [CrossRef]
20. Pakravan, H.A.; Yaghoubi, M. Combined thermophoresis, Brownian motion and Dufour effects on natural convection of nanofluids. *Int. J. Therm. Sci.* **2011**, *50*, 394–402. [CrossRef]
21. Anbuchezhian, N.; Srinivasan, K.; Chandrasekaran, K.; Kandasamy, R. Thermophoresis and Brownian motion effects on boundary layer flow of nanofluid in presence of thermal stratification due to solar energy. *Appl. Math. Mech.* **2012**, *33*, 765–780. [CrossRef]
22. Kandasamy, R.; Muhaimin, I.; Mohamad, R. Thermophoresis and Brownian motion effects on MHD boundary-layer flow of a nanofluid in the presence of thermal stratification due to solar radiation. *Int. J. Mech. Sci.* **2013**, *70*, 146–154. [CrossRef]
23. Hayat, T.; Asad, S.; Alsaedi, A. Analysis for flow of Jeffrey fluid with nanoparticles. *Chin. Phys. B* **2015**, *24*, 044702. [CrossRef]
24. Mabood, F.; Ibrahim, S.; Khan, W. Framing the features of Brownian motion and thermophoresis on radiative nanofluid flow past a rotating stretching sheet with magnetohydrodynamics. *Results Phys.* **2016**, *6*, 1015–1023. [CrossRef]
25. Sulochana, C.; Ashwinkumar, G.; Sandeep, N. Transpiration effect on stagnation-point flow of a Carreau nanofluid in the presence of thermophoresis and Brownian motion. *Alex. Eng. J.* **2016**, *55*, 1151–1157. [CrossRef]
26. Astanina, M.; Abu-Nada, E.; Sheremet, M. Combined effects of thermophoresis, brownian motion, and nanofluid variable properties on CuO-water nanofluid natural convection in a partially heated square cavity. *J. Heat Transf.* **2018**, *140*, 082401. [CrossRef]
27. Awan, S.E.; Raja, M.A.Z.; Mehmood, A.; Niazi, S.A.; Siddiqa, S. Numerical treatments to analyze the nonlinear radiative heat transfer in MHD nanofluid flow with solar energy. *Arab. J. Sci. Eng.* **2020**, *45*, 4975–4994. [CrossRef]
28. Rekha, M.; Sarris, I.E.; Madhukesh, J.; Raghunatha, K.; Prasannakumara, B. Impact of thermophoretic particle deposition on heat transfer and nanofluid flow through different geometries: An application to solar energy. *Chin. J. Phys.* **2022**, *80*, 190–205. [CrossRef]
29. Sharma, B.; Kumar, A.; Gandhi, R.; Bhatti, M. Exponential space and thermal-dependent heat source effects on electro-magneto-hydrodynamic Jeffrey fluid flow over a vertical stretching surface. *Int. J. Mod. Phys. B* **2022**, *36*, 2250220. [CrossRef]
30. Sharma, B.; Khanduri, U.; Mishra, N.K.; Mekheimer, K.S. Combined effect of thermophoresis and Brownian motion on MHD mixed convective flow over an inclined stretching surface with radiation and chemical reaction. *Int. J. Mod. Phys. B* **2022**, 2350095. [CrossRef]
31. Sharma, B.; Gandhi, R.; Mishra, N.K.; Al-Mdallal, Q.M. Entropy generation minimization of higher-order endothermic/exothermic chemical reaction with activation energy on MHD mixed convective flow over a stretching surface. *Sci. Rep.* **2022**, *12*, 17688. [CrossRef] [PubMed]
32. Parvin, S.; Nasrin, R.; Alim, M. Heat transfer and entropy generation through nanofluid filled direct absorption solar collector. *Int. J. Heat Mass Transf.* **2014**, *71*, 386–395. [CrossRef]
33. Khan, M.I.; Hayat, T.; Khan, M.I.; Waqas, M.; Alsaedi, A. Numerical simulation of hydromagnetic mixed convective radiative slip flow with variable fluid properties: A mathematical model for entropy generation. *J. Phys. Chem. Solids* **2019**, *125*, 153–164. [CrossRef]
34. Wang, W.W.; Cai, Y.; Wang, L.; Liu, C.W.; Zhao, F.Y.; Sheremet, M.A.; Liu, D. A two phase closed thermosyphon operated with nanofluids for solar energy collectors: Thermodynamic modeling and entropy generation analysis. *Sol. Energy* **2020**, *211*, 192–209. [CrossRef]
35. Naz, R.; Noor, M.; Shah, Z.; Sohail, M.; Kumam, P.; Thounthong, P. Entropy generation optimization in MHD pseudoplastic fluid comprising motile microorganisms with stratification effect. *Alex. Eng. J.* **2020**, *59*, 485–496. [CrossRef]
36. Farooq, U.; Munir, S.; Malik, F.; Ahmad, B.; Lu, D. Aspects of entropy generation for the non-similar three-dimensional bioconvection flow of nanofluids. *AIP Adv.* **2020**, *10*, 075110. [CrossRef]
37. Yusuf, T.A.; Mabood, F.; Prasannakumara, B.; Sarris, I.E. Magneto-bioconvection flow of Williamson nanofluid over an inclined plate with gyrotactic microorganisms and entropy generation. *Fluids* **2021**, *6*, 109. [CrossRef]
38. Li, Y.X.; Khan, M.I.; Gowda, R.P.; Ali, A.; Farooq, S.; Chu, Y.M.; Khan, S.U. Dynamics of aluminum oxide and copper hybrid nanofluid in nonlinear mixed Marangoni convective flow with entropy generation: Applications to renewable energy. *Chin. J. Phys.* **2021**, *73*, 275–287. [CrossRef]

39. Sharma, B.; Gandhi, R.; Bhatti, M. Entropy analysis of thermally radiating MHD slip flow of hybrid nanoparticles (Au-Al$_2$O$_3$/Blood) through a tapered multi-stenosed artery. *Chem. Phys. Lett.* **2022**, *790*, 139348. [CrossRef]
40. Khanduri, U.; Sharma, B.K. Entropy Analysis for MHD Flow Subject to Temperature-Dependent Viscosity and Thermal Conductivity. In *Nonlinear Dynamics and Applications*; Springer: Berlin/Heidelberg, Germany, 2022; pp. 457–471.
41. Salawu, S.; Obalalu, A.; Shamshuddin, M. Nonlinear solar thermal radiation efficiency and energy optimization for magnetized hybrid Prandtl–Eyring nanoliquid in aircraft. *Arab. J. Sci. Eng.* **2022**, 1–12. [CrossRef]
42. Avramenko, A.; Kuznetsov, A. Stability of a suspension of gyrotactic microorganisms in superimposed fluid and porous layers. *Int. Commun. Heat Mass Transf.* **2004**, *31*, 1057–1066. [CrossRef]
43. Avramenko, A.; Kuznetsov, A. The onset of bio-thermal convection in a suspension of gyrotactic microorganisms in a fluid layer with an inclined temperature gradient. *Int. J. Numer. Methods Heat Fluid Flow* **2010**, *20*, 111–129. [CrossRef]
44. Mutuku, W.N.; Makinde, O.D. Hydromagnetic bioconvection of nanofluid over a permeable vertical plate due to gyrotactic microorganisms. *Comput. Fluids* **2014**, *95*, 88–97. [CrossRef]
45. Acharya, N.; Das, K.; Kundu, P.K. Framing the effects of solar radiation on magneto-hydrodynamics bioconvection nanofluid flow in presence of gyrotactic microorganisms. *J. Mol. Liq.* **2016**, *222*, 28–37. [CrossRef]
46. Saleem, S.; Rafiq, H.; Al-Qahtani, A.; El-Aziz, M.A.; Malik, M.; Animasaun, I. Magneto Jeffrey nanofluid bioconvection over a rotating vertical cone due to gyrotactic microorganism. *Math. Probl. Eng.* **2019**, *2019*, 3478037. [CrossRef]
47. Sohail, M.; Naz, R.; Abdelsalam, S.I. On the onset of entropy generation for a nanofluid with thermal radiation and gyrotactic microorganisms through 3D flows. *Phys. Scr.* **2020**, *95*, 045206. [CrossRef]
48. Song, Y.Q.; Hamid, A.; Khan, M.I.; Gowda, R.P.; Kumar, R.N.; Prasannakumara, B.; Khan, S.U.; Khan, M.I.; Malik, M. Solar energy aspects of gyrotactic mixed bioconvection flow of nanofluid past a vertical thin moving needle influenced by variable Prandtl number. *Chaos Solitons Fractals* **2021**, *151*, 111244. [CrossRef]
49. Naidu, K.K.; Babu, D.H.; Reddy, S.H.; Narayana, P.S. Radiation and partial slip effects on magnetohydrodynamic Jeffrey nanofluid containing gyrotactic microorganisms over a stretching surface. *J. Therm. Sci. Eng. Appl.* **2021**, *13*, 031011. [CrossRef]
50. Sharma, B.K.; Khanduri, U.; Mishra, N.K.; Chamkha, A.J. Analysis of Arrhenius activation energy on magnetohydrodynamic gyrotactic microorganism flow through porous medium over an inclined stretching sheet with thermophoresis and Brownian motion. *Proc. Inst. Mech. Eng. Part E J. Process. Mech. Eng.* **2022**, 09544089221128768. [CrossRef]
51. Bhatti, M.; Arain, M.; Zeeshan, A.; Ellahi, R.; Doranehgard, M. Swimming of Gyrotactic Microorganism in MHD Williamson nanofluid flow between rotating circular plates embedded in porous medium: Application of thermal energy storage. *J. Energy Storage* **2022**, *45*, 103511. [CrossRef]
52. Hussain, A.; Farooq, N. Gyrotactic micro-organisms swimming under the Hyperbolic Tangent Blood Nano Material and Solar biomimetic system over the Esophagus. *Int. Commun. Heat Mass Transf.* **2023**, *141*, 106579. [CrossRef]
53. Shahzad, F.; Jamshed, W.; Nisar, K.S.; Khashan, M.M.; Abdel-Aty, A.H. Computational analysis of Ohmic and viscous dissipation effects on MHD heat transfer flow of Cu-PVA Jeffrey nanofluid through a stretchable surface. *Case Stud. Therm. Eng.* **2021**, *26*, 101148. [CrossRef]
54. Ali, A.; Maqsood, M.; Anjum, H.; Awais, M.; Sulaiman, M. Analysis of heat transfer on MHD Jeffrey nanofluid flow over nonlinear elongating surface of variable thickness. *ZAMM-J. Appl. Math. Mech. Angew. Math. Mech.* **2022**, *102*, e202100250. [CrossRef]
55. Muhammad, T.; Waqas, H.; Manzoor, U.; Farooq, U.; Rizvi, Z.F. On doubly stratified bioconvective transport of Jeffrey nanofluid with gyrotactic motile microorganisms. *Alex. Eng. J.* **2022**, *61*, 1571–1583. [CrossRef]

nanomaterials

Article

Comparative Analysis of LiMPO$_4$ (M = Fe, Co, Cr, Mn, V) as Cathode Materials for Lithium-Ion Battery Applications— A First-Principle-Based Theoretical Approach

Sayan Kanungo [1,2], Ankur Bhattacharjee [1], Naresh Bahadursha [1] and Aritra Ghosh [3,*]

1 Department of Electrical and Electronics Engineering, Birla Institute of Technology and Science-Pilani, Hyderabad Campus, Hyderabad 500078, India
2 Materials Center for Sustainable Energy & Environment, Birla Institute of Technology and Science-Pilani, Hyderabad Campus, Hyderabad 500078, India
3 Faculty of Environment, Science and Economy (ESE), Renewable Energy, Electric and Electronic Engineering, University of Exeter, Penryn TR10 9FE, UK
* Correspondence: a.ghosh@exeter.ac.uk

Abstract: The rapidly increasing demand for energy storage has been consistently driving the exploration of different materials for Li-ion batteries, where the olivine lithium-metal phosphates (LiMPO$_4$) are considered one of the most potential candidates for cathode-electrode design. In this context, the work presents an extensive comparative theoretical study of the electrochemical and electrical properties of iron (Fe)-, cobalt (Co)-, manganese (Mn)-, chromium (Cr)-, and vanadium (V)-based LiMPO$_4$ materials for cathode design in lithium (Li)-ion battery applications, using the density-functional-theory (DFT)-based first-principle-calculation approach. The work emphasized different material and performance aspects of the cathode design, including the cohesive energy of the material, Li-intercalation energy in olivine structure, and intrinsic diffusion coefficient across the Li channel, as well as equilibrium potential and open-circuit potential at different charge-states of Li-ion batteries. The results indicate the specification of the metal atom significantly influences the Li diffusion across the olivine structure and the overall energetics of different LiMPO$_4$. In this context, a clear correlation between the structural and electrochemical properties has been demonstrated in different LiMPO$_4$. The key findings offer significant theoretical and design-level insight for estimating the performance of studied LiMPO$_4$-based Li-ion batteries while interfacing with different application areas.

Keywords: lithium-ion battery; cathode material; LiMPO$_4$; olivine structure; lithium transport; first-principle calculations; density functional theory

Citation: Kanungo, S.; Bhattacharjee, A.; Bahadursha, N.; Ghosh, A. Comparative Analysis of LiMPO$_4$ (M = Fe, Co, Cr, Mn, V) as Cathode Materials for Lithium-Ion Battery Applications—A First-Principle-Based Theoretical Approach. *Nanomaterials* 2022, 12, 3266. https://doi.org/10.3390/nano12193266

Academic Editors: M. M. Bhatti, Kambiz Vafai and Sara I. Abdelsalam

Received: 26 August 2022
Accepted: 15 September 2022
Published: 20 September 2022

Publisher's Note: MDPI stays neutral with regard to jurisdictional claims in published maps and institutional affiliations.

1. Introduction

Over the last two decades, the increasing concerns about global warming, rapidly depleting international reserves of fossil fuels, and a growing presence of portable electronics in different application areas have primarily driven the ever-increasing requirement for clean and efficient energy-storage systems [1–4]. In this context, the lithium (Li)-ion battery demonstrates steadily growing commercial footprints as an efficient energy-storage element owing to its high energy density, low self-discharge property, high open-circuit voltage, and long lifespan [3,4]. Specifically, the high-energy density of Li-ion batteries holds considerable promise as the power source of hybrid or fully electric vehicles in the background of reducing CO$_2$ and other greenhouse gas emissions and cutting down fossil-fuel consumption in the automotive industry [3,5]. However, like any other energy-storage device, the performance of Li-ion batteries also significantly depends on the material properties of their constituent elements, specifically electrodes.

Some of the significant challenges involving electrode design—and thereby lifespan degradation of high-energy-density Li-ion batteries—include damage, fracture, and diffusion-induced stress in the electrodes with the Li-ion intercalation/de-intercalation [6,7] chemo-mechanical degradation of the electrodes [8], the co-existence of electrochemical reaction, and irradiation in the extreme environment leading to compromise of operational integrity of the electrodes [9]. Therefore, a renewed research interest has been observed in enhancing the performance of Li-ion batteries by integrating novel or emerging materials and their hybrids in the cathode and anode design [2,3,10]. In this context, the density-functional-theory-based first-principle-calculation approach has emerged as a highly efficient computational technique for estimating the potential of different emerging/novel material systems for the Li-ion battery designs, which can efficiently guide as well as complement experimental exploration in this direction [4,6]. The material specifications of cathode materials significantly influence the overall performance of Li-ion batteries, leading to growing research interest in theoretically exploring different material systems as well as material engineering strategies for cathode-design optimization [4,11].

The olivine $LiMPO_4$ material family from the Pnma space group presents some of the most promising cathode materials for Li-ion battery applications, which are also commercially adopted (such as $LiFePO_4$ and $LiCoPO_4$) for battery design [12]. In this context, considerable research has recently been observed to optimize the relevant material properties of different $LiMPO_4$ materials, where the explorations based on the first-principle calculation remain at the forefront of such research. In one of the early works in this direction, the Li diffusion was theoretically analyzed in different $LiMPO_4$ (M = Mn, Fe, Co, and Ni) structures, where the diffusivity is typically underestimated compared to experimental reports [13]. The incorporation of atomistic defects in the $LiMPO_4$ (M = Mn, Fe, Co, and Ni) structure reproduces a more reliable assessment of Li diffusion and establishes a curved Li-diffusion pathway across the Li channel along the [10] crystallographic directions [14,15]. Specifically, the presence of Li/M antisite defects in the lattice significantly affects the Li diffusion in $LiMPO_4$ (M = Fe) [14] and may lead to inter-channel hopping [16]. The bulk diffusion in Li is affected by the presence of intrinsic strain as well as Li concentration in $LiMPO_4$ (M = Fe) lattices [17]. Furthermore, the formation of the interface in $Li_{1-x}MPO_4$ (M = Fe) plays an important role in the faster charging/discharging in Li-ion batteries [5]. In this effect, the bulk and surface Li diffusion in the Fe-doped $LiMPO_4$ (M = Co) structure has been theoretically studied, which reveals that the (010) surface demonstrates the lowest surface energy, wherein the Li-ion diffusion barrier can be notably reduced by Fe doping [18]. The doping in olivine $LiMPO_4$ structures has been identified as a highly efficient approach for engineering the material properties of different $LiMPO_4$ materials. In this context, the theoretical investigation suggests that the introduction of Ni and Fe co-doping presents a promising route to optimizing the electrochemical properties of $LiMPO_4$ (M = Mn) [19]. A similar exploration suggests that the Mn doping in $LiMPO_4$ (M = Co) can significantly influence the electrochemical activities of the material [20]. Apart from doping, the formation of hybrid $LiM_xN_{1-x}PO_4$ with different metals is another promising material engineering strategy for optimizing the electrochemical properties. In this context, a recent theoretical investigation reveals that $LiFe_xMn_{1-x}PO_4$ possesses considerable potential as a cathode for Li-ion battery design [21,22]. On the other hand, very recently, an extensive DFT-based investigation has been performed to investigate the different reaction routes, intrinsic defect formations, and influence of a wide variety of doping in $LiMPO_4$ (M = Fe), offering a vital theoretical understanding of the synthesis process [23].

However, the majority of the theoretical investigations studied the electronic and electrochemical properties of one or few $LiMPO_4$, emphasizing specific electrochemical aspects of these materials. To date, no systematic theoretical investigation has been observed on the different olivine phosphate-based cathode materials on their electrochemical and electrical properties. Consequently, in this work, a detailed first-principle-based theoretical investigation has been performed on the $LiMPO_4$ (M = Fe, Co, Cr, Mn, and V), emphasizing the structural, electrochemical, and electrical properties. In this context, the comparative cohesive energy, M-O (L-O) bond length, M-O (L-O) charge transfer, diffusion activation barrier, hopping distance, diffusion coefficient, equilibrium potential, and open-circuit voltage as a function of state-of-charge for different cathode materials are emphasized in this work.

2. Materials and Computational Methodology

2.1. Battery Mechanism

In a conventional lithium-ion battery, the battery cell includes a positive electrode (cathode), a negative electrode (anode), and a non-aqueous liquid electrolyte. Conventionally, graphite is adopted for the anode material, whereas the non-aqueous liquid electrolyte constitutes of $LiPF_6$ salt in suitable organic solvents [4]. The choice of cathode materials greatly influences the overall performance of the Li-ion battery, and, subsequently, a wide number of materials have been explored in this context [4,5,24,25]. In this work, different $LiMPO_4$ materials are considered for Li-ion batteries while keeping the specifications of other constituents of the battery cell unchanged.

During the charging cycle of the battery, a potential is applied across the electrodes leading to Li-ion extraction from the $LiMPO_4$ cathode and subsequent ionic diffusion through electrolytes towards the anode. After reaching the anodes, the Li ions are intercalated between the layers of graphite, as shown in Figure 1a. Next, the Li ions return to the cathode from the anode during the discharging cycle. At the same time, electrons pass from the anode to the cathode through an external circuit acting as a power source for externally attached devices with the battery, as depicted in Figure 1b. The overall efficiency of the battery depends on the discharging cycle, whence the electrode reactions can be given as [4]:

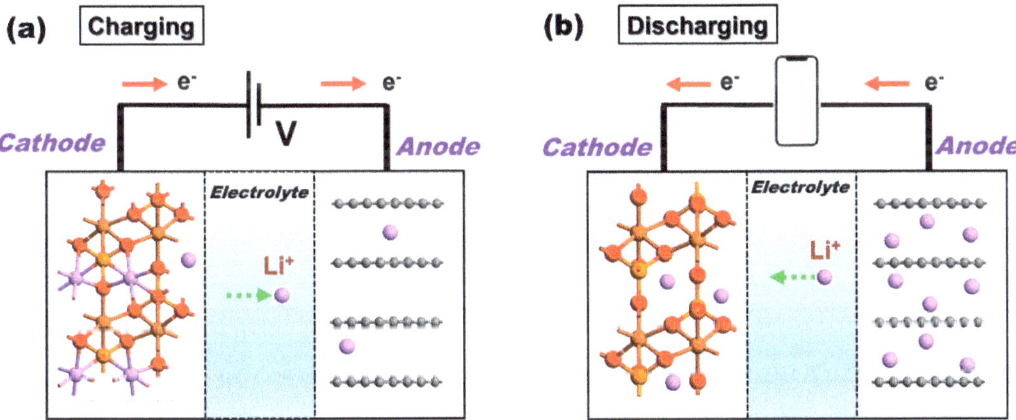

Figure 1. Schematic representation of Li-ion battery (**a**) charging and (**b**) discharging mechanisms.

At anode:

$$C_6Li \rightarrow 6C \text{ (Graphite)} + x \cdot Li^+ + x \cdot e^- \tag{1}$$

At cathode:

$$Li_{1-x}MPO_4 + x \cdot Li^+ + x \cdot e^- \rightarrow LiMPO_4 \tag{2}$$

2.2. Material Specifications

The $LiMPO_4$ family of phosphate materials demonstrates olivine structures and belongs to the orthorhombic Pnma space group [16,25]. In these three-dimensional structures, lithium (Li) is bonded with six oxygen (O) atoms to form LiO_6 octahedra, which share corners with four equivalent MO_6 octahedra and four equivalent PO_4 tetrahedra, and, at the same time, share edges with two equivalent LiO_6 octahedra and two equivalent MO_6 octahedra, two equivalent PO_4 tetrahedra, as depicted in Figure 2. A closer inspection of the crystal structure of $LiMPO_4$ indicates that the LiO_6 octahedra are corner-shared with each other and bridged by PO_4 tetrahedra in parallel to the B-crystallographic axis, which creates a Li channel along the B-axis of $LiMPO_4$. Furthermore, the individual channels are separated along the A-crystallographic axis by the MO_6 octahedra, sharing both the edge and corner with LiO_6 octahedra. Hence, there are two possible Li diffusion pathways through the B- and A-crystallographic axes. However, a number of previous theoretical reports conclusively established that the Li diffusions across individual channels along the B-crystallographic axis of $LiMPO_4$ materials are more favorable compared to inter-channel hopping-dominated Li diffusion from one channel to another along the C-crystallographic axis [4,5,9,14,16,24,26].

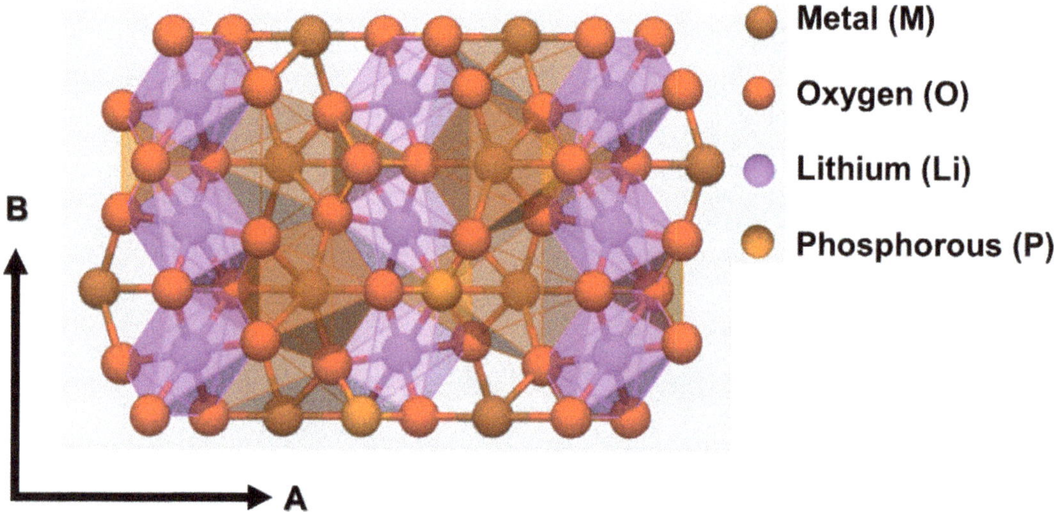

Figure 2. Schematic representation of the generic structure of $LiMPO_4$.

2.3. Simulation Framework

In this work, the first principle calculations are performed using the commercially available Atomixtix Tool Kit (ATK) and Virtual Nano Lab (VNL) simulation packages from Synopsys Quantum Wise [27]. For the first-principle calculations, the linear combination of atomic orbitals (LCAO) PseudoDojo basis sets with the $3 \times 5 \times 6$ Monkhorst-Pack grid for sampling the k points in the Brillouin zone and a density-mesh cutoff energy of 125 Hartree is considered [27]. The LCAO calculator of ATK provides a description of electronic structure using density functional theory (DFT) and the norm-conserving pseudopotentials, where the single-particle wave functions are expanded on the basis of numerical atomic orbitals [27]. It should be noted that the default energy-mesh cutoff significantly varies from one $LiMPO_4/MPO_4$ material system to another. Consequently, a uniform energy-mesh cutoff is considered in this work. The energy-mesh cutoff is taken as 125 Hartree, which is 1.25 times the highest default energy-mesh cutoff value (~100 Hartree) to ensure a proper trade-off between the reliability of the calculation and the overall computational cost. Moreover, the energy-mesh cutoff in the range of 100–200 Hartree in the LCAO basis set of ATK has also been reported previously in the literature [18,28–30].

The unit cells of bulk $LiMPO_4$ (M = Fe, Co, Cr, Mn, and V) and MPO_4 (M = Fe, Co, Cr, Mn, and V) are first geometry optimized using the Generalized Gradient Approximation (GGA) method with the Perdew–Burke–Ernzerhof (PBE) Exchange-Correlation Functional and the Limited-Memory Broyden–Fletcher–Goldfarb–Shanno (LBFGS) Algorithm with a force and pressure tolerance of 0.001 eV/Å and 0.0001 eV/Å^3, respectively [27]. The relaxed unit cells of different $LiMPO_4$ (M = Fe, Co, Cr, Mn, and V) and MPO_4 (M = Fe, Co, Cr, Mn, and V) are demonstrated in Figure 3.

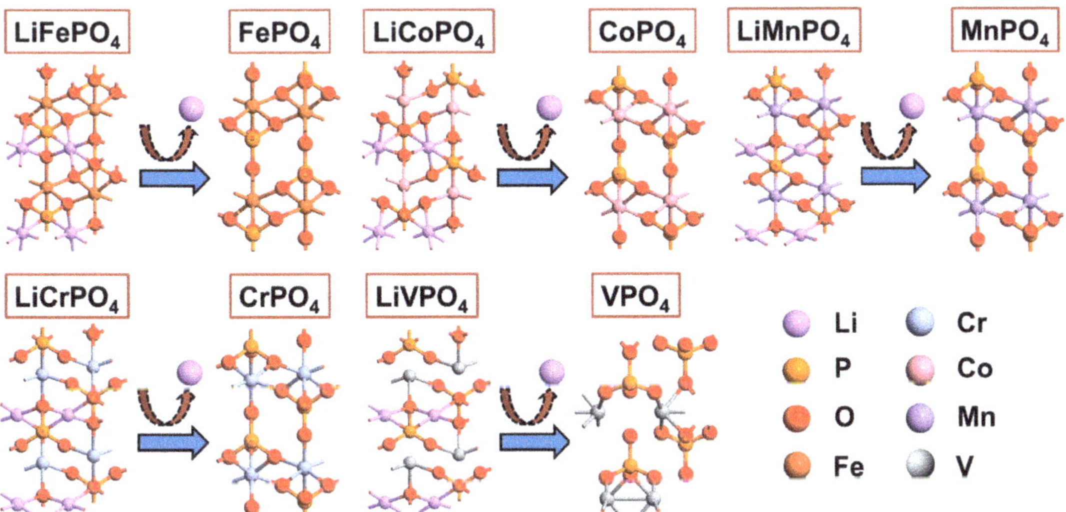

Figure 3. Schematic representation of optimized unit cells of different $LiMPO_4$ (M = Fe, Co, Cr, Mn, and V) and MPO_4 (M = Fe, Co, Cr, Mn, and V).

The relaxed unit cells of $LiMPO_4$ are modified to design initial and final material configurations for Li diffusion across a preferred crystalline axis and are subjected to further geometry optimization following the aforementioned procedure. The initial and final configurations are utilized for calculating Li diffusion across any preferred crystalline axis using the Ab-Initio Molecular Dynamics (AIMD)-based Nudge Elastic Band (NEB) simulation with the climbing image method by considering a 0.5 Å interval across the diffusion path between intermediate images [27]. It should be noted that in this work, the Li diffusion is studied across the Li channel (parallel to the B-crystallographic axis) of different $LiMPO_4$ (M = Fe, Co, Cr, Mn, and V) materials, which is considered as the predominant Li transport direction. In this context, a number of existing reports argued that the GGA-PBE method could not accurately account for the electron-correlation effect, and the Hubbard GGA-PBE + U method improves the accuracy of the Li diffusion calculation for $LiMPO_4$ materials [1,16,24]. The GGA + U method involves an empirical Hubbard U correction term for the individual material systems, where the results of the simulation are significantly influenced by the choice of U. However, choosing a set of appropriate empirical correction terms for each $LiMPO_4$ material poses a significant challenge for the reliability of comparative performance estimations. Consequently, in this work, the GGA-PBE method is considered for both ground-state energy and AIMD-NEB calculations to ensure a uniform simulation platform for reliable qualitative performance estimations of different materials.

2.4. Parameter Definitions

The structural stability of the cathode materials in completely lithiated and completely de-lithiated phases are assessed from the cohesive energies ($E_{cohesive\_lithiated}$ and $E_{cohesive\_de-lithiated}$), which are defined as [31,32]

$$E_{cohesive\_lithiated} = [E_{LiMPO4} - (4 \times E_{Lithium} + 4 \times E_{Phosphorus} + 16 \times E_{Oxygen} + 4 \times E_{Metal})]/28 \tag{3}$$

$$E_{cohesive\_de-lithiated} = [E_{MPO4} - (4 \times E_{Phosphorus} + 16 \times E_{Oxigen} + 4 \times E_{Metal})]/24 \tag{4}$$

where E_{LiMPO4} and E_{MPO4} are the ground-state energies of the $LiMPO_4$ and MPO_4 lattices. Furthermore, the $E_{Lithium}$, $E_{Phosphorus}$, E_{Oxygen}, and E_{Metal} are the ground-state energies of isolated lithium, phosphorus, oxygen, and metal (iron, cobalt, chromium, manganese, and vanadium) atoms, respectively. The higher cohesive energies imply higher structural stability of the $LiMPO_4$ and MPO_4 lattices. The Li intercalation energy ($E_{Intercallation}$) in Li_xMPO_4 is another essential performance matrix for cathode materials in Li-ion battery applications, which is defined between completely lithiated and single Li-ion de-lithiated phases of $LiMPO_4$ lattices as follows [4]:

$$E_{Intercallation} = [E_{Li4MPO4} - E_{Li3MPO4} - E_{Lithium}] \tag{5}$$

where $E_{Li4MPO4}$ and $E_{Li3MPO4}$ are the ground-state energies of the completely lithiated and single Li-ion de-lithiated (averaged over Li position) $LiMPO_4$ lattices.

The Li diffusion within olivine $LiMPO_4$ structures is assessed in terms of intrinsic diffusion coefficient (D), which can be calculated from the atomic scale behavior using the framework of transition state theory under one-dimensional transport as follows [5,9]:

$$D = a^2 \times v^* \times \exp(-E_{act}/K_BT) \tag{6}$$

where a, v^*, E_{act}, K_B, and T are the hopping distance, attempt frequency, diffusion activation energy, Boltzmann constant, and absolute temperature, respectively. In this context, it should be noted that the diffusion activation energy represents the potential barrier that has to be surmounted by the Li-ion during diffusion in $LiMPO_4$. Furthermore, several definitions of attempt frequency in the context of diffusion exist in the literature [33,34]. However, the attempt frequency can be simply obtained by considering every vibration of a diffusing atom as an attempt, which allows defining the attempt frequency as the averaged vibrational frequency [33]. In this work, the diffusion coefficient is calculated at room temperature (T = 300 K), and a standard and uniform attempt frequency of $v^* = 10^{12}$ Hz is considered from literature in the context of $LiMPO_4$ materials [13]. Furthermore, both the hopping distance and activation barriers are calculated from the NEB simulated Li diffusion between the neighboring atomic sites across the specified crystallographic axis.

The equilibrium cell potential (V_{eq}) of a Li battery with a specific $LiMPO_4$ cathode material can be reasonably estimated from the first-principle calculation by ignoring the contributions of vibrational and configurational entropy to the cell potential at room temperature as follows [4]:

$$V_{eq} = [-(E_{LiMPO4} - E_{MPO4} - 4 \times E_{Lithium})]/4q \tag{7}$$

where q is the electronic charge. The equilibrium cell potential in this work is calculated considering the completely lithiated and completely de-lithiated phases of cathode materials [4].

Finally, the open-circuit voltage (V_{OCV}) of the Li-ion battery cell is calculated for different state-of-charge (SOC) levels using the Nernst Potential Equation as follows:

$$V_{OCV} = V_{eq} + (2 \times R \times T/F) \times \ln[SOC/(1 - SOC)] \tag{8}$$

where R is the molar gas constant, and F is Faraday's number.

Equation (8) offers a quantitative estimation of the electrical characteristics of the battery, wherein the V_{OCV} can be determined under the variation of SOC. The V_{OCV}-SOC characteristics indicate the available potential of a battery at different charging states (SOC). In this work, the range of SOC has been considered from 10% to 90% limit, which is a standard for Li-ion batteries for ensuring safe operation, i.e., avoiding the over-charge and over-discharge conditions.

3. Results and Discussion

This work discusses the comparative structural, electrochemical, and electrical properties of different $LiMPO_4$ materials in the following subsections.

3.1. Structural Properties of LiMPO$_4$ (M = Fe, Co, Cr, Mn, and V)

First, the relative structural stability of individual LiMPO$_4$ and MPO$_4$ materials are analyzed from the cohesive energies and illustrated in Figure 4. The results indicate that, generally, the cohesive energy is considerably higher in the delithiated phase compared to the lithaited phase. Furthermore, the highest and lowest structural stability can be found for vanadium- and iron-based olivine phosphates, both in their lithiated and delithiated phases. However, the smaller difference in the cohesive energies between the lithiated and delithiated phases indicates the suitability of Li insertion in the olivine phosphate matrices [35]. Consequently, the LiCoPO$_4$ and LiFePO$_4$ suggest a superior cathode performance, whereas the least favorable performance can be expected from LiVPO$_4$.

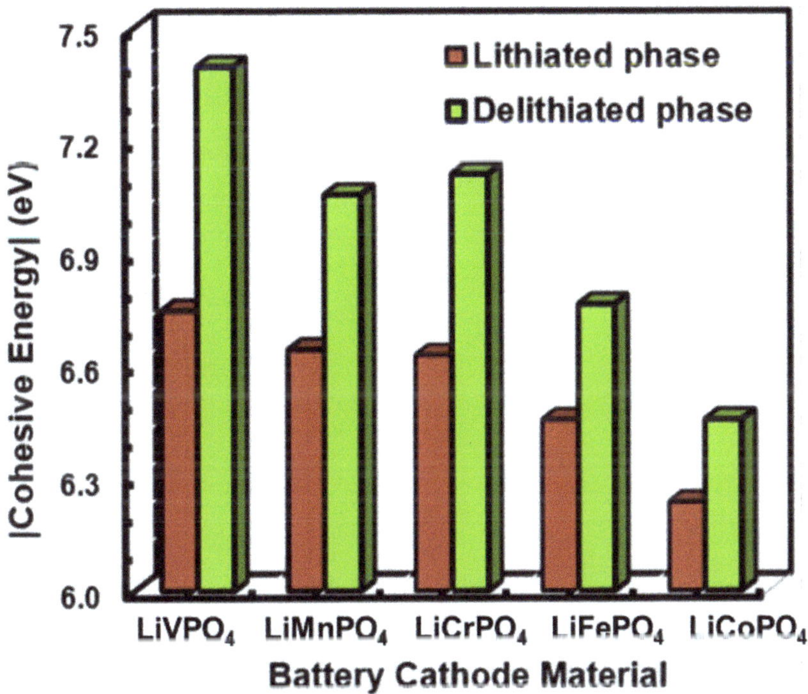

Figure 4. Comparative plots of cohesive energies in lithiated and de-lithiated phases for different LiMPO$_4$ cathode materials.

Next, the structural properties of LiMPO$_4$ materials are estimated in terms of lattice vectors, lithium (Li) oxygen (O) bond length, and Li atomic charge, which is tabulated in Tables 1 and 2.

Table 1. The Lattice Constants of Different Olivine LiMPO$_4$.

Materials	Lattice Constant, a (Å)	Lattice Constant, b (Å)	Lattice Constant, c (Å)
LiFePO$_4$	9.851	5.773	4.672
LiCoPO$_4$	9.885	5.882	4.702
LiMnPO$_4$	9.953	5.830	4.689
LiCrPO$_4$	10.041	5.882	4.707
LiVPO$_4$	10.276	5.981	4.709

Table 2. The Li-O bond length and Mulliken Charge on Li in Olivine $LiMPO_4$.

Materials	Li-O Bond Length Range (Å)	Mulliken Charge on Li (e$^-$)
$LiFePO_4$	2.07–2.13	0.009
$LiCoPO_4$	2.09–2.13	0.009
$LiMnPO_4$	2.07–2.17	0.052
$LiCrPO_4$	2.07–2.20	0.054
$LiVPO_4$	2.09–2.21	0.083

Table 1 indicates that the largest unit-cell lattice vector can be found for $LiVPO_4$. In contrast, the smallest and the most comparable lattice vectors can be found in $LiFePO_4$ and $LiCoPO_4$. This trend can be attributed to the fact that the largest atomic radius of Vanadium (1.71 Å [36]) spatially expands the VO_6 octahedra and thereby increases the volume of the entire unit cell of $LiVPO_4$. In contrast, the smaller atomic radii of cobalt (1.52 Å [36]) and iron (1.56 Å [36]) restrict the volume of CoO_6 and FeO_6 octahedra, which leads to the smaller unit cell lattice vectors in $LiFePO_4$/$LiCoPO_4$. On the other hand, intermediate atomic radii of manganese (1.61 Å [36]) and chromium (1.66 Å [36]) result in intermediate lattice vector values in $LiCrPO_4$ and $LiMnPO_4$ compared to $LiVPO_4$ and $LiFePO_4$/$LiCoPO_4$, as depicted in Table 1. Such distinctly different lattice vectors in $LiMPO_4$ are expected to significantly influence the octahedral structure of LiO_6 and thereby the Li-O bond strengths in different $LiMPO_4$ materials, which are illustrated in Table 2. The larger values of unit cell lattice vector expand the LiO_6 octahedra, and a relatively larger bond length (2.21 Å) can be found in $LiVPO_4$. In contrast, the smaller unit cell lattice vector leads to relatively smaller bond lengths (2.13 Å) in $LiFePO_4$ and $LiCoPO_4$. Interestingly, the Mulliken charge accumulation on the Li atom strongly depends on the lattice vector and Li-O bond length in LiO_6 octahedra, where an expanded bond length increases the Mulliken charge on Li, suggesting a weaker covalent bond formation. In this context, $LiVPO_4$ exhibits the weakest Li-O bond strength, whereas the strongest bond strength can be found in $LiFePO_4$ and $LiCoPO_4$. In essence, the structural analysis offers crucial insight into the Li-O bond strength of LiO_6 octahedra in different olivine $LiMPO_4$ materials, which can be exploited for analyzing the comparative electrochemical properties of $LiMPO_4$. It should be noted that the experimentally observed lattice constants of $LiFePO_4$ (a = 10.35 Å, b = 6.02 Å, c = 4.70 Å) [37], $LiCoPO_4$ (a = 10.20 Å, b = 5.92 Å, c = 4.70 Å) [38], and $LiMnPO_4$ (a = 10.44 Å, b = 6.09 Å, c = 4.75 Å) [39] are in reasonable agreement with theoretically calculated values in this work.

3.2. Electrochemical Properties of $LiMPO_4$ (M = Fe, Co, Cr, Mn, and V)

The electrochemical properties of $LiMPO_4$-based cathode materials are appreciated from the Li intercalation energies in the lattice, as well as the Li diffusion process along the most favorable diffusion path, which is across the Li channels parallel to the B-crystallographic axis.

First, the relative Li intercalation energies for different cathode materials are depicted in Figure 5. Subsequently, a notably small (<3 eV) and large (>4 eV) Li intercalation energy can be found in $LiVPO_4$ and $LiFePO_4$/$LiCoPO_4$, respectively. This trend can satisfactorily be co-related with the Li-O bond strength analysis that was performed in the previous subsection. Specifically, the relatively stronger covalent bonding between Li and O in $LiFePO_4$ and $LiCoPO_4$ makes it less favorable to remove one Li ion from these lattices, resulting in comparatively larger intercalation energies, as discussed previously. In contrast, the relatively weaker covalent Li-O bonding ensures a smaller Li intercalation energy in $LiVPO_4$. In this line, the intermediate Li-O bond strength results in intermediate intercalation energies in $LiMnPO_4$ and $LiCrPO_4$. It should be noted that the calculated Li intercalation energy of 4.05 eV in $LiFePO_4$ is slightly overestimated compared to the experimentally observed 3.50 eV value [5]. However, the theoretical prediction can be further improved by adopting more reliable DFT methods like the GGA-PBE + U [4].

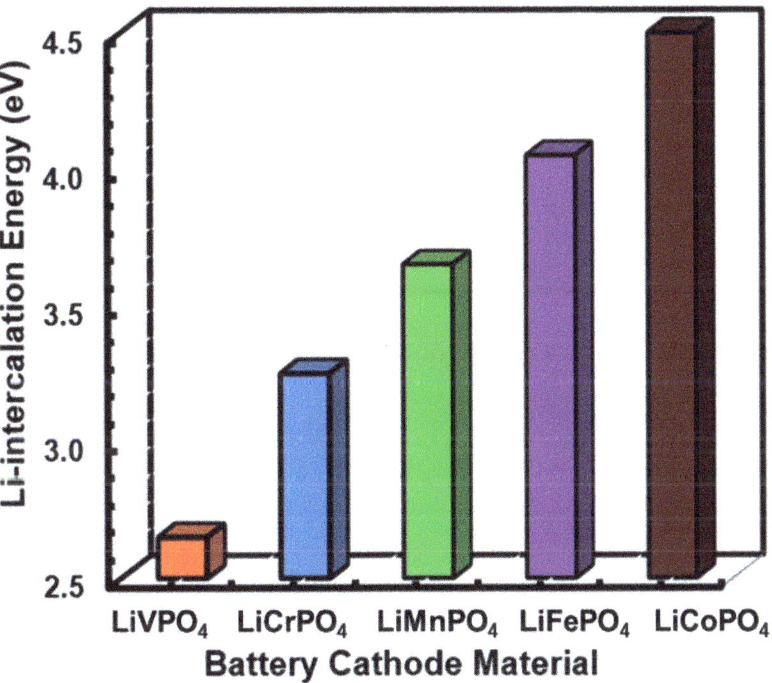

Figure 5. Comparative plots of lithium intercalation energies defined between completely lithiated and single Li-ion de-lithiated phases of the LiMPO$_4$ lattice.

Next, the Li diffusion energy profiles and corresponding Li migration pathways between two neighboring Li sites in different LiMPO$_4$ are analyzed and are depicted in Figure 6. Figure 6 indicates that, in general, the Li diffusion between two neighboring Li-sites followed a curved migration pathway across the Li channel (as depicted in Figure 2). The results predict Li diffusion is essentially one-dimensional with a zigzag pathway across the AB-crystallographic plane of the lattices. It should be noted that the shapes of Li diffusion trajectories and the Li diffusion barriers in different LiMPO$_4$ is consistent with the literature [4,5,9,14,16,24,26]. From the result, it is apparent that the atomic and chemical environment of the Li diffusion path significantly influences the Li diffusion activation barrier and hopping distance between two neighboring Li atom positions. Consequently, the structural configurations of different LiMPO$_4$, which represent the maximum Li diffusion activation barrier have also been considered in Figure 6. In the structural configuration leading to the maximum Li diffusion activation barrier height, the Li forms three bonds with the one neighboring metal atom and two neighboring oxygen atoms, which is distinctly different from any other structural configurations in Li diffusion path. Interestingly, it has been found that the Li-O bond length remains comparable, i.e., in the range of 1.85 Å–1.87 Å and 1.89 Å–1.90 Å, irrespective of the particular olivine structure. On the other hand, the Li-M bond length more noticeably varies within the range of 2.53 Å–2.63 Å for different LiMPO$_4$. This suggests that the M atom significantly influences the Li diffusion process in each LiMPO$_4$ material. Correspondingly, a distinctly different Li diffusion activation barrier height (across the B-crystallographic axis) and energy-distance curvature of the diffusion barrier can be found in each olivine structure.

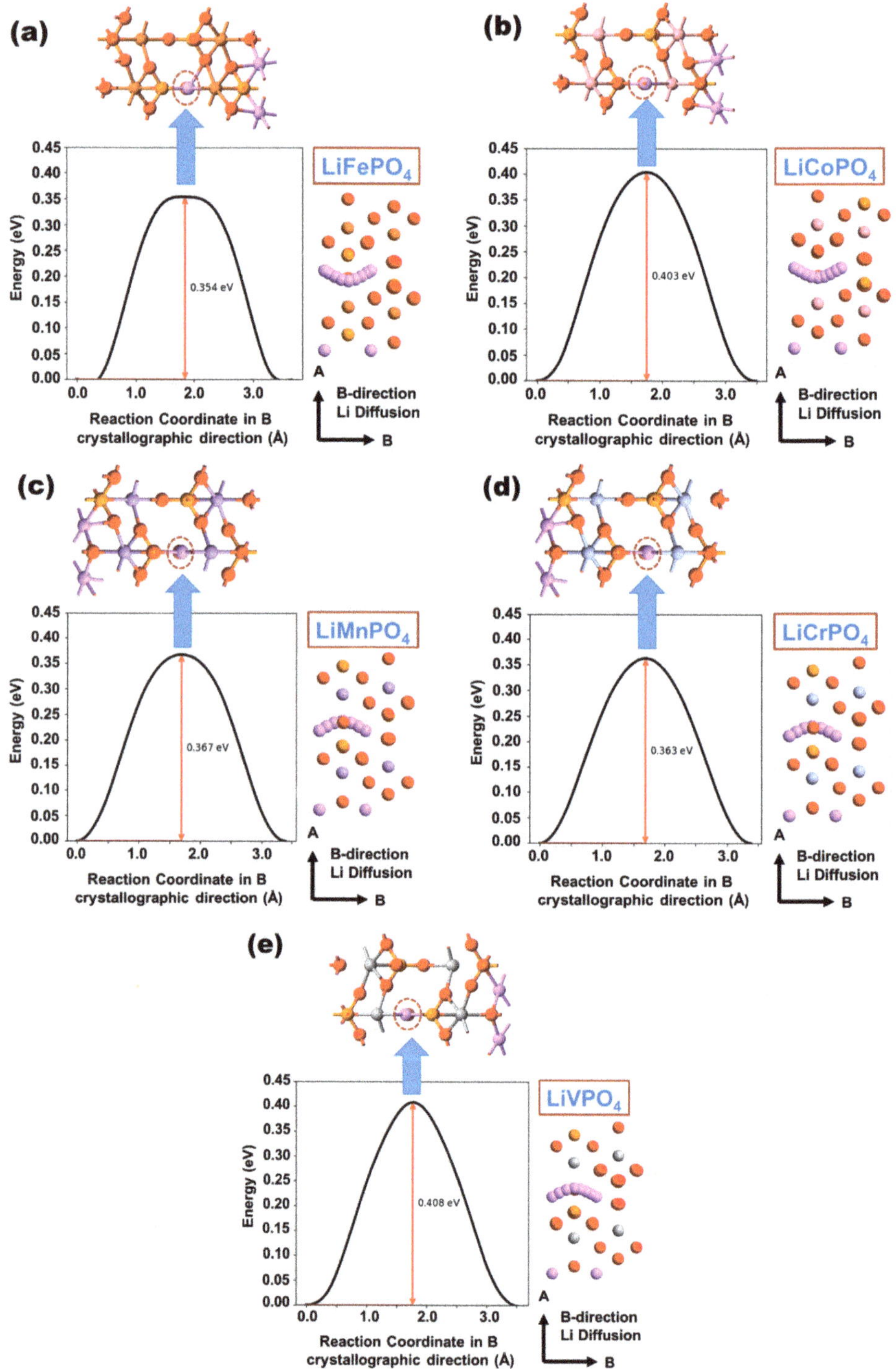

Figure 6. Plots of Li diffusion activation barrier along B-crystallographic axis and corresponding overlay image of Li diffusion for (**a**) LiFePO$_4$, (**b**) LiCoPO$_4$, (**c**) LiMnPO$_4$, (**d**) LiCrPO$_4$, and (**e**) LiVPO$_4$.

The calculated Li activation energy of 0.35 eV in LiFePO$_4$ is in reasonable agreement with the previously GGA-PBE-based theoretically reported values in the range of 0.27–0.29 eV [5,9]. In contrast, the theoretically calculated value is notably underestimated compared to the experimentally observed range of 0.55–0.65 eV [40]. In this context, the adoption of the GGA-PBE + U method can further improve the theoretical estimation of activation energy compared to the GGA-PBE. However, it is worth mentioning that compared to experimental reports, a similar underestimation of activation energy has also been observed in previous theoretical reports, which is often correlated with the presence of non-idealities and their subsequent influence on Li-atom diffusion in the LiMPO$_4$ structures [4].

Next, the Li atom diffusion across the B-crystallographic axis is quantified in terms of diffusion activation barrier height, hopping distance, and intrinsic diffusion coefficient, and are depicted in Figure 7a–c, respectively. The LiFePO$_4$ exhibits the lowest diffusion activation barrier height, which is slightly lesser than the barrier height of LiCrPO$_4$ and LiMnPO$_4$. Moreover, notably larger diffusion activation barrier height can be found in LiCoPO$_4$ and LiVPO$_4$. In contrast, the hopping distance indicates a different trend, where the smallest and largest hopping distance B-crystallographic axes can be found in LiCrPO$_4$ and LiFePO$_4$, respectively. However, the intrinsic diffusion coefficient (within the range of 0.354 eV–0.408 eV) exhibits an exponential dependence over activation barrier height, as depicted in Equation (6). Moreover, a relatively smaller range of variation in the hopping distances (3.55 Å–3.86 Å) can be found in different LiMPO$_4$. Consequently, the activation barrier height effectively dominates over the diffusion coefficients. Subsequently, the smallest and largest intrinsic diffusion coefficient can be found in LiVPO$_4$ and LiFePO$_4$, where nearly an order of magnitude difference (1.97×10^{-8} cm^2/s–1.55×10^{-7} cm^2/s) can be found between these two olivine structures. Moreover, the LiCrPO$_4$ and LiMnPO$_4$ also demonstrate slightly smaller diffusion coefficients (1.03×10^{-8} cm^2/s and 8.94×10^{-8} cm^2/s) compared to LiFePO$_4$, whereas LiCoPO$_4$ exhibits a diffusion coefficient (2.60×10^{-8} cm^2/s) that is comparable to LiVPO$_4$.

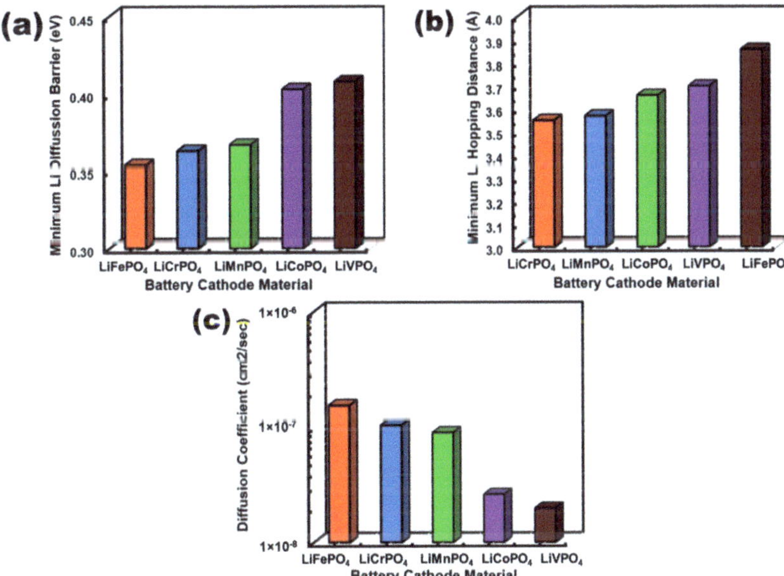

Figure 7. Comparative plots of (**a**) minimum activation barrier, (**b**) minimum hopping distance, and (**c**) intrinsic diffusion co-efficient for Li diffusion across Li channel (parallel to B-crystallographic axis) for different LiMPO$_4$ cathode materials.

3.3. Electrical Characteristics of LiMPO$_4$ (M = Fe, Co, Cr, Mn, and V)

In this work, five different and compatible metal (M = Fe, Co, Cr, Mn, and V) based samples of Li-ion battery cells have been simulated to analyze their electrical characteristics. It is important to analyze the battery's electrical characteristics [41] for further estimation of their performance while interfacing with different useful applications. As mentioned in Equation (7), the Nernst potential equation provides the battery open-circuit voltage (V_{OCV}) as a function of the equilibrium potential (V_{eq}). The V_{eq} significantly varies with the cathode material. Therefore, in this paper, to examine the electrical characteristics of 5 different olivine phosphate material-based cathodes, the V_{eq} has been presented in Figure 8a. Further, based on the principle of the Nernst equation, the V_{OCV} profile over the range of SOC (10–90%) has been determined for different LiMPO$_4$ cathode materials, as shown in Figure 8b. It is to be noted from the simulation results that the V_{eq} and V_{OCV} of LiCoPO$_4$ cell is the highest (V_{OCV} around 4.9 V/cell), and that of LiVPO$_4$ attains the lowest value (V_{OCV} around 2.7 V/cell). In the presence of Vanadium, the Li diffusion barrier increases, resulting in the battery voltage attaining a lesser value. Interestingly, it is found in Figure 8 that for the LiFePO$_4$ sample, the V_{OCV} attains a value of 4.6 V/cell, which is close to the maximum value possessed by LiCoPO$_4$. Also, the diffusion barrier for LiFePO$_4$ is lesser compared to that of the other LiMPO$_4$ battery samples. Therefore, due to these properties of LiFePO$_4$, it has been reasonably popular among the other LiMPO$_4$ batteries in electrical power system applications where high energy density and fast charge-discharge capacity are concerned. It should be noted that the experimentally demonstrated cell voltage of LiFePO$_4$ typically lies between 3.5 V (10% SOC) to 4.2 V (90% SOC) [42]. However, the theoretically calculated V_{OCV} exhibits a range of 4.51 V (10% SOC) to 4.72 V (90% SOC). This overestimation of V_{OCV} in this work can be attributed to both the choice of the GGA-PBE DFT method instead of GGA-PBE + U as well as inherent theoretical limitations in accounting for practical factors like self-discharge drop.

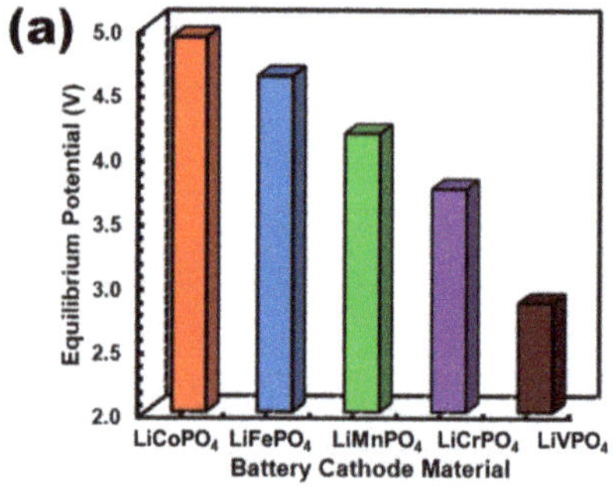

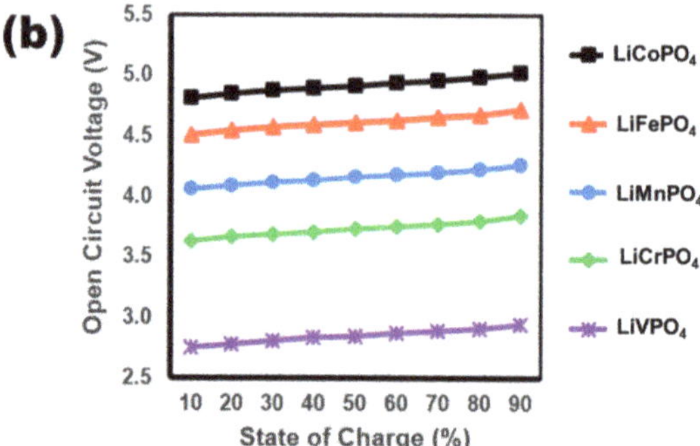

Figure 8. Comparative plots of (**a**) equilibrium potential and (**b**) open-circuit voltage as a function of state of charge for different LiMPO$_4$ cathode materials.

4. Conclusions

The work presents an extensive first principle calculation based on theoretical investigations of the comparative electrochemical and electrical properties of different olivine lithium metal phosphates for battery-storage applications. The results indicate that the presence of iron (Fe), cobalt (Co), manganese (Mn), chromium (Cr), and vanadium (V) atoms in the olivine phosphate structure distinctly influence structural stabilities and Li-atom diffusion across the Li channel, as well as battery open-circuit potential. It has been found that the lattice vector and, thereby, the Li-O bond specifications in LiO_6 octahedral primarily determine the Li intercalation energies in different $LiMPO_4$ materials. Furthermore, the chemical-bond formation of Li with specific metal atoms in the lattice during Li-atom diffusion effectively determines the lithium activation barrier height and curvature. Typically, $LiVPO_4$ demonstrates a notably smaller Li intercalation energy of 2.65 eV compared to $LiFePO_4$ and $LiCoPO_4$, which exhibits intercalation energies in the range of 4.00–4.50 eV. Specifically, despite its lowest intercalation energy, the potential of $LiVPO_4$ is compromised by its notably smaller Li intrinsic diffusion coefficient as well as low equilibrium voltage. In contrast, the notably higher Li intrinsic diffusion coefficient of 1.55×10^{-7} cm^2/s and equilibrium voltage of 4.62 V can be achieved in $LiFePO_4$, in contrast to the Li intrinsic diffusion coefficient in the range of 1.97×10^{-8}–2.60×10^{-8} cm^2/s in $LiVPO_4$ and $LiCoPO_4$ and equilibrium voltage of 2.85 V in $LiVPO_4$. However, the relatively higher intercalation energy remains a limiting factor in $LiFePO_4$. On the other hand, except for the higher equilibrium voltage of 4.92 V in $LiCoPO_4$, both the higher intercalation energy and lower Li atom diffusivity present a severe bottleneck for this material for cathode design. However, both $LiCrPO_4$ and $LiMnPO_4$ represent a reasonable trade-off between intercalation energy (in the range of 3.25–3.65 eV), intrinsic diffusion coefficient (in the range of 1.03×10^{-7}–8.94×10^{-8} cm^2/s), and equilibrium voltage (in the range of 3.74–4.16 V). The trend suggests that none of the studied $LiMPO_4$ is simultaneously co-optimizing both intercalation energy and diffusion coefficient to achieve an unequivocally faster charging/discharging in the cathode. Subsequently, a dedicated material engineering approach is necessary to optimize the overall charge-discharge property of $LiMPO_4$ materials for Li-ion battery storage.

Author Contributions: Conceptualization, S.K. and A.B.; methodology, S.K. and A.B.; software, S.K. and N.B.; validation, A.B., N.B. and A.G.; formal analysis, S.K. and A.B.; investigation, N.B. and A.B.; resources, S.K.; data curation, A.B. and A.G.; writing—original draft preparation, S.K. and A.B.; writing—review and editing, A.G.; visualization, A.G.; supervision, S.K. and A.G.; project administration, A.G.; funding acquisition, S.K. All authors have read and agreed to the published version of the manuscript.

Funding: This work has been supported in part by an Outstanding Potential for Excellence in Research and Academics (OPERA) Grant by BITS-Pilani, India (Grant No. BITS/OPERA/0765) and Start-up Research Grant (SRG) by DST-SERB (Grant No. SRG/2020/000547) awarded to Sayan Kanungo.

Data Availability Statement. The raw/processed data required to reproduce the above findings cannot be shared at this time, as the data also forms part of an ongoing study.

Conflicts of Interest: The authors declare no conflict of interest.

References

1. Dunn, B.; Kamath, H.; Tarascon, J.M. Electrical Energy Storage for the Grid: A Battery of Choices. *Science* **2011**, *334*, 928–935. [CrossRef] [PubMed]
2. Nitta, N.; Wu, F.; Lee, J.T.; Yushin, G. Li-ion battery materials: Present and future. *Mater. Today* **2015**, *18*, 252–264. [CrossRef]
3. Kim, T.; Song, W.; Son, D.Y.; Ono, L.K.; Qi, Y. Lithium-ion batteries: Outlook on present, future, and hybridized technologies. *J. Mater. Chem. A* **2019**, *7*, 2942–2964. [CrossRef]
4. Islam, M.S.; Fisher, C.A. Lithium and sodium battery cathode materials: Computational insights into voltage, diffusion and nanostructural properties. *Chem. Soc. Rev.* **2014**, *43*, 185–204. [CrossRef] [PubMed]
5. Nakayama, M.; Yamada, S.; Jalem, R.; Kasuga, T. Density functional studies of olivine-type $LiFePO_4$ and $NaFePO_4$ as positive electrode materials for rechargeable lithium and sodium ion batteries. *Solid State Ion.* **2016**, *286*, 40–44. [CrossRef]

6. Chen, Y.; Sang, M.; Jiang, W.; Wang, Y.; Zou, Y.; Lu, C.; Ma, Z. Fracture predictions based on a coupled chemo-mechanical model with strain gradient plasticity theory for film electrodes of Li-ion batteries. *Eng. Fract. Mech.* **2021**, *253*, 107866. [CrossRef]

7. Ma, Z.S.; Xie, Z.C.; Wang, Y.; Zhang, P.P.; Pan, Y.; Zhou, Y.C.; Lu, C. Failure modes of hollow core-shell structural active materials during the lithiation-delithiation process. *J. Power Sources* **2015**, *290*, 114–122. [CrossRef]

8. Chen, C.F.; Barai, P.; Mukherjee, P.P. An overview of degradation phenomena modelling in lithium-ion battery electrodes. *Curr. Opin. Chem. Eng.* **2016**, *13*, 82–90. [CrossRef]

9. Ma, Z.; Wu, H.; Wang, Y.; Pan, Y.; Lu, C. An electrochemical-irradiated plasticity model for metallic electrodes in lithium-ion batteries. *Int. J. Plast.* **2017**, *88*, 188–203. [CrossRef]

10. Tian, Y.; Zeng, G.; Rutt, A.; Shi, T.; Kim, H.; Wang, J.; Koettgen, J.; Sun, Y.; Ouyang, B.; Chen, T.; et al. Promises and challenges of next-generation "beyond Li-ion" batteries for electric vehicles and grid decarbonization. *Chem. Rev.* **2020**, *121*, 1623–1669. [CrossRef]

11. Urban, A.; Seo, D.H.; Ceder, G. Computational understanding of Li-ion batteries. *NPJ Comput. Mater.* **2016**, *2*, 16002. [CrossRef]

12. Padhi, A.; Nanjundaswamy, K.S.; Goodenough, J.B. Phospho-Olivines as Positive-Electrode Materials for Rechargeable Lithium Batteries. *J. Electrochem. Soc.* **1997**, *144*, 1188–1194. [CrossRef]

13. Morgan, D.; Ven, A.; Ceder, G. Li conductivity in Li_xMPO_4 (M = Mn, Fe, Co, Ni) olivine materials. *Electrochem. Solid-State Lett.* **2004**, *7*, 4032–4037. [CrossRef]

14. Islam, M.S.; Driscoll, D.J.; Fisher, C.A.; Slater, P.R. Atomic-scale investigation of defects, dopants, and lithium transport in the $LiFePO_4$ olivine-type battery material. *Chem. Mater.* **2005**, *17*, 5085–5092. [CrossRef]

15. Ficher, C.A.J.; Prieto, V.M.H.; Islam, M.S. Lithium battery materials $LiMPO_4$ (M= Mn, Fe, Co and Ni): Insights into defect association, transport mechanisms and doping behaviour. *Chem. Mater.* **2008**, *20*, 5907–5915. [CrossRef]

16. Yang, J.; Tse, J.S. Li ion diffusion mechanisms in $LiFePO_4$: An ab initio molecular dynamics study. *J. Phys. Chem. A* **2011**, *115*, 13045–13049. [CrossRef]

17. Dathar, G.K.P.; Sheppard, D.; Stevenson, K.J.; Henkelman, G. Calculations of Li-ion diffusion in olivine phosphates. *Chem. Mater.* **2011**, *23*, 4032–4037. [CrossRef]

18. Wu, K.C.; Hsieh, C.H.; Chang, B.K. First principles calculations of lithium diffusion near the surface and in the bulk of Fe-doped $LiCoPO_4$. *Phys. Chem. Chem. Phys.* **2022**, *24*, 1147–1155. [CrossRef]

19. Oukakhou, S.; Maymoun, M.; Elomrani, A.; Sbiaai, K.; Hasnaoui, A. Enhancing the Electrochemical Performance of Olivine $LiMnPO_4$ as Cathode Materials for Li-ion Batteries by Ni-Fe codoping. *ACS Appl. Energy Mater.* **2022**. [CrossRef]

20. Ping, L.Z.; Ming, Z.Y.; Jun, Z.Y. First-principles studies of Mn-doped $LiCoPO_4$. *Chin. Phys. B* **2011**, *20*, 018201. [CrossRef]

21. Zhang, H.; Gong, Y.; Li, J.; Du, K.; Cao, Y.; Li, J. Selecting substituent element for $LiMnPO_4$ cathode materials combined with density functional theory (DFT) calculations and experiments. *J. Alloys Compd.* **2019**, *793*, 360–368. [CrossRef]

22. Yang, L.; Deng, W.; Xu, W.; Tian, Y.; Wang, A.; Wang, B.; Zou, G.; Hou, H.; Deng, W.; Ji, X. Olivine $LiMn_xFe_{1-x}Po_4$ cathode materials for lithium ion batteries: Restricted factors of rate performances. *J. Mater. Chem. A* **2021**, *9*, 14214–14232. [CrossRef]

23. Kuganathan, N.; Chroneos, A. Formation, doping, and lithium incorporation in $LiFePO_4$. *AIP Adv.* **2022**, *12*, 045225. [CrossRef]

24. Alfaruqi, M.H.; Kim, S.; Park, S.; Lee, S.; Lee, J.; Hwang, J.Y.; Sun, Y.K.; Kim, J. Density functional theory investigation of mixed transition metals in olivine and tavorite cathode materials for Li-ion batteries. *ACS Appl. Mater. Interfaces* **2020**, *12*, 16376–16386. [CrossRef]

25. Molenda, J.; Kulka, A.; Milewska, A.; Zając, W.; Świerczek, K. Structural, transport and electrochemical properties of $LiFePO_4$ substituted in lithium and iron sublattices (Al, Zr, W, Mn, Co and Ni). *Materials* **2013**, *6*, 1656–1687. [CrossRef] [PubMed]

26. Ouyang, C.; Shi, S.; Wang, Z.; Huang, X.; Chen, L. First-principles study of Li ion diffusion in $LiFePO_4$. *Phys. Rev. B* **2004**, *69*, 104303. [CrossRef]

27. "Quantum ATK". Available online: https://www.synopsys.com/silicon/quantumatk.html (accessed on 1 January 2022).

28. Berdiyorov, G.R.; Madjet, M.E.; Mohmoud, K.A. First-Principles Density Functional Theory Calculations of Bilayer Membranes Heterostructures of $Ti_3C_2T_2$ (MXene)/Graphene and AgNPs. *Membranes* **2021**, *11*, 543. [CrossRef] [PubMed]

29. Stanley, J.C.; Mayr, F.; Gagliardi, A. Machine Learning Stability and Bandgaps of Lead-Free Perovskites for Photovoltaics. *Adv. Theory Simul.* **2020**, *3*, 1900178. [CrossRef]

30. Palepu, J.; Anand, P.P.; Parshi, P.; Jain, V.; Tiwari, A.; Bhattacharya, S.; Chakraborthy, S.; Kanungo, S. Compartitive analysis of strain engineering on the electronic properties of homogenous and heterostructure bilayers od MoX_2 (X = S, Se, Te). *Micro Nanostruct.* **2022**, *168*, 207334. [CrossRef]

31. Tiwari, A.; Palepu, J.; Choudhury, A.; Bhattacharya, S.; Kanungo, S. Theoretical analysis of the NH_3, NO, and NO_2 adsorption on boron-nitrogen and boron-phosphorous co-doped monolayer graphene-A comparative study. *FlatChem* **2022**, *34*, 100392. [CrossRef]

32. Yang, C.C.; Li, S. Cohesive energy: The intrinsic dominant of thermal stability and structural evolution in Sn from size scales of bulk to dimer. *J. Phys. Chem. C* **2009**, *113*, 14207–14212. [CrossRef]

33. Klerk, N.J.J.de.; Maas, E.V.D.; Wagemaker, M. Analysis of Diffusion in Solid-State Electrolytes through MD Simulations, improvement of the Li-Ion Conductivity in β-Li_3PS_4 as an Example. *ACS Appl. Energy Mater.* **2018**, *1*, 3230–3242. [CrossRef] [PubMed]

34. Koettgen, J.; Zacherle, T.; Grieshammer, S.; Martin, M. Ab initio calculation of the attempt frequency of oxygen diffusion in pure and samarium doped ceria. *Phys. Chem. Chem. Phys.* **2017**, *19*, 9957–9973. [CrossRef] [PubMed]

35. Gardiner, G.R.; Islam, M.S. Anti-site defects and ion migration in the LiFe$_{0.5}$Mn$_{0.5}$PO$_4$ mixed-metal cathode material. *Chem. Mater.* **2010**, *22*, 1242–1248. [CrossRef]
36. Clementi, E.; Raimondi, D.L.; Reinhardt, W.P. Atomic Screening Constants from SCF Functions. II. Atoms with 37 To 86 Electrons. *J. Chem. Phys.* **2004**, *47*, 1300. [CrossRef]
37. Fan, C.L.; Lin, C.R.; Han, S.C.; Chen, J.; Li, L.F.; Bai, Y.M.; Zhang, K.H.; Zhange, X. Structure, conductive mechanisn and electrochemical of LiFePO$_4$/C doped with Mg^{2+}, Cr^{3+} and Ti^{4+} by a carbothermal reduction method. *New J. Chem.* **2014**, *38*, 795–801. [CrossRef]
38. Strobridge, F.C.; Clement, R.J.; Leskes, M.; Middlemiss, D.S.; Borekwicz, O.J.; Wiaderek, K.M.; Chapman, K.W. Identifying the structure of the Intermediate, Li$_{2/3}$CoPO$_4$, formed during Electrochemical cycling of LiCoPO$_4$. *Chem. Mater.* **2014**, *26*, 6193–6205. [CrossRef]
39. Sgroi, M.F.; Lazzaroni, R.; Beljonne, D.; Pullini, D. Doping LiMnPO$_4$ with Cobalt and Nickel: A First Principle study. *Batteries* **2017**, *3*, 11. [CrossRef]
40. Satyavani, T.V.S.L.; Kiran, B.R.; Kumar, V.R.; Kumar, A.S.; Naidu, S.V. Effect on particle size on dc conductivity, activation energy and diffusion coefficient of lithium iron phosphate in Li-ion cells. *Eng. Sci. Technol. Int. J.* **2016**, *19*, 40–44. [CrossRef]
41. Chen, S.; Zhao, Z.; Gu, X. The Research on Characteristics of Li-NiMnCo Lithium-Ion Batteries in Electric Vehicles. *J. Energy* **2020**, *2020*, 3721047. [CrossRef]
42. Zhang, R.; Xia, B.; Li, B.; Cao, L.; Lai, Y.; Zheng, W.; Wang, H.; Wang, W.; Wang, M. A Study on the Open Circuit Voltage and State of the Charge Characterization of High Capacity Lithium-Ion Battery Under Different Temperatures. *Energies* **2018**, *11*, 2408. [CrossRef]

nanomaterials

Article

Biochar-Mediated Zirconium Ferrite Nanocomposites for Tartrazine Dye Removal from Textile Wastewater

Shazia Perveen [1], Raziya Nadeem [1], Farhat Nosheen [2], Muhammad Imran Asjad [3], Jan Awrejcewicz [4] and Tauseef Anwar [5,*]

[1] Department of Chemistry, University of Agriculture, Faisalabad 38000, Pakistan
[2] Institute of Chemistry, University of Sargodha, Sargodha 40100, Pakistan
[3] Department of Mathematics, University of Management and Technology, Lahore 54000, Pakistan
[4] Department of Automation, Biomechanics and Mechatronics, Lodz University of Technology, Żeromskiego 116, 90-924 Łódź, Poland
[5] Department of Physics, The University of Lahore, Lahore 54000, Pakistan
* Correspondence: tauseef.anwar@phys.uol.edu.pk

Abstract: To meet the current challenges concerning the removal of dyes from wastewater, an environmentally friendly and efficient treatment technology is urgently needed. The recalcitrant, noxious, carcinogenic and mutagenic compound dyes are a threat to ecology and its removal from textile wastewater is challenge in the current world. Herein, biochar-mediated zirconium ferrite nanocomposites (BC-ZrFe$_2$O$_5$ NCs) were fabricated with wheat straw-derived biochar and applied for the adsorptive elimination of Tartrazine dye from textile wastewater. The optical and structural properties of synthesized BC-ZrFe$_2$O$_5$ NCs were characterized via UV/Vis spectroscopy, Fourier transform Infra-red (FTIR), X-Ray diffraction (XRD), Energy dispersive R-Ray (EDX) and Scanning electron microscopy (SEM). The batch modes experiments were executed to explore sorption capacity of BC-ZrFe$_2$O$_5$ NCs at varying operative conditions, i.e., pH, temperature, contact time, initial dye concentrations and adsorbent dose. BC-ZrFe$_2$O$_5$ NCs exhibited the highest sorption efficiency among all adsorbents (wheat straw biomass (WSBM), wheat straw biochar (WSBC) and BC-ZrFe$_2$O$_5$ NCs), having an adsorption capacity of (mg g^{-1}) 53.64 ± 0.23, 79.49 ± 0.21 and 89.22 ± 0.31, respectively, for Tartrazine dye at optimum conditions of environmental variables: pH 2, dose rate 0.05 g, temperature 303 K, time of contact 360 min and concentration 100 mg L^{-1}. For the optimization of process variables, response surface methodology (RSM) was employed. In order to study the kinetics and the mechanism of the adsorption process, kinetic and equilibrium mathematical models were used, and results revealed 2nd order kinetics and a multilayer chemisorption mechanism due to complexation of hydroxyl, Fe and Zr with dyes functional groups. The nanocomposites were also recovered in five cycles without significant loss (89 to 63%) in adsorption efficacy. This research work provides insight into the fabrication of nanoadsorbents for the efficient adsorption of Tartrazine dye, which can also be employed for practical engineering applications on an industrial scale as efficient and cost effective materials.

Keywords: adsorption; biochar; BC-ZrFe$_2$O$_5$ NCs; batch study; response surface methodology

Citation: Perveen, S.; Nadeem, R.; Nosheen, F.; Asjad, M.I.; Awrejcewicz, J.; Anwar, T. Biochar-Mediated Zirconium Ferrite Nanocomposites for Tartrazine Dye Removal from Textile Wastewater. *Nanomaterials* **2022**, *12*, 2828. https://doi.org/10.3390/nano12162828

Academic Editor: George Z. Kyzas

Received: 14 July 2022
Accepted: 9 August 2022
Published: 17 August 2022

Publisher's Note: MDPI stays neutral with regard to jurisdictional claims in published maps and institutional affiliations.

1. Introduction

Environmental pollution has been increasing in the current global scenario owing to different toxicants discharged from such industrial sectors as textile, fertilizer, cosmetics and pharmaceutical. The textile industries have been found to be the largest generator of colored effluents because of greater consumption of water during various processing operations. However, 20% of the dye is lost during the process of dying due to poor levels of dye fixation to fiber [1]. It is projected that approximately 7×10^5 tons of dyes are being produced by textile industries annually. The synthetic dyes extensively used in textile industries are non-biodegradable because of their complex aromatic structure. Tartrazine dye, whose International Union of Pure and Applied Chemistry (IUPAC) name is trisodium 1-(4-sulfonatophenyl)-4-(4-sulfonatophenylazo)-5-pyrazolone-3-carboxylate (Figure 1), is a typical synthetic, water-soluble anionic dye. Tartrazine dye has a chromophore (-N=N-) entity in its molecular structure, and is an example of azo dye that is utilized in the textile industry to color wool, cotton, polyamide and silk [2].

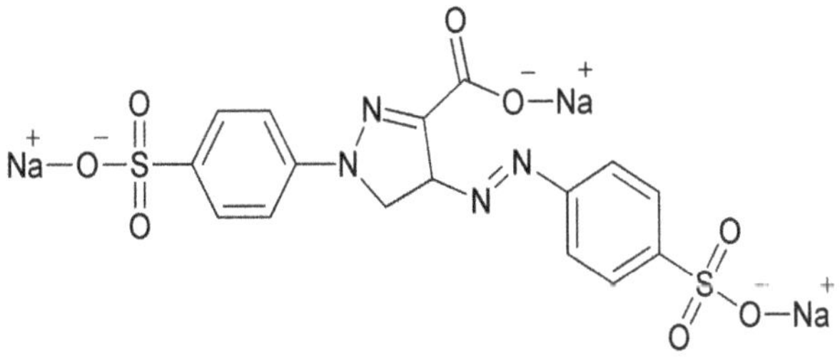

Figure 1. Tartrazine dye structure.

The chromophoric azo group in dye molecules is known to pose a serious threat to the biosphere when discharged with waste effluents. The potential health risks associated with dyes are hyperactivity, allergic reactions particularly among asthmatics and those with aspirin intolerance, as well as having a bio-accumulative, carcinogenic and mutagenic nature [3]. Once dyes enter into the environment they are rarely removed; hence, their discharge into water reservoirs results in the disruption of aquatic biota, reduction in sunlight penetration and the retardation of photosynthetic potential of some organisms [4]. Dyes also exhibit toxicity in fish, with a lethal 50% concentration (LC_{50}) in test animals of more than 100 mg g^{-1}. Therefore, the treatment of wastewater containing dyes is important before its disposal otherwise it is a threat to ecology [5]. Concerns about environmental protection have been stimulated around the globe, prompting researchers and scientists to focus their attention on the remediation of wastewater [3]. As dyes are stable against chemical and biological degradation as well as the conventional treatment, approaches have some limitations, including being cost-prohibitive, time-consuming, producing secondary wastes, membrane fouling, incomplete mineralization and sludge formation [6,7]. Thus, the use of conventional methodologies are not recommended for dye removal, such as filtration, settling and others, as these methods are often inefficient in the removal of this class of pollutants [8]. Therefore, some alternative methods are distinguished by their ability to remove dyes from an aqueous medium, such as membrane separation [9], reverse osmosis [10,11], coagulation [11], oxidative remediation [12] and adsorption [13,14], among others.

Thus, the adsorption of dyes from contaminated water has become a popular topic recently due to its environmental and ecological importance [15]. Adsorption is an effective physico-chemical treatment technique due to ease of design, simplicity of the procedure, environmentally friendly approaches and also has the potential to use activated carbon, biological wastes, mineral oxides or polymer materials for the elimination of contaminants from aqueous solution [7]. The strategy for improving the adsorptive efficacy of industrial effluents is the implication of nanotechnology. In this context, metal nanoparticles are extensively studied owing to their catalytic, chemical, mechanical magnetic and other properties. In addition to a decrease in size, the enhanced surface area improves the capability to react, interact and adsorb other compounds [16]. Biochar is an emerging low-cost carbonaceous material synthesized from cheap agro-waste biomass residue. Its production on an industrial scale is already feasible as its propitious potential for environmental applications attracts considerable interest. Furthermore, biochar is considered a promising candidate for the remediation of inorganic and organic contaminants from water, though the removal efficiency is highly dependent on its chemical and physical attributes [16,17]. The native biochar possesses the limited capability to remove toxicants from water reservoirs, particularly for severely contaminated water. Alternatively, owing to the well-defined porous structure and abundant surface functional groups, biochar displays fascinating properties for the rational design of functional materials. Usually, biochar could be exploited to stabilize and disperse nanoparticles in order to enhance their catalytic reactivity for reactions.

More importantly, biochar-mediated catalysts/adsorbents can exert beneficial impacts to control water pollution when employed for the sequestration of organic pollutants. The biochar nanocomposites are cost-effective and efficient adsorbents to meet the stringent quality criteria of healthy and pure water availability [18–20]. Although biochar is an interesting material, it must be modified with metal nanoparticles [21,22] in order to increase its sorption capacity towards dyes and other organic and inorganic pollutants [23,24]. Therefore, many research teams have been working to develop alternative nanoadsorbents that can replace activated carbon in the use for pollution control. In comparison with activated carbons, carbon nanotubes (CNTs), biochar and biochar-based nanocomposites are more attractive because of their favorable thermal/chemical stabilities, high selectivity and structural diversity [25,26]. The reported surface area is ($340 \text{ m}^2 \text{ g}^{-1}$) and porosity (0.21 $\text{cm}^3 \text{ g}^{-1}$) for biochar pyrolyzed at 600 °C [25,26], while for the activated carbon (AC) of Mango seed (MS) husk, the highest specific surface area is $1943 \text{ m}^2 \text{ g}^{-1}$ and average pore volume $0.397 \text{ cm}^3 \text{ g}^{-1}$ [27]. Additionally, the recent progress in their large-scale production makes them better for use as ideal organic and inorganic contaminants nanoadsorbents [28]. Extensive experimental work has been conducted on the adsorption of different dyes onto biochar-mediated nanocomposites, such as Congo Red dye [27], Reactive Red 24 [29], Tartrazine dye [30], Tartrazine dye [31], Reactive Blue 19 dye [32], Cr^{3+} and Cd^{2+} [33], Zn^{+2} and Fe^{+3} [34] and Tartrazine dye [35].

The objective of the present work was to determine the fabrication of biochar $ZrFe_2O_5$ nanocomposites through a facile process using biochar as supportive material for adsorbent materials in the removal of Tartrazine dye from aqueous solutions. The effect of contact time, pH and initial concentration on adsorption characteristics of biochar-based nanocomposites was studied, and the experimental data obtained from the equilibrium studies were fitted to the Langmuir and Freundlich adsorption models. In addition, kinetics of the adsorption process was also studied. The adsorption process optimization was carried out using Response Surface Methodology (RSM). The main goal of the present work is to discover new possibilities of combining biochar from cheaper agro-waste biomass with nanomaterials. Recently emerging $ZrFe_2O_5$ nanocomposites with biochar will be an important step towards wastewater treatment and its purification.

2. Experimental

2.1. Synthesis of Wheat Straw Biochar

Wheat straw biomass was obtained from Jhang, Pakistan. It was washed using distilled water to eliminate debris and dust particles and allowed to dry in sunlight and in the oven at 60 °C for 24 h. Then, biomass grounded to fine particle size by food processor (Moulinex., Paris, France) was sieved using an Octagon nano-sieve (OCT-DIGITAL 4527-01) to the size of the 200 mm mesh. The sample in the crucible was kept in a muffle furnace at 600 °C for 1 h. The N_2 gas flow was maintained continuously from the inlet, forcing the pyrolysis fumes to pass via the outlet pipe and was kept underwater in order to avoid the direct discharge of fumes into the air. After 1 h of heating, the furnace was switched off and allowed to cool to room temperature. Then, the biochar was taken out and stored in an airtight bottle for further experiments.

2.2. Synthesis of Zirconium Ferrite (ZrFe₂O₅) Nanoparticles

The 0.91 g zirconium nitrate (Zr(NO$_3$)$_4$) and 2.73 g ferric chloride hexahydrate (FeCl$_3$.6H$_2$O) were mixed into 100 mL of distilled water. The suspension was oscillated ultrasonically for 8 min to avoid zirconium particle conglomeration. The pH of the suspension was maintained at 9.0 by adding 1 N NaOH solution. The suspension was stirred vigorously to avoid the precipitation of zirconium, and the particles were separated by centrifugation after 5 h stirring. The precipitates were washed with distilled water and then placed in a heating oven for 5 h at 160 °C to dry [36].

2.3. Synthesis of Biochar Based Nanocomposites

The biochar-supported ZrFe$_2$O$_5$ NCs was synthesized according to the method described by [20] with little modification. In a 250 mL conical flask, 10 g of sieved biochar and 0.1 g of ZrFe$_2$O$_5$ nanoparticles were mixed using 150 mL of distilled water under reflux conditions for 24 h. After this time duration, the conical flask was taken out and allowed to settle down for the next 12 h. When the nanocomposites had settled, the supernatant was removed. The obtained composites were dried at 50 °C and grounded into fine powder form.

2.4. Characterization of Nanocomposites

The functional group of BC-ZrFe$_2$O$_5$ NCs was interpreted and analyzed using the FTIR spectrometer (Bruker Tensor 27, Bruker, Hamburg, Germany) (sample prepared as KBr disc), while the surface structure of BC-ZrFe$_2$O$_5$ NCs was evaluated via SEM (JEOL JMT 300, Chicago, IL, USA). The elemental composition was determined by EDX analysis, and XRD was used to find the crystal nature of composites using Cu K radiations (D8 ADVAHCL, Bruker, Hamburg, Germany).

2.5. Point of Zero Charge (pH$_{pzc}$)

The sorbents' (BM, BC and BC-ZrFe$_2$O$_5$ NCs) point of zero charge (pH$_{pzc}$) was measured using the salt addition process. To determine pH$_{pzc}$, the series of 50 mL potassium nitrate (KNO$_3$) solutions (0.1 M each) were made, and pH was adjusted in a range from 1–12 by adding 0.1 N NaOH and HCl. Then, 0.1 g of adsorbent was added to each solution and was kept for a period of 48 h with intermittent shaking. The final pH (pH$_f$) of the solution was noted and the difference between final and initial pH (ΔpH) (*Y*-axis) was plotted versus initial pH (*X*-axis). The intersection point of the curve yield was pH$_{pzc}$ [37].

2.6. Preparation of Dye Solutions

Tartrazine dye was utilized in experiments without purification. To prepare 1000 mg L^{-1} stock solution 1 g of dye was dissolved in 1000 mL of deionized water. The solution of various concentrations (ranged from 10–200 mg L^{-1}) was prepared by adequate dilution of stock solution for experimental work. Tartrazine dye being anionic had λ_{max} 430 nm.

2.7. Batch Experimental Studies

Optimization of imperative environmental variables as pH (1–10), sorbent dosage (0.05–0.1 g), contacts time (0–480 min), initial dye concentration (10–200 mg L^{-1}) and temperature (303–333 K) for the elimination of Tartrazine dye was conducted via classical approach. The conical flasks (250 mL) comprising 50 mL of the dye solution of known pH, adsorbent dosage and concentration, were shaken in an orbital shaking incubator (PA250/25H) at 200 rpm for 480 min. The pH was adjusted by adding 0.1 N NaOH and HCl solutions. The blank solutions were also carried out under identical experimental conditions, excepting sorbent addition. All experiments were executed in triplicate and findings were determined as mean standard deviation ($\pm$SD). After a specified interval of time, the sample solutions were taken out and concentrations of residual dye solutions were determined using UV-Vis spectrophotometer (Schimadzu Corporation, Tokyo, Japan).

Percent removal R% and the equilibrium sorption uptake, q_e (mg g^{-1}), was calculated using the following relationships:

$$\text{Removal } \% = \frac{(C_o - C_e)}{C_o} \times 100 \tag{1}$$

$$q_e = \frac{(C_o - C_e)\ V}{W} \tag{2}$$

where C_o is the initial dye concentration (mg L^{-1}), C_e is the equilibrium dye concentration (mg L^{-1}), W is the mass of the sorbent (g) and V is solution volume (L).

2.8. Optimization of Parameters Using Response Surface Methodology (RSM)

The Response Surface Methodology (RSM) was applied for an interactive study of important influential parameters using "design expert" software. Central Composite Design (CCD) gives an idea about the fitness of experimental data with comparatively lesser numbers of run/design point, thus, reduced overall cost of experiments [38]. The adsorbent dose, pH, contact time and initial dye concentration were selected as independent variables and adsorption capacity (q_e) was the dependent (response) variable. There was a total of thirty experimental runs, comprising three central points, eight factorial points and six axial points. $X_i = -1$, 0 and 1 were three levels of each input variable [39]. The least-square followed by the second-order differential model was taken into consideration for interpretation of correlation among independent and dependent (response) variables.

$$Y = \beta_o + \sum\nolimits_{i=1}^{k} \beta_i x_i + \sum\nolimits_{i=1}^{k} \beta_{ii} x_i^2 + \sum\nolimits_{i=1}^{k} \sum\nolimits_{i \neq i=i}^{k} \beta_{ij} x_i x_j + \varepsilon \tag{3}$$

where Y is the adsorption capacity (response), β_o belongs to coefficients possessing specific numerical values, β_i, β_{ii} and β_{ij} are coefficients about linear, quadratic and interaction effects, respectively, while ε (Epsilon) is random error and k is a number of independent variables. The suitability and validity of polynomial equations were monitored to check significance by computing statistically the values of the regression coefficient (R^2) by means of analysis of variance (ANOVA) and F test at 0.05 probability (p) [40].

2.9. Kinetic and Adsorption Isotherm Models

Sorption kinetics experiments were carried out by using pseudo-1st order (Lagergren, 1898), pseudo-2nd order (Ho et al., 2000) and intraparticles diffusion (Webers and Morris, 1963) kinetic models.

The Freundlich (Freundlich, 1906) and Langmuir adsorption isotherms were investigated in the current study for exploring the adsorption mechanism in Tartrazine dye removal.

3. Results and Discussions

3.1. Characterization of BC-ZrFe$_2$O$_5$ NCs

Point of zero charge (pH$_{pzc}$) is the pH at which charge on a sorbent's surface is neutral, i.e., the amount of electric positive and negative charges is equal. The pH$_{pzc}$ of wheat straw biomass (WSBM), wheat straw biochar (WSBC) and BC-ZrFe$_2$O$_5$ NCs was observed to be 3.6, 2.5 and 3.8, respectively, as presented in Figure 2. Below this pH value, the adsorbents attain net positive charge because of functional group protonation resulting in strong electrostatic attractions among anionic dye and adsorbents. Beyond this pH value, the sorbent's surface acquires a negative charge. Thus, the sorption of anionic pollutants (dye) was preferred at pH < pH$_{pzc}$ where sorbent surfaces become positively charged [23]. These results are in accordance with the findings of [41].

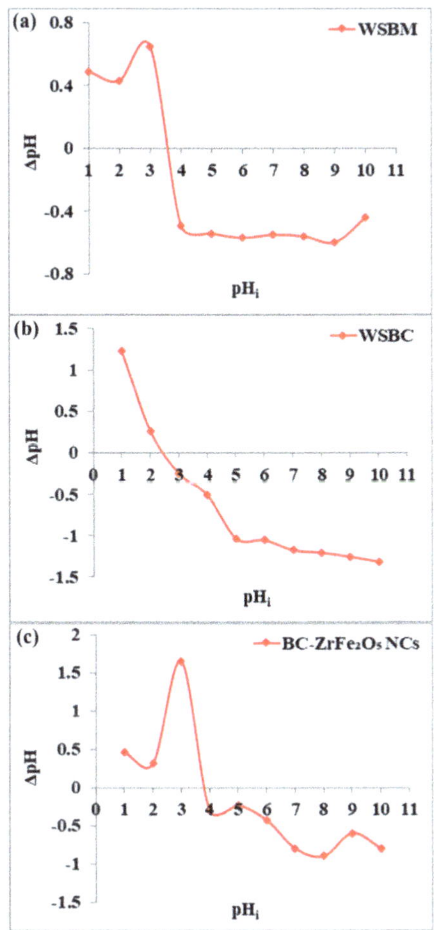

Figure 2. pH$_{pzc}$ of adsorbents: (**a**) wheat straw biomass (WSBM), (**b**) wheat straw biochar (WSBC) and (**c**) biochar-based zirconium ferrite nanocomposite (BC-ZrFe$_2$O$_5$ NCs).

The Fourier Transform Infra-red (FTIR) spectroscopy is an important analytical technique that analyzes vibrations characteristic of each functional group in a molecule. The FTIR spectra of biochar, BC-ZrFe$_2$O$_5$ NCs and Tartrazine dye-loaded BC-ZrFe$_2$O$_5$ NCs were explored in a range from 400–4000 cm^{-1} as given in Figure 3b. The FTIR spectra of unloaded BC-ZrFe$_2$O$_5$ NCs showed the presence of peaks in the range from 3512–3113, 1559, 1384, 1048 and 762 cm^{-1}. The broad band from 3512–3113 cm^{-1} was attributed to the O–H vibration of ZrO-H on the surface of the sorbent. The vibration peak at 1595 cm^{-1} indicated the presence of a –C=C– group of aromatic rings. The peak at a region of 1384 cm^{-1} represented the presence of the alkyl group in the structure of BC-ZrFe$_2$O$_5$ NCs. A distinctive sharp peak at 1061 cm^{-1} was assigned to C–O–C stretching vibrations while the Zr–Fe bond was responsible for vibration at 762 cm^{-1}. In the FTIR spectra of biochar (BC), the band from 3371–2825 cm^{-1} was attributed to the O–H group (carboxylic acids, phenols and alcohols) on the surface of sorbents such as cellulose, pectin and lignin. The peak at 1584 cm^{-1} indicated the stretching vibration of the –C=C–group. The peaks at 1184 cm^{-1} embodied the presence of C–O–C functional groups in BC. In the case of spectra of dye-loaded BC-ZrFe$_2$O$_5$ NCs, there is distinctive vanishing and broadening of some bands owing to their involvement in the adsorption process. Thus, FTIR spectra specified the functional groups and exchanging active sites on which adsorption occurred [42,43].

The X-ray diffraction (XRD) pattern of BC-ZrFe$_2$O$_5$ NCs before and after dye loading was recorded for the determination of the crystalline nature of adsorbents as shown in Figure 3a. The peaks 2θ at position 28.9°, 36° and 40° correspond to (012) (200) (311) and (220) planes for BC-ZrFe$_2$O$_5$ NCs matched with standard card no. 96-152-3795. These planes could be attributed to a tetragonal crystal structure. The size of the crystal was calculated by using the Debye–Scherrer formula as given below:

$$D = \frac{K\lambda}{\beta cos\theta} \tag{4}$$

where in D is the size of the crystal, λ is the X-ray wavelength (Cu Kα radiations), K is the Scherrer constant, θ is the angle of diffraction and β is full with half maximum. The size of the BC-ZrFe$_2$O$_5$ NCs crystal was calculated to be 40.72 nm. There was an obvious change in XRD spectra after dye sorption owing to electrostatic and complex formation being a major mechanism in dye removal. These findings are in agreement to the XRD pattern reported by [18].

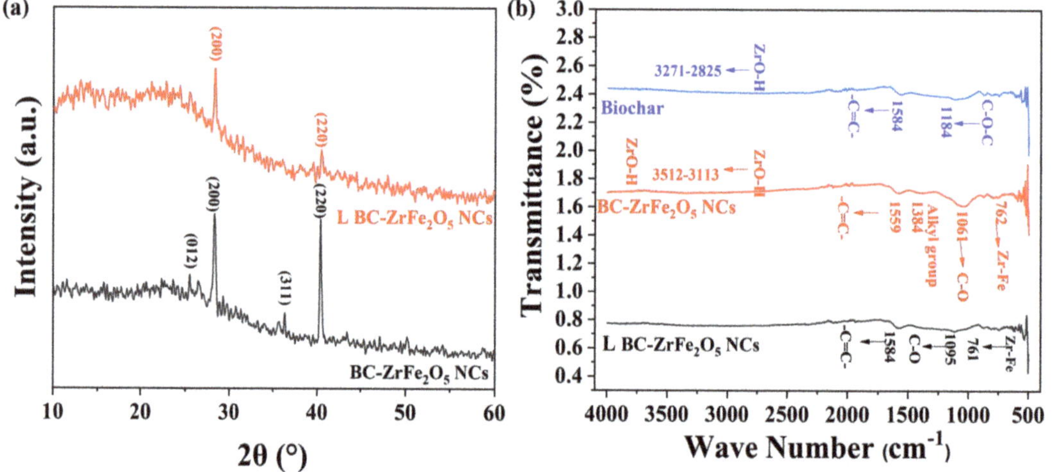

Figure 3. (a) XRD pattern of native BC-ZrFe$_2$O$_5$ NCs and dye-loaded BC-ZrFe$_2$O$_5$ NCs, (b) FT-IR spectrum of Biochar, BC-ZrFe$_2$O$_5$ NCs and dye-loaded BC-ZrFe$_2$O$_5$ NCs.

The morphological characteristics and surface features, i.e., shape, size, pore properties and arrangements of particles of the adsorbent were studied using a Scanning Electron Microscope (SEM). The higher the number of pores, the higher the dye sorption onto the adsorbent surface. The SEM images of free and Tartrazine dye-loaded wheat straw biochar (WSBC) are shown in SI Figure S2, while the SEM micrographs of free BC-ZrFe$_2$O$_5$ NCs and Tartrazine dye-loaded adsorbents are depicted in Figure 4a–f at different magnification levels. These images clearly show porous, rough, fibrous, regular and rod-like textures of sorbents facilitating the sorption of dye. In the SEM micrographs, after Tartrazine sorption, the regular rod-like structure seems to be cramped by dye molecules, thus, minimizing the size of pores as well as causing surface roughness, corroborating that the WSBC and BC-ZrFe$_2$O$_5$ NCs represent suitable morphological profiles for dye take up. These results are related to the reported results of [44] with minor changes due to the difference in experimental conditions.

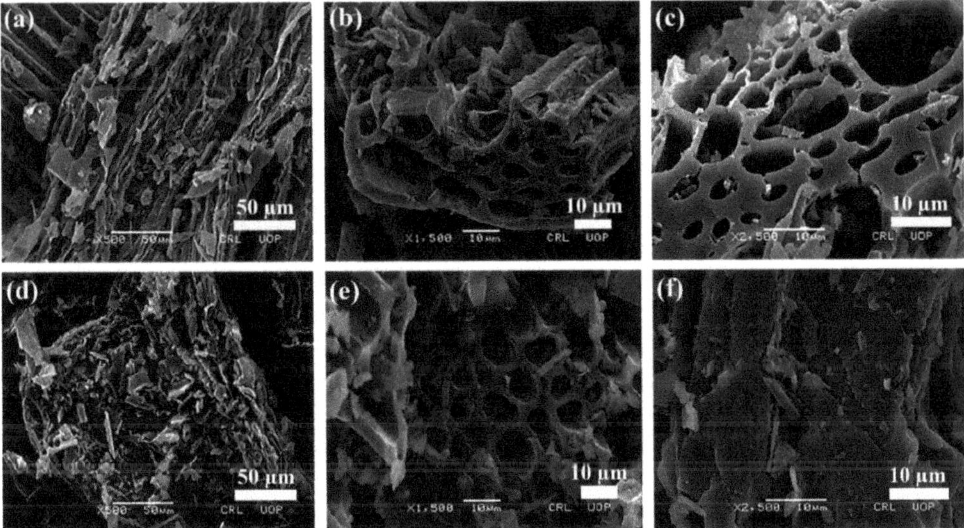

Figure 4. SEM micrographs of BC ZrFe$_2$O$_5$ NCs (**a–c**) and Tartrazine dye-loaded BC-ZrFe$_2$O$_5$ NCs (**d–f**) at three different magnification levels.

The Energy Dispersive X-ray (EDX) studies documented the elemental composition of sorbents. The EDX pattern of free and loaded BC-ZrFe$_2$O$_5$ NCs is depicted in Figure S1a,b. The EDX spectrum verified strong signals for Zr and Fe presence. In addition, Na, Ca, Si, C, O, S, Al and Cl signals were assigned to molecules present in biochar as an integral part. However, in the EDX spectrum of the dye-loaded sorbent, the clear vanishing of Zr, Fe, Na, Ca, C, O and Al and peaks were observed, supporting electrostatic and ionic exchange interaction as a major mechanism for adsorption phenomena. These findings are analogous to the results of [43].

3.2. Optimization of Environmental Variables

3.2.1. Impact of pH

Solution pH is a substantial factor that monitors the process of sorption as the pH affects the functional groups' activity on the adsorbent surfaces. The influence of pH on the remediation of Tartrazine dye from aqueous media is presented in Figure 5a. The sorption was carried out by varying dye solutions pH from 2 to 10 and the findings revealed that the rise in pH culminates in the decline of dye adsorptive removal. The maximum sorptive removal of Tartrazine dye by BM, BC and BC- ZrFe$_2$O$_5$ NCs was achieved at pH 2. This was attributed to the fact that at acidic conditions, functional groups (binding sites) on adsorbent surfaces get protonated owing to a net increment of a positive charge. The electrostatic attractions among positively charged sorbents and anionic dye surfaces were responsible for the extent of dye elimination from the solution. An increment in medium pH had a contrary influence on sorbent removal capacity for dye because of the increase in negative charges onto the adsorbent surfaces resulting in net electrostatic repulsion among sorbent functional groups and dye anions subsequently causing the reduced sorption of dye [45]. These findings are also in concord to the point of zero charges of BM, BC and BC-ZrFe$_2$O$_5$ NCs that were found to be 3.6, 2.5 and 3.8, respectively. Thus, below the aforementioned pH values, adsorbents hold net positive charge resulting in greater sorption of anionic Tartrazine dye. Maximum removal (88.09 mg g^{-1}) of dye was obtained by utilizing BC-ZrFe$_2$O$_5$ NCs. These findings were similar to the results reported by [46].

3.2.2. Impact of Contact Time

For cost-effective and large scale wastewater treatment, the contact time course studies between the sorbate and sorbent are imperative in the sorption system. To explain the influence of contact time on the removal of Tartrazine, the experiments were performed by varying time intervals from 15–480 min while keeping the other environmental variables such as pH, concentration, temperature and adsorbent dose constant. Figure 5b illustrates results of time-dependent experimental data points. The outcomes show that during the adsorption process, initially the rate of reaction was rapid, followed by slow removal and, finally, the trend becomes constant on the achievement of equilibrium. Within 90 min, BM, BC and BC-ZrFe$_2$O$_5$ NCs adsorbents showed up to 60, 71 and 86 mg g^{-1} removal for Tartrazine dye. Approximately 80% of dye uptake took place within 90 min; therefore, this time (90 min) was deemed to be adequate in successive experiments to establish equilibrium. During sorption phenomena, the dye molecules initially come into contact with the boundary of the adsorbents, then adsorb onto the adsorbent surface and finally diffuse into the porous and permeable sorbent that requires longer contact time. Hence, the sorption process arises in two steps, the initial (rapid) stage and final (equilibrated) stage. At the start of the reaction, rapid sorption might be due to the vacancy of all the active binding sites onto the sorbent surface and dye molecules easily occupy the binding sites. As the reaction progressed, the adsorption process slowed down due to the saturation of active sites resulting in the slower movement of adsorbate molecules from the boundary of the adsorbent into the interior of the adsorbent [47,48]. In one study chitosan/geopolymer beads were used for Crystal violet dye sorption and similar results were obtained from the study [24].

3.2.3. Impact of Initial Dye Concentration

The initial dye concentration appears to be of paramount importance as it has a pronounced influence on sorption phenomena. The concentration dependence efficacy of sorbents was recorded by changing Tartrazine concentrations from 10 to 200 mg L^{-1} at pre-optimized conditions of pH (2), sorbent dose (0.05 g) and contact time (120 min). Figure 5c represents the finding regarding the impact of initial dye concentration on sorptive uptake of BM, BC and BC-ZrFe$_2$O$_5$ NCs. Results revealed that on enhancing the concentration of dye (from 10–200 mg/L) the sorption (mg g^{-1}) improved from 5.23–113.24, 5.66–137.88 and 6.69–162.67 for BM, BC and BC-ZrFe$_2$O$_5$ NCs, respectively. The initial dye concentration offers the main driving force for the collision of all dye molecules between the solid (adsorbent) and aqueous (adsorbate) phases so yielding greater uptake of dye. At lower concentration, the ratio of active sites of the adsorbent surface to available adsorbate molecules was lower; consequently, fractional adsorption no longer dependent upon initial dye concentration. However, at greater concentration, the binding sites accessible for adsorption might be fewer in comparison to the molecules of dye present; therefore, removal strongly depends upon adsorbate concentration. Contrary to this, the % removal decreased on increasing initial concentration because of the saturation of adsorption sites at a fixed sorbent dose [49]. These results are concordant to studies reported by [50] with the analogous trend of increased dye removal with an augmented concentration of dye.

3.2.4. Impact of Sorbent Dosage

The sorbents amount also plays an imperative role in the process of biosorption. To explore the influence of the amount of adsorbent on the removal of Tartrazine, the dose of BM, BC and BC- ZrFe$_2$O$_5$ NCs was raised from 0.05 to 0.1 g/50 mL of Tartrazine dye solution while keeping initial concentration (100 mg L^{-1}), and findings in Figure 5d depicted that the dye uptake was reduced by augmenting the sorbent concentration. Almost 60% reduction in sorption was recorded on enhancing the sorbent and the maximum removal of dye was attained by 0.05 g sorbent dose. It was attributed to the aggregation of particles of sorbents resulting in the lower surface area, hence, the decrease in the active sites to dye molecules ratio [36]. These results closely relate to the literature cited by other researchers [51].

3.2.5. Impact of Temperature

The temperature is an imperious parameter as it is an indicator to determine whether the sorption process is exothermic or endothermic. The experimental data regarding temperature influence on sorption of Tartrazine dye from aqueous solution was studied at a different temperature ranging from 303–333 K as shown in Figure 5e. It was examined from the results that the adsorption of Tartrazine dye onto BM, BC and BC-ZrFe$_2$O$_5$ NCs was an exothermic process, i.e., rising temperature caused a reduction in dye sorption. This can be elucidated as the rise in temperature weakened binding forces that caused detachment of the dye molecule from the adsorbent's surface [14]. The highest removal was observed at 303 K with BM, BC and BC-ZrFe$_2$O$_5$ NCs. It was also examined that the sorption of textile dye onto activated carbon was of exothermic nature [52].

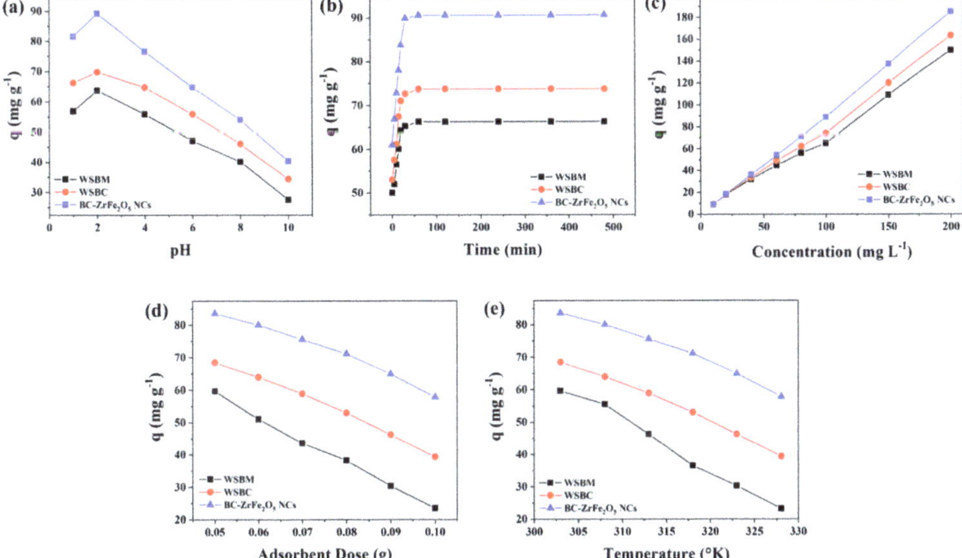

Figure 5. Influence of (**a**) pH, (**b**) time of contact, (**c**) initial concentration, (**d**) sorbent dosage, (**e**) temperature on the adsorption of Tartrazine dye using, WSBM, WSBC and BC-ZrFe$_2$O$_5$ NCs.

3.3. Optimization of BC-ZrFe$_2$O$_5$ NCs Using Response Surface Methodology

Response Surface Methodology (RSM) was employed for the optimization of process parameters that assisted in identifying the maximum possible interactions among various parameters. Central Composite Design (CCD) of 30 experimental runs was performed to examine the interactive influence of four independent variables including pH, sorbent dose, contact time and the initial concentration of dye for the response (adsorption capacity). To analyze the statistical significance of factors and their interactions, the Fischer's test for ANOVA was employed. The probability values (p values) for model terms were calculated at a 95% confidence level. p-values of each input variable proposed whether it was significant. This multiple regression analysis on experimental data was performed and gave a 2nd order polynomial equation representing the relation among response Y (adsorption capacity) and input variables (pH, dye concentration, contact time and sorbent dose) [38,39]. The summarized results of regression coefficients and ANOVA have been tabulated in Table S1 (for Tartrazine dye).

The model adequacy for the presentation of the results of experimental data was justified by greater values of R^2 and adj R^2. The non-significant lack of fit ($p > 0.05$) guarantees the goodness of the suggested model. Various input variables combinations (sorbent dose (g), contact time (min), pH and initial dye concentration (mg L^{-1}) were adjusted to perform adsorption experiments using CCD design. Response in terms of adsorption capacity (mg/g) is described with contour and 3D response surface plots (Figures 6 and 7). The response ranged from 11–147 mg g^{-1}. The 2nd order regression polynomial equation signified the relation as:

$$Y = 80.9 - 5.67A + 59.29B - 0.93C + 31.83 - 8.21AB + 0.82AC - 2.91AD \\ - 1.50BC + 31.50BD - 1.25CD + 3.39A^2 - 4.31B^2 - 5.49C^2 - 31.49D^2 \tag{5}$$

where Y = response (adsorption capacity) and A, B, C and D designated input variables, i.e., pH, dye initial concentrations, sorbent dose and contact time, respectively. To assess model significance, the coefficients were calculated. These embody the individual impacts of linear, i.e., A, B, C and D and quadratic terms such as A^2, B^2, C^2 and D^2 as well as effects of 1st order interactions such as AB, AC, AD, BC, BD and CD. The values of adjusted R^2 nearer to 1 showed a greater correlation among predicted and observed values [53,54]. It was obvious from the results that all parameters (A, B, C, D) individually as well as in combinations had substantial influence on dye adsorption.

Optimization of Input Variables

The two-dimensional (2D) contour plots and corresponding three-dimensional (3D) response surfaces were employed to comprehend the influence of the input variables and optimization on the adsorption process [38]. The findings have been depicted in Figures 6 and 7. The contour and 3D surface plots to examine the mutual impact of concentration and pH have been presented in Figure 6a,b. In these plots, initial concentrations of pH and dye were varied, keeping sorbent amount and contact time constant. As noticed from the plots, the adsorption of Tartrazine dye enhanced with increasing concentration up to a certain limit, but pH had a contrary effect; upon increment in pH, the adsorption of respective dye declined (from 3–10 pH). Very high pH imparts lower adsorption efficiency even if the amount of dye concentration is high due to functional group deprotonation resulting in electrostatic repulsion among sorbent and anionic sorbate [54]. The maximum removal capacity was exhibited at 100 mg/L concentration and 2 pH. In addition, the nearly linear contour plots implied relatively weak interaction among concentration and pH. The interactive influence of adsorbent dosage and pH on dye elimination are displayed in Figure 6c,d. It is evident from the figures that lower levels of sorbent and high levels of concentration support the attainment of a higher removal percentage of dye (while time and concentration were held constant). Higher sorbent concentration causes the aggregation of sorbent molecules, thus, lowering the surface area and active binding sites. The response surface and contour plot due to the combined effect of contact time and pH is shown in Figure 6e,f. It was evident from the graph that time and pH had a paramount impact on the response (sorption of dye). Maximum sorption was detected at lower pH and greater contact time owing to maximum interaction among sorbate and sorbent at greater contact time. The combined effect of contact time and pH on dye removal are described in Figure 7a,b, while keeping initial dye concentration and sorbent dose fixed. The % removal was augmented with increments in contact and reduced with pH that is accredited to the overlapping of binding sites. In addition, from the contour plots, it was evident that elliptical lines implied significant association between contact time and pH. The interaction between contact time and sorbent dose was not so influential (Figure 7c,d); the mutual influence of contact time and sorbent dose has been illustrated in Figure 7e,f. The circular lines of the contour plots signified the interaction among contact time and initial concentration. In order to get higher removal efficacy, we need to maximize contact time and minimize the sorbent dosage. Moreover, the significance of the interaction of the combined effect of pH and sorbent dose could be evident from the linear nature of the contour plots. Combined impacts of input variables in numerical values were also set in terms of the coefficient of the polynomial quadratic equation [54,55].

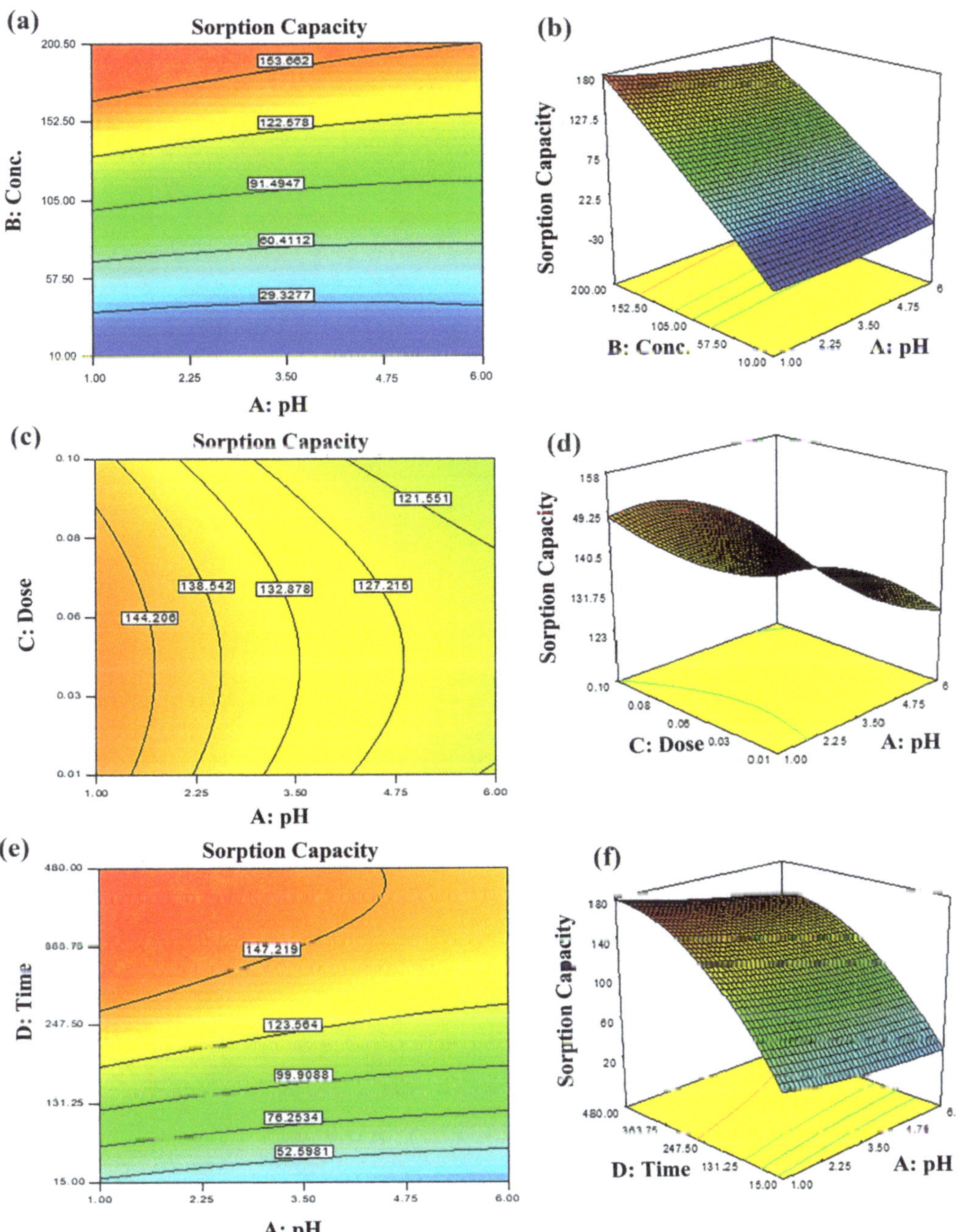

Figure 6. Predicted dye removal contour plots under influence of: (**a**) pH–initial concentration, (**b**) pH–sorbent dose, (**c**) contact time–pH and the related response surface plots in (**d**–**f**) at central point values of other parameters.

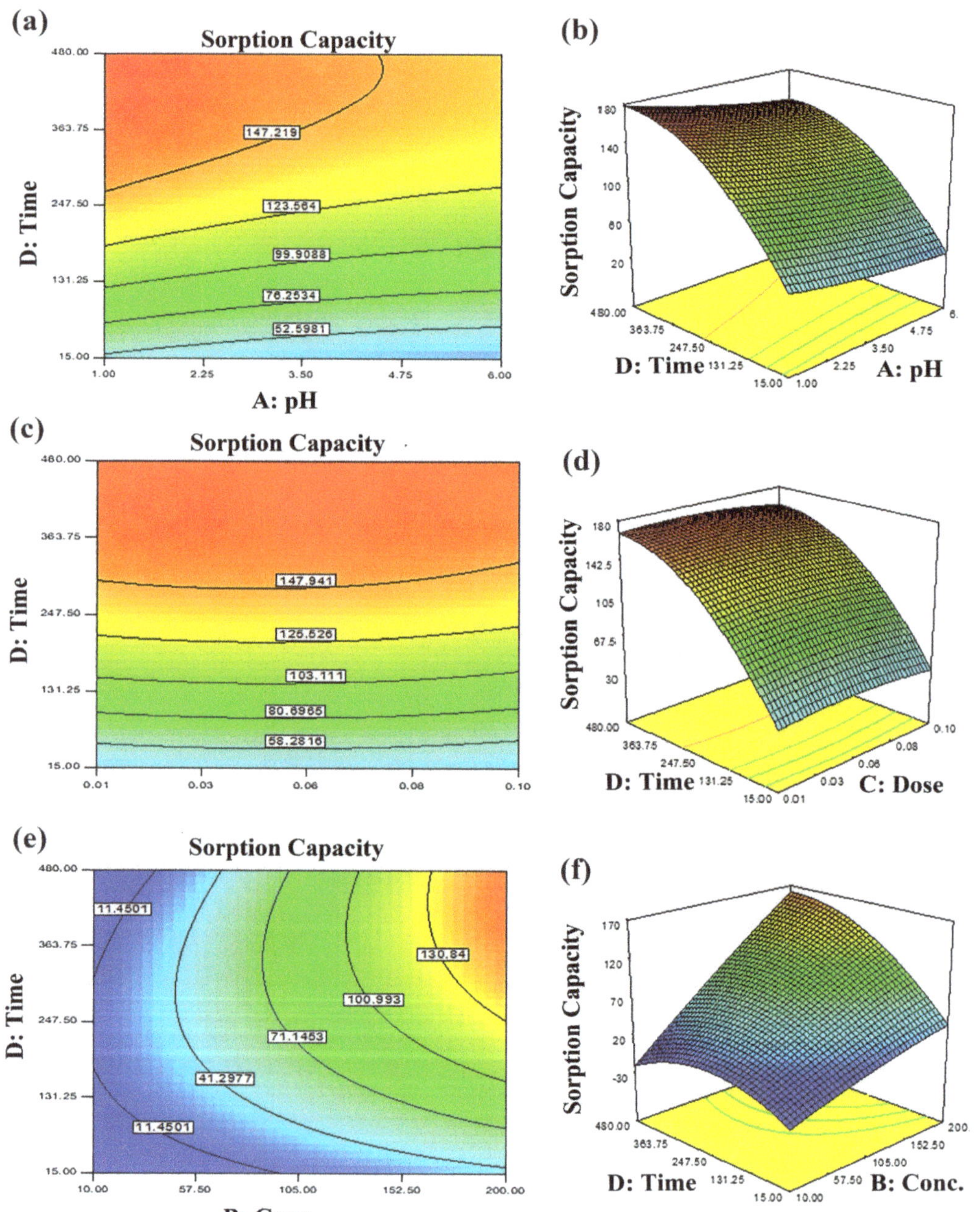

Figure 7. Predicted dye removal contour plots under influence of: (**a**) pH–contact time, (**b**) contact time–sorbent dose and (**c**) contact time–initial concentration and the related response surface plots in (**d**–**f**) at central point values of other parameters.

3.4. Adsorption Kinetic Modeling

Kinetic models describe the reaction order of the sorption process based on adsorption capacity of the adsorbent including Lagergren's first order equation and Ho's second-order expression. Adsorption kinetic modeling is essential to describe the rate of the transfer of dye from aqueous solution to the surface of the adsorbent, which is important to measure the retention time of the adsorption process necessary for the process of optimization in industry. In the present study, the Tartrazine dye kinetics was investigated at various time intervals to understand the behavior of adsorbents such as WSBM, WSBC and BC-ZrFe$_2$O$_5$ NCs.

3.4.1. Pseudo-First Order

The rate of reaction is proportional to the empty sites of adsorbents. The pseudo-first order equation is given as:

$$\log(q_e - q_t) = \log q_e - \frac{K_{1t}}{2.303} \tag{6}$$

where K_1 (per min) is the adsorption rate constant for the pseudo-first order reaction, q_t (mg g^{-1}) is the amount of adsorption at time "t", q_e (mg g^{-1}) is the amount of adsorption of dye at equilibrium. The K_1 and q_e values were attained from the plot of log ($q_e - q_t$) verses t. The slopes and intercepts of plots of log ($q_e - q_t$) verses t was used to compute the K_1 and q_e. In Figure 8a, the nonlinear plot of Log ($q_e - q_t$) versus t represented a non-significant regression coefficient (R^2) which indicated that pseudo-first order kinetics might be inadequate to understand the adsorption mechanism for Tartrazine dye. Moreover, [56] investigated the biosorption kinetics and equilibrium of RO 22 using peels and fruits of *Trapa bispinosa* and observed that R^2 values for the pseudo-first order kinetic plots were lesser than pseudo-second order kinetic plots.

3.4.2. Pseudo-Second Order Equation

In the pseudo-second order kinetic model, the rate of adsorption is proportional to the square of number of the empty sites of the adsorbent, i.e., it is based on the adsorption capacity of the adsorbent. The pseudo-second order differential equation is given as:

$$\frac{dq_t}{t} = K_2 \left(q_e - q_t\right)^2 \tag{7}$$

By applying and integrating boundary conditions q = 0 to q = q_t and t = 0 to t = t the pseudo-second order kinetics model may be expressed as:

$$\frac{t}{q_t} = \frac{1}{K_2 q^2{}_e} + \frac{1}{q_e} \tag{8}$$

where K_2 (mg g^{-1} min^{-1}) is the adsorption rate constant for the pseudo-second order reaction. The value of K_2 and q_e can be calculated from the intercept and slope of the plot of t/q_t. In Figure 8b, the plots of t/q_t versus t showed a linear relationship with a higher regression coefficient (R^2) than the pseudo first-order model. Kinetics parameters are shown in Table S2 showing chemical adsorption as the rate limiting step for the mechanism of adsorption. Data presented shows there is an acquiescence between pseudo-second order theoretical and experimental q_e values for Tartrazine dye. Moreover, the values of R^2 for the pseudo-second order was almost equal to 1 and significantly higher than the pseudo-first order, suggesting that the system of adsorbents is well designated by the pseudo-second order kinetic model. All the evidence predicted that the adsorption of Tartrazine dye follows the pseudo-second order model that is based on the statement that adsorption may be a rate limiting step. The isotherm depicted that how the molecules were distributed between the liquid and solid phases at equilibrium [57,58].

3.4.3. Interparticles Diffusion Model

The solute molecules transfer from aqueous media to sorbent surfaces in several steps, as continuously fast stirring is involved in the batch study, thus, the rate monitoring step might contain intra-particles or film diffusions or both mechanisms. The intraparticles diffusion model equation is given as:

$$q_t = K_{pi}t^{1/2} + C_i \tag{9}$$

where C_i = the intercept describing boundary layer thickness and K_{pi} (mg/g min)$^{1/2}$ represents the rate constant for intraparticle diffusion. The plot q_t versus $t^{1/2}$ could be linear only if intraparticle diffusion is involved in the sorption phenomenon, but if the line passes through the origin, then the intraparticle diffusion could be the rate determining step. When the plot did not pass via origin, it indicated boundary layer control at some extent and it showed that the intraparticle diffusion was not the only rate controlling step, but other kinetics models might also control the rate of sorption, all of which could operate simultaneously [32,57,59]. The findings of the intraparticle diffusion model are tabulated in Table S2 and Figure S3 (SI). The lower values of R^2 as compared to the 2nd order kinetic model indicate that Tartrazine adsorption did not follow this model.

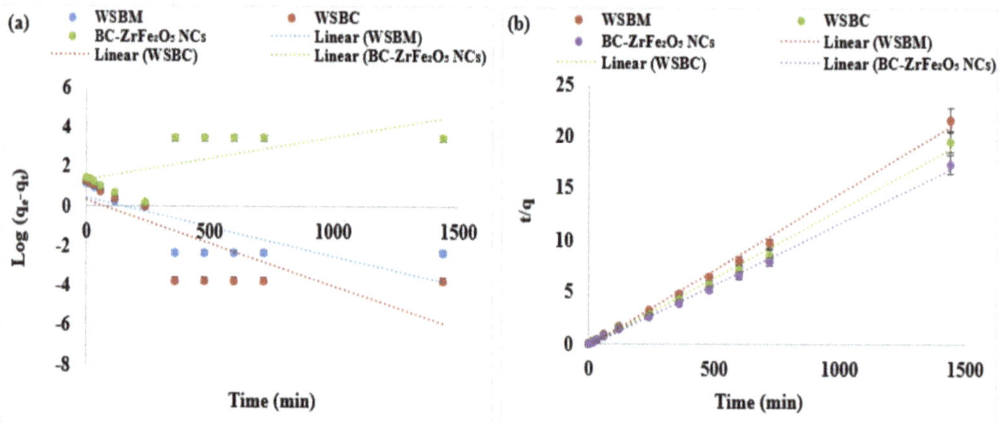

Figure 8. Adsorption kinetic models: (**a**) first order, (**b**) second-order plots of adsorption of Tartrazine dye by WSBM, WSBC, and ZrFe$_2$O$_5$ NCs.

3.5. Adsorption Equilibrium Studies

Adsorption equilibrium models are used to examine the adsorption mechanism or the interaction between adsorbate and adsorbent at equilibrium and used to deduct the maximum adsorption capacity for adsorbents.

3.5.1. Langmuir Isotherm

The Langmuir isotherm model is effective for the monolayer adsorption process with a limited number of energetically equivalent identical active sites. According to this model, the maximum adsorption takes place when the monolayer of adsorbent was saturated with adsorbate molecules and there is no passage for molecules of adsorbate to pass over the surface of the adsorbent. The maximum monolayer adsorption capacity (q_e mg/g) and other factors were examined by the following equation:

$$\frac{C_e}{q_e} = \frac{1}{X_m K_L} + \frac{C_e}{X_m} \tag{10}$$

where q_e (mg g^{-1}) is equilibrium sorption capacity, C_e (mg L^{-1}) is concentration of dye at equilibrium, X_m (mg g^{-1}) is complete monolayer or adsorption capacity, K_L is Langmuir constant as apparent energy of adsorption. A non-linear plot was found when C_e/q_e is plotted against C_e. Thus, a non-linear relation between C_e/q_e and C_e, and the lower value of R^2 predicted that the adsorption of Tartrazine dye and does not follow the Langmuir isotherm model that is based on the statement that biosorption may be a monolayer. The Langmuir isotherm could also be represented as a dimensionless constant partitioning element for the equilibrium parameters, R_L calculated by equation:

$$R_L = \frac{1}{1 + bC_o} \tag{11}$$

where C_o = initial adsorbate concentration and b = Langmuir constant. R_L values indicated isotherm type; favorable ($0 < R_L < 1$), irreversible ($R_L = 0$), unfavorable ($R_L > 1$) or linear ($R_L = 1$) as shown in Figure 9a [56].

3.5.2. Freundlich Isotherm

The Freundlich isotherm is based on the concept that adsorption occurs on heterogeneous surfaces and the adsorption capacity depends on the concentration of dye at equilibrium. The Freundlich equation is an empirical equation, and its linearized form can be given as:

$$\text{Log } q_e = \log K_F + \frac{1}{n} \log C_e \tag{12}$$

where the K_F (mg g^{-1}) and $1/n$ indicates adsorption capacity and adsorption intensity, respectively. The Freundlich constants KF and $1/n$ can be calculated from the slope and intercept of the linear plot, with log q_e versus log C_e as shown in Table S3. The value of $1/n$ less than one indicated the simple separation of dye from aqueous solution. The lesser value of $1/n$ showed that the adsorption was better and a comparatively stronger bond was developed between the adsorbent and adsorbate. The good depiction of adsorption equilibrium by the Freundlich model recommended that adsorption of multilayers occurred for Tartrazine dye which predicts many active sites. The results showed that Freundlich isotherm was best fitted to Tartrazine dye than the Langmuir model [56,59,60].

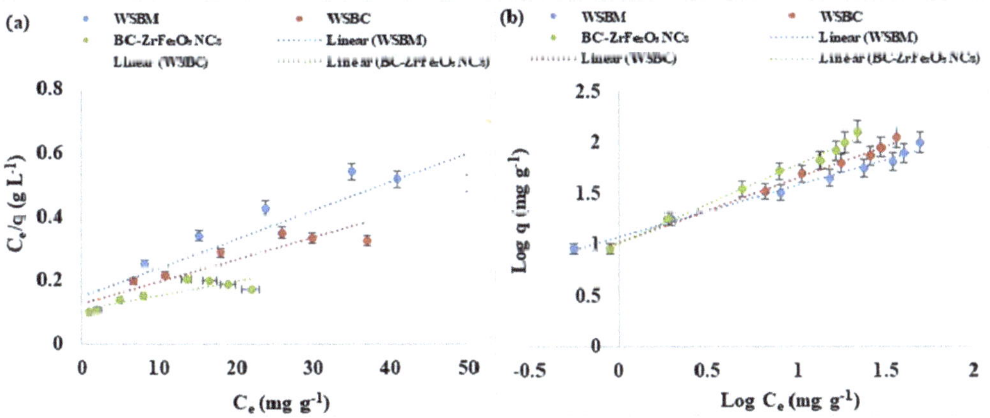

Figure 9. Adsorption equilibrium models: (**a**) Langmuir, (**b**) linearized Freundlich of adsorption of Tartrazine dye by WSBM, WSBC, and BC-ZrFe$_2$O$_5$ NCs.

3.6. Adsorption Mechanism

Based on the results of FTIR, EDX, SEM and XRD and the adsorption models in this study, the adsorption mechanism can be proposed. The surface hydroxyl, carbonyl and Zr-Fe groups on BC-ZrFe$_2$O$_5$ NCs are the active adsorption sites for dye in solution, which may interact by complex formation or ion-exchange with dye molecules. The mechanism of Tartrazine dye adsorption by the BC-ZrFe$_2$O$_5$ NCs be elucidated as follows:

$$MOH + H^+ \leftrightarrow MOH_2^+ \equiv MOH_2 + TD^- \leftrightarrow \equiv MTD + H_2O \equiv MOH + TD^- \leftrightarrow \equiv MTD + OH^-$$

where $\equiv$M represents the surface of the adsorbent and TD represent Tartrazine dye.

The ZrFe$_2$O$_5$ NCs, which are entrapped with the biochar matrix, can form an aqua complex with water and develop a charged surface through amphoteric dissociation. When the pH of the medium is acidic, positively charged surface sites are developed, which attract the negatively charged Tartrazine dye (anionic dye) by electrostatic attraction, resulting in the enhanced dye removal in acidic pH values. With a neutral pH of the solution, dye adsorption can be due to the ion-exchange and complexation reaction between anionic dye and hydroxyl, carbonyl and Zr-Fe ions. The adsorption mechanism has been reported as being able to remove dyes with high selectivity and adsorption capacity The excess hydroxyl ions will compete with dye anions on the active adsorption sites, which lead to the decrease in the adsorption capacity with increasing solution pH.

Based on the FTIR analysis, the complex formation mechanism would play an important role in the dye removal. The hydroxyl groups on the surface of the BC-ZrFe$_2$O$_5$ NCs could form metal complex dye ions during the removal process. Specifically, at high solution pH, the surface of the adsorbent becomes negatively charged, and stronger electrostatic repulsion between active sites and dye exists. The dye removal would be mainly due to complexation by coordination of Fe and Zr with the dyes' functional groups.

3.7. Catalyst Regeneration and Comparison

Catalyst stability, recyclability and reusability are very important for commercial-scale application. After completion of adsorption, the synthesized BC-ZrFe$_2$O$_5$ NCs catalyst was separated by centrifugation and washed by using 0.1 N solution of NaOH, then deionized with distilled water and desiccated in the oven at 80 °C and saved for subsequent reaction. It was also seen that the catalytic activity of biochar-based composite remained constant up to five cycles of operation so the recycled BC-ZrFe$_2$O$_5$ NCs could be reused at least five times with no significant loss of catalytic activity as given in Figure 10a. The removal efficacy declined from 89–63%. This sorbent's membrane fouling owing to dye sorption could be recovered by centrifuging the sorbents in 1 N solution of NaOH at the completion of each cycle. The larger portion of sorbent could be recovered that was useful as well as economic for the regeneration of sorbents for practical applications [24]. The removal efficacy of sorbents viz. WSBM, WSBC and BC-ZrFe$_2$O$_5$ NCs in batch experimental mode have been depicted in Figure 10b. The findings revealed that sorption efficacy (mg g^{-1}) of WSBM, WSBC and BC-ZrFe$_2$O$_5$ NCs was 67, 75 and 89 for Tartrazine dye, respectively. The extraordinarily high remediation potential of BC-ZrFe$_2$O$_5$ NCs for dye was attributed to its larger surface area, smaller size, greater surface functional groups and lower resistance of mass transfer [57,61]. The comparison of current study with already reported research work has been compiled in Table 1.

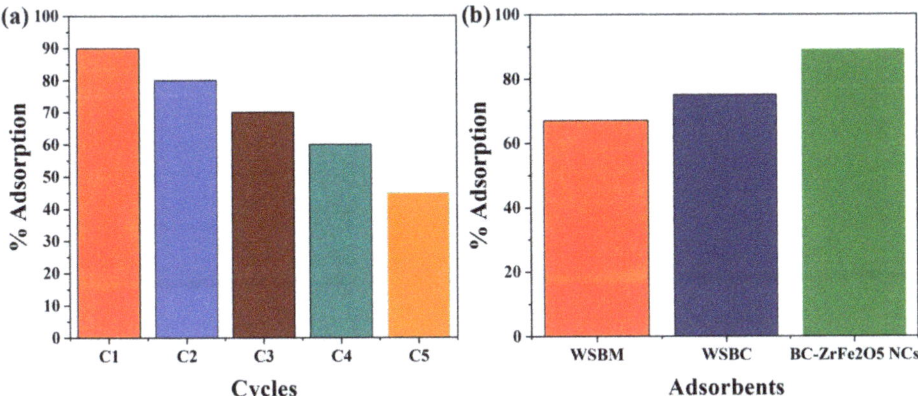

Figure 10. (a) Catalyst recyclability and (b) comparison among different catalysts (WSBM, WSBC and BC-ZrFe$_2$O$_5$ NCs) for sorption of Tartrazine dye (at conditions 100 mg g^{-1} dye concentration, 30 °C temperature, 0.5 g sorbent dose).

Table 1. Comparison of BC-ZrFe$_2$O$_5$ NCs to previously reported nano-sorbents for different contaminants.

Adsorbent	Pollutant	Removal Capacity	References
Agro-waste-derived biochars impregnated with ZnO	As(III) Pd(II)	25.9 mg/g 25.8 mg/g	[25]
ZnO/biochar composites	Sulfamethoxazole Methyl Orange	90.8% 88.3%	[26]
ZnO/cotton stalks biochar	Congo Red dye	89.65%	[27]
Agricultural waste-derived biochar modified with ZnO nanoparticles	Reactive red 24	71%	[29]
Polystyrene/magnetite nanocomposite (PS-DVB/Fe$_3$O$_4$)	Tartrazine dye	90%	[30]
Iron-loaded natural zeolite (NZ-A-Fe)	Tartrazine dye	90%	[31]
Biochar caged zirconium ferrite nanocomposites	Reactive Blue 19 dye	88.8%	[32]
Layered double hydroxide-based material	Cr^{3+} Cd^{2+}	56.95 mg/g 198.34 mg/g	[33]
Etidronic acid-functionalized layered double hydroxide	Zn^{2+} Fe^{+3}	281.36 mg/g 206.03 mg/g	[34]
Biochar-supported zinc oxide (BC-ZnO NCs) and Graphene oxide/zinc oxide (BC-GO/ZnO NCs)	Tartrazine dye	78.85 mg/g (87%) 88.8 mg/g (94%)	[35]
ZnO modified with nanobiochar	Phenol	99.8%	[59]
α-MnO$_2$ photocatalyst	RhB Organic Dye	95–100%	[62]
Biochar-mediated zirconium ferrite Nanocomposites	Tartrazine dye	89.22 mg/g	**This study**

4. Conclusions

The current research aimed to appraise the sorption efficacy of biochar-mediated ZrFe$_2$O$_5$ nanocomposites for the remediation of Tartrazine dye containing textile wastewater. Batch mode experimentations were executed for this purpose. BC-ZrFe$_2$O$_5$ NCs depicted high sorption potential for Tartrazine dye. The Response Surface Methodology was used for the optimization of process parameters. For future practical applications of biochar-based nanocomposites on an industrial scale as efficient and cost effective materials, much work is required. Regarding these aspects, very little literature is available, thus, further studies are required. The application of biochar-mediated nanocomposites is usually concentrated at the laboratory scale only, however, its application for treatment of simulated and real wastewater is still lacking. Real wastewater commonly consists of diverse contaminants, hence, resource recycling and selective elimination of the pollutants is of great importance. The main goal of the present work is to discover novel, efficient and ecofriendly nanomaterials for application in wastewater treatment and new possibilities of combining biochar from cheaper agro-waste biomass with nanomaterials.

Supplementary Materials: The following supporting information can be downloaded at: https://www.mdpi.com/article/10.3390/nano12162828/s1, Figure S1: EDX spectrum of (a) BC-ZrFe$_2$O$_5$ NCs and (b) dye loaded BC-ZrFe$_2$O$_5$ NCs, with an elemental composition; Figure S2: SEM micrographs of WSBC (a–c) and Tartrazine dye loaded WSBC (d–f) at three different magnification levels; Figure S3: Intraparticles diffusion adsorption kinetic model plot for adsorption of Tartrazine dye by WSBM, WSBC, and BC-ZrFe$_2$O$_5$ NCs evaluation of the most common gases during biomass pyrolysis under inert conditions, title; Table S1: ANOVA results for Response Surface Quadratic Model of Tartrazine dye using BC-ZrFe$_2$O$_5$ NCs; Table S2: Comparison between pseudo-first order, pseudo-second-order and intraparticles diffusion kinetic models for Tartrazine Dye; Table S3: Equilibrium isotherm parameters for Tartrazine Dye.

Author Contributions: Conceptualization, S.P. and R.N.; methodology, S.P.; software, M.I.A. and S.P.; validation, T.A., R.N. and J.A.; formal analysis, T.A.; investigation, S.P.; resources, S.P.; data curation, R.N. and T.A.; writing—original draft preparation, S.P.; writing—review and editing, F.N. and T.A.; visualization, F.N.; supervision, R.N.; project administration, R.N. and J.A.; funding acquisition, J.A. All authors have read and agreed to the published version of the manuscript.

Funding: This work was financially supported by the National Science Centre, Poland, under the grant OPUS 14 No. 2017/27/B/ST8/01330.

Institutional Review Board Statement: Not applicable.

Informed Consent Statement: Not applicable.

Data Availability Statement: Not applicable.

Acknowledgments: The authors gratefully acknowledge to the Department of Chemistry, University of Agriculture, Faisalabad for providing nice environment and facilities to carry out this research work.

Conflicts of Interest: The authors declare no conflict of interest.

References

1. Nath, J.; Ray, L.; Bera, D. Continuous removal of malachite green by calcium alginate immobilized *Bacillus cereus* M116 in packed bed column. *Environ. Technol. Innov.* **2016**, *6*, 132–140. [CrossRef]
2. Ismail, M.; Khan, M.I.; Khan, S.B.; Khan, M.A.; Akhtar, K.; Asiri, A.M. Green synthesis of plant supported CuAg and CuNi bimetallic nanoparticles in the reduction of nitrophenols and organic dyes for water treatment. *J. Mol. Liq.* **2018**, *260*, 78–91. [CrossRef]
3. De Lima Barizao, A.C.; Silva, M.F.; Andrade, M.; Brito, F.C.; Gomes, R.G.; Bergamasco, R. Green synthesis of iron oxide nanoparticles for tartrazine and bordeaux red dye removal. *J. Environ. Chem. Eng.* **2020**, *8*, 103618. [CrossRef]
4. Kausar, A.; Naeem, K.; Hussain, T.; Bhatti, H.N.; Jubeen, F.; Nazir, A.; Iqbal, M. Preparation and characterization of chitosan/clay composite for direct Rose FRN dye removal from aqueous media: Comparison of linear and non-linear regression methods. *J. Mater. Res. Technol.* **2019**, *8*, 1161–1174. [CrossRef]
5. Nasrollahzadeh, M.S.; Hadavifar, M.; Ghasemi, S.S.; Chamjangali, M.A. Synthesis of ZnO nanostructure using activated carbon for photocatalytic degradation of methyl orange from aqueous solutions. *Appl. Water Sci.* **2018**, *8*, 1–12. [CrossRef]
6. Asghar, A.; Raman, A.A.A.; Daud, W.M.A.W. Advanced oxidation processes for in-situ production of hydrogen peroxide/hydroxyl radical for textile wastewater treatment: A review. *J. Clean. Prod.* **2015**, *87*, 826–838. [CrossRef]
7. Sampurnam, S.; Muthamizh, S.; Dhanasekaran, T.; Latha, D.; Padmanaban, A.; Selvam, P.; Stephen, A.; Narayanan, V. Synthesis and characterization of Keggin-type polyoxometalate/zirconia nanocomposites—Comparison of its photocatalytic activity towards various organic pollutants. *J. Photochem. Photobiol. A Chem.* **2019**, *370*, 26–40.
8. Ali, M.A.; Mubarak, M.F.; Keshawy, M.; Zayed, M.A.; Ataalla, M. Adsorption of Tartrazine anionic dye by novel fixed bed Core-Shell-polystyrene Divinylbenzene/Magnetite nanocomposite. *Alex. Eng. J.* **2022**, *61*, 1335–1352. [CrossRef]

9. Essandoh, M.; Garcia, R.A.; Palochik, V.L.; Gayle, M.R.; Liang, C. Simultaneous adsorption of acidic and basic dyes onto magnetized polypeptidylated-Hb composites. *Sep. Purif. Technol.* **2021**, *255*, 117701. [CrossRef]

10. Joseph, J.; Radhakrishnan, R.C.; Johnson, J.K.; Joy, S.P.; Thomas, J. Ion-exchange mediated removal of cationic dye-stuffs from water using ammonium phosphomolybdate. *Mater. Chem. Phys.* **2020**, *242*, 122488. [CrossRef]

11. Yao, J.; Zhang, Y.; Dong, Z. Enhanced degradation of contaminants of emerging concern by electrochemically activated peroxymonosulfate: Performance, mechanism, and influencing factors. *Chem. Eng. J.* **2021**, *415*, 128938. [CrossRef]

12. Ayodele, B.V.; Alsaffar, M.A.; Mustapa, S.I.; Cheng, C.K.; Witoon, T. Modeling the effect of process parameters on the photocatalytic degradation of organic pollutants using artificial neural networks. *Process Saf. Environ. Prot.* **2021**, *145*, 120–132. [CrossRef]

13. Ebrahimpoor, S.; Kiarostami, V.; Khosravi, M.; Davallo, M.; Ghaedi, A. Optimization of tartrazine adsorption onto polypyrrole/SrFe12O19/graphene oxide nanocomposite using central composite design and bat inspired algorithm with the aid of artificial neural networks. *Fibers Polym.* **2021**, *22*, 159–170. [CrossRef]

14. Ouassif, H.; Moujahid, E.M.; Lahkale, R.; Sadik, R.; Bouragba, F.Z.; Diouri, M. Zinc-Aluminum layered double hydroxide: High efficient removal by adsorption of tartrazine dye from aqueous solution. *Surf. Interfaces* **2020**, *18*, 100401. [CrossRef]

15. Naseem, K.; Farooqi, Z.H.; Begum, R.; Irfan, A. Removal of Congo red dye from aqueous medium by its catalytic reduction using sodium borohydride in the presence of various inorganic nano-catalysts: A review. *J. Clean. Prod.* **2018**, *187*, 296–307. [CrossRef]

16. Burakov, A.E.; Galunin, E.V.; Burakova, I.V.; Kucherova, A.E.; Agarwal, S.; Tkachev, A.G.; Gupta, V.K. Adsorption of heavy metals on conventional and nanostructured materials for wastewater treatment purposes: A review. *Ecotoxicol. Environ. Saf.* **2018**, *148*, 702–712. [CrossRef]

17. Shakya, A.; Núñez-Delgado, A.; Agarwal, T. Biochar synthesis from sweet lime peel for hexavalent chromium remediation from aqueous solution. *J. Environ. Manag.* **2019**, *251*, 109570. [CrossRef]

18. Ito, D.; Nishimura, K.; Miura, O. Removal and recycle of phosphate from treated water of sewage plants with zirconium ferrite adsorbent by high gradient magnetic separation. *J. Phys. Conf. Ser.* **2009**, *156*, 012033. [CrossRef]

19. Tan, X.F.; Liu, Y.G.; Gu, Y.L.; Xu, Y.; Zeng, G.M.; Hu, X.J.; Liu, S.B.; Wang, X.; Liu, S.M.; Li, J. Biochar-based nano-composites for the decontamination of wastewater: A review. *Bioresour. Technol.* **2016**, *212*, 318–333. [CrossRef]

20. Banerjee, S.; Joshi, S.; Mandal, T.; Halder, G. Application of zirconium caged activated biochar alginate beads towards deionization of Cr (VI) laden water in a fixed bed column reactor. *J. Environ. Chem. Eng.* **2018**, *6*, 4018–4029. [CrossRef]

21. Vikrant, K.; Kim, K.H.; Ok, Y.S.; Tsang, D.C.; Tsang, Y.F.; Giri, B.S.; Singh, R.S. Engineered/designer biochar for the removal of phosphate in water and wastewater. *Sci. Total Environ.* **2018**, *616*, 1242–1260. [CrossRef] [PubMed]

22. Imran, M.; Khan, Z.U.H.; Iqbal, M.M.; Iqbal, J.; Shah, N.S.; Munawar, S.; Ali, S.; Murtaza, B.; Naeem, M.A.; Rizwan, M. Effect of biochar modified with magnetite nanoparticles and HNO3 for efficient removal of Cr (VI) from contaminated water: A batch and column scale study. *Environ. Pollut.* **2020**, *261*, 114231. [CrossRef] [PubMed]

23. Aichour, A.; Zaghouane-Boudiaf, H. Highly brilliant green removal from wastewater by mesoporous adsorbents: Kinetics, thermodynamics and equilibrium isotherm studies. *Microchem. J.* **2019**, *146*, 1255–1262. [CrossRef]

24. Zhang, J.; Ge, Y.; Li, Z.; Wang, Y. Facile fabrication of a low-cost and environmentally friendly inorganic-organic composite membrane for aquatic dye removal. *J. Environ. Manag.* **2020**, *256*, 109969. [CrossRef]

25. Cruz, G.J.; Mondal, D.; Rimaycuna, J.; Soukup, K.; Gómez, M.M.; Solis, J.L.; Lang, J. Agrowaste derived biochars impregnated with ZnO for removal of arsenic and lead in water. *J. Environ. Chem. Eng.* **2020**, *8*, 103800. [CrossRef]

26. Gonçalves, M.G.; da Silva Veiga, P.A.; Fornari, M.R.; Peralta-Zamora, P.; Mangrich, A.S.; Silvestri, S. Relationship of the physicochemical properties of novel ZnO/biochar composites to their efficiencies in the degradation of sulfamethoxazole and methyl orange. *Sci. Total Environ.* **2020**, *748*, 141381. [CrossRef]

27. Wickramaarachchi, W.K.P.; Minakshi, M.; Gao, X.; Dabare, R.; Wong, K.W. Hierarchical porous carbon from mango seed husk for electro-chemical energy storage. *Chem. Eng. J. Adv.* **2021**, *8*, 100158. [CrossRef]

28. Iqbal, M.M.; Imran, M.; Hussain, T.; Naeem, M.A.; Al-Kahtani, A.A.; Shah, G.M.; Ahmad, S.; Farooq, A.; Rizwan, M.; Majeed, A.; et al. Effective sequestration of Congo red dye with ZnO/cotton stalks biochar nanocomposite: MODELING, reusability and stability. *J. Saudi Chem. Soc.* **2021**, *25*, 101176. [CrossRef]

29. Van, H.T.; Nguyen, L.H.; Dang, N.V.; Chao, H.P.; Nguyen, Q.T.; Nguyen, T.H.; Nguyen, T.B.L.; Van Thanh, D.; Nguyen, H.D.; Thang, P.Q.; et al. The enhancement of reactive red 24 adsorption from aqueous solution using agricultural waste-derived biochar modified with ZnO nanoparticles. *RSC Adv.* **2021**, *11*, 5801–5814. [CrossRef]

30. Ali, S.A.M. Networks of Effectiveness? The Impact of Politicization on Bureaucratic Performance in Pakistan. *Eur. J. Dev. Res.* **2022**, *34*, 733–753. [CrossRef]

31. Jafari-Arvari, H.; Saei-Dehkordi, S.S.; Farhadian, S. Evaluation of interactions between food colorant, tartrazine, and Apo-transferrin using spectroscopic analysis and docking simulation. *J. Mol. Liq.* **2021**, *339*, 116715. [CrossRef]

32. Perveen, S.; Nadeem, R.; Ali, S.; Jamil, Y. Biochar caged zirconium ferrite nanocomposites for the adsorptive removal of Reactive Blue 19 dye in a batch and column reactors and conditions optimizaton. *Z. Für Phys. Chem.* **2021**, *235*, 1721–1745. [CrossRef]

33. Zhu, S.; Khan, M.A.; Kameda, T.; Xu, H.; Wang, F.; Xia, M.; Yoshioka, T. New insights into the capture performance and mechanism of hazardous metals Cr^{3+} and Cd^{2+} onto an effective layered double hydroxide based material. *J. Hazard. Mater.* **2022**, *426*, 128062. [CrossRef]

34. Zhu, S.; Chen, Y.; Khan, M.A.; Xu, H.; Wang, F.; Xia, M. In-depth study of heavy metal removal by an etidronic acid-functionalized layered double hydroxide. *ACS Appl. Mater. Interfaces* **2022**, *14*, 7450–7463. [CrossRef]

35. Perveen, S.; Nadeem, R.; Nosheen, F.; Tongxiang, L.; Anwar, T. Synthesis of biochar-supported zinc oxide and graphene oxide/zinc oxide nanocomposites to remediate tartrazine dye from aqueous solution using fixed-bed column reactor. *Appl. Nanosci.* **2022**, *12*, 1491–1505. [CrossRef]

36. Wang, L.; Wang, J.; He, C.; Lyu, W.; Zhang, W.; Yan, W.; Yang, L. Development of rare earth element doped magnetic biochars with enhanced phosphate adsorption performance. *Colloids Surf. A Physicochem. Eng. Asp.* **2019**, *561*, 236–243. [CrossRef]

37. Maqbool, M.; Bhatti, H.N.; Sadaf, S.; Zahid, M.; Shahid, M. A robust approach towards green synthesis of polyaniline-Scenedesmus biocomposite for wastewater treatment applications. *Mater. Res. Express* **2019**, *6*, 055308. [CrossRef]

38. Ali, M.E.M.; Assirey, E.A.; Abdel-Moniem, S.M.; Ibrahim, H.S. Low temperature-calcined TiO$_2$ for visible light assisted decontamination of 4-nitrophenol and hexavalent chromium from wastewater. *Sci. Rep.* **2019**, *9*, 19354. [CrossRef]

39. Tabasum, A.; Zahid, M.; Bhatti, H.N.; Asghar, M. Fe$_3$O$_4$-GO composite as efficient heterogeneous photo-Fenton's catalyst to degrade pesticides. *Mater. Res. Express* **2018**, *6*, 015608. [CrossRef]

40. Sadaf, S.; Bhatti, H.N. Response surface methodology approach for optimization of adsorption process for the removal of Indosol Yellow BG dye from aqueous solution by agricultural waste. *Desalinat. Water Treat.* **2016**, *57*, 11773–11781. [CrossRef]

41. Dawodu, F.A.; Onuh, C.U.; Akpomie, K.G.; Unuabonah, E.I. Synthesis of silver nanoparticle from Vigna unguiculata stem as adsorbent for malachite green in a batch system. *SN Appl. Sci.* **2019**, *1*, 346. [CrossRef]

42. Sajjadi, S.A.; Meknati, A.; Lima, E.C.; Dotto, G.L.; Mendoza-Castillo, D.I.; Anastopoulos, I.; Alakhras, F.; Unuabonah, E.I.; Singh, P.; Hosseini-Bandegharaei, A. A novel route for preparation of chemically activated carbon from pistachio wood for highly efficient Pb (II) sorption. *J. Environ. Manag.* **2019**, *236*, 34–44. [CrossRef]

43. Huang, H.; Wang, Y.; Zhang, Y.; Niu, Z.; Li, X. Amino-functionalized graphene oxide for Cr (VI), Cu (II), Pb (II) and Cd (II) removal from industrial wastewater. *Open Chem.* **2020**, *18*, 97–107. [CrossRef]

44. Zhang, M.; Liu, Y.-H.; Shang, Z.-R.; Hu, H.-C.; Zhang, Z.-H. Supported molybdenum on graphene oxide/Fe$_3$O$_4$: An efficient, magnetically separable catalyst for one-pot construction of spiro-oxindole dihydropyridines in deep eutectic solvent under microwave irradiation. *Catal. Commun.* **2017**, *88*, 39–44. [CrossRef]

45. Cruz, J.C.; Nascimento, M.A.; Amaral, H.A.; Lima, D.S.; Teixeira, A.P.C.; Lopes, R.P. Synthesis and characterization of cobalt nanoparticles for application in the removal of textile dye. *J. Environ. Manag.* **2019**, *242*, 220–228. [CrossRef] [PubMed]

46. Melo, R.; Neto, E.B.; Nunes, S.; Dantas, T.C.; Neto, A.D. Removal of Reactive Blue 14 dye using micellar solubilization followed by ionic flocculation of surfactants. *Sep. Purif. Technol.* **2018**, *191*, 161–166. [CrossRef]

47. Naeem, H.; Bhatti, H.N.; Sadaf, S.; Iqbal, M. Uranium remediation using modified Vigna radiata waste biomass. *Appl. Radiat. Isot.* **2017**, *123*, 94–101. [CrossRef]

48. Ooi, G.T.; Tang, K.; Chhetri, R.K.; Kaarsholm, K.M.; Sundmark, K.; Kragelund, C.; Litty, K.; Christensen, A.; Lindholst, S.; Sund, C.; et al. Biological removal of pharmaceuticals from hospital wastewater in a pilot-scale staged moving bed biofilm reactor (MBBR) utilising nitrifying and denitrifying processes. *Bioresour. Technol.* **2018**, *267*, 677–687. [CrossRef]

49. Benally, C.; Messele, S.A.; El-Din, M.G. Adsorption of organic matter in oil sands process water (OSPW) by carbon xerogel. *Water Res.* **2019**, *154*, 402–411. [CrossRef]

50. Salam, K.A. Assessment of Surfactant Modified Activated Carbon for Improving Water Quality. *J. Encapsulat. Adsorpt. Sci.* **2019**, *9*, 13. [CrossRef]

51. Abbas, R.; Hami, H.; Mahdi, N. Removal of doxycycline hyclate by adsorption onto cobalt oxide at three different temperatures: Isotherm, thermodynamic and error analysis. *Int. J. Environ. Sci. Technol.* **2019**, *16*, 5439–5446. [CrossRef]

52. Noorimotlagh, Z.; Mirzaee, S.A.; Martinez, S.S.; Alavi, S.; Ahmadi, M.; Jaafarzadeh, N. Adsorption of textile dye in activated carbons prepared from DVD and CD wastes modified with multi-wall carbon nanotubes: Equilibrium isotherms, kinetics and thermodynamic study. *Chem. Eng. Res. Des.* **2019**, *141*, 290–301. [CrossRef]

53. Sakdasri, W.; Sawangkeaw, R.; Ngamprasertsith, S. Response surface methodology for the optimization of biofuel production at a low molar ratio of supercritical methanol to used palm olein oil. *Asia-Pac. J. Chem. Eng.* **2016**, *11*, 539–548. [CrossRef]

54. Bezerra, M.A.; Santelli, R.E.; Oliveira, E.P.; Villar, L.S.; Escaleira, L.A. Response surface methodology (RSM) as a tool for optimization in analytical chemistry. *Talanta* **2008**, *76*, 965–977. [CrossRef]

55. Hemmati, A.; Rashidi, H. Optimization of industrial intercooled post-combustion CO$_2$ absorber by applying rate-base model and response surface methodology (RSM). *Process Saf. Environ. Prot.* **2019**, *121*, 77–86. [CrossRef]

56. Saeed, M.; Nadeem, R.; Yousaf, M. Removal of industrial pollutant (Reactive Orange 122 dye) using environment-friendly sorbent Trapa bispinosa's peel and fruit. *Int. J. Environ. Sci. Technol.* **2015**, *12*, 1223–1234. [CrossRef]

57. Siyal, A.A.; Shamsuddin, M.R.; Khan, M.I.; Rabat, N.E.; Zulfiqar, M.; Man, Z.; Siame, J.; Azizli, K.A. A review on geopolymers as emerging materials for the adsorption of heavy metals and dyes. *J. Environ. Manag.* **2018**, *224*, 327–339. [CrossRef]

58. Ullah, A.; Rahman, L.; Hussain, S.Z.; Yazdani, M.B.; Jilani, A.; Iqbal Khan, D.; Nasir, M.Z.; Khan, W.S.; Hussain, I.; Rehman, A. Tin Oxide Supported on Nanostructured MnO$_2$ as Efficient Catalyst for Nitrophenol Reduction: Kinetic Analysis and Their Application as Heterogeneous Catalyst. *Mater. Innov.* **2022**, *2*, 83–91. [CrossRef]

59. Tho, P.T.; Van, H.T.; Nguyen, L.H.; Hoang, T.K.; Tran, T.N.H.; Nguyen, T.T.; Nguyen, T.B.H.; Le Sy, H.; Tran, Q.B.; Sadeghzadeh, S.M.; et al. Enhanced simultaneous adsorption of As (iii), Cd (ii), Pb (ii) and Cr (vi) ions from aqueous solution using cassava root husk-derived biochar loaded with ZnO nanoparticles. *RSC Adv.* **2021**, *11*, 18881–18897. [CrossRef]

60. Dar, A.; Ashfaq, H.; Sherin, L.; Salman, M.; Anwar, J. Synthesis, Characterization and Applications of Plant Based Silver for Effective Removal of Chromium(VI) from Contaminated Water. *Mater. Innov.* **2021**, *1*, 56–63. [CrossRef]

61. Ikram, A.; Jamil, S.; Fasehullah, M. Green Synthesis of Copper Oxide Nanoparticles from Papaya/Lemon tea Extract and its Application in Degradation of Methyl Orange. *Mater. Innov.* **2022**, *2*, 115–122. [CrossRef]
62. Baral, A.; Das, D.P.; Minakshi, M.; Ghosh, M.K.; Padhi, D.K. Probing Environmental Remediation of RhB Organic Dye Using α-MnO$_2$ under Visible-Light Irradiation: Structural, Photocatalytic and Mineralization Studies. *ChemistrySelect* **2016**, *1*, 4277–4285. [CrossRef]

Article

Hybrid Plasma–Liquid Functionalisation for the Enhanced Stability of CNT Nanofluids for Application in Solar Energy Conversion

Ruairi J. McGlynn *, Hussein S. Moghaieb, Paul Brunet, Supriya Chakrabarti, Paul Maguire and Davide Mariotti *

Nanotechnology and Integrated Bio-Engineering Centre (NIBEC), Ulster University, Newtownabbey BT37 0QB, UK

* Correspondence: r.mcglynn@ulster.ac.uk (R.J.M.); d.mariotti@ulster.ac.uk (D.M.); Tel.: +44-7548493594 (R.J.M.); +44-28-9536-5266 (D.M.)

Abstract: Macroscopic ribbon-like assemblies of carbon nanotubes (CNTs) are functionalised using a simple direct-current-based plasma–liquid system, with oxygen and nitrogen functional groups being added. These modifications have been shown to reduce the contact angle of the ribbons, with the greatest reduction being from 84° to 35°. The ability to improve the wettability of the CNTs is of paramount importance for producing nanofluids, with relevance for a number of applications. Here, in particular, we investigate the efficacy of these samples as nanofluid additives for solar–thermal harvesting. Surface treatments by plasma-induced non-equilibrium electrochemistry are shown to enhance the stability of the nanofluids, allowing for full redispersion under simulated operating conditions. Furthermore, the enhanced dispersibility results in both a larger absorption coefficient and an improved thermal profile under solar simulation.

Keywords: solar–thermal; plasma functionalisation; carbon nanotubes

check for
updates

Citation: McGlynn, R.J.; Moghaieb, H.S.; Brunet, P.; Chakrabarti, S.; Maguire, P.; Mariotti, D. Hybrid Plasma–Liquid Functionalisation for the Enhanced Stability of CNT Nanofluids for Application in Solar Energy Conversion. *Nanomaterials* **2022**, *12*, 2705. https://doi.org/10.3390/nano12152705

Academic Editors: M.M. Bhatti, Kambiz Vafai and Sara I. Abdelsalam

Received: 4 July 2022
Accepted: 1 August 2022
Published: 6 August 2022

Publisher's Note: MDPI stays neutral with regard to jurisdictional claims in published maps and institutional affiliations.

1. Introduction

Zero-carbon renewable energy sources such as solar energy provide an alternative method to satisfy the continued increase in energy demand without the concomitant environmental degradation caused by burning fossil fuels. One method to obtain thermal energy from solar sources is to utilise direct-absorption solar collectors (DASCs) [1], where the working fluid acts both to absorb incident solar radiation and to transfer the captured energy to a hot-water tank or other storage systems. As typical heat-transfer fluids (e.g., water, ethylene glycol, oil) are poor absorbers in the visible-light region [1–3], it is necessary to introduce additives to effectively capture a significant portion of this high-energy region. Early additives included micron-scale carbonaceous particles, which whilst increasing the absorption of light in the visible region, also had the negative effects of fouling the transparent covers and clogging the pumping systems, and therefore limited the lifetime and efficiency of the overall system [1,2]. Reduction in system fouling has been achieved by adding nanoscale particles to produce nanofluids, whilst also providing enhanced optical absorption with lower particle concentrations, as demonstrated for aluminium [4]. Beyond this, other nanoparticles have been tested, including gold [5,6], silver [7,8], and copper oxide nanoparticles [9–11], providing the opportunity to tune the optical absorption of the working fluid by blending a carefully selected range of these additives.

By comparison to these noble- or transition-metal additives, carbon-based materials are relatively Earth-abundant and eco-friendly. Additionally, some of these materials—such as graphene/graphene oxide [12,13] and carbon nanotubes (CNTs) [14–18]—demonstrate extremely high optical absorption, approaching blackbody levels [19], which when paired with exceptional thermal conductivity [20,21] present attractive options for nanofluid additives. CNTs are potentially one of the most attractive options, as they undergo a phenomenon called "shear thinning", which is beneficial to their application in DASCs. Where the fluid is constantly pumped, the CNTs can align in the direction of the liquid flow, which at low CNT concentrations can result in a working fluid with a viscosity even lower than that of the base fluid alone [22]. At greater shear rates, entangled and agglomerated CNTs can be deagglomerated, enhancing the stability of the nanofluid [23].

Initial studies have demonstrated the high suitability of functionalised CNTs as nanofluid additives, with much of the thermal conductivity and optical properties of the CNTs lent to the nanofluid even at low CNT contents. At 150 parts per million carboxylate-functionalised CNTs, Karami et al. obtained an extinction coefficient of 4.1 cm^{-1} and a 32.2% thermal conductivity increase when using water as the base fluid [15]. Furthermore, the authors noted that higher dispersions were possible, although these absorbed too much light to perform any optical measurements. Hordy et al. [18] addressed the suitability of water- and glycol-based nanofluids for usage at 85 °C and 170 °C, respectively, with oxygen-grafted plasma-functionalised CNTs. Beyond this, they also confirmed a storage stability of over eight months in glycol-based fluids, losing only 8.6% of the effective concentration. It was noted that the water-based fluid lost approximately 5% of the effective concentration per month. This is echoed in other similar works, where even ultralow contents of CNT additives result in near-perfect absorption (0.2 volume percent) and substantially improved thermal conductivity over the base fluid alone (0.508 W m^{-1}K^{-1}) [24]. Studies of CNTs within DASC devices have also yielded promising results, with carboxyl-functionalised CNTs in a 30:70 ethylene glycol:water mixture demonstrating an improvement in collector efficiency of between 10 and 29% compared to the base fluid alone, with increasing CNT contents and flow rates giving greater enhancements [14]. The improvement was attributed to the 54.4% increase in thermal conductivity at 100 ppm CNTs coupled with an extinction coefficient of over 5 cm^{-1}. In a tubular collector configuration, similar benefits of CNT additives were found [25], with a 0.01 weight percentage of surfactant-stabilised CNTs yielding a thermal efficiency of 80%, which was 37.9% greater than that of a comparable opaque receiver plate. The full-scale system was operated for 45 days and 15 h per day, with no degradation to the pumping system noted, although there was evidence of some deposition of materials near the pump impeller.

CNTs are inherently hydrophobic and, thus, require modification of the surface chemistry to enable superior dispersion and stability within a fluid matrix [26], as is the case with each of the above examples. Two main methods exist to remedy this: coating with surfactants, and chemical treatments. Surfactants can provide a repulsive force, due to zeta potential, by creating a surface charge, which results in improved stability and lifetime of the nanomaterials in solution [27]. Whilst these have been demonstrated to provide high stability, there are three main limiting factors in the use of surfactants for stabilisation: (1) the surfactant can degrade at temperatures above 100 °C [15,16,27–29], although some are capable of working up to 300 °C, such as sodium dodecylbenzenesulfonate [16]; (2) the coated surfactant can lower the thermal conductivity of the individual carbon additives, though in some cases this may be overcome by the enhanced distribution of the absorbing additives in the fluid [28,29]; and (3) the surfactants can modify the refractive index and scattering of the particles, inducing a spectral shift [30]. As an alternative to surfactants, the surface functional groups of the CNTs can be modified by using strong oxidants such as sulfuric or nitric acids [15,16], or weak oxidants such as potassium hydroxide [15], to add hydroxyl and carboxylic groups. However, these chemical methods often require protracted treatment times and the use of harsh, often hazardous chemicals. This highlights the need for an alternative pathway to produce CNTs that are stable in aqueous solutions and at higher temperatures.

One option for this is to utilise plasma treatments, which can both modify the surface chemistry and remove excess amorphous carbons [26]. With plasma-based treatments, reduced reaction times and the avoidance of aggressive chemicals—such as concentrated acids—are key improvements over traditional covalent chemical methodologies [26]. A range of low-pressure plasma systems have been utilised to provide oxygen- and nitrogen-based functional groups, including microwave-source [31,32] and radio-frequency systems [33–37], with the radio-frequency systems also reporting the additional benefit of the removal of excess amorphous carbons. Eliminating the vacuum equipment provides the benefits of reducing the capital investment involved in establishing a process, as well as simplifying the processing itself by allowing direct processing of liquids and colloids, such as in plasma-induced non-equilibrium electrochemistry [38–40]. The use of plasma-induced non-equilibrium electrochemistry for the functionalisation of nanocellulose has been demonstrated in water-based suspensions, with both oxygen and nitrogen functional groups added depending on the plasma conditions and electrolytes [41].

In this work, an atmospheric-pressure plasma treatment is demonstrated for the formation of oxygen and nitrogen functional groups on the sidewalls of macroscopic CNT assemblies. This system utilises a direct-current (DC) source between a submerged electrode (CNT ribbon) and a microplasma counter-electrode, across an electrolyte solution with or without a nitrogen precursor [42].

2. Materials and Methods

2.1. CNT Synthesis

The CNTs in this work were produced in a floating-catalyst chemical vapour deposition system that produces ribbon-like macroscopic assemblies of continuous length [43]. Ferrocene, thiophene, and methane were flown in a furnace [43] with hydrogen as a carrier gas, and where the iron catalyst nanoparticles were formed by the thermal decomposition of ferrocene powder. After the hot zone of the furnace, the material began to cool, and a dense web of dark carbon material formed in the centre of the work tube. This aerogel was sufficiently strong that it could be drawn out with a rod to form a long black "sock".

To produce the ribbon-like CNT assemblies, gas flows of 115.5 standard cubic centimetres per minute (sccm) of hydrogen through the ferrocene bubbler, 90 sccm of hydrogen through the thiophene bubbler, a methane flow rate of 160 sccm, and a carrier hydrogen flow rate of 1470 sccm were flown through a tube furnace maintained at 1290 °C. In the context of this work, the term "ribbon" is used to describe the material obtained by this process after pressing the macroscopic CNT aerogel between two glass microscope slides for 1 min under a 2 kg force to create a 3–4 cm long, 0.5–1.0 cm wide, and 1–10 μm thick flat ribbon-like material.

2.2. CNT Functionalisation

The CNT ribbon samples were subjected to different treatments and compared to the pristine samples. Prior to plasma functionalisation, all samples discussed in this manuscript were annealed for 15 min at 350 °C in a Carbolite VMF 10/15 furnace with a standard air atmosphere. Samples not pre-treated with this annealing step are documented in the Supplementary Materials, and show a slightly lower overall performance. A plasma-induced non-equilibrium electrochemistry system was used for the functionalisation of the CNT ribbons (Figure 1). In this system, direct-current microplasma is generated in helium between a nickel capillary tube (inner diameter of 0.7 mm and outer diameter of 1 mm) and the electrolyte surface. The helium gas flow of 25 sccm (controlled by a mass flow controller, MKS Instruments, UK) ensures that the plasma is largely formed in helium gas, with a small amount of turbulent mixing with the surrounding air expected. A voltage of 1.3 kV was applied to the CNT ribbons using a Matsusada AU-10 * 15 power supply, and all samples of CNTs were treated at a constant current of 10 mA for 15 min. Over the treatment period, the applied voltage dropped from approximately 1.3 kV to approximately 1 kV in order to maintain the constant current. The CNT ribbon was secured in place by a PTFE support frame (Figure 1b), which prevented the ribbon from moving during the treatment. An exposed surface of 1.5 cm in length and between 0.5 cm and 1 cm in width was exposed to the electrolyte solution. We used two different solutions for the plasma–liquid treatment. The first comprised ethanol and water at a ratio of 1:9, with a volume of 18 mL, hereafter referred to as "Plasma–liquid". The second, "Plasma–liquid with EDA", included a 10% volume of ethylenediamine with the same stock solution of 1:9 ethanol:water—again with a volume of 18 mL. After the plasma treatments, the ribbons were submerged in a beaker of distilled water for 5 min.

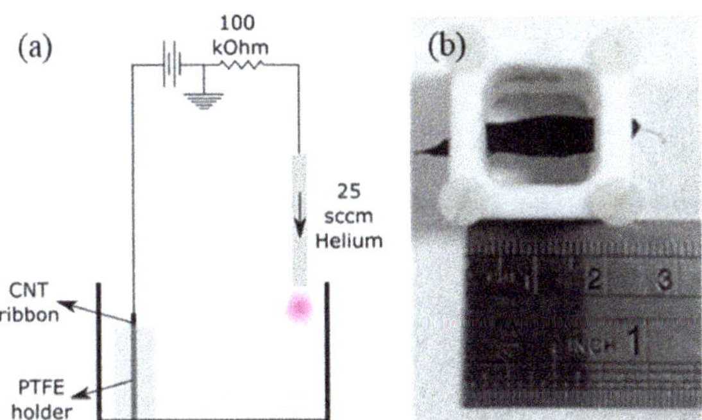

Figure 1. (**a**) The configuration of the plasma-induced non-equilibrium electrochemistry system used to treat the ribbons, and (**b**) a photograph illustrating the PTFE holder used to secure the carbon nanotubes.

2.3. CNT Ribbon Characterisation

X-ray photoelectron spectroscopy (XPS) was used in this study to analyse the elemental composition and the chemical bonding. The spectrometer was an ESCALAB XI⁺ instrument, (Thermo Fisher, East Grinstead, United Kingdom). The base pressure during spectral acquisition was above 5×10^{-7} mbar, achieved using an Edwards E2M28 rotary vane pump. The main background gas in the analysis chamber was argon at all points during the loading, pumping, and measurements. The excitation source was a monochromated aluminium anode with an excitation energy of 1486.68 eV operated at approximately 15 kV and 15 mA, giving a source power of 225 W. The work function of the spectrometer was determined by the software as 4.68 eV. The recorded spectra included C 1s, O 1s, N 1s, Fe 2p, S 2p, and Pt 4f (reference), which were acquired sequentially, with a total acquisition time of 662 s. With the selected scan parameters, the energy resolution was 0.1 eV for high-resolution spectra and 1 eV for survey spectra. The size of the analysed sample area was 650 μm, taking the form of an elongated circle. The samples were stored overnight in standard lab conditions before loading into the spectrometer. The transfer procedure within the fast-entry lock of the spectrometer included exposure to 2×10^{-6} mbar, achieved with an Edwards RV5 rotary vane pump (Edwards, West Sussex, United Kingdom) in less than 10 min prior to XPS analysis. Charge compensation, by means of an electron beam, was applied via a flood gun operated at 100 μA. The charge referencing method was performed by shifting the asymmetric Pt $4f_{7/2}$ peak of freshly sputter-cleaned Pt foil to 71.2 eV. The foil was pressed onto the surface of the CNT ribbons with a copper clip. The sputter-cleaning was performed with a monoatomic argon ion gun held at 4000 eV and 15 mA at an angle of 30° to the sample over a 1.5 mm wide square. This was expected to give a sputter rate of 0.97 nm s^{-1} for tantalum(V) oxide.

A Horiba LabRAM 300 spectrometer with a 632.81 nm helium–neon fixed-wavelength laser was used to obtain Raman spectra. Measurements were taken 1 cm apart for a total of five measurements per sample to assess the variance of the graphitisation within the sample. An 8th-order polynomial baseline subtraction was performed using the system software (LabSpec, Horiba). The D band was considered the peak at 1335 cm^{-1}, with the G band at 1580 cm^{-1}. The third peak at 1622 cm^{-1} was considered as the D′ band, and was fitted to exclude it from the G-band area. The area enclosed by each of these components was calculated by integrating the peaks with OriginPro 9 software, where a peak was defined as having a minimum height of 15% of the data and a width between

5% and 15% of the data. Finally, the area ratio of G:D was calculated by dividing the two integrated areas.

The effect of the treatments on the hydrophobicity of the CNT ribbons was assessed by contact angle measurements with a KSV Instruments CAM 200 system. A drop of distilled water was placed on the surface of the ribbons and allowed to settle for 15 s before an image was taken. The contact angle was then calculated by the system software and recorded. This was performed 5 times per sample to generate an average and a standard deviation.

2.4. Nanofluid Preparation and Optical Characterisation

The nanofluids were prepared by weighing the CNT ribbons on a microbalance and then adding the ribbon and the necessary volume of ethylene glycol to a sample vial to produce a concentration of 25 mg L^{-1}. The CNT ribbon was then dispersed as separate CNTs using a Sonics VCX 130PB system with a 3 mm probe diameter, operated at 70% power for 300 s, giving an average tip power of 10 W over the duration.

The optical properties of the produced nanofluids were tested with a PerkinElmer Lambda 650 system with a 150 mm integrating sphere. This system allows for the determination of the constituent parts of light attenuation by the sample: transmittance, scattering, and absorption. Furthermore, the absorption, scattering, and extinction coefficients can be calculated. A quartz cuvette with a path length of 1 cm was filled with pure ethylene glycol and used as a reference. Measurements were carried out both in standard transmittance mode and within the integrating sphere. The calculations for the absorption, scattering coefficients, and power are provided in the Supplementary Materials.

For UV–Vis data collected from day 120 onwards, the instrument was upgraded to a PerkinElmer Lambda 1050+, although the sample integrating sphere unit was migrated to the new system.

2.5. Methods for the Assessment of the Solar–Thermal Conversion

Solar simulation experiments were performed with a 150 W xenon arc lamp with a 25 mm diameter uniform beam (LS0106, Lot-QuantumDesign). This lamp was calibrated to an intensity of one sun (1000 W m^{-2}) across the wavelength range of 400–1100 nm. As illustrated in Figure 2, a small-volume transparent vessel (>92% transmittance in visible light [44]) was produced from poly(methyl methacrylate), with the following dimensions: internal diameter of 21 mm, base thickness of 1.7 mm, and sidewall thickness of 2.5 mm. A cover made from a 1.8 mm thick quartz disc of 21 mm in diameter was used as a lid for the vessel to minimise heat loss by natural convection during the solar–thermal conversion (STC) experiments. The transmittance of the cover was experimentally measured and found to be ≈94% (see Supplementary Materials) over a wavelength range of 300–2500 nm, i.e., a net intensity of 940 W m^{-2} reached the nanofluid surface. Then, 800 μL of each nanofluid was added, giving a fluid height of approximately 2.3 mm from the base of the vessel. This small volume allowed for a more uniform heat distribution within the sample. T-type thermocouples were used to measure the sample and ambient air temperatures, with the data recorded onto a PC via a TC-08 thermocouple data acquisition module (Omega). A black plate was positioned 3 cm below the sample vessel to capture transmitted light and prevent reflection back into the sample. No thermal insulation was used in these experiments, in order to allow for more accurate calculation of the various parameters. This setup allowed for temperature measurements of static-state nanofluid samples exposed to solar irradiation. On this basis, we were able to then calculate the total power absorbed, which was the sum of the stored thermal power within the nanofluid and the heat loss to the surroundings. Full details of the calculations are provided in the Supplementary Materials.

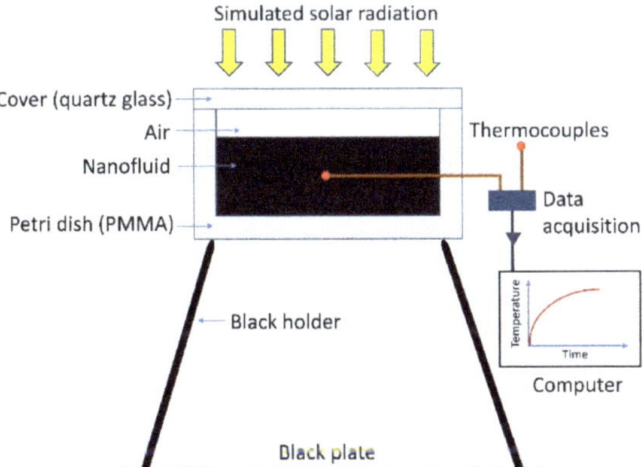

Figure 2. Schematic of the experimental solar thermal conversion setup used to measure the time-dependent temperature variation of nanofluids exposed to simulated solar irradiation.

3. Results

3.1. CNT Ribbon Chemical Analysis

XPS was used to assess the functionalisation of the CNTs by the plasma–liquid treatments, with a summary of the atomic percentages given in Figure 3. The oxygen content was 2.5 at% for the pristine sample, and was slightly decreased after the annealing step to 2.2 at%. Both the pristine and annealed samples showed no nitrogen. When CNTs were subjected to either plasma treatment, an increase in oxygen content was observed, and the presence of nitrogen was then detected. The oxygen content increased to 7.4 at% with the plasma–liquid treatment, with a lesser increase to 4.8 at% with the plasma–liquid with EDA treatment. The nitrogen content was 5 at% for the plasma–liquid sample and 7 at% when EDA was also included. Whilst nitrogen was not seen in the pristine or annealed-only materials, the plasma treatment in both electrolyte solutions resulted in the inclusion of nitrogen within the sample. Reactive nitrogen species such as nitrate, nitrite, and the corresponding nitric acid can be generated in the plasma due to the interaction of the plasma and the surrounding atmospheric nitrogen. These reactive species can then be dissolved within the liquid, where they may interact with the CNTs if they are sufficiently long-lived [45]. As a result, even in the absence of a deliberately dissolved nitrogen precursor, the plasma liquid samples presented a relatively high content of nitrogen. Sulphur and iron were present due to the catalysts, and were within the range 0.4–0.7 at% and 0.6–1.1 at%, respectively, for all samples.

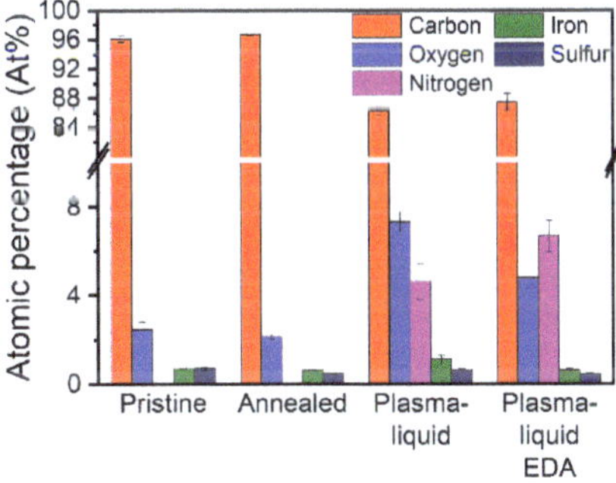

Figure 3. X-ray photoelectron spectroscopy summary of atomic percentages by element for oxygen and nitrogen. The remaining percentage is composed of carbon.

Figure 4 shows the high-resolution spectra for oxygen and nitrogen. All of the O 1s spectra present two features corresponding to the carbon–oxygen bonding: C–O and C=O. Annealing results in an increase in the C=O peak area (by 6%) and an increase in the C–O peak area (by 29%) compared to the pristine sample. Both peak areas are much larger for both the plasma–liquid-treated samples. In particular, the plasma–liquid sample exhibits the most oxygen functionalisation, with the largest peak areas for C=O and C–O species—increased by 204% and 311%, respectively, with respect to the pristine sample. Whilst the plasma–liquid with EDA treatment also enhances the C=O peak area (61% increase) and C–O peak area (383% increase) components significantly compared to the pristine material, its total carbon–oxygen bonding is lower than that of the plasma–liquid-treated material without EDA. The peak that is linked to Fe_xO_y develops after annealing, and is further enhanced after either plasma treatment. This could be due to the removal of some of the surface carbons by the plasma treatment, which exposes more catalysts to the XPS measurements. It is also possible that the catalysts are subject to a degree of oxidation during the different treatments. However, we should note that there was some variability in the catalyst density in the sampled areas; for instance, the plasma–liquid sample appeared to have a higher atomic concentration of iron (Figure 3), which would justify the higher contribution to the Fe_xO_y component in the O 1s signal in Figure 4. Figure 4b includes the N 1s spectra for all four samples; however, the annealed and pristine samples clearly indicate that no nitrogen was detected in these cases. However both of the plasma treatments resulted in the inclusion and development of nitrogen bonds with features including pyridines, amines, pyrroles, and graphitic nitrogen groups [46–51]. Whilst the graphitic nitrogen contents and pyridinic contents were comparable for the plasma–liquid and plasma–liquid with EDA samples, the peak areas of the pyrollic and amine components were vastly different. The pyrollic peak area was shown to increase from 3812 CPSeV to 12,099 CPSeV, and the amine peak area increased from 2879 CPSeV to 12,020 CPSeV when EDA was included in the electrolyte solution. This demonstrates that pyrollic and amine bonding are directly impacted by adding EDA to the electrolyte.

Raman spectroscopy was employed to assess the G-band to D-band area ratio (G:D ratio), which gives an indication of the graphitisation of the CNTs and whether the treatment results in defects (reduction in G:D). We should note that the synthesis method of the CNT ribbons introduces an intrinsic variability of the G:D ratio, which has an average value of 1.3 with a 0.2 standard deviation (determined over five spots per sample for 108 samples). Figure 5 shows a reduction in the G:D ratio of the annealed samples from 1.1 to 0.6. The G:D area ratio recovered after either plasma–liquid treatment, to approximately 0.8. This can be explained by the inclusion of graphitic nitrogen in the carbon lattice, which could reduce the number of missing atoms in the graphitic lattice and, therefore, give a strong lower D-band peak area.

The change in the hydrophobicity of the CNT ribbons after each plasma treatment was assessed using contact angle measurements (Figure 6). The untreated sample showed a contact angle of 84°, with a large drop to 46° for the samples treated with plasma–liquid in the presence of EDA. A remarkable decrease to 35° was measured for the plasma–liquid-treated samples in the base electrolyte of ethanol and water only. This represents a substantial shift towards a hydrophilic regime, and is suggested to be due to the functionalisation of the plasma–liquid treatment, as the intermediary annealing step results in a contact angle of only 72°.

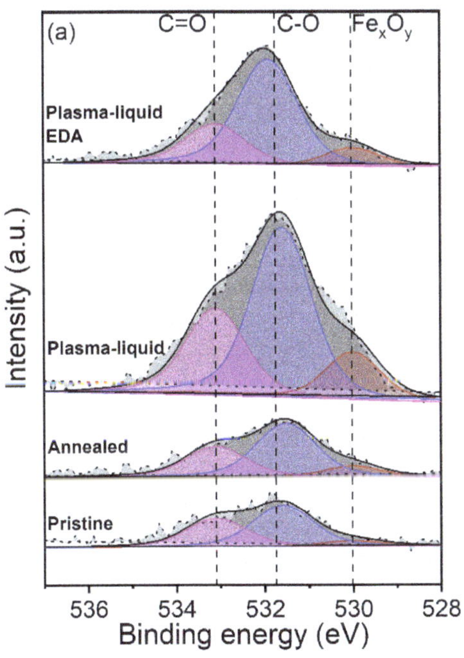

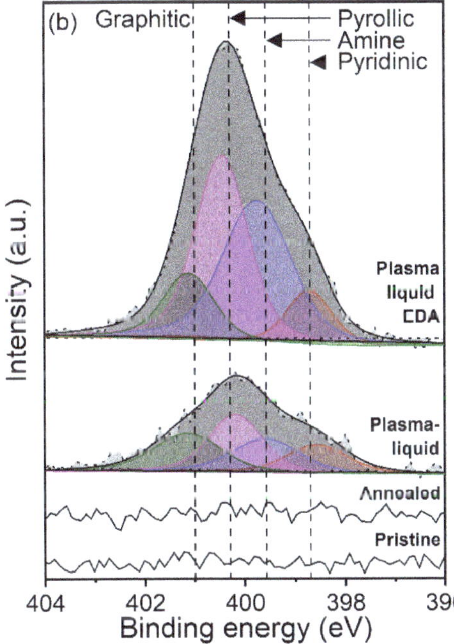

Figure 4. High-resolution X-ray photoelectron spectra of (**a**) oxygen and (**b**) nitrogen for pristine and pre-annealed carbon nanotube ribbons.

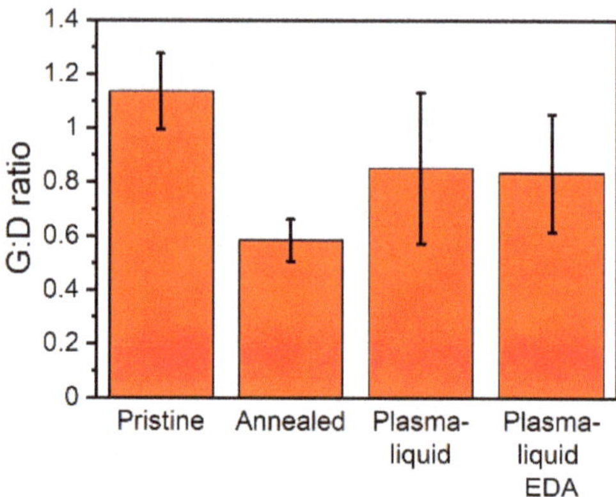

Figure 5. Summary of the G-band to D-band area ratios determined by Raman spectroscopy.

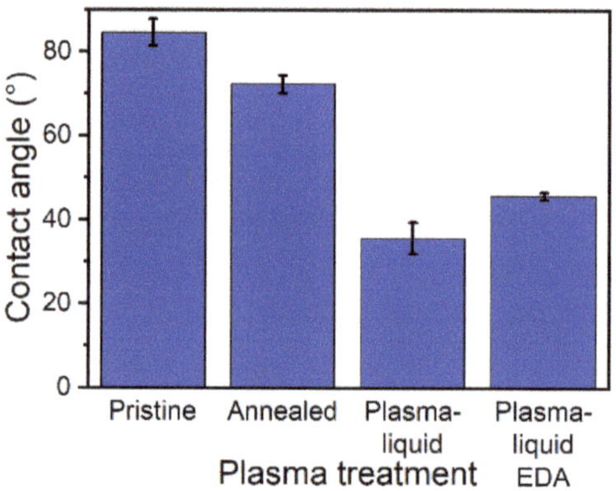

Figure 6. Summary of the contact angle measurement results for the 6 different carbon nanotube treatments.

3.2. Nanofluid Optical Characterisation and Solar–Thermal Conversion

The sedimentation and redispersion of the CNTs in nanofluids were monitored over a period of more than 2 years (809 days exactly). The nanofluids were allowed to settle in a dark cupboard with no disturbances (Figure 7). The pristine sample showed clear signs of flocculation. The same, to a lesser degree, was observed in the annealed sample and the plasma–liquid EDA sample. The plasma–liquid sample was the one that appeared to be the least impacted by long-term storage. All four vials were then shaken with no more than two vigorous shakes by hand (Figure 7). Shaking improved the dispersion of all samples, although clear differences remained, with the plasma–liquid sample still visually presenting the best dispersion, as it did before shaking. This was consistent with the contact angle measurements, which showed the best hydrophilicity for the CNTs treated with the plasma–liquid sample. More results at different timing and for non-annealed samples are documented in the Supplementary Materials.

These observations are useful to understand the mechanisms of sedimentation, but they are a qualitative assessment and are not fully representative of real-life applications, where the nanofluids are subject to circulation, with continued opportunities for mixing and gaining the advantages of shear-thinning mechanisms [22,23]. Hence, before UV–Vis spectroscopy, each sample was shaken vigorously by hand twice to redisperse the CNTs. UV–Vis spectroscopy measurements of the CNT nanofluids showed that all samples exhibited the same optical features and, therefore, that the treatment type did not appear to affect the qualitative nature of the light's interaction with the nanofluids (Figure 8). The nanofluid samples presented strong attenuation across the scanned range and a Pi-plasmon absorbance in the UV region at approximately 250–260 nm [52,53]. However, this region was also dominated by the absorption of the ethylene glycol base fluid; as such, it was not assessed here. The differences between the samples lie in the intensity of the optical coefficients and the stability over time, and it should be noted that the scattering coefficients in each case are an order of magnitude smaller than the absorption coefficients. Over the time period assessed, the treated samples possessed higher absorption coefficients across the full spectral range when compared to the pristine and annealed samples. The absorption coefficients of both of the plasma-treated samples showed very little change over time, while the pristine and annealed samples exhibited clear reductions in the absorption coefficients over the ~2-year period. The annealed sample, however, was slightly better in terms of both absorptivity and stability when compared to the pristine sample. Similarly, the plasma–liquid sample showed better absorption and stability when compared to the plasma–liquid EDA sample. For the plasma–liquid sample, we observed an increase in the absorption coefficient across the spectral range, and we attributed this to the progressive disentanglement of the CNTs, which might have improved over time and after shaking at day 60. This was corroborated by a decrease in the scattering coefficient over the same period, suggesting an improved dispersion. These results are consistent with the trends observed in the contact angle measurements (Figure 6) and photos (Figure 7). The scattering coefficients are somewhat more difficult to interpret, as their values, which are much lower than the absorption coefficients, are more susceptible to experimental variations, and such differences could be considered negligible. This is particularly the case when comparing the scattering curves over time for the same samples. In general, we can confirm that the scattering coefficient is higher for samples with a higher absorption coefficient, confirming that flocculation and sedimentation are at least in part responsible for the different optical properties of the four samples.

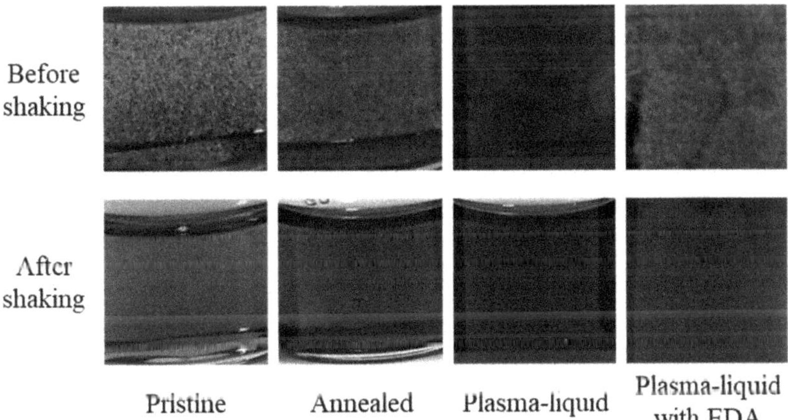

Figure 7. Photographs of the nanofluids after 809 days of storage before shaking, and after 2 vigorous shakes by hand. Note that the photographs have been modified to enhance the contrast and brightness to highlight flocculation.

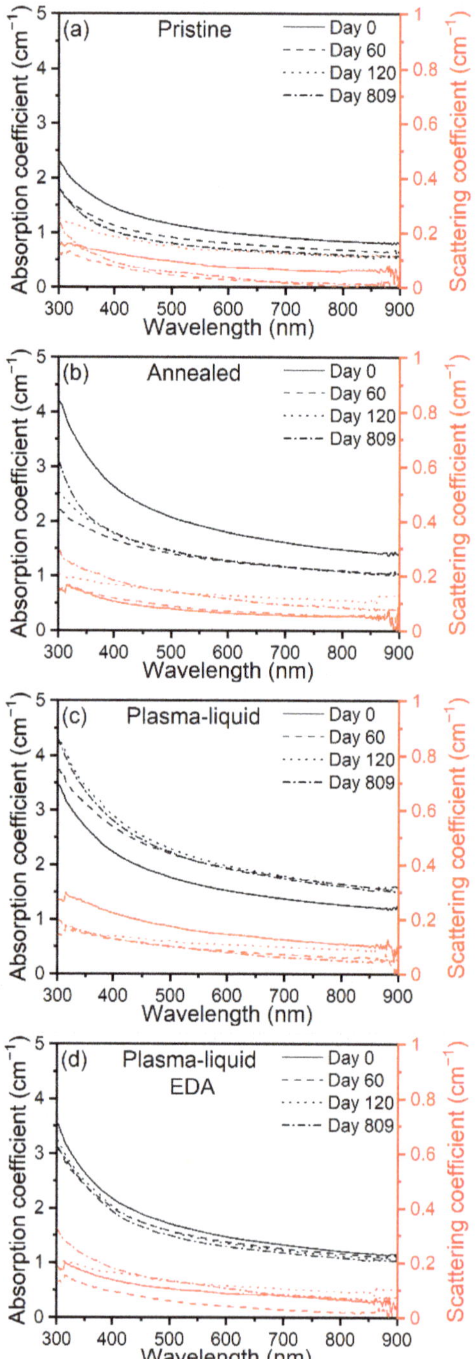

Figure 8. *Cont.*

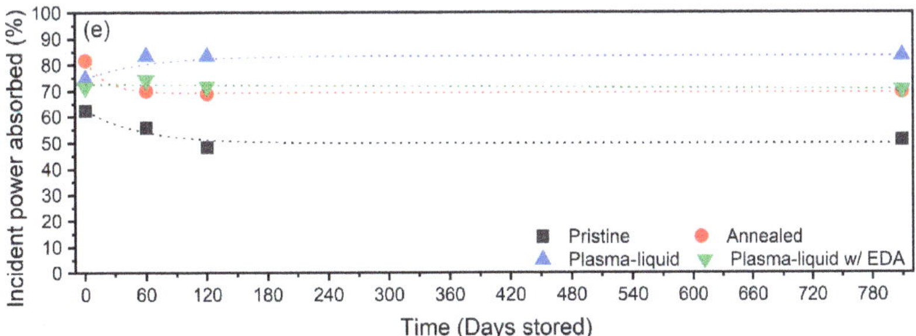

Figure 8. The values of absorption and scattering coefficients obtained from ultraviolet–visible spectroscopy via the transmittance port and an integrating sphere (**a–d**). The percentage of incident power absorbed for each carbon-nanotube-based nanofluid over 120 days (**e**).

The absorption coefficient measured for each nanofluid at different times was then used to calculate the percentage of power absorbed from a spectrum of solar light at AM1.5G.ach and with a nanofluid, for a theoretical fluid depth of 1 cm (see the Supplementary Materials and Equation (16)). On day 0, the pristine sample absorbed 63% of the solar power, whilst the plasma–liquid with EDA and the plasma–liquid samples absorbed 72% and 75%, respectively. The annealed sample also exhibited a very high percentage of absorbed power (82%). After 60 days, a substantial decrease in the absorptive properties of the pristine and annealed samples was found, dropping to 56% and 70%, respectively. Consistent with the measured absorption coefficients, the absorbed power increased to 84% and 74% for the plasma–liquid and plasma–liquid EDA samples, respectively. By day 120, the power absorbed by the pristine sample continued to fall, plummeting to 48%. After more than 2 years of storage, it could be seen that the plasma–liquid and plasma–liquid EDA samples retained their absorptive properties, with a calculated power absorbed above 80% and above 70%, respectively. The pristine CNT nanofluid remained below 50%, and the annealed samples remained below 70%. These results strongly reinforce that the "plasma–liquid" treatment can produce nanofluids with superior optical properties and enhanced stability.

These observations clearly show that both plasma-treated samples gained in terms of absorbed power compared to the pristine sample, likely due to the added functional groups. Hence, these not only play a role in the stability of the CNTs, but also can partially improve light absorption. The difference in stability was also obvious, as both plasma-treated samples were very little affected after more than 2 years. We also observed that the plasma–liquid sample outperformed the plasma–liquid with EDA sample. This could also be linked to differences in nitrogen- and oxygen-based functional groups. This suggests that oxygen-based groups appear to be more effective in promoting light absorption and, therefore, should be preferred for enhancing the overall absorption.

Solar–thermal conversion efficiency was assessed for each of the treated CNT types; this was carried out on day 60 (Figure 9). In terms of the recorded temperatures of the nanofluids (Figure 9a), the pristine CNT sample showed the lowest temperature rise, with an increase of 12.3 °C over 20 min. Improving upon this, the plasma–liquid with EDA sample demonstrated a temperature increase of 13.1 °C. As expected from the previous UV–Vis measurements and absorbed power calculations, the plasma–liquid sample demonstrated the greatest temperature rise under the solar simulator, with a temperature increase of 16.7 °C. For each CNT additive, the performance of the nanofluid was enhanced over pure ethylene glycol, which only provided a 9.6 °C temperature increase. These results align with the incident power absorbed, as shown in Figure 8e. Figure 9b shows a substantial enhancement in the STC efficiency—in particular with the "Plasma–liquid" CNT nanofluid, where an efficiency of approximately 50% was achieved. In addition to the greatest achieved STC efficiency, the "Plasma–liquid" sample also retained this over the greatest temperature range, surpassing 40 °C before dipping. By comparison, the "Pristine" and "Plasma–liquid with EDA" samples showed a decrease in the STC efficiency at much lower temperatures, even below 30 °C. The ethylene glycol did not demonstrate a decrease in efficiency with temperature, although this can be ascribed to the substantially lower maximum temperature recorded during the solar–thermal experiments.

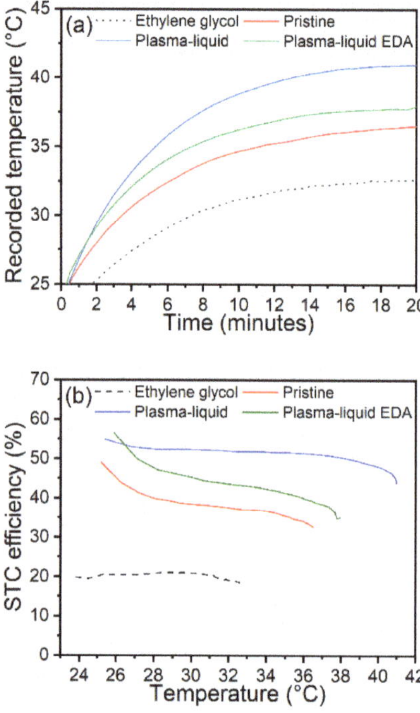

Figure 9. (**a**) Variation in temperature of the nanofluids and ethylene glycol over 20 min of exposure to simulated solar radiation, and (**b**) their resulting solar–thermal conversion efficiency after accounting for heat lost.

4. Conclusions

We used plasma-induced non-equilibrium electrochemistry to functionalise the surface of CNT ribbons and enhance their dispersion within ethylene glycol for application in solar-to-thermal energy conversion. In particular, the nanofluids with CNTs that showed a combined nitrogen-based and high oxygen-based functionalisation—the "plasma–liquid" samples—performed exceptionally well. These showed a reduction in the contact angle from 84° to 35°, along with corresponding improvements in hydrophilicity and dispersion, leading to the enhancement of the absorption coefficient, reaching values greater than 2 cm^{-1} at 600 nm, with much greater values above 5 cm^{-1} observed at the logarithmic Van Hove peak at 256 nm. Solar simulations showed that this "plasma–liquid" additive results in the greatest temperature increase—to above 40 °C—as well as a solar–thermal conversion efficiency of 50%. The long-term performance of the treated nanofluids was assessed, with values of 80% incident radiation absorption retained after 67 months of storage, with only a brief manual shake required to fully redisperse the material prior to UV–Vis measurements, highlighting the readiness for industrial application.

Supplementary Materials: The following supporting information can be downloaded at: https://www.mdpi.com/article/10.3390/nano12152705/s1, Figure S1: UV–Vis transmittance measurement of the fused quartz cover used to seal the Petri dish for solar–thermal experiments; Figure S2: Schematic diagram showing (a) the heat loss mechanism from the nanofluid to the ambient surroundings, and (b) the thermal circuit of the thermal resistances during heat loss; Figure S3: Scanning electron micrographs of the different carbon nanotube ribbons after sonication and drop-casting onto a silicon wafer; Figure S4: Photographs of the non-annealed nanofluids after 809 days of storage (a) before shaking and (b) after 2 vigorous shakes by hand; Figure S5: Scanning electron micrographs of the different carbon nanotube ribbons after sonication and drop-cast onto a silicon wafer; Figure S6: X-ray photoelectron spectroscopy summary of atomic percentages by element for oxygen and nitrogen; Figure S7: High-resolution X-ray photoelectron spectra of oxygen and nitrogen for pristine and pre-annealed carbon nanotube ribbons; Figure S8: Raw spectral data of the carbon nanotubes prior to dispersion; Figure S9: Summary of the G-band to D-band area ratios determined by Raman spectroscopy; Figure S10: Summary of the contact angle measurement results for the 6 different carbon nanotube treatments; Figure S11: Photographs highlighting the short-term flocculation effect on stored nanofluids without disturbance; Figure S12: The values of absorption and scattering coefficients obtained from ultraviolet–visible spectroscopy via the transmittance port and an integrating sphere. The percentage of incident power absorbed for each carbon-nanotube-based nanofluid over 120 days is presented in the fourth chart; Figure S13: (a) Variation in temperature of the nanofluids and ethylene glycol over 20 min of exposure to simulated solar radiation, with small improvements noted for the plasma-treated and annealed samples. (b) The resultant solar thermal conversion efficiency after accounting for heat lost; Figure S14: Photographs taken during the solar–thermal conversion experiments; (a) shows the "pristine" sample, whereas (b) is a photograph of the "plasma–liquid" conditions; Table S1: A table of symbols and constants used in the solar simulation calculations. References [3,14,54–61] are cited in the Supplementary Materials.

Author Contributions: Conceptualisation, R.J.M.; methodology, R.J.M. and D.M.; formal analysis, R.J.M., H.S.M. and P.B.; investigation, R.J.M., H.S.M. and P.B.; resources, D.M.; data curation, R.J.M.; writing—original draft preparation, R.J.M.; writing—review and editing, D.M.; visualisation, R.J.M.; supervision, D.M., P.M. and S.C.; project administration, D.M.; funding acquisition, D.M. and P.M. All authors have read and agreed to the published version of the manuscript.

Funding: The authors would like to acknowledge the EPRSC for supporting this work through EP/M015211/1 and EP/R008841/1.

Institutional Review Board Statement: Not applicable.

Informed Consent Statement: Not applicable.

Data Availability Statement: All relevant data can be made available upon request to the corresponding author.

Conflicts of Interest: The authors declare no conflict of interest.

References

1. Minardi, J.E.; Chuang, H.N. Performance of a "Black" Liquid Flat-Plate Solar Collector. *Sol. Energy* **1975**, *17*, 179–183. [CrossRef]
2. Bertocchi, R.; Karni, J.; Kribus, A. Experimental Evaluation of a Non-Isothermal High Temperature Solar Particle Receiver. *Energy* **2004**, *29*, 687–700. [CrossRef]
3. *ASTM G173-03*; ASTM Standard Tables for Reference Solar Spectral Irradiances: Direct Normal and Hemispherical on 37° Tilted Surface 2012. ASTM International: West Conshohocken, PA, USA, 2020.
4. Tyagi, H.; Phelan, P.; Prasher, R. Predicted Efficiency of a Low-Temperature Nanofluid-Based Direct Absorption Solar Collector. *J. Sol. Energy Eng.* **2009**, *131*, 041004. [CrossRef]
5. Sharaf, O.Z.; Rizk, N.; Joshi, C.P.; Abi Jaoudé, M.; Al-Khateeb, A.N.; Kyritsis, D.C.; Abu-Nada, E.; Martin, M.N. Ultrastable Plasmonic Nanofluids in Optimized Direct Absorption Solar Collectors. *Energy Convers. Manag.* **2019**, *199*, 112010. [CrossRef]
6. Beicker, C.L.L.; Amjad, M.; Bandarra Filho, E.P.; Wen, D. Experimental Study of Photothermal Conversion Using Gold/Water and MWCNT/Water Nanofluids. *Sol. Energy Mater. Sol. Cells* **2018**, *188*, 51–65. [CrossRef]
7. Taylor, R.A.; Phelan, P.E.; Otanicar, T.P.; Adrian, R.; Prasher, R. Nanofluid Optical Property Characterization: Towards Efficient Direct Absorption Solar Collectors. *Nanoscale Res. Lett.* **2011**, *6*, 225. [CrossRef]
8. Otanicar, T.P.; Phelan, P.E.; Prasher, R.S.; Rosengarten, G.; Taylor, R.A. Nanofluid-Based Direct Absorption Solar Collector. *J. Renew. Sustain. Energy* **2010**, *2*, 033102. [CrossRef]

9. Karami, M.; Akhavan-Behabadi, M.A.; Raisee Dehkordi, M.; Delfani, S. Thermo-Optical Properties of Copper Oxide Nanofluids for Direct Absorption of Solar Radiation. *Sol. Energy Mater. Sol. Cells* **2016**, *144*, 136–142. [CrossRef]

10. Menbari, A.; Alemrajabi, A.A.; Rezaei, A. Heat Transfer Analysis and the Effect of CuO/Water Nanofluid on Direct Absorption Concentrating Solar Collector. *Appl. Therm. Eng.* **2016**, *104*, 176–183. [CrossRef]

11. Moghaieb, H.S.; Padmanaban, D.B.; Kumar, P.; Ul Haq, A.; Maddi, C.; McGlynn, R.; Arredondo, M.; Singh, H.; Maguire, P.; Mariotti, D. Efficient Solar-Thermal Energy Conversion with Surfactant-Free Cu-Oxide Nanofluids. *Nano Energy Rev.* 2022. *Under Review.*

12. Rose, B.A.J.; Singh, H.; Verma, N.; Tassou, S.; Suresh, S.; Anantharaman, N.; Mariotti, D.; Maguire, P. Investigations into Nanofluids as Direct Solar Radiation Collectors. *Sol. Energy* **2017**, *147*, 426–431. [CrossRef]

13. Vakili, M.; Hosseinalipour, S.M.; Delfani, S.; Khosrojerdi, S.; Karami, M. Experimental Investigation of Graphene Nanoplatelets Nanofluid-Based Volumetric Solar Collector for Domestic Hot Water Systems. *Sol. Energy* **2016**, *131*, 119–130. [CrossRef]

14. Delfani, S.; Karami, M.; Akhavan-Behabadi, M.A. Performance Characteristics of a Residential-Type Direct Absorption Solar Collector Using MWCNT Nanofluid. *Renew. Energy* **2016**, *87*, 754–764. [CrossRef]

15. Karami, M.; Akhavan Bahabadi, M.A.A.; Delfani, S.; Ghozatloo, A. A New Application of Carbon Nanotubes Nanofluid as Working Fluid of Low-Temperature Direct Absorption Solar Collector. *Sol. Energy Mater. Sol. Cells* **2014**, *121*, 114–118. [CrossRef]

16. Mesgari, S.; Taylor, R.A.; Hjerrild, N.E.; Crisostomo, F.; Li, Q.; Scott, J. An Investigation of Thermal Stability of Carbon Nanofluids for Solar Thermal Applications. *Sol. Energy Mater. Sol. Cells* **2016**, *157*, 652–659. [CrossRef]

17. Hordy, N.; Coulombe, S.; Meunier, J. Plasma Functionalization of Carbon Nanotubes for the Synthesis of Stable Aqueous Nanofluids and Poly(Vinyl Alcohol) Nanocomposites. *Plasma Process. Polym.* **2013**, *10*, 110–118. [CrossRef]

18. Hordy, N.; Rabilloud, D.; Meunier, J.L.; Coulombe, S. High Temperature and Long-Term Stability of Carbon Nanotube Nanofluids for Direct Absorption Solar Thermal Collectors. *Sol. Energy* **2014**, *105*, 82–90. [CrossRef]

19. Yang, Z.-P.; Ci, L.; Bur, J.A.; Lin, S.-Y.; Ajayan, P.M. Experimental Observation of an Extremely Dark Material Made by a Low-Density Nanotube Array. *Nano Lett.* **2008**, *8*, 446–451. [CrossRef]

20. Samani, M.K.; Khosravian, N.; Chen, G.C.K.; Shakerzadeh, M.; Baillargeat, D.; Tay, B.K. Thermal Conductivity of Individual Multiwalled Carbon Nanotubes. *Int. J. Therm. Sci.* **2012**, *62*, 40–43. [CrossRef]

21. Kim, P.; Shi, L.; Majumdar, A.; McEuen, P.L. Thermal Transport Measurements of Individual Multiwalled Nanotubes. *Phys. Rev. Lett.* **2001**, *87*, 215502. [CrossRef]

22. Shi, Q.; Liu, Y.; Chen, F.; Dong, S. Investigation on Rheological Properties of Carbon Nanotube Nanofluids. *Phys. Chem. Liq.* **2019**, *57*, 37–42. [CrossRef]

23. Ruan, B.; Jacobi, A.M. Ultrasonication Effects on Thermal and Rheological Properties of Carbon Nanotube Suspensions. *Nanoscale Res. Lett.* **2012**, *7*, 127. [CrossRef] [PubMed]

24. Tan, N.; Zhang, Y.; Wei, B.; Zou, C. Experimental Investigation on Optical and Thermal Properties of Propylene Glycol–Water Based Nanofluids for Direct Absorption Solar Collectors. *Appl. Phys. A Mater. Sci. Process.* **2018**, *124*, 569. [CrossRef]

25. Struchalin, P.G.; Yunin, V.S.; Kutsenko, K.V.; Nikolaev, O.V.; Vologzhannikova, A.A.; Shevelyova, M.P.; Gorbacheva, O.S.; Balakin, B.V. Performance of a Tubular Direct Absorption Solar Collector with a Carbon-Based Nanofluid. *Int. J. Heat Mass Transf.* **2021**, *179*, 121717. [CrossRef]

26. Williams, J.; Broughton, W.; Koukoulas, T.; Rahatekar, S.S. Plasma Treatment as a Method for Functionalising and Improving Dispersion of Carbon Nanotubes in Epoxy Resins. *J. Mater. Sci.* **2013**, *48*, 1005–1013. [CrossRef]

27. Ghadimi, A.; Saidur, R.; Metselaar, H.S.C. A Review of Nanofluid Stability Properties and Characterization in Stationary Conditions. *Int. J. Heat Mass Transf.* **2011**, *54*, 4051–4068. [CrossRef]

28. Mingzheng, Z.; Guodong, X.; Jian, L.; Lei, C.; Lijun, Z. Analysis of Factors Influencing Thermal Conductivity and Viscosity in Different Kinds of Surfactant Solutions. *Exp. Therm. Fluid Sci.* **2012**, *36*, 22–29. [CrossRef]

29. Cakmak, N.K. The Impact of Surfactants on the Stability and Thermal Conductivity of Graphene Oxide de-Ionized Water Nanofluids. *J. Therm. Anal. Calorim.* **2020**, *139*, 1895–1902. [CrossRef]

30. Movsesyan, A.; Marguet, S.; Muravitskaya, A.; Béal, J.; Adam, P.-M.; Baudrion, A.-L. Influence of the CTAB Surfactant Layer on Optical Properties of Single Metallic Nanospheres. *J. Opt. Soc. Am. A* **2019**, *36*, C78. [CrossRef]

31. Chen, C.; Liang, B.; Ogino, A.; Wang, X.; Nagatsu, M. Oxygen Functionalization of Multiwall Carbon Nanotubes by Microwave-Excited Surface-Wave Plasma Treatment. *J. Phys. Chem. C* **2009**, *113*, 7659–7665. [CrossRef]

32. Kalita, G.; Adhikari, S.; Aryal, H.R.; Afre, R.; Soga, T.; Sharon, M.; Umeno, M. Functionalization of Multi-Walled Carbon Nanotubes (MWCNTs) with Nitrogen Plasma for Photovoltaic Device Application. *Curr. Appl. Phys.* **2009**, *9*, 346–351. [CrossRef]

33. Zhao, B.; Zhang, L.; Wang, X.; Yang, J. Surface Functionalization of Vertically-Aligned Carbon Nanotube Forests by Radio-Frequency Ar/O2 Plasma. *Carbon* **2012**, *50*, 2710–2716. [CrossRef]

34. Hussain, S.; Amade, R.; Jover, E.; Bertran, E. Nitrogen Plasma Functionalization of Carbon Nanotubes for Supercapacitor Applications. *J. Mater. Sci.* **2013**, *48*, 7620–7628. [CrossRef]

35. Gohel, A.; Chin, K.C.; Zhu, Y.W.; Sow, C.H.; Wee, A.T.S. Field Emission Properties of N2 and Ar Plasma-Treated Multi-Wall Carbon Nanotubes. *Carbon* **2005**, *43*, 2530–2535. [CrossRef]

36. Ham, S.W.; Hong, H.P.; Kim, J.H.; Min, S.J.; Min, N.K. Effect of Oxygen Plasma Treatment on Carbon Nanotube-Based Sensors. *J. Nanosci. Nanotechnol.* **2014**, *14*, 8476–8481. [CrossRef]

37. Park, O.-K.; Kim, W.Y.; Kim, S.M.; You, N.-H.; Jeong, Y.; Su Lee, H.; Ku, B.-C. Effect of Oxygen Plasma Treatment on the Mechanical Properties of Carbon Nanotube Fibers. *Mater. Lett.* **2015**, *156*, 17–20. [CrossRef]
38. Mitra, S.; Švrček, V.; Mariotti, D.; Velusamy, T.; Matsubara, K.; Kondo, M. Microplasma-Induce Liquid Chemistry for Stabilizing of Silicon Nanocrystals Optical Properties in Water. *Plasma Process. Polym.* **2014**, *11*, 158–163. [CrossRef]
39. Mitra, S.; Cook, S.; Švrček, V.; Blackley, R.A.; Zhou, W.; Kovač, J.; Cvelbar, U.; Mariotti, D. Improved Optoelectronic Properties of Silicon Nanocrystals/Polymer Nanocomposites by Microplasma-Induced Liquid Chemistry. *J. Phys. Chem. C* **2013**, *117*, 23198–23207. [CrossRef]
40. Liu, Y.; Sun, D.; Askari, S.; Patel, J.; Macias-Montero, M.; Mitra, S.; Zhang, R.; Lin, W.F.; Mariotti, D.; Maguire, P. Enhanced Dispersion of TiO$_2$ Nanoparticles in a TiO$_2$/PEDOT:PSS Hybrid Nanocomposite via Plasma-Liquid Interactions. *Sci. Rep.* **2015**, *5*, 15765. [CrossRef]
41. Panaitescu, D.M.; Vizireanu, S.; Nicolae, C.A.; Frone, A.N.; Casarica, A.; Carpen, L.G.; Dinescu, G. Treatment of Nanocellulose by Submerged Liquid Plasma for Surface Functionalization. *Nanomaterials* **2018**, *8*, 467. [CrossRef]
42. McGlynn, R.J.; Brunet, P.; Chakrabarti, S.; Boies, A.; Maguire, P.; Mariotti, D. High Degree of N-Functionalization in Macroscopically Assembled Carbon Nanotubes. *J. Mater. Sci.* **2022**, *57*, 13314–13325. [CrossRef]
43. Sundaram, R.M.; Koziol, K.K.K.; Windle, A.H. Continuous Direct Spinning of Fibers of Single-Walled Carbon Nanotubes with Metallic Chirality. *Adv. Mater.* **2011**, *23*, 5064–5068. [CrossRef] [PubMed]
44. Perspex Sheet. Available online: https://www.perspexsheet.uk/clear-000-perspex/ (accessed on 7 July 2019).
45. Vlad, I.-E.; Anghel, S.D. Time Stability of Water Activated by Different On-Liquid Atmospheric Pressure Plasmas. *J. Electrostat.* **2017**, *87*, 284–292. [CrossRef]
46. Stöhr, B.; Boehm, H.P.; Schlögl, R. Enhancement of the Catalytic Activity of Activated Carbons in Oxidation Reactions by Thermal Treatment with Ammonia or Hydrogen Cyanide and Observation of a Superoxide Species as a Possible Intermediate. *Carbon* **1991**, *29*, 707–720. [CrossRef]
47. Pels, J.R.; Kapteijn, F.; Moulijn, J.A.; Zhu, Q.; Thomas, K.M. Evolution of Nitrogen Functionalities in Carbonaceous Materials during Pyrolysis. *Carbon* **1995**, *33*, 1641–1653. [CrossRef]
48. Liu, H.; Zhang, Y.; Li, R.; Sun, X.; Désilets, S.; Abou-Rachid, H.; Jaidann, M.; Lussier, L.-S. Structural and Morphological Control of Aligned Nitrogen-Doped Carbon Nanotubes. *Carbon* **2010**, *48*, 1498–1507. [CrossRef]
49. Liu, Z.-Q.; Cheng, H.; Li, N.; Ma, T.Y.; Su, Y.-Z. ZnCo$_2$O$_4$ Quantum Dots Anchored on Nitrogen-Doped Carbon Nanotubes as Reversible Oxygen Reduction/Evolution Electrocatalysts. *Adv. Mater.* **2016**, *28*, 3777–3784. [CrossRef]
50. Ito, Y.; Cong, W.; Fujita, T.; Tang, Z.; Chen, M. High Catalytic Activity of Nitrogen and Sulfur Co-Doped Nanoporous Graphene in the Hydrogen Evolution Reaction. *Angew. Chem. Int. Ed.* **2015**, *54*, 2131–2136. [CrossRef]
51. Carolan, D.; Rocks, C.; Padmanaban, D.B.; Maguire, P.; Svrcek, V.; Mariotti, D. Environmentally Friendly Nitrogen-Doped Carbon Quantum Dots for next Generation Solar Cells. *Sustain. Energy Fuels* **2017**, *1*, 1611–1619. [CrossRef]
52. Landi, B.J.; Ruf, H.J.; Evans, C.M.; Cress, C.D.; Raffaelle, R.P. Purity Assessment of Single-Wall Carbon Nanotubes, Using Optical Absorption Spectroscopy. *J. Phys. Chem. B* **2005**, *109*, 9952–9965. [CrossRef]
53. Attal, S.; Thiruvengadathan, R.; Regev, O. Determination of the Concentration of Single-Walled Carbon Nanotubes in Aqueous Dispersions Using UV–Visible Absorption Spectroscopy. *Anal. Chem.* **2006**, *78*, 8098–8104. [CrossRef] [PubMed]
54. Pak, B.C.; Cho, Y.I. HYDRODYNAMIC AND HEAT TRANSFER STUDY OF DISPERSED FLUIDS WITH SUBMICRON METALLIC OXIDE PARTICLES. *Exp. Heat Transf.* **1998**, *11*, 151–170. [CrossRef]
55. Zhou, S.-Q.; Ni, R. Measurement of the Specific Heat Capacity of Water-Based Al$_2$O$_3$ Nanofluid. *Appl. Phys. Lett.* **2008**, *92*, 093123. [CrossRef]
56. Incropera, F.P.; Dewitt, D.P.; Bergman, T.L.; Lavine, A.S. *Fundamentals of Heat and Mass Transfer*; Wiley: Hoboken, NJ, USA, 2007.
57. Chen, M.; He, Y.; Zhu, J.; Wen, D. Investigating the Collector Efficiency of Silver Nanofluids Based Direct Absorption Solar Collectors. *Appl. Energy* **2016**, *181*, 65–74. [CrossRef]
58. Gorji, T.B.; Ranjbar, A.A. A Numerical and Experimental Investigation on the Performance of a Low-Flux Direct Absorption Solar Collector (DASC) Using Graphite, Magnetite and Silver Nanofluids. *Sol. Energy* **2016**, *135*, 493–505. [CrossRef]
59. Jeon, J.; Park, S.; Lee, B.J. Analysis on the Performance of a Flat-Plate Volumetric Solar Collector Using Blended Plasmonic Nanofluid. *Sol. Energy* **2016**, *132*, 247–256. [CrossRef]
60. Moncrieff, D.A.; Barker, P.R. Secondary Electron Emission in the Scanning Electron Microscope. *Scanning* **1978**, *1*, 195–197. [CrossRef]
61. Inada, H.; Su, D.; Egerton, R.F.; Konno, M.; Wu, L.; Ciston, J.; Wall, J.; Zhu, Y. Atomic Imaging Using Secondary Electrons in a Scanning Transmission Electron Microscope: Experimental Observations and Possible Mechanisms. *Ultramicroscopy* **2011**, *111*, 865–876. [CrossRef]

nanomaterials

Review

Progress and Recent Trends in the Application of Nanoparticles as Low Carbon Fuel Additives—A State of the Art Review

Jeffrey Dankwa Ampah [1], Abdulfatah Abdu Yusuf [2], Ephraim Bonah Agyekum [3], Sandylove Afrane [1], Chao Jin [1], Haifeng Liu [4,*], Islam Md Rizwanul Fattah [5,6], Pau Loke Show [7], Mokhtar Shouran [8], Monier Habil [8,*] and Salah Kamel [9]

1. School of Environmental Science and Engineering, Tianjin University, Tianjin 300072, China; jeffampah@live.com (J.D.A.); sandfran20@gmail.com (S.A.); jinchao@tju.edu.cn (C.J.)
2. Department of Mechanical and Automobile Engineering, Sharda University, Knowledge Park III, Greater Noida 201310, UP, India; abdulfatahabduyusuf@gmail.com
3. Department of Nuclear and Renewable Energy, Ural Federal University Named after the First President of Russia Boris Yeltsin, 19 Mira Street, 620002 Ekaterinburg, Russia; agyekum@urfu.ru
4. State Key Laboratory of Engines, Tianjin University, Tianjin 300072, China
5. Centre for Green Technology, Faculty of Engineering and IT, University of Technology Sydney, Ultimo, NSW 2007, Australia; IslamMdRzwanul.Fattah@uts.edu.au
6. Department of Mechanical Engineering, College of Engineering, Universiti Tenaga Nasional, Kajang 43000, Selangor Darul Ehsan, Malaysia
7. Department of Chemical and Environmental Engineering, Faculty of Science and Engineering, University of Nottingham Malaysia, Jalan Broga, Semenyih 43500, Selangor Darul Ehsan, Malaysia; Pauloke.show@nottingham.edu.my
8. Wolfson Centre for Magnetics, School of Engineering, Cardiff University, Cardiff CF24 3AA, UK; shouranma@cardiff.ac.uk
9. Electrical Engineering Department, Faculty of Engineering, Aswan University, Aswan 81542, Egypt; skamel@aswu.edu.eg
* Correspondence: haifengliu@tju.edu.cn (H.L.); habilmm@cardiff.ac.uk (M.H.)

check for updates

Citation: Ampah, J.D.; Yusuf, A.A.; Agyekum, E.B.; Afrane, S.; Jin, C.; Liu, H.; Fattah, I.M.R.; Show, P.L.; Shouran, M.; Habil, M.; et al. Progress and Recent Trends in the Application of Nanoparticles as Low Carbon Fuel Additives—A State of the Art Review. *Nanomaterials* **2022**, *12*, 1515. https://doi.org/10.3390/nano12091515

Academic Editor: Diego Cazorla-Amorós

Received: 22 March 2022
Accepted: 26 April 2022
Published: 29 April 2022

Publisher's Note: MDPI stays neutral with regard to jurisdictional claims in published maps and institutional affiliations.

Abstract: The first part of the current review highlights the evolutionary nuances and research hotspots in the field of nanoparticles in low carbon fuels. Our findings reveal that contribution to the field is largely driven by researchers from Asia, mainly India. Of the three biofuels under review, biodiesel seems to be well studied and developed, whereas studies regarding vegetable oils and alcohols remain relatively scarce. The second part also reviews the application of nanoparticles in biodiesel/vegetable oil/alcohol-based fuels holistically, emphasizing fuel properties and engine characteristics. The current review reveals that the overall characteristics of the low carbon fuel–diesel blends improve under the influence of nanoparticles during combustion in diesel engines. The most important aspect of nanoparticles is that they act as an oxygen buffer that provides additional oxygen molecules in the combustion chamber, promoting complete combustion and lowering unburnt emissions. Moreover, the nanoparticles used for these purposes exhibit excellent catalytic behaviour as a result of their high surface area-to-volume ratio—this leads to a reduction in exhaust pollutants and ensures an efficient and complete combustion. Beyond energy-based indicators, the exergy, economic, environmental, and sustainability aspects of the blends in diesel engines are discussed. It is observed that the performance of the diesel engine fuelled with low carbon fuels according to the second law of efficiency improves under the influence of the nano-additives. Our final part shows that despite the benefits of nanoparticles, humans and animals are under serious threats from the highly toxic nature of nanoparticles.

Keywords: nanoparticles; biodiesel; vegetable oil; alcohol; research hotspots; fuel properties; engine characteristics

1. Introduction

The emergence of the industrial revolution coupled with modernization in lifestyle and vehicular population globally has led to a significant increase in energy demand, and according to Joshi et al. [1], global car ownership will double by the end of 2040 compared to 2016, by which conventional energy sources will power 80% of these cars. This trend has put excessive pressure on the global energy demand and supply market. Transportation uses 30% of the world's total supplied energy, with road transport accounting for 80% of it. This sector is thought to account for approximately 60% of global oil demand and will continue to be the fastest expanding demand sector in the future [2]. Moreover, because of the increasing expansion of vehicles, demand for petroleum products is anticipated to climb to more than 240 million metric tonnes by 2021–2022 and to about 465 million metric tonnes by 2031–2032, assuming strong output growth [3]. The high energy demand from this market, together with other sectors, will eventually contribute to the exhaustion of the available petroleum reserves. In addition to resource depletion from excessive use of conventional fuels in the transport sector is their environmental emissions. Between 2007 and 2020, an estimated 4.1 billion metric tonnes of carbon dioxide were emitted into the atmosphere. Furthermore, between 2020 and 2035, an extra 8.6 billion metric tons of carbon dioxide is expected to be emitted into the environment [4,5]. For the aforementioned predicted timeframe, this is estimated to represent a 43 percent raise. Therefore, engine manufacturers are being forced to develop technologies and investigate cleaner alternative fuel sources without having to worry about engine changes due to strict pollution laws and rising energy demands.

Alcohols, biodiesel, and vegetable oils are among the most promising and popular liquid biofuels studied for their application in internal combustion engines (ICE). Though liquid biofuels such as biodiesels, vegetable oils, and primary alcohols have the potential of solving the world energy crisis, sometimes their direct application in conventional diesel engines is limited. For instance, biodiesels tend to oxidize quickly due to the presence of unsaturated fatty acids, which is the main disadvantage of biodiesel [6,7]. The presence of 11 wt% oxygen reduces its heating value compared to neat diesel fuel [8,9]. Vegetable oils are about 10–20 times more highly viscous [10,11]. The highly viscous nature of vegetable oils presents poor fuel atomization, negative cold flow characteristics, incomplete combustion, ring sticking, and carbon deposit in the combustion chamber, among many others [10,11]. The drawback with low carbon alcohols (i.e., methanol and ethanol) is that their combustion in diesel engines is characterized by lower efficiency as a result of their inherent inferior physico-chemical properties such as high latent heat of vaporization, low ignition qualities, and relatively poor calorific value [12,13]. Moreover, they are very hygroscopic, and thus, they have poor miscibility with diesel. These problems with the direct application of neat biofuels in engines are often circumvented by forming blends with diesel fuel. Many researchers acknowledge the fuel blending approach for achieving certain fuel characteristics in order to increase the performance and emission control of a diesel engine without modifying the present engine. Several researchers have focused on improving the quality of fuels with additives and emulsification [14–17]. With microscale additives, sedimentation, aggregation, and non-uniform size distribution are issues [18]. Particle sizes smaller than 100 nm may now be easily produced and utilized as additives in engines because of the progress made in nanoscience, resolving the aforementioned issues [19].

One of the most important and novel themes in ICE is nanotechnology. The extant literature in the field has shown that, under the influence of nanoparticles (NP), the aforementioned liquid biofuels and their blends with conventional fuels exhibit overall improved fuel properties and combustion characteristics. This significant improvement is as a result of the excellent thermophysical properties of NPs and their high reactivity characteristics, which are suitable for combustion in ICE. In addition, the high thermal conductivity of these NPs provides them with optimal heat and mass transport features [20–23]. For these NPs to be considered suitable additives for fuel combustion, Ribeiro et al. [24] outline the key requirements and features that should be present or exhibited by the NPs: (1) exhaust emissions should minimize after the addition of these additives to base fuels, (2) the presence of these additives should ensure that the oxygen concentration in the particle filter and the combustion chamber of the engine is boosted, (3) the stability of the nanofluids should not be a problem over a wide range of conditions, (4) the presence of the additives should see to it that the viscosity index of the resulting fuel blend is increased, (5) the additives should be able to produce an increased rate of ignition, i.e., the flash point and ignition delay period of the resulting fuel should reduce, and (6) wearing, friction loss, and corrosivity should not be a problem after introducing these additives to the base fuel.

The addition of nanoparticles to liquid fuels (biofuels and diesel) as a secondary energy carrier has enhanced combustion, performance, and emission properties. Numerous scientists have investigated the possibility of using these modified fuels in diesel engines. Several researchers have comprehensively reviewed the application of nano-additives in biofuel–diesel blends, including but not limited to Kegl et al. [25], Kumar et al. [18], Venkatesan et al. [26], Hoang [27], Shaafi et al. [28]; Khond and Kriplani [20], Dewangan et al. [29], Nanthagopal et al. [30], and Soudagar et al. [31]. Though these studies offer significant contributions to the corpus of literature, there exist some gaps that need to be filled;

(1) To the best of our knowledge, studies that holistically review all three biofuels (alcohols, biodiesel, and vegetable oil) in the context of nanoparticles and engine characteristics are scarce; most of these studies typically consider only one type of the biofuels, especially biodiesel, with limited review specifically dedicated for alcohols or vegetable oils in the broader spectrum.

(2) When doing a literature review on the evolution of any theory or concept over time, it is critical to include the development component by posing questions such as, "What are the evolutionary trends in the research field?", "What future research areas have been emphasized in significant research articles?", and "What are the major research areas?" [32]. The existing reviews clearly lack these aspects, and it is very imperative to systematically analyse the broad literature body, which could help structure the existing knowledge and identify future research gaps [33].

(3) Energy-based indicators of ICE such as brake thermal efficiency (BTE), brake specific fuel consumption (BSFC), and emission characteristics are usually the most used assessment criteria for nanofuels [34]. However, an assessment based on these energetic indicators alone is not enough to describe an all-round performance of the diesel engine [35]. In addition, it is difficult to examine the renewability and sustainability of an energy resource using energy analysis since this indicator fails to consider the effects of the second law's limitation on energy conversion [36]. Exergy analysis bridges this gap as it is a combination of both first and second law of thermodynamics and is closely linked to the renewability and sustainability nexus. In order to achieve a better understanding of the irreversibility or resource destruction, one could employ exergy analysis as it is a powerful technique for investigating the imperfections in an energy conversion system [36,37]. Despite its tremendous ability to optimize energy systems, conventional exergy analysis is often criticized for overlooking the economics and environmental aspects of the thermal system being considered. In nutshell, for an overall performance of any fuel in a thermal system, the energy and exergy indicators are very important, but the addition of the economic and environmental analysis is also key in determining the profitability and sustainability of an improvement in process through exergo-economic and exergo-environmental analysis [35]. A number of studies on the aforementioned aspects related to nano-low carbon fuels in diesel engines have been conducted [34–36,38–42]—however, these generalized discussions are missing in the extant literature review papers on the current subject.

(4) It is worth noting that, besides the engine emissions, performance, and combustion characteristics, most of the existing reviews have only focused on the dispersion stability, wear and friction loss, corrosion, and cost-related issues with nanoparticles,

with limited discussion on a very important aspect of these nano-additives, which is their toxicity and health impacts when they come into contact with humans and animals over a period of exposure. There is numerous evidence supporting how toxic these nanoparticles are and how detrimental they could be to an individual's health [43–48]. It will therefore be prudent to augment the existing literature with these findings.

The review paper reports the impact of potential nanofuel additives on properties of fuel, engine performance, exhaust emissions, and combustion characteristics at different operating conditions. The past and present state of research of this field is also presented in the current work to reveal key research hotspots and ignored areas for future development. The exergy, economic, environmental, and sustainability of these nanofuels in low carbon fuelled-engines are reviewed. We conclude the current study with the toxicity and health impacts of nanoparticles based on results from literary sources.

2. Discussion on Zero Carbon Ecology and Circular Economy

Several organizations and experts around the world have emphasized that efforts to mitigate and adapt to climate change must be accelerated. Approximately 80% of the energy produced in the world comes from fossil fuels [49], with global fossil carbon emissions on the rise since the start of the last century. In this context, the transportation industry consumes approximately 21% of global energy, with oil accounting for 94% of that consumption and 8.0 Gt of direct carbon dioxide (CO_2) emissions from fuel burning, accounting for almost a quarter of global totals [50,51]. This trend in emissions from the transportation sector has aroused significant attention from the scientific community in recent years as efforts are being made to attain a carbon neutral future.

Carbon neutrality, or achieving a carbon-free society, has piqued the interest of scholars, researchers, and policymakers throughout the last three decades. Countries around the globe are converting to renewable energy to reach carbon neutrality, with the goal of keeping global warming below 2 °C compared to pre-industrial levels [52,53]. To achieve a sustainable, low carbon, and resource efficient environment, modern concepts such the circular economy indeed have a significant role to play [54].

Recently, the principle of circular economy is gaining momentum in climate change mitigation measures, and it is believed to have an important role to play in reaching carbon neutral targets. Circular economy hinges on three main components, i.e., reduce, reuse, and recycle. In relation to carbon mitigation strategies particularly from major CO_2 emitting sectors such as the transportation sector, circular economy can be translated to circular carbon economy through the use of alternative fuels in the following ways: (1) to reduce the carbon that must be managed in the first place, (2) to reuse carbon as an input to create feedstocks and fuels, (3) to recycle carbon through the natural carbon cycle with bioenergy, and (4) unique to circular carbon economy, to remove excess carbon and store it [55]. Based on the nature and characteristics of biofuels, they fit the bill in all four pathways of circular carbon economy. Hence, by increasing the share of biofuels in the transport sector, carbon emissions can be dramatically decreased, and chances of reaching carbon neutral targets are increased. In this context, nanoparticles indeed have a role to play in simultaneously promoting cleaner and efficient combustion of low carbon fuels in diesel engines.

3. Research Hotspots and Evolutionary Trends

Bibliometric analysis is one of the modern tools researchers have adopted in ascertaining the research focus and trend of a topic of interest. It is defined as applying mathematics and statistical methods to books and other media of communication [56]. It is a type of research study that provides the basis for what has been achieved and what needs to be investigated [57]. The methodology consists of descriptive and exploratory techniques deemed worthy tools to analyse the relevant literature's recent trends [58]. Many researchers have used the tool to help identify research hotspots in different fields of science. To mention a few, in 2015, Mao et al. [59] conducted a bibliometric analysis on various renewable energy sources. Their study revealed that biogas and biodiesel were the two main areas researchers focused on between 1994 and 2013 regarding bioenergy. In another study, the same set of tools Min and Hao [60] employed between 1990 and 2017 to evaluate the research on biofuels; they found that the three most researched biofuels were biogas, vegetable oil, and bioethanol. Jin et al. [61] also comprehensively reviewed the past and current state of research on ethanol and methanol fuel combustion in ICE between 2000 and 2021. According to Zhang et al. [62], bibliometric investigation, Jatropha curcas, algae, waste cooking oil, and vegetable oil were the most hotspot-related papers for biofuel generation between 1991 and 2015.

As already mentioned in our introduction, to the best of our knowledge, the use of these tools in the area of nano-additives in liquid fuels for ICE does not exist in the literature. The 'Web of Science' Core Collection database was used to determine the historical and present research paradigms of the issue under consideration by utilizing the strategy and techniques of comparable bibliometric studies. Our search strategy is described as follows: "TOPIC: (nanoparticle* or nanoadditive* or 'metal additive*' or 'nano emulsion' or 'nano material') AND TOPIC: ('diesel blend*') Timespan: 2000–2021. The use of 'diesel blend*' is purposeful to reveal all related studies for diesel engine combustion. However, we also looked through the preliminary findings for publication titles and abstracts that were purely relevant to the present research. Thus, papers relating to any other biofuel either than biodiesel, alcohol, or vegetable oils were not included. Papers outside the scope of combustion, performance, and emission characteristics were also excluded. A total of 689 documents was finally retrieved and analysed with an R-statistical package (Biblioshiny) for identifying the core research focus for the subject of the current work.

Figure 1 depicts the 50 most commonly used words or phrases in the subject area under consideration. PERFORMANCE, COMBUSTION, BLENDS, BIODIESEL, NANOPARTI-CLES, EMISSION CHARACTERISTICS, FUEL, and METHYL-ESTER are the words or phrases having at least 100 occurrences. These words suggest that the interest of the investigators in this research field lies in the application of nanoparticles as additives for improving the performance, combustion, and emission characteristics of liquid fuels. Besides these characteristics, STABILITY (rank 22) of liquid fuel–nano-additive blends is another area of interest in this field. One of the main concerns related to the application of NPs as additives for low carbon fuels is their stability aspects [31]. By virtue of their high surface activity and large surface, NPs are prone to aggregation-causing stability problems within the base fuel they are present in. Hence, more work is being carried out in this area to address the situation. Amongst the three biofuels under review in this work, biodiesel seems to be the most investigated fuel as far as nano-additive blending in liquid fuels is concerned. Other fuels in the top 50 keywords are ETHANOL, WASTE COOKING OIL, DI ETHYL ETHER, N-BUTANOL, and JATROPHA METHYL ESTER. Although 'nanoparti-cles' is a term representing several investigated nanoparticles or nano-additives, those to distinctively appear in the top 50 keywords were CARBON NANOTUBE, ALUMINIUM OXIDE, and ZINC OXIDE. It also appears that NOx emissions was the most frequently used environmental-related keyword.

Following that, we add to the discourse by noting quantitative developments in nanoparticles as additives for biodiesel/vegetable oil/alcohol–diesel blends. Figure 2 shows a graphical representation of year-to-year research patterns. Prior to 2018, researchers were heavily involved in using CARBON NANOTUBES as additives mostly for controlling emissions, as seen in trend topics such as PARTICULATE MATTER, SOOT, and PARTICLE SIZE DISTRIBUTION. Some of the popular works completed in that period (according to citations) include: (1) Hosseini et al. [63], who studied the blends of carbon nanotubes and diesel-biodiesel and revealed that carbon monoxide (CO), unburned hydrocarbon (UHC), and soot emissions were dropped by 65.7, 44.98, and 29.41%, respectively; (2) Sadhik Basha and Anand [64], who used Jatropha biodiesel in the presence of carbon nanotubes to conduct their experiment. Their findings showed that smoke opacity and NOx emissions were 69% and 1282 ppm for the neat biodiesel, while the nano-emulsified fuel was 910 ppm and 49%, respectively; and (3) Heydari-Maleney et al. [65] analysed and investigated diesohol–B2 fuels under the influence of carbon nanotubes. Their results indicated that 6.69%, 31.72%, and 5.47% of soot, unburned hydrocarbons, and carbon monoxide were recorded, respectively.

Figure 1. Top 50 keywords of research on nanoparticles as additives for biodiesel/vegetable oil/alcohol–diesel blends.

Furthermore, from the same Figure 2, it is seen that between 2018 and 2021, the attention shifted towards other nanoparticles such as CERIUM OXIDE (CeO_2), TITANIUM DIOXIDE (TiO_2), and ALUMINUM OXIDE (Al_2O_3). Kegl et al. [25] attempted to rank several nanoparticles and their base fuels under two main criteria; the first criteria considered Criterion A as representative of exhaust emissions and engine performance, whereas Criterion B, which was the second criterion, denoted only emission characteristics. Results from both criteria revealed that blends with Al_2O_3 delivered the most optimal feasibility for use in diesel engine. It is therefore not surprising that this nanoparticle has begun to attract the most interest in recent years, as seen in Figure 2.

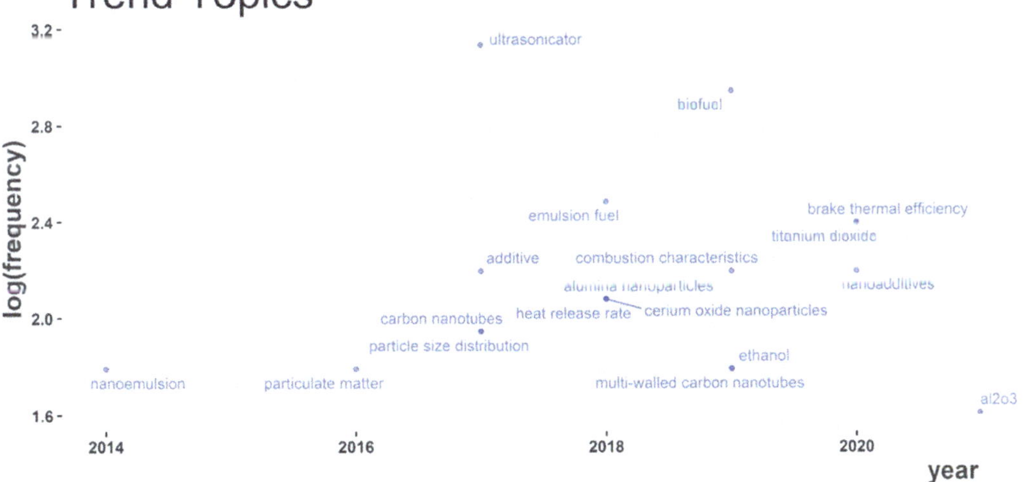

Figure 2. Analysis of topic trends for research on nanoparticles as additives for biodiesel/vegetable oil/alcohol–diesel blends between 2014 and 2021.

In addition, the Multiple Correspondence Analysis (MCA) in R Biblioshiny was used to visualize the conceptual structure of the investigated topic. By using Porter's stemming approach, this technique extracts terms from the papers' title, keywords, and abstract—and in order to make these terms consistent throughout their usage, they are reduced to their base/root/word stem. Moreover, common themes can be identified by using K-means clustering technique to extract and group themes according to clusters. To conclude, the MCA takes into account the distribution of words according to their degree of similarity to construct a two-dimensional graphical map [58]. If two or more words focus on the same theme (common research theme), they are likely to appear closer to each other on the map away from unrelated themes. If a particular cluster is identified in the red region, then these themes have been paid attention to the most by the scientific community, while relatively less attention (relatively ignored) has been given to themes in green and blue clusters (Figure 3). The closer the dots on the graph representing each phrase are, the more similar the keyword distribution is, meaning that they co-occur more frequently in the articles. Furthermore, the proximity of a term to the centre point shows its importance in the study subject, whereas those at the margin are less relevant to other research topics.

The conceptual structure map aided in identifying the important research topics, their connections to other areas, and the topics that had attracted the least attention.

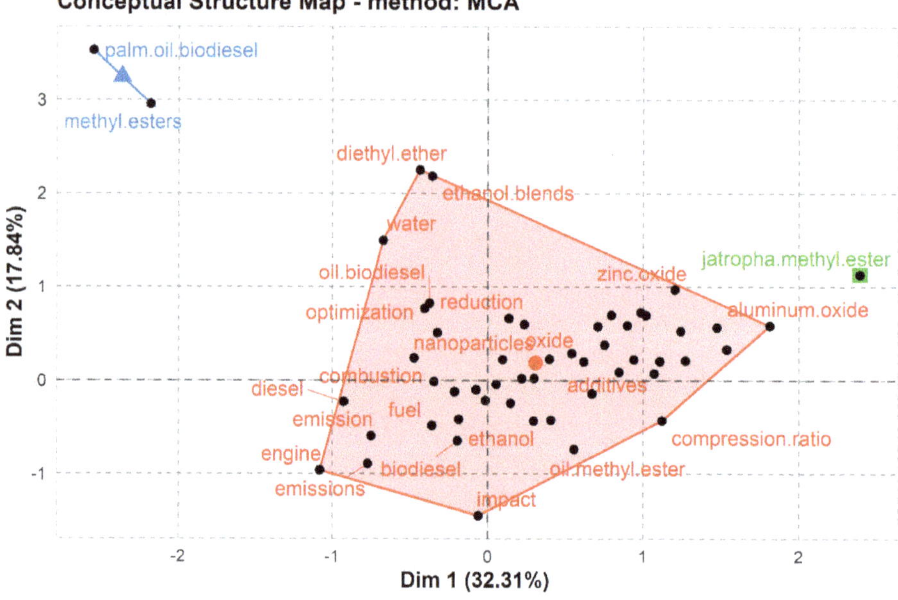

Figure 3. Conceptual structure plot using Multiple Correspondence Analysis (MCA).

The strategic diagram, as illustrated in Figure 4a,b, is a two-dimensional diagram that identifies two properties ("centrality" and "density") that describe the themes. The degree to which one network interacts with another is measured by centrality. The centrality of a theme's external ties to other subjects is measured and may be used as an indication to quantify the theme's influence across the overall academic area. Density measures the strength of internal linkages among all keywords within a topic. As a result, the richness of a subject reflects its progression. The themes are divided into four quadrants based on their centrality and density. In the last decade, scholars have also improved their interpretation of this figure [66]; the interpretation is that, the first quadrant (central and developed) represents motor themes, the second quadrant (central and undeveloped) represents basic themes, the third quadrant (peripheral and developed) represents niche themes, and the fourth quadrant (peripheral and undeveloped) represents emerging or declining themes. Figure 4 is divided into two different periods; Figure 4a represents the themes of this research field during 2000–2010, whereas those of 2011–2021 is represented by Figure 4b.

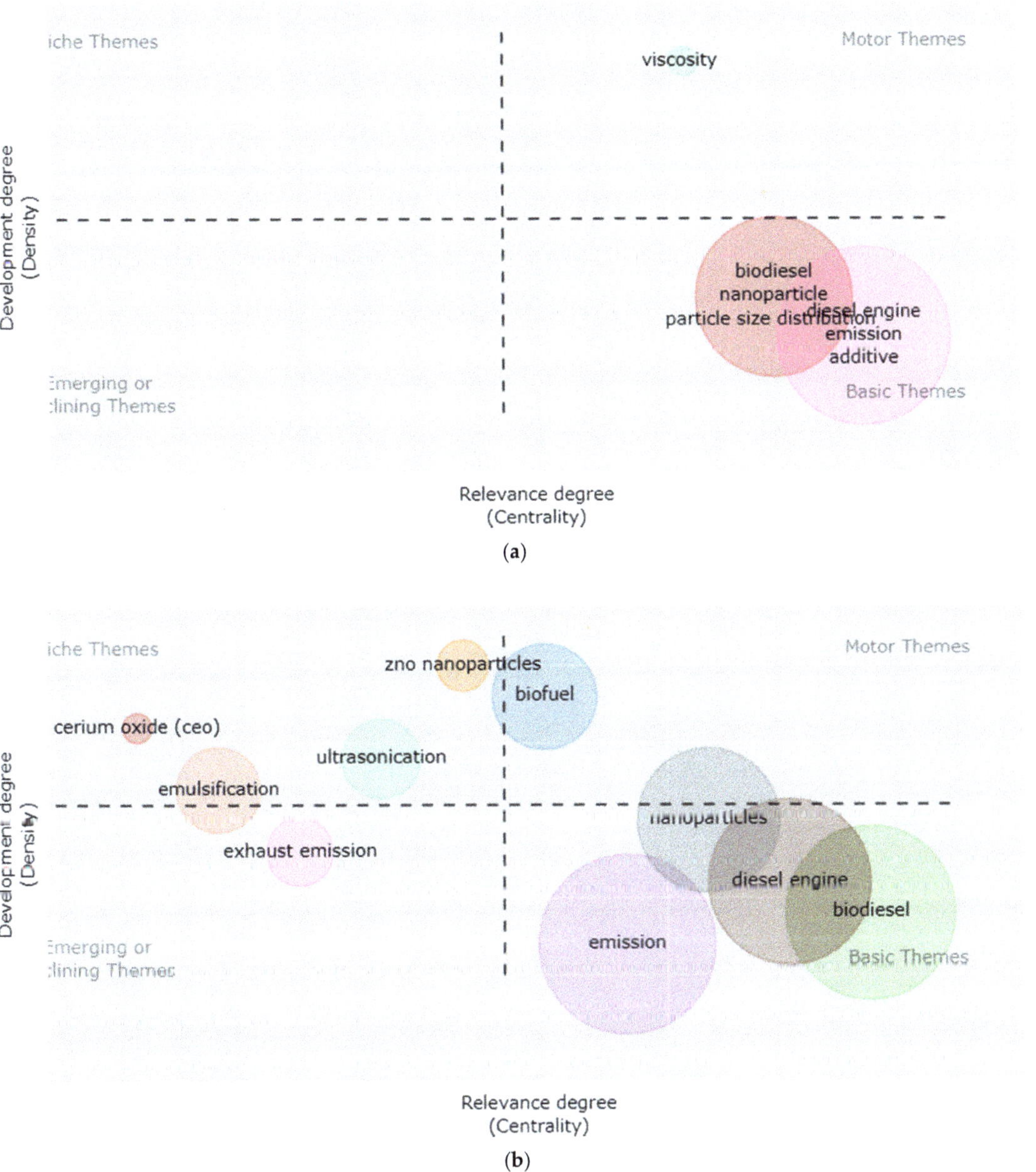

Figure 4. Thematic map of a research field from 2000 to 2010 (**a**); Thematic map of a research field from 2011 to 2021 (**b**).

The key themes with varied levels of density and centrality throughout the first half of the two decades (2000–2010) may be seen in the strategy diagram developed in Figure 4a, and they are VISCOSITY, BIODIESEL, NANOPARTICLE, PARTICLE SIZE DISTRIBUTION, DIESEL ENGINE, EMISSION, and ADDITIVE. Most of these themes were classified as important but undeveloped (according to their quadrant). However, since the Euro VI vehicle emission standard, Vehicular Emission Scheme, Bharat Stage IV, Paris Agreement, Sustainable Development Goals, and other global emission regulations came into effect in the last decade (2011–2021), several efforts have been made to make liquid fuel combustion cleaner and more efficient. Therefore, it is not surprising that the themes of this research field greatly intensified in the last decade, i.e., 2011–2021 (Figure 4b), compared to that of the first decade, i.e., 2000–2010 (Figure 4a). There has been a general increase in research interest in different nanoparticles and biofuels.

Figure 5 shows the geographical distribution of the active researchers in this field of research. It can be seen that the field is largely driven by contributions from Asia, mainly by India (59.85%), China (9.57%), Malaysia (8.70%), and Iran (8.55%). Egypt (6.09%), Turkey (4.35%), USA (2.32%), and Brazil (1.45%) are the key contributors from Africa, Europe, North America, and South America, respectively. Studies from Africa, South America, and Oceania have been heavily underrepresented.

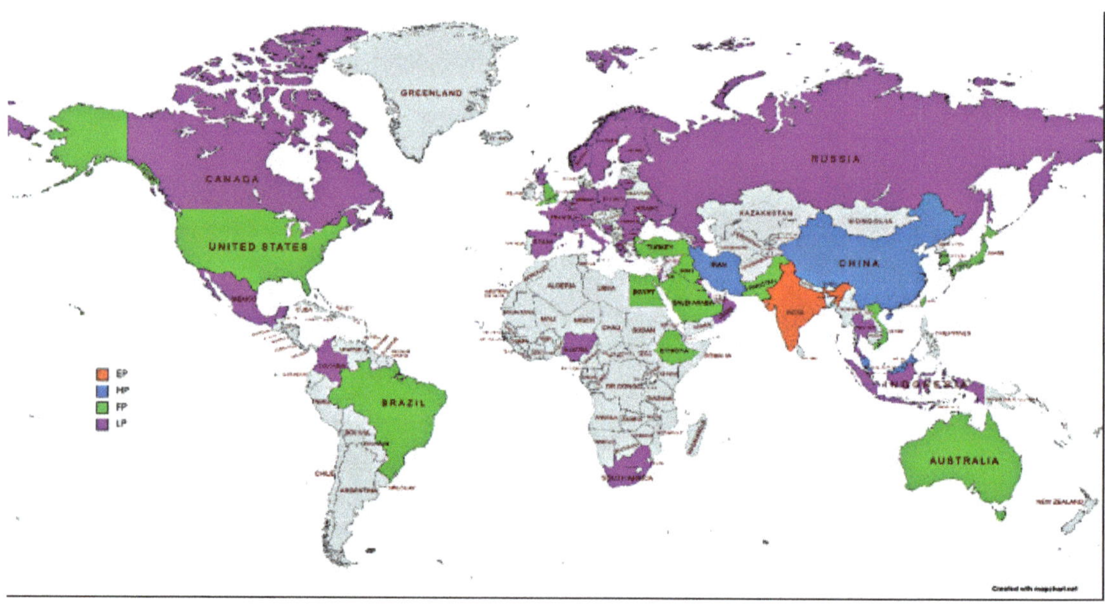

Figure 5. Geographical distribution of research on nanoparticles blending in alcohols/biodiesel/vegetable oil-based fuels (Note—EP: productivity is excellent; HP: productivity is high; FP: productivity is fair/average; LP: productivity is low).

4. Fuel Properties, Emissions, Performance, and Combustion Characteristics

In this section, different fuel combinations from literary sources are reviewed, and the main results from these studies with respect to the effect of nanoparticles on the fuel properties, performance, emission, and combustion characteristics of diesel engines fuelled with alcohol/vegetable oil/biodiesel-based fuels are presented.

Table 1 shows the elemental composition (carbon, oxygen, and hydrogen) of the base fuels considered in the current review. It is seen that for the low carbon fuels reviewed, the highest share of carbon is 77% while the lowest is 37.8%. Similarly, the hydrogen content and oxygen content ranges between 12 and 13.61% and 11 and 49.93%, respectively. The low carbon and oxygen content of these fuels relative to that of conventional diesel (87% carbon, 13% hydrogen, no oxygen) makes them cleaner for combustion in ICE. According to Low

Carbon Technology Partnerships initiative (LCTPi), a low carbon fuel should have a CO_2 performance significantly better than conventional fossil transport fuels by at least 50% [67].

The main fuel properties reviewed in this section are density, kinematic viscosity, cetane number, calorific value, and flash point. Under performance characteristics, we review evidence from the literature relating to brake specific fuel consumption (BSFC), brake thermal efficiency (BTE), brake torque (BT), and brake power (BP). However, there are currently limited experimental data on the brake torque and brake power after the inclusion of the nanoparticles for vegetable oils; thus, only BTE and BSFC are reviewed for this particular fuel. For combustion characteristics, we looked at in-cylinder pressure rise rate, ignition delay, and heat release rate (HRR). Finally, carbon monoxide, oxides of nitrogen, and hydrocarbon pollutants are reported under the emission characteristics.

4.1. Effect of Nanoparticles on Fuel Properties of Low-Carbon Fuels

The physico-chemical properties of a fuel tell how much influence it will have on the emission, performance, and combustion characteristics when fuelled in a diesel engine. For example, viscosity and density impact the duration of fuel atomization penetration; calorific value influences fuel consumption; and cetane number influences fuel ignition quality, resulting in more complete combustion. When these fuel properties are optimized, the resulting fuel provides better fuel performance, combustion, and emission characteristics. Nanoparticles have excellent characteristics, making them suitable as additives for various fuels. Nanthagopal et al. [30] summarized these excellent features of nanoparticles as shown in Figure 6. Several researchers have thus investigated the adjustment to fuel properties such as viscosity, density, flash point, cetane number/index, calorific value, etc., upon the addition of nanoparticles. Sections 4.1.1–4.1.3 are a summary of the effect of nanoparticles on fuel properties of alcohol, vegetable oil, and biodiesel-based fuels, respectively.

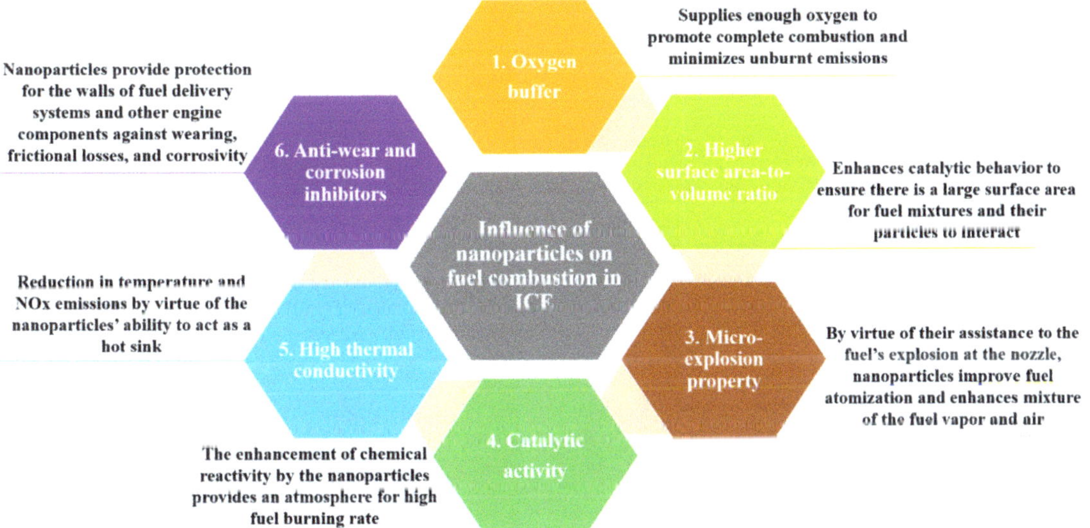

Figure 6. Characteristics of nanoparticles in CI Engines (Reprinted with permission from Ref. [30] Copyright © 2020 Elsevier).

4.1.1. Alcohol-Based Fuels

The inferior properties of alcohol-based fuels (especially low carbon alcohols, i.e., methanol and ethanol) such as poor ignition quality and lower heating value generally improve under the influence of nanoparticles. Carbon-based nanomaterials such as multi-walled nanocarbon tubes (MWCNT), graphene nanoplatelets [68], Fe_2O_3 [69], Al_2O_3 [70,71], and TiO_2 [72] are generally better nanoparticle candidates for boosting the cetane number and calorific value of alcohol-based fuels. However, viscosity and density adjustments are dependent on the blend components. For example, zinc oxide (ZnO) [73], silicon dioxide (SiO_2) [74], CeO_2 [75] increases the densities of the original fuels. However, SiO_2 [76], Fferric oxide (Fe_2O_3) [69], and Al_2O_3 [71] decrease the densities of the original fuels. The viscosity of the neat fuels worsens upon the addition of SiO_2 [74], CeO_2 [75], and TiO_2 [72] but improves under the influence of the following nanoparticles; graphene oxide [77], multi-walled carbon nanotubes [68], graphene quantum dot [78], and Al_2O_3 [79]. Most of the authors added the nanoparticles in concentrations between 10 and 250 ppm (or mg/L). It is worth mentioning that the concentration of the added nanoparticles also had an effect of varying the fuel properties. For instance, when the concentration of SiO_2 in neat methanol increases from 25 to 100 ppm, it negatively affects the density and viscosity of methanol [74]. Furthermore, when the concentration of Fe_2O_3 in N-amyl ternary fuel is increased from 40 to 120 ppm, the corresponding calorific value increases from 41.73 to 42.97 MJ/kg [69]. Table 1 summarizes the effect of various nanoparticles and their dosages on the properties of alcohol-based fuels from literary sources.

Table 1. Summary of nanoparticles' effect on fuel properties of alcohol-based fuels.

Alcohol	Case #	Fuel	Nanoparticle (DOSAGE)	Density (kgm^{-3})	Viscosity (mm^2/s)	Flash Point (°C)	Calorific Value (MJ/kg)	Cetane Number
Ethanol [73]	1a	D40B30E30	Absent	828.5	2.42	10	39.90	57
	1b	D40B30E30	ZnO (250 ppm)	836.3	2.32	16	36.89	55
Methanol [74]	2a	M100	Absent	790	0.59	-	20.3	-
	2b	MSN25	SiO_2 (25 ppm)	793	0.62	-	21.9	-
	2c	MSN50	SiO_2 (50 ppm)	798	0.65	-	22.4	-
	2d	MSN100	SiO_2 (100 ppm)	804	0.71	-	23.2	-
Pentanol [76]	3a	TF	Absent	841	3.3	3	41.62	48
	3b	TF40	SiO_2 (40 ppm)	839	3.37	2.8	41.73	48.5
	3c	TF80	SiO_2 (80 ppm)	837	3.21	4	41.96	55
	3d	TF120	SiO_2 (120 ppm)	830	3.01	3	42.97	47.4
N-amyl [69]	4a	TF	Absent	841	3.3	3	41.62	48
	4b	TF40	Fe_2O_3 (40 ppm)	839	3.37	2.8	41.73	48.5
	4c	TF80	Fe_2O_3 (80 ppm)	837	3.21	4	41.96	55
	4d	TF120	Fe_2O_3 (120 ppm)	830	3.01	3	42.97	47.4
Ethanol [70]	5a	DF90E10	Absent	821.5	2.7	-	41.7	52.44
	5b	DF90E10	Al_2O_3 (100 ppm)	821.6	2.8	-	42.5	53.68
	5c	DF90E10	TiO_2 (100 ppm)	821.6	2.8	-	42.3	53.24
Ethanol [71]	6a	TF	Absent	852	3.18	59	43.18	48.4
	6b	TF10	Al_2O_3 (10 ppm)	849	3.07	60	43.41	48.6
	6c	TF20	Al_2O_3 (20 ppm)	848	3.02	63	43.85	48.7
	6d	TF30	Al_2O_3 (30 ppm)	845	3.1	62	43.58	48.4
Ethanol [80]	7a	BDE	Absent	840.2	2.86	20	39.98	53
	7b	BDE	Al_2O_3 (25 ppm)	837.2	2.57	22	39.14	54
Methanol [75]	8a	M100	Absent	790	0.59	-	20.3	-
	8b	MCN25	CeO_2 (25 ppm)	800	0.62	-	20.8	-
	8c	MCN100	CeO_2 (100 ppm)	810	0.66	-	22.1	-
Isopropanol, Butanol [79]	9a	B20	Absent	847 *	3.70	-	43	42
	9b	D80SBD15E4S1	Al_2O_3 (100 mg/L)	840 *	3.37	-	42.59	52
Butanol [72]	10a	J50D10Bu	Absent	848 *	4.49	-	44.99	52.5
	10b	J50D10Bu25TiO$_2$	TiO_2 (25 mg/L)	849 *	4.51	-	45.11	53.5
	10c	J50D10Bu50TiO$_2$	TiO_2 (50 mg/L)	849 *	4.55	-	45.14	54.5
Butanol [81]	11a	B20But10	Absent	840.1	2.62	46.75	39.96	-
	11b	B20But10	TiO_2 (0.01% by mass)	840.2	2.63	45	39.84	-

Table 1. *Cont.*

Alcohol	Case #	Fuel	Nanoparticle (DOSAGE)	Density (kgm^{-3})	Viscosity (mm^2/s)	Flash Point (°C)	Calorific Value (MJ/kg)	Cetane Number
Ethanol [65]	12a	B2	Absent	820.7	2.31	-	42.66	-
	12b	B2E2C20	Carbon nanotubes (20 ppm)	821.8	2.39	-	42.23	-
	12c	B2E2C60	Carbon nanotubes (60 ppm)	821.8	2.38	-	42.27	-
	12d	B2E2C100	Carbon nanotubes (100 ppm)	821.9	2.39	-	42.23	-
	12e	B2E4C20	Carbon nanotubes (20 ppm)	820.7	2.31	-	42.66	-
	12f	B2E4C60	Carbon nanotubes (60 ppm)	820.8	2.31	-	42.68	-
	12g	B2E4C100	Carbon nanotubes (100 ppm)	820.9	2.31	-	42.62	-
	12h	B2E6C20	Carbon nanotubes (20 ppm)	819.6	2.24	-	43.11	-
	12i	B2E6C60	Carbon nanotubes (40 ppm)	819.7	2.24	-	43.13	-
	12j	B2E6C100	Carbon nanotubes (100 ppm)	819.9	2.25	-	43.03	-
Ethanol [78]	13a	B10	Absent	835	3.33	70	-	-
	13b	B10E2GQD30	Graphene quantum dot (30 ppm)	834	3.11	<28	-	-
	13c	B10E4GQD30	Graphene quantum dot (30 ppm)	834	2.99	<28	-	-
	13d	B10E6GQD30	Graphene quantum dot (30 ppm)	834	2.94	<28	-	-
	13e	B10E8GQD30	Graphene quantum dot (30 ppm)	834	2.83	<28	-	-
Heptanol [68]	14a	H20D	Absent	839.5 *	3.34	-	34.65	48.5
	14b	H40D	Absent	838.1 *	3.33	-	43.11	45.5
	14c	H20DMWCNT	Multi-walled carbon nanotubes (50 mg/L)	842.2 *	3.16	-	44.79	51.5
	14d	H20DGNP	Graphene nanoplatelets (50 mg/L)	842.1 *	3.11	-	44.79	50.5
	14e	H20DGO	Graphene oxide (50 mg/L)	842.3 *	3.12	-	44.80	51
	14f	H40DMWCNT	Multi-walled carbon nanotubes (50 mg/L)	841 *	3.16	-	43.60	49.5
	14g	H40DGNP	Graphene nanoplatelets (50 mg/L)	840.5 *	3.13	-	43.59	50
	14h	H40DGO	Graphene oxide (50 mg/L)	840.7 *	3.13	-	43.60	50.5
Butanol [77]	15a	JME40B	Absent	849.9 *	3.73	-	37.53	43.53
	15b	JME40B50GO	Graphene oxide (50 mg/L)	851.0 *	3.65	-	37.55	48.10
	15c	JME40BGNPs	Graphene nanoplatelets (50 mg/L)	851.1 *	3.68	-	37.56	47.95
	15d	JME40BMWCNTs	Multi-walled nanocarbon nanotubes (50 mg/L)	851.1 *	3.69	-	37.56	47.98

* Specific gravity.

4.1.2. Vegetable Oil-Based Fuels

The application of nanoparticles in vegetable oil-based fuels, to some extent, follows a similar pattern as that of alcohol-based fuels. Annamalai et al. [82] added cerium oxides in the concentration of 30 ppm to an emulsion of lemongrass oil. It was observed that the presence of the nanoparticle increased the densities and viscosities of the emulsion fuel. Similar observations have been made by Dhinesh et al. [83], where CeO$_2$ was blended in

Cymbopogon Flexuosu oil. With increase in the concentration of the nanoparticles, it was observed that the viscosity and density of the Cymbopogon Flexuosu oil became worse. Results from several works show that CeO_2 may not be an ideal nanoparticle when the goal is to address the viscous and dense nature of vegetable oils [83–87]. On the other hand, Al_2O_3 had a positive effect on the density and viscosity of a pyrolyzed biomass oil when 50 ppm and 100 ppm of the nanoparticle was added to the base fuel [88]. Increasing the concentration of CeO_2 in orange peel oil and lemon peel oil results in an increase in calorific value, but the trend reverses if CeO_2 is replaced with carbon nanotubes [86]. Nano ferrocene shows excellent cetane-enhancing abilities in vegetable oil-based fuels than CeO_2 [89]. Similarly, CeO_2 also provides more energy content to oil-containing fuels than carbon nanotubes [90]. Some nanoparticles also had marginal or no effect on the properties of the based fuels. It is worth mentioning that, though nanoparticles can offer improvements to fuel properties, their concentrations in the blend should be moderated. Excessive addition of nanoparticles, especially CeO_2, could defeat the original purpose of their inclusion in the base fuels. Table 2 summarizes the effect of various nanoparticles and their dosages on the properties of vegetable oil-based fuels from literary sources.

Table 2. Summary of nanoparticles' effect on fuel properties of vegetable oil/pure bio-oil-based fuels.

Vegetable Oil	Case #	Fuel	Nanoparticle (Dosage)	Density (kgm^{-3})	Viscosity (mm^2/s)	Flash Point (°C)	Calorific Value (MJ/kg)	Cetane Number
Polanga seed oil [91]	1a	Neat polanga	Absent	937.4 *	57.8	-	-	-
	1b	Diesel + polanga	Fe_2O_3 (100 ppm)	835.3 *	3.49	-	44.08	-
	1c	Diesel + polanga	Fe_2O_3 (200 ppm)	837.3 *	3.62	-	44.03	-
	1d	Diesel + polanga	Fe_2O_3 (300 ppm)	837.5 *	3.39	-	44.00	-
Tyre oil ** [84]	2a	B10	Absent	820	6.59	49	42.90	-
	2b	B10D85	CeO_2 (50 ppm)	822	6.65	50	42.94	-
	2c	B10D80	CeO_2 (100 ppm)	824	6.72	51	42.98	-
Lemongrass oil [85]	3a	LGO25	Absent	870 *	3.48	53	41.69	-
	3b	LGO25 + WE + CE	CeO_2 (50 ppm)	910 *	4.16	58	41.06	-
Pyrolyzed biomass oil ** [88]	4a	PBO20	Absent	845	4.24	96	41.1	-
	4b	PB020	Al_2O_3 (50 ppm)	839	4.08	94	41.2	-
	4c	PBO40	Absent	862	4.86	108	39.5	-
	4d	PBO40	Al_2O_3 (100 ppm)	852	4.72	104	41.3	-
Lemon peel oil [86]	5a	LPO20	CeO_2 (50 ppm)	856	2.43	44	41.20	-
	5b	LPO20	CeO_2 (100 ppm)	856	2.56	40	42.44	-
	5c	LPO20	CNT (50 ppm)	856	2.38	42	42.11	-
	5d	LPO20	CNT (100 ppm)	856	2.64	44	41.88	-
Orange peel oil [86]	6a	OPO20	CeO_2 (50 ppm)	858	2.54	46	42.48	-
	6b	OPO20	CeO_2 (100 ppm)	858	2.80	42	42.32	-
	6c	OPO20	CNT (50 ppm)	858	2.72	44	42.41	-
	6d	OPO20	CNT (100 ppm)	858	3.01	43	42.17	-
Nerium olender [87]	7a	ENOB	Absent	906	4.67	74	35.8	-
	7b	NENOB	CeO_2 (30 ppm)	916.4	4.99	67	36.2	-
Lemongrass oil [82]	8a	Neat LGO	Absent	905	4.60	55	37	48
	8b	LGO emulsion	Absent	906	4.67	74	35.8	46.3
	8c	LGO nano emulsion	CeO_2 (30 ppm)	916.4	4.99	67	36.2	48.8
Hydrotreated vegetable oil [89]	9a	B7 + 10%HVO	Absent	828.5	2.73	59	-	55.2
	9b	B7 + 10%HVO	CeO_2 (1:4000)	828.3	2.73	60	-	53.1
	9c	B7 + 10%HVO	Nano ferrocen (1:1000)	828.1	2.72	59	-	57.7
Tyre pyrolysis oil ** [90]	10a	JME90TPO10	Absent	868.7	6.39	-	9962.7 ***	-
	10b	JME90TPO10	CeO_2 (100 ppm)	868.3	6.39	-	9537.5 ***	-
	10c	JME90TPO10	CNT (100 ppm)	872.6	5.25	-	9311.5 ***	-
	10d	JME80TPO20	Absent	874.1	6.36	-	10,001.43 ***	-
	10e	JME80TPO20	CeO_2 (100 ppm)	873.5	6.40	-	9630.2 ***	-
	10f	JME80TPO20	CNT (100 ppm)	878.1	5.35	-	9482.6 ***	-
	10g	JME70TPO30	Absent	880.4	6.48	-	10062 ***	-
	10h	JME70TPO30	CeO_2 (100 ppm)	880.3	6.39	-	9726.8 ***	-
	10i	JME70TPO30	CNT (100 ppm)	881.8	5.29	-	9656.5 ***	-

Table 2. *Cont.*

Vegetable Oil	Case #	Fuel	Nanoparticle (Dosage)	Density (kgm^{-3})	Viscosity (mm^2/s)	Flash Point (°C)	Calorific Value (MJ/kg)	Cetane Number
Cymbopogon flexuosus biofuel [83]	11a	C20D80	Absent	843	3.21	49	42.19	-
	11b	C20D80	CeO$_2$ (10 ppm)	844.1	3.28	47	42.14	-
	11c	C20D80	CeO$_2$ (20 ppm)	844.5	3.31	46	41.88	-
	11d	C20D80	CeO$_2$ (30 ppm)	844.9	3.37	45	41.62	-

* Specific gravity; ** not typical vegetable oil, but these neat oils share similar characteristics with vegetable oil; *** gross calorific value (cal/gm).

4.1.3. Biodiesel-Based Fuels

El-Seesy et al. [92] blended multi-walled carbon nanotubes of concentrations between 10 and 50 mg/L into Jatropha biodiesel blends. The density of the original fuel remained unchanged, viscosity and cetane number increased, while a negligible difference was recorded in its calorific value. In a similar study, Alenezi et al. [93] increased the concentration of the multi-walled carbon nanotubes to 100 ppm in Palm oil biodiesel blends. The density and cetane number of the base fuel reduced, but its viscosity and calorific value increased. The density, viscosity, and calorific value of Jojoba biodiesel blends increased upon the addition of cupric oxide (CuO) (25–75 ppm) in the work of Rastogi et al. [94], but its flash point kept decreasing with an increase in the dosage of the nanoparticle. Al$_2$O$_3$ (0.2–0.04 ppm) has a positive effect on the density adjustment of Madhuca Indica, but it will decrease the methyl ester's flash point and marginally/negligibly increase its calorific value, as shown by Rastogi et al. [95]. The addition of titanium oxide of 300 ppm to canola biodiesel produces a fuel that has improved density, viscosity, cetane number, and sulfur content than the neat biodiesel in the study of Nithya et al. [96]. Venu and Madhavan [80] show that Al$_2$O$_3$ (25 ppm) in Jatropha biodiesel-containing fuel will follow a similar trend observed in the work of Rastogi et al. [95] for density and viscosity, but assume an opposite trend in the flash point and calorific value. Janakiraman et al. [97] compared the fuel adjustment abilities of three different nanoparticles, namely TiO$_2$, Zirconium dioxide (ZrO$_2$), and CeO$_2$, in Garcinia gummi-gutta methyl esters; in general, TiO$_2$ and ZrO$_2$ showed better fuel modification compared to CeO$_2$. Following this observation, CeO$_2$ generally offered negative fuel modifications when it was added to waste cooking oil according to Khalife et al. [98], but Karthikeyan et al. [99] stipulates otherwise; the addition of CeO$_2$ to rice bran biodiesel had a positive effect on the density, viscosity, and flash point of the original fuel. Hajjari et al. [100] investigated the effect of CeO$_2$ nanoparticles on the oxidative stability of biodiesel. Their results revealed that upon the addition of the nanoparticle at 50 ppm, the oxidative stability of the neat biodiesel worsened, but caused slight improvement when the concentration of CeO$_2$ was increased to 200 ppm. However, at that high concentration, the oxidative stability of the resulting fuel still failed to meet the ASTM/EN requirement of 6 h induction period for biodiesel. Table 3 summarizes the effect of various nanoparticles and their dosages on properties of biodiesel-based fuels from literary sources.

Table 3. Summary of nanoparticles' effect on fuel properties of biodiesel-based fuels.

Biodiesel	Case #	Fuel	Nanoparticle (Dosage)	Density (kgm^{-3})	Viscosity (mm^2/s)	Flash Point (°C)	Calorific Value (MJ/kg)	Cetane Number
Jatropha [92]	1a	JB20D	Absent	847.1 *	4.06	-	45.43	52
	1b	JB20D	MWCNT (10 mg/L)	847.1 *	4.1	-	45.43	52.7
	1c	JB20D	MWCNT (20 mg/L)	847.1 *	4.19	-	45.45	53.5
	1d	JB20D	MWCNT (30 mg/L)	847.1 *	4.25	-	45.46	54.2
	1e	JB20D	MWCNT (40 mg/L)	847.1 *	4.31	-	45.46	55.4
	1f	JB20D	MWCNT (50 mg/L)	847.1 *	4.35	-	45.46	56
Canola biodiesel [101]	2a	Canola biodiesel	Absent	886.5	5.38	172	38.76	48
	2b	Canola emulsion	CeO$_2$ (50 ppm)	906.8	17.2	185	33.54	38
Jatropha [80]	3a	BDE	Absent	840.2	2.86	20	39.98	53
	3b	BDE	Al$_2$O$_3$ (25 ppm)	837.2	2.57	22	39.14	54

Table 3. *Cont.*

Biodiesel	Case #	Fuel	Nanoparticle (Dosage)	Density (kgm^{-3})	Viscosity (mm^2/s)	Flash Point (°C)	Calorific Value (MJ/kg)	Cetane Number
Jojoba [94]	4a	JB20	Absent	845.36	3.59	71	41.93	-
	4b	JB20CN25	CuO (25 ppm)	858.15	3.68	66	41.22	-
	4c	JB20CN50	CuO (50 ppm)	864.56	3.76	64	41.43	-
	4d	JB20CN75	CuO (75 ppm)	871.17	3.87	63	41.66	-
Rice bran [99]	5a	B20	Absent	828	6.62	39	38.96	-
	5b	B20	CeO_2 (50 ppm)	830	6.16	35	39.44	-
	5c	B20	CeO_2 (100 ppm)	826	5.96	45	39.25	-
Madhuca Indica [95]	6a	B100	Absent	889	5.21	173	40.30	-
	6b	B10A0.2	Al_2O_3 (0.2 gm)	848	4.38	65	41.78	-
	6c	B10A0.4	Al_2O_3 (0.4 gm)	853	4.35	63	41.82	-
	6d	B20A0.2	Al_2O_3 (0.2 gm)	858	4.49	59	41.91	-
	6e	B20A0.4	Al_2O_3 (0.4 ppm)	862	4.42	56	41.92	-
Palm oil [93]	7a	B100	Absent	860	4.61	-	38.6	62.5
	7b	B30C100	MWCNT (100 ppm)	852	5.12	-	40.3	52.2
Waste cooking oil [98]	8a	B5W3	Absent	-	3.6	78	44.35	-
	8b	B5W5	Absent	-	3.57	76	42.84	-
	8c	B5W7	Absent	-	3.92	74	42.49	-
	8d	B5W3$_m$	CeO_2 (90 ppm)	-	3.82	80	43.48	-
	8e	B5W5$_m$	CeO_2 (90 ppm)	-	3.82	78	42.73	-
	8f	B5W7$_m$	CeO_2 (90 ppm)	-	3.88	77	42.38	-
Neem oil [102]	9a	NBD	Absent	830	4.1	-	38.96	53
	9b	NBDCNT 50	Carbon nanotubes (50 ppm)	820	3.8	-	39.15	54
	9c	NBDCNT100	Carbon nanotubes (100 ppm)	810	3.5	-	39.56	55
Canola oil [96]	10a	B20	Absent	915	4.8	-	-	42
	10b	B20	TiO_2 (300 ppm)	840	3.4	-	-	56
Kapok oil [103]	11a	B100	Absent	931	4.2	170	38	48
	11b	B20	Cobalt chromite (50 ppm)	845	3.8	145	39	49
Used cooking oil [104]	12a	B20	Absent	843.2	3.19	76	43.33	52.5
	12b	B20	MWCNT (25 ppm)	843.9	3.15	74	43.37	52.9
	12c	B20	MWCNT (50 ppm)	845.2	3.09	71	43.4	53.4
	12d	B20	MWCNT (75 ppm)	846.9	2.97	69	43.45	54.1
	12e	B20	MWCNT (100 ppm)	848.1	2.95	67	43.62	55.3
Garcinia gummi-gutta [97]	13a	B20	Absent	863	4.51	90.7	40.81	50.7
	13b	B20	TiO_2 (25 ppm)	864	4.39	96.8	41.06	51.62
	13c	B20	CeO_2 (25 ppm)	863	4.54	90.2	40.68	50.85
	13d	B20	ZrO_2 (25 ppm)	866	4.51	93.1	41.31	50.91
Karanja oil/waste cooking oil [105]	14a	KBD20	Graphene oxide (60 ppm)	839	3.66	80	41.82	-
	14b	KBD20	Graphene nanoplatelets (60 ppm)	837	3.65	81	41.8	-
	14c	WBD20	Graphene oxide (60 ppm)	838	3.57	79	41.7	-
	14d	WBD20	Graphene nanoplatelets (60 ppm)	837	3.56	81	41.7	-
	14e	KBD20	Absent	836	3.65	81	41.8	-
	14f	WBD20	Absent	836.6	3.55	80	41.7	-
Orange peel oil [106]	15a	OOME	Absent	850.7	4.83	94	38.1	47
	15b	OOMET50	TiO_2 (50 ppm)	856.5	5.17	96	35.98	50
	15c	OOMET100	TiO_2 (100 ppm)	861.3	5.42	99	36.1	53
Waste frying oil [107]	16a	WFOME	Absent	898	4.21	160	43.85	-
	16b	WFOME	MWCNT (25 ppm)	830	4.75	57	43.73	-
	16c	WFOME	MWCNT (50 ppm)	831.1	4.45	65	43.93	-
Camelina oil [108]	17a	B20	Absent	836	5.67	-	44.09	-
	17b	B20G60	Graphene oxide (60 ppm)	832	5.53	-	44.49	-
Honge oil [109]	18a	HOME	Absent	-	5.6	170	36.02	-
	18b	HOME25CNT	MWCNT (25 ppm)	-	5.7	166	34.56	-
	18c	HOME50CNT		-	5.8	164	35.1	-
Sardine oil [110]	19a	SOME	Absent	890	4.5	58	37.41	45
	19b	SOME	CeO_2 (25 ppm)	894	5.6	191	43.37	56

Table 3. *Cont.*

Biodiesel	Case #	Fuel	Nanoparticle (Dosage)	Density (kgm^{-3})	Viscosity (mm^2/s)	Flash Point (°C)	Calorific Value (MJ/kg)	Cetane Number
Calophyllum	20a	CIB20	Absent	843.3	3.56	69	40.92	53.85
inophyllum [111]	20b	CIB20ANP40	Al$_2$O$_3$ (40 ppm)	858	3.64	64	41.44	54.58

* Specific gravity.

4.2. Effect of Nanoparticles on Engine Performance/Emission/Combustion Characteristics of Low Carbon Fuels

4.2.1. Engine Performance Characteristics of Nanoparticles in Alcohol-Based Fuels

Brake Thermal Efficiency

Prabakaran and Udhoji [73] revealed that the enhancement of surface area-to-volume ratio by zinc oxide nanoparticles led to an improvement in the BTE of diesel–ethanol–biodiesel blends at 100% load conditions and increased the average combustion temperature, as well as an exhibited an increase in the exhaust gas temperature along with the increase in load. Wei et al. [74] achieved a similar result for their blends of silicon dioxide nanoparticles and methanol. As the load increased, BTE increased, as well as being a result of the increased fuel injection quantity. Moreover, the increase in BTE became more obvious at higher concentrations of the silicon dioxide nanoparticle. In the study of Ramachander et al. [76], with a 40–120 ppm increase in silicon dioxide nanoparticles in a ternary fuel containing pentanol, the corresponding BTE increased by 1.58% to 2.34%. The authors explained that the catalytic activity of the nanoparticles may have resulted in finer combustion characteristics, thus positively influencing the BTE. Ağbulut et al. [70] revealed that the process of adding oxides of aluminum and titanium of 100 ppm concentration to diesel–bioethanol blends results in an increase of BTE of 5.70% and 5.15% for DF90E10 + A100 and DF90E10 + T100, respectively, compared to the DF90E10 test fuel. The increase in BTE of the base fuel upon the addition of the nanoparticles was attributed to the catalyst activity role thereof, micro exploits in primary droplets, the higher energy content of nanoparticles, their higher surface area to volume ratio, oxygen-buffer role, and superior thermal properties. According to Venu et al. [71], the addition of aluminum oxide nano-additives increased the BTE of ternary fuel (diesel–biodiesel–ethanol) by 2.48%, 7.8%, and 1.42% for doping concentrations of 10 ppm, 20 pm, and 30 ppm, respectively. The reasons behind this surge in BTE for the ternary fuel post doping were similar to those described in the work of Ağbulut et al. [70]. Another possible explanation was that the Al$_2$O$_3$ positively influenced the heat transfer rate due to its enhanced conductive, radiative, and heat mass transfer. The presence of Al$_2$O$_3$ ensured that the mixture of air with fuel vapor is enhanced, thereby promoting complete combustion.

Brake Specific Fuel Consumption

Shaafi and Velraj [79] added alumina nanoparticles (100 mg/L) to ethanol and iso-propanol as additives with diesel–soybean biodiesel blend (D80SBD15E4S1 + alumina fuel). A minimum BSFC was recorded for D80SBD15E4S1 + alumina fuel blend at 75 and 100% load conditions. The BSFC of B20, D80SBD15E4S1 + alumina fuel blend, and neat diesel was 0.312, 0.309, and 0.349 kWh, respectively. The large surface area of the alumina nanoparticles enhanced the combustion process of D80SBD15E4S1 + alumina fuel blend. El-Seesy and Hassan [72] shows that TiO$_2$ nanoparticles' presence in the Jatropha biodiesel–diesel–n-butanol blend (J50D10Bu) leads to a significant reduction in BSFC. The investigators explained that the high surface area of TiO$_2$ nanoparticles resulted in a more reactive surface area with air, which improved evaporation rate and reduced ignition delay; thus, the combustion process enhanced. The presence of titanium dioxide nanoparticles in J50D10Bu reduced the BSFC up to 18% in contrast to that of the pure J50B10Bu blend. Diesohol fuel and B2 blends were doped with carbon nanotubes to investigate the effect on engine performance in the work of Heydari-Maleney et al. [65]. Carbon nanotubes were in concentrations of 20, 60, and 100 ppm. Results revealed that the addition of the carbon nanoparticle reduces BSFC of the base fuel. The addition of carbon nanotubes and ethanol in diesel fuel resulted in an average decrease in BSFC by 8.86%. The effect of alumina nano methanol fluid on the performance, combustion, and emission characteristics of a diesel engine fuelled with diesel methanol dual fuel was investigated by Zenghui et al. [112]. Three methanol-based nanofluids with Al_2O_3 of mass fractions 25, 50, and 100 ppm were prepared. Results from their investigation revealed that the methanol-based nanofluids recorded lower BSFC compared to the other test fuels without the nanoparticle. The lowest BSFC was recorded by the blend with the highest dosage of the nanoparticle (100 ppm). They claimed the accelerated evaporation mixing which led to an enhanced combustion process was a result of Al_2O_3 nanoparticles' ability to decrease fuel droplet size and also positively influence the mixtures' thermal conductivity. These processes eventually reduced the fuel requirement. In the work of El-Seesy and Hassan [77], the investigators reveal that by adding carbon nanomaterial, i.e., graphene oxide, multi-walled carbon nanotubes, and graphene nanoplatelets, to the blend of Jatropha biodiesel–butanol fuel (JME40B), a significant reduction in BSFC could be achieved. The carbon nanomaterial and the blends' BSFC was reduced by approximately 35% compared to the pure JME40B. Some of the reasons the authors ascribed to this observation were that the reduction in consumed fuel and improved combustion process was as a result of an increase in the engine's carbon oxidation rate by virtue of subjecting the JME40B to carbon nanomaterials doping. Moreover, these carbon nanomaterials shorten burnout time, which enhances the expansion work on the piston. As a result, a lower BSFC was obtained.

Brake Power and Brake Torque

Örs et al. [81] performed a study to experiment with a ternary blend consisting of n-butanol, diesel, and biodiesel fuelled in a diesel engine under the influence of TiO_2. The average BT of B20 + TiO_2 increased around 10.20% compared to B20. Furthermore, the BT of B20But10 + TiO_2 was approximately 9.74% higher than B20But10. The presence of n-butanol decreased the BT of the fuel blends but interestingly, doping the blend with TiO_2 (B20But10 + TiO_2) increased the BT value relative to B20. The corresponding maximum BP values at 2800 rpm were 9.17 kW by B20 + TiO_2 and 8.59 kW for B20But10 + TiO_2 in comparison with 8.16 kW for B20 and 7.69 kW for B20But10, respectively. The increase in BT and BP of the fuels after the addition of TiO_2 was attributed to the high energy content of TiO_2, which is about 100–150 MJ/kg. In addition, the high surface–volume proportion of the TiO_2 provided better oxidation of fuel; hence, high combustion enthalpy and energy density were released so that the maximum engine BT and BP increased. The exhaust emissions and engine performance of a single cylinder diesel engine fuelled with diesel–biodiesel–ethanol (DBE) ternary blend in the presence of nano-biochar was modelled by Mirbagheri et al. [113]. The average BT was remarkably increased by approximately 11.7% after the addition of the nano-biochar particles. The maximum BP of approximately 7.6 kW was recorded by DBE blends with a nano-biochar concentration of 113 ppm. According to the authors, the improved combustion process and atomization of the dispersed nano-organic particles resulted in an efficient conversion of the fuel blend's chemical energy into mechanical work, which consequently boosted the BP and BT values. Diesohol–B2 blends were mixed with carbon nanotubes for evaluation on the characteristics of a diesel engine in the work of Heydari-Maleney et al. [65]. In this work, the fuel blends with carbon nanotubes produced the maximum BT while B2 and D100 produced the minimum BT. The increase in BT became very spontaneous by increasing the dosage of the carbon nanotubes. By doing this, the investigators claimed that the energy generated by the combustion in the cylinder is more complete, and the quality of combustion improves. Hence, the average pressure becomes greater, causing an increase in the piston force and torque. The results for BP are analogous to that of BT. The fuel blends with the carbon nanotubes recorded the highest BP, while B2 and D100 recorded the lowest BP. At higher concentrations of the carbon nanotubes, combustion improves and energy conversion to useful work becomes more effective. This observation could be due to the increase in heat transfer co-efficient attributed to the carbon nanotubes' high surface area-to-volume ratio. Ethanol–biodiesel blends doped with Graphene Quantum Dot (GQD) nanomaterials were researched by Heidari-Maleni et al. [78]. By virtue of GQD's influence, the brake power and brake torque of the oxygenated blends increased by 28.18% and 12.42%, respectively. Additionally, according to Safieddin Ardebili et al. [114], the presence of nano-biochar (SNB) slightly increased the brake power of fusel oil–diesel fuel by ~3%. The researchers explained that the catalytic activity of the SNB particles contributed to reducing ignition delay, which resulted in higher peak cylinder pressure and a better combustion process. At 100 ppm, full load condition, and 2000 rpm engine speed, the highest BT value of 7.8 Nm was recorded for 10% fusel oil. When the test conditions were kept constant, the corresponding engine torque was approximately 7.4% without the SNB particles.

4.2.2. Engine Emission Characteristics of Nanoparticles in Alcohol-Based Fuels
Carbon Monoxide

Mardi K. et al. [115] created three nano emulsions fuel, namely, BD.CNT.DEE.E, BD.ALO.EHN.M, and BD.TIO.GLC.B, with nanoparticles of CNT, Al_2O_3, and TiO_2, respectively of concentration 50 ppm. BD.CNT.DEE.E showed the lowest CO emissions compared with the other emulsion fuels and a 26% decrease from biodiesel values, whereas BD.TIO.GLC.B and BD.ALO.EHN.M showed 20% and 12% CO reduction, respectively. CNT has better combustion attributes and higher oxygen content of ethanol, improved fuel atomization of DEE, and better formation of air–fuel mixture by micro-explosions of water led to complete oxidation of the fuel mixture, and hence, the highest reduction in CO emissions for the BD.CNT.DEE.E emulsion fuel. In the work of Soudagar et al. [116], biodiesel was blended with octanol under the influence of 3% of functionalized MWCNTs. The authors claimed that MWCNT nanoparticles were inefficient for promoting combustion. The CO emission of MWCNT-containing fuels increased by an average of 38.4% more than diesel at all loads. According to Venu et al. [71], the presence of 10–30 ppm of alumina nanoparticles in a ternary fuel made up of diesel–biodiesel–ethanol reduces CO emissions by 2.81–11.24% compared to the neat ternary fuels. According to the authors, alumina nanoparticles have the ability to act as an oxygen donating catalyst and buffer for CO molecules' oxidation. In addition, the chemical reactivity enhances leading to a decrease in the ignition delay period by virtue of the nanoparticle's large surface area-to-volume ratio. These processes promote complete combustion and reduce emissions of CO. It is worth mentioning that Al_2O_3 nanoparticles dissociate to Al_2O and O at elevated temperatures. Inside the combustion chamber, 'Al_2O' is very unstable at those extreme temperatures, and this further decomposes it to 2Al and $\frac{1}{2}O_2$. As seen in Equations (1)–(3), CO_2 is produced from further reaction of this oxygen molecule with the CO. The above-mentioned mechanism contributes to a much-lowered CO emission.

$$Al_2O_3 \rightarrow Al_2O + 2O \tag{1}$$

$$Al_2O \rightarrow 2Al + \frac{1}{2}O_2 \tag{2}$$

$$O + CO \rightarrow CO_2 \tag{3}$$

According to El-Seesy and Hassan [72], the presence of titanium dioxide in the blend of Jatropha biodiesel–diesel-n-butanol blends ensured a significant decrease in CO emissions. The reduced ignition delay period and ignition characteristics enhancement by the action of the TiO_2 nanoparticles was responsible for this observation. Moreover, these nanoparticles improve fuel–air mixing inside the combustion chamber as a result of their high catalytic activity—and the process aids in the reduction of CO emissions. Nutakki et al. [69] prepared a blend of n-amyl alcohol/biodiesel/diesel blend with(out) the influence of iron oxide nanoparticles, whose dosages were 40 ppm (TF40), 80 ppm (TF80), and 120 ppm (TF120). The CO emissions in TF40, TF80, and TF120 were 7.89%, 11.23%, and 23.26% lower than the ternary fuels without the nanoparticles. The researchers supported their results with the explanation that iron oxide nanoparticles are inherently high oxygen-bearing in nature, which aids in the oxidation of CO molecules as a result of the nanoparticles' high catalytic activity. The reduction in CO emissions, according to the authors, is also attributed to the improved combustion process due to the high surface area per volume of the nanoparticles, causing ignition delay period to shorten.

Hydrocarbons

Pan et al. [75] investigated the impacts of adding cerium dioxide nanoparticles to methanol on the combustion, performance, and emission of a dual-fuel diesel engine. Without the CeO_2, HC emissions in methanol mode increased significantly, particularly for the cases with higher methanol concentrations. However, with the addition of CeO_2, HC emissions are effectively reduced irrespective of the operating conditions. Compared to methanol mode, the maximum reduction in HC emissions were 47.8%, 56.3%, 31%, and 41.1% for M10Ce25, M10Ce100, M30Ce25, and M30Ce100, respectively. The authors provided explanations for the trend in HC emissions of methanol obtained under the influence of CeO_2: (1) CeO_2 nanoparticles as oxygen buffers provide oxygen atoms to improve combustion and hence reduce HC emissions. (2) CeO_2 exists as an oxidation catalyst that can promote the oxidation of hydrocarbons in which CeO_2 is converted to Ce_2O_3, according to Equation (4). (3) The catalytic activity of CeO_2 nanoparticles can lower the combustion activation temperature of carbon and promote more complete combustion.

$$(2X + Y)\, CeO_2 + CxH_y \; \rightarrow \; \left(\frac{(2X + Y)}{2} \right) Ce_2O_3 + \frac{X}{2} CO_2 + \frac{Y}{2} H_2O \qquad (4)$$

Silicon dioxide nanoparticles were blended as additives to methanol in the work of Wei et al. [74]. Their results revealed that the increase in the concentration of the SiO_2 resulted in a significant reduction in HC emissions regardless of the engine loads and methanol substitution ratios. The catalytic action of the SiO_2 nanoparticles may have played a role in the HC reduction by lowering the combustion activation temperature of carbon to promote combustion. Furthermore, SiO_2 nanoparticles provided additional oxygen molecules to help promote combustion. A maximum reduction in HC emission of 74.2 could be possible due to the presence of the SiO_2 over the tested conditions. Heidari-Maleni et al. [78] experimented on an ethanol–biodiesel blend using graphene quantum dot (GQD) nanoparticles. Due to the high catalytic activity of the nanoparticles, their surface area-to-volume ratio increases and thus produces more energy inside the cylinder to obtain more complete fuel combustion and reduce emission of pollutants. By adding GQD to ethanol–biodiesel blends, HC emissions reduced by 33.12%. Three different nanoparticles viz graphene oxide (GO), graphene nanoplatelets (GNP), and multi-walled carbon nanotubes (MWCNT) were added as fuel additives to n-butanol–Jatropha biodiesel by El-Seesy and Hassan [77] to investigate its performance on a diesel engine. Under the influence of the nanoparticles, the HC emissions of JME40B were significantly reduced. These nanoparticles can shorten ignition delay and improve ignition characteristics of the fuel they are blended into. Moreover, because these nanoparticles have a higher surface area to volume ratio, they exhibit high catalytic activity, and this attribute helps promote fuel–air mixing during the combustion process. The above-mentioned factors may be the reason for the nanoparticles' positive impact on the HC emissions according to the researchers. The results from their work showed that an approximately 50% reduction in HC emissions could be achieved for the blends of the JME40B + nanoparticles. Venu and Madhavan [80] compared two different additives for biodiesel–diesel–ethanol blends, i.e., diethyl ether and alumina nanoparticles, for their combustion, performance, and emission characteristics. Their results showed that the blends with diethyl ether recorded more unburned HC. However, the alumina-containing blends exhibited lower HC emissions throughout the engine load except at full load conditions. The catalytic combustion activity of Al_2O_3 was well recognized for lower and part loads and may have improved the combustion process, thereby lowering the HC emissions.

Nitrogen Oxides

Khan et al. [117] prepared a nanofluid involving Nigella Sativa biodiesel, diesel, n-butanol, and graphene oxide nanoparticles with the aim of enhancing the performance, combustion, and symmetric characteristics and reducing the emissions from a diesel engine. They concluded that NO_x emissions from the nanofluid were higher than that of neat diesel and the diesel–biodiesel blends. Per the explanations given by the researchers, the presence of the graphene oxide added more oxygen molecules to what n-butanol and biodiesel had already added. The excess oxygen molecules in the nanofluid may have contributed to higher NO_x emissions. Mehregan and Moghiman [118] numerically investigated the effect of nano aluminum on NO_x and CO pollutants emission in liquid fuels combustion. Their analysis revealed that the mass fraction of NO_x pollutants decreases by adding the aluminum nanoparticles to ethanol and n-decane liquid fuels. Their results confirmed that aluminum nanoparticles, due to their enhanced thermal conductivity, led to improved combustion features of ethanol and n-decane liquid fuels. El-Seesy et al. [68] showed that the addition of carbon nanomaterials (multi-walled carbon nanotubes, graphene oxide, and graphene nanoplatelets) to a blend of n-heptanol–diesel leads to an increase in NO_x level at various engine loads, except at high loads. El-Seesy and his team explained that at lower and part loads, the increase in NO_x emissions is attributable to the positive effect of the carbon nanomaterials and n-heptanol that lead to an increase in peak pressure increases NO_x emissions (Zeldovich Mechanism). However, at higher loads, the presence of the additives may have reduced combustion duration; therefore, there was not sufficient time for the formation of NO_x. By adding aluminum oxide and titanium oxide to diesel–bioethanol blends, Ağbulut et al. [70] showed that the process resulted in a 6.40% and 4.99% drop in NO_x emission for DF90E10 + A100 and DF90E10 + T100, respectively, compared to neat DF90E10. The investigators explained that the main reason behind this drop in NO_x emission might be due to the increase in thermal conductivity of the blends after the addition of the nanoparticles, which ensured a rapid heat transfer for the resulting fuel. Thus, the proper elevated temperature required for the formation of NO_x was not highly reached, and NO_x emission was seen lesser with the nanoparticle-doped fuels. In the work of Nour et al. [119], Al_2O_3 was added to diesterol (70% diesel+ 20% ethanol+ 10% Jojoba biodiesel) blends. Without the nanoparticles, JE20D exhibited a higher NO_x emission especially at lower loads. The addition of the Al_2O_3 nanoparticles to JE20D caused no impact in NO_x emissions at lower loads. However, at high engine loads, lower NO_x was reported for JE20D25A, JE20D75A, and JE20D100A in comparison to pure diesel and JE20D blends. The authors ascribed this trend to the high catalytic behaviour of Al_2O_3 nanoparticles that led to a more complete combustion, forming the final products with a minimum thermal breakdown of the hydrocarbon compounds. Hence, per the existence of lower active radicals, the possibility of forming thermal NO_x was lowered.

4.2.3. Effect of Nano-Additives and Diesel–Alcohol Fuels on Engine Combustion

The adverse effect of diesel fuel usage in CI engines has created significant interest in prospective renewable additives, such as alcohol, including butanol, ethanol, and methanol. These fuels can be used as emulsion, dual, or blend with diesel and biodiesel, to enhance fuel properties and stability. Small changes in combustion might not yield a significant improvement in cylinder chamber. However, it is necessary to create the correct mixture conditions, in particular, to control the air movement and turbulence [120]. Table 4 summarizes the recent experiments on the variation in combustion characteristics from CI engines fuelled with various nano-additives and alcohol fuels.

The effect of Al_2O_3 NPs with a dosing range of 10 ppm–30 ppm in ethanol fuel on combustion characteristics was investigated by Venu et al. [71]. It was found that the in-cylinder pressure decreases by 2.33% with TF20 as compared to diesel fuel. This trend absolutely matches with results found by Ağbulut et al. [70] with TiO_2 and Wei et al. [121] with Al_2O_3, they observed a significant reduction in the peak cylinder pressure for ethanol and methanol fuels due to the high specific heat of alcohol fuels and high latent heat of vaporization, which could be reason for the drop-in peak [121]. On the contrary, the in-cylinder pressure significantly improved with the presence of CeO_2 [75], Fe_2O_3 [69], and ZnO [73] nano-additives with different alcohol-based fuels. This is linked to the shortened ignition delay using nanofluids due to high thermal conductivity and surface area to volume ratio.

El-Seesy and Hassan [72], Örs et al. [81], and Yaşar et al. [122] assessed the impact of butanol with nanoparticles TiO_2 as a diesel engine catalyst. They discovered that the fuel with 25 ppm and 50 ppm TiO_2 NPs produced superior combustion efficiency with better emission reduction as compared to diesel–butanol fuels without TiO_2. Heidari-Maleni et al. [78] found that when GQD NPs concentration is elevated (added to ethanol fuel), the peak HRR is reduced by ~14.35% compared to that of diesel, consequently demonstrating a less combustible mixture formed at low cylinder temperature. Similar evidence in reduction with nano-biochar/diesel–ethanol was observed using a dosing range of 25 ppm–125 ppm [113]. For diesel–methanol fuel, the maximum peak HRR is obtained at 100 ppm dosage of nano-additives with up to 7.79% increase with CeO_2 [75] and 8.6% increase with SiO_2 [74], respectively.

Table 4. Summary of the most recent experiments on different nanoparticles addition in alcohol fuels used in CI engines.

Type of NPs Used	Alcohol Based Fuel	Blends	Size of NPs/NPs Concentration	Engine Sp.	Combustion		Performance		Gaseous Emission			Observation	Reference
					HRR	ICP	BFSC	BTE	CO	HC	NOx		
	Ethanol	DF90E10 + A100, DF90E10 and DF	18 nm / 100 ppm	Lombardini 15 LD 350, CI, CR 20.3:1, RP 7.5HP, RS3600 rpm, IP 207 bar.	—	↓ with DF90E10 + A100 compared to DF	↓ by 2.25% with DF90E10 + A100 compared to DF	↑ by 3.48% with DF90E10 + A100 compared to DF	↓ by 25% with DF90E10 + A100 compared to DF	↓ by 30.15% with DF90E10 + A100 compared to DF	↓ by 3.02% with DF90E10 + A100 compared to DF	• NPs-doped to DF90E10 acted as an oxygen-donating catalyst and ensured more oxygen atoms which in turn increase complete combustion.	[70]
	Ethanol	TF (TF + 10, TF + 20 and TF + 30) and DF	28 nm-30 nm/10 ppm, 20 ppm and 30 ppm	Kirloskar TAF 1, 1C, 4S, CI, CR 17.5:1, RP 4.4 kW, RS1500 rpm, IT 23° bTDC, IP 220 bar.	↑ highly with DF100 compared to other fuels	↓ by 2.33% with TF + 20 compared to DF	BSEC ↓ by 4.93% with TF + 20 compared to TF and DF	↑ by 7.8% with TF + 20 compared to TF and DF	↓ by 11.25% with TF + 20 compared to TF and DF	↓ by 5.63% with TF + 20 compared to TF and DF	↓ by 9.39% with TF + 20 compared to TF	• TF + 20 was found to produce excellent combustion and emission behaviour. • Al_2O_3 additives help to improve the catalytic combustion and shortened ID which in turn led to better air-fuel interaction.	[71]
Al_2O_3	Methanol	MDFs (M10, M30, and M50); MD-NFs (M10A25, M10A50, M10A100, M30A25, M30A50, M30A100, M50A25, M50A50, and M50A100) and DF	30 nm/25 ppm, 50 ppm and 100 ppm	Kirloskar, 1C, 4S, CI, CR17.5:1, RP11.32 kW, RS2200 rpm, IT 20° bTDC, IP 18 MPa.	↓ with MDFs-NFs compared to DF and MDF at high load.	↑ by 30–50% with MD-NFs compared to MDFs and DF	—	—	↓ with MD-NFs highly compared to MDFs and DF	↓ with MD-NFs highly compared to MDFs and DF	↑ with MD-NFs significantly compared to MDFs and DF	• NFs as additive leads to a definite reduction in CO and HC with an increase in NOx emissions. • High LVH of methanol lead to reduction in temperature of in-cylinder charge • In addition, MD-NFs lessen pre-combustion reactivity with increase in ID.	[123]
	Methanol	MDFs (M5 and M15); MD-NFs (M5A50, M5A100, M15A50 and M15A100) and DF	30 nm/50 ppm and 100 ppm	Kirloskar, 1C, 4S, WC, CI, CR17.5:1, RP11.32 kW, RS2200 rpm, IT 20° bTDC, IP 18 MPa.	Improved by 16.1% with ↓ of 6.9% in ID via 100 ppm	Improved by 2.5% with ↓ of 16% in CD via 100 ppm	↓ by ~3.7% with M5A100 and M15A100 compared to MDFs	↑ by ~3.6% with M5A100 and M15A100 compared to MDFs	↓ by 83.3% with MD-NFs compared to MDFs	↓ by ~40.9% with MD-NFs compared to MDFs	↑ slightly by 14.4% with MD-NFs compared to MDFs	• Addition of NP in MDFs helps in improving the fuel cetane number which lead to improvement in HRR and ICR while reduction in ID and CD. • Emissions were reduced compared to MDFs except NOx emission.	[121]

Table 4. *Cont.*

Type of NPs Used	Alcohol Based Fuel	Blends	Size of NPs/NPs Concentration	Engine Sp.	Combustion		Performance		Gaseous Emission			Observation	Reference
					HRR	ICP	BFSC	BTE	CO	HC	NO$_x$		
	Ethanol	D45EB10; D45EB10 + Al$_2$O$_3$ (D45EB10A50, D45EB10A75 and D45EB10A100)	30 nm/50 ppm, 75 ppm and 100 ppm	Kirloskar TVI, 1C, 4S, WC, CI, CR17.5:1, RP7 kW, RS1500 rpm, IT 23° bTDC, IP 220 kgf/cm².	↑ highly with D45EB10A100 compared to D45EB10	↑ highly with D45EB10A100	—	—	↓ slightly by 0.02% with D45EB10A100 at 100% load.	—	↑ with an ↑ in Al$_2$O$_3$ rate and load.	• Results showed that Al2O3 advances the initiation of combustion decreasing and shorten the ignition delay.	[124]
	Methanol and Ethanol	E.M.BioD.Al (5%Eth, 3%Meth, 86%BioD and 50 ppm)	20 nm/50 ppm	KIPOR KM186FA, 1C, 4S, AC, CI, CR19:1, RP5.7 kW, RS3000 rpms	Significantly improved	↑ highly with E.M.BioD.Al compared to other fuels	↓ with E.M.BioD.Al compared to BioD.	↑ by 6% with E.M.BioD.Al compared to BioD.	↓ by 12% with E.M.BioD.Al compared to BioD.	↓ slightly with E.M.BioD.Al compared to BioD.	↓ by 12.3% with E.M.BioD.Al compared to BioD.	• Addition of Meth and NPs to the blend enhances the cetane number, which then results in an efficient combustion. • The combustion of E.M.BioD.Al fuel provides more time for the oxidation of soot.	[115]
	Ethanol	JE20D; JE20D + Al$_2$O$_3$ (JE20D25A, JE20D50A, JE20D75A, JE20D100A) and DF	20 nm–50 nm/25 ppm, 50 ppm, 75 ppm and 100 ppm	HATZ-1B30-2, 1C, AC, CI, CR21.5:1, RP5.4 kW, RS3600 rpm, IT 20° bTDC, IP 18 MPa.	↑ highly with JE20D + Al$_2$O$_3$ compared to JE20D and DF	↑ highly with JE20D + Al$_2$O$_3$ compared to JE20D.	↓ by 17–25% with JE20D + Al$_2$O$_3$ compared to DF	↑ highly with JE20D + Al$_2$O$_3$ compared to JE20D	↓ by 20% with JE20D + Al$_2$O$_3$	↓ by 60% with JE20D + Al$_2$O$_3$	↓ by 30–50% with JE20D + Al$_2$O$_3$	• The JE20D25A and JE20D75A blends improved the combustion process and resulted in lowered emissions compared to JE20D.	[119]
CeO$_2$	Methanol	M10 and M30; MCN (M10C25, M10C100, M30C25 and M30C100) and DF	25 ppm and 100 ppm	Kirloskar, 1C, 4S, WC, CI, CR17.5:1, RP11.32 kW, RS2200 rpm, IT 20° bTDC, IP 18 MPa.	↑ by 7.9% slightly with MCN	↑ with MCN	↓ by 5.7–8.1% with MCN compared to M10 and M30	Improved by adding CeO$_2$ to M10 and M30 with 5.2–108%	↓ by 79.8% with MCN compared to DF, M10 and M30.	↓ by 56.3% with MCN compared to DF, M10 and M30.	↓ by 70–90% with MCN compared to DF, M10 and M30.	• Methanol and CeO$_2$ NPs proved to be a promising technique for dual fuel in CI engines.	[75]
CNTs	Ethanol	D100, B2C20, B2C60, B2C100, B2E2C20, B2E2C60, B2E2C100, B2E4C20, B2E4C60, B2E4C100, B2E6C20, B2E6C60, and B2E6C100	4 nm–8 nm/20 ppm, 60 ppm and 100 ppm	DICOM 50.1 15/5, 1C, 4S, AC, DI, CI, RP9kW, RS3000 rpm.	—	—	↓ by 11.73% with an ↑ in CNTs NPs.	↑ by 13.97% with an ↑ in CNTs NPs.	↓ by ~5.47% with CNTs NPs	↓ by 31.72% with CNTs NPs	↑ by 12.22% with CNTs NPs.	• B2E4C60 has the optimal performance and emissions. • The negative effects of NPs on CI engine need to be investigated.	[65]

Table 4. *Cont.*

Type of NPs Used	Alcohol Based Fuel	Blends	Size of NPs/NPs Concentration	Engine Sp.	Combustion		Performance		Gaseous Emission			Observation	Reference
					HRR	ICP	BFSC	BTE	CO	HC	NO_x		
Fe_2O_3	Pentanol	TF (P10B20D70); TF40, TF80 and TF120	40 ppm, 80 ppm and 120 ppm	Kirloskar TVI, 1C, 4S, CEDI, CI, CR18.1:1, RP3.7 kW, RS3000 rpm, IP 250–500 kgf. cm⁻².	↑ with an ↑ in Fe_2O_3 NPs	Significantly improve with the presence of Fe_2O_3 NPs	↓ significantly by 4.93% with TF80 and TF120 compared to TF	↑ by 7.8% with TF120 compared to TF.	↓ significantly by 5.69% with TF120 compared to other fuels	↓ significantly by 11.24% with TF120 compared to other fuels	↓ significantly by 9.39% with TF120 compared to other fuels	Result showed that the chemical reactivity in combustion takes place very fast by decreasing the ID while completing the combustion rate.	[69]
GQD	Ethanol	BEGQD (B10E2GQD30, B10E4GQD30, B10E6GQD30 and B10E8GQD30) and DF	30 ppm	DICCM, 1C, 4S, AC, CI.	↓ by ~14.35% with BEGQD compared to DF.	—	—	—	↓ by ~29.54% with BEGQD compared to DF.	↓ by ~31.12% with BEGQD compared to DF.	↓ with DF compared to BEGQD.	GQD NPs improve the performance and emission behaviour of the CI engine fuelled with diesel-bioethanol-biodiesel blends.	[78]
SiO_2	Methanol	MSN (M10); M10Si (M10Si25, M10Si50 and M10Si100) and DF	20 nm–30 nm/25 ppm, 50 ppm and 100 ppm	Kirloskar, 1C, 4S, CI, CR17.5:1, RP11.32 kW, RS2200 rpm, IT 20° bTDC, IP 18 MPa.	↑ by ~8.6% max with M10Si100.	—	↓ by 6.2% with an ↑ in SiO_2 NPs.	↑ by 5.1% with an ↑ in SiO_2 NPs.	↓ by 55.4% with SiO_2 NPs.	↓ by 38.5% with SiO_2 NPs.	↓ by 5.2% with SiO_2 NPs.	Reductions in the ID and CD can be found in the cases of high NPs dosage.	[74]
Nano-biochar	Ethanol	DB2E2, DB4E4, DB6E6, DB8E8; DBE (with 25–125 ppm) and DF	25 ppm–125 ppm	CT-159 1C, 4S, CI, CR 21:1	↓ by ~3% with DBE compared to DF.				↓ by ~0.03–0.015% with an ↑ in nano-biochar DBE.	↓ by ~28% with 125 ppm fuels compared to other fuels.	↓ by ~15% with 100 ppm compared to other fuels.	Nano-biochar NPs lead to an improved combustion efficiency and reduced pollutant emissions.	[113]
TiO_2	Ethanol	DF90E10 + T100, DF90E10 and DF	43 nm/100 ppm	Lombardin 15 LD 350, CI, CR20.3:1, RP7.5HP, RS3600 rpm, IP 207 bar.	—	↓ with DF90E10 + T100 compared to DF	↓ by 1.26% with DF90E10 + T100 compared to DF	↑ by 2.94% with DF90E10 + T100 compared to DF	↓ by 21.43% with DF90E10 + T100 compared to DF	↓ by 26.47% with DF90E10 + T100 compared to DF	↓ by 1.57% with DF90E10 + T100 compared to DF	Result showed that biofuel worsen the emission, performance and combustion as compared to DF. The addition of TiO_2-based NPs allows these worsened to drawback.	[70]
	Butanol	J50Bu10; JBu + TiO_2 (J50Bu10T25 and J50Bu10T50) and DF	25 ppm and 50 ppm	HATZ-B30-2, 1C, 4S, WC, VVA, CI, CR8.39:1, RS1000 rpm, IT 6° bTDC, IP 150 bar.	↑ with JBu + TiO_2	↑ with JBu + TiO_2	↑ by 15% highly with JBu + TiO_2	↑ by 17% highly with JBu + TiO_2	↓ by 30% significantly with JBu + TiO_2	~ by 50% significantly with JBu + TiO_2	↑ with an ↑ in TiO_2 NPs.	With TiO_2 NPs, no negative effects were recorded on CI engine components.	[72]

Table 4. *Cont.*

Type of NPs Used	Alcohol Based Fuel	Blends	Size of NPs/NPs Concentration	Engine Sp.	Combustion		Performance		Gaseous Emission				Observation	Reference
					HRR	ICP	BFSC	BTE	CO	HC	NO$_x$			
	Butanol	B20 and B100; B20Bu20; B + TiO$_2$ (B20 + TiO$_2$ and B20Bu10 + TiO$_2$) and DF	0.1689 g	3 LD 510, 1C, 4S, WC, CI, CR17.5:1, RP9 kW, RS3300 rpm, IP 190 bar.	—	↑ with B + TiO$_2$	↓ by 27.73–28.37% with B + TiO$_2$ compared to all other fuels.	↑ by 0.34–0.66% with B + TiO$_2$ compared to other fuels.	↓ by 14–38% with B + TiO$_2$ compared to all other fuels except B100.	↓ by 22.38–34.39% with B + TiO$_2$ compared to all other fuels except B100.	↑ by 1.20–3.94% with B + TiO$_2$ compared to other fuels.	• n-butanol improved cold flow properties of fuel blends. Adding TiO$_2$ in fuels has positively effect on engine performance.	[81]
	Butanol	B5 and B10; BTiO$_2$ (B5T25, B5T50, B10T25 and B10T50) and DF	25 ppm and 50 ppm	Kirloskar, 1C, 4S, WC, CI, CR18:1, RP3.5 kW, RS1500 rpm, IT 25° bTDC.	—	↑ slightly with an ↑ in engine load.	↓ by 2.87–6.47% with all BTiO$_2$ except B5T25 which ↑ by 7.91% compared toDF.	—	↓ by 22.34–36.17% with BTiO$_2$ compared to DF.	—	↑ by 0.89–0.7.78% with B5T25, B10T25 and B10T50 while B5T50 ↓ by 2.69%.	• Butanol and TiO$_2$-based additives can be used as fuel without engine modification.	[122]
ZnO	Ethanol	D40B30E30; D40B30E30Z250; TFu (D40B30E30C6 and D40B30E30Z250C6) and DF	30 nm/250 ppm	Kirloskar TAF 1, 1C, 4S, CI, CR17.5:1, RP4.41 kW, RS1500 rpm, IT 23° bTDC, IP 200 bar.	↑ significantly with TFu compared to DF.	↑ by 8% and 13% with TFu compared to DF.	↑ by 14–39% with TFu compared to DF.	↓ by 9–21% with TFu compared to DF.	↓ by 62–92% with TFu compared to DF.	↑ by 21% with D40B30E30C6 and ↓ by 9% with D40B30E30Z 250C6.	↓ by 16–35% with TFu.	• Fuel solubility played a vital role for limiting the emissions effect while improving the combustion performance of the engine.	[73]

Note: All engines considered are research-based, solely for testing purposes; ↑ = increase; ↓ = decrease; AC = air-cooled; bDTC = before top dead centre; BSEC = brake specific energy consumption; C = cylinder; CD = combustion duration; CI = compressive ignition; CR = compression ratio; CRDI = common rail direct injection; DF = diesel fuel; DI = direct injection; GQD = graphene quantum dot; HRR: heat release rate; ICP = in-cylinder pressure; ID = ignition delay; IT = ignition timing; IP = injection pressure; LVH = latent vaporization heat; Max. = Maximum; MDF = methanol-diesel fuel; NPs = nanoparticles; RS = rated speed; RP = rated power; S = stroke; Sp. = specification; TF = ternary fuel; TFu = stable fuels; VVA = variable valve actuation; WC = water cooled.

4.2.4. Engine Performance Characteristics of Nanoparticles in Vegetable Oil-Based Fuels
Brake Thermal Efficiency

Polanga oil-diesel blends were doped with iron oxide nanoparticles for evaluation on a CI engine by Santhanamuthu et al. [91]. It was seen that the BTE of the blends with iron oxide nanoparticles was on par with that of neat diesel. According to the authors, the improvement in BTE due to the presence of iron oxide can be attributed to the enhancement of thermal properties such as thermal conductivity, thermal diffusivity, and convective heat transfer co-efficient that the nanoparticles present. Purushothaman et al. [125] added 25–100 ppm of Al_2O_3 and TiO_2 nanoparticles alternatively to mahua oil. It was reported in this work that the BTE value of 100 ppm Al_2O_3 and TiO_2 blended emulsified mahua oil were 29.2% and 28.4%, respectively, compared to 23.8% of neat mahua oil. From the authors' perspective, the nanoparticles acted as a heat source which shortened the ignition delay and also enhanced the combustion due to the higher surface area to volume ratio. In the work of Ramesh et al. [126], canola oil blended diesel with Al_2O_3 nanoparticles was optimized through single and multi-objective optimization techniques. Results from their experiment showed that 18.8% of canola blends with 30 ppm of nanoparticles had BTE of 33.81%, which was a 16% increase with reference to pure diesel. Additionally, 10–30 ppm of alumina nanoparticle was added to lemongrass oil by Balasubramanian et al. [127]. The BTE of B20A20 was higher than any other test fuel at low and medium engine load conditions. At medium load, BTE of B20A20 increased by 12.24% and 4.08% over B20 blend and neat diesel, respectively. Moreover, at 100% load, BTE of B20A20 increased by 2.71% over B20 fuel. The authors mentioned that the presence of the alumina nanoparticle provided more oxygen molecules that boosted the combustion inside the cylinder. This was possible due to a higher area to volume ratio, improved atomization, quick evaporation, and greater mixing of fuel and air brought about by the alumina nanoparticle. Dhinesh et al. [83] showed that the addition of cerium oxide (10–30 ppm) to Cymbopogon flexuous biofuel was blended with diesel fuel positively impacts BTE. C20-D80 + 20 ppm CeO_2 resulted in higher BTE than C20-D80 blend. The authors explained that the presence of CeO_2 in the base fuels acted as a catalyst and oxygen buffer for combustion enhancement.

Brake Specific Fuel Consumption

The impact of rice husk nanoparticles on a diesel engine running on pine oil–diesel blends was investigated by Panithasan et al. [128]. It appears that the pine oil blends with 0.1% rice husk nanoparticles consume less fuel than the blends without nanoparticles. They explained that the rice husk nanoparticle acted as an oxygenated additive which enhanced the combustion process. Sathiyamoorthi et al. [85] studied the combined effect of nano emulsion and EGR on the characteristics of neat lemongrass oil–diethyl ether–diesel blend. BSFC of the cerium oxide-based nano emulsified LGO25 with EGR mode was increased by 10.8% compared to LGO25. They attributed this rise in BSFC to the lower calorific value of the nano emulsified LGO25, although the high cetane number and oxygen content of diethyl ether could partially reduce BSFC while operating in EGR mode. According to Dhinesh and Annamalai [87], utilizing cerium oxide nanoparticles mixed with an emulsion of Nerium oleander biofuel results in lower energy consumption when compared to neat Nerium oleander. The energy consumption of the nano emulsified fuel was 13.33 MJ/kWh whereas the neat Nerium oleander was 14.21 MJ/kWh. In a similar study, Annamalai et al. [82] dispersed 30 ppm of ceria nanoparticles into lemongrass oil (LGO) emulsion fuel. The process resulted in a nano emulsified fuel with energy consumption of 12.99 MJ/kWh, whereas that of neat LGO and diesel were both 13.8 MJ/kWh. In both studies, the researchers supported this observation claiming that, by introducing cerium oxide to the emulsion of Nerium oleander biofuel/lemongrass oil, the secondary atomization and micro-explosion improved, which in turn resulted in a heightened evaporation rate and mixing of the fuel. In the study by Dhinesh et al. [83] involving *Cymbopogon flexuous* biofuel was blended with diesel fuel under the influence of cerium oxide nanoparticles, lower energy consumption was achieved in the nano-blended fuels than the non-nano blended fuels. The researchers attributed cerium oxide's catalytic ability and oxygen buffer which promotes complete combustion as the reason behind the obtained results.

4.2.5. Engine Emission Characteristics of Nanoparticles in Vegetable Oil-Based Fuel

Carbon Monoxide

Panithasan et al. [128] experimented with rice husk nanoparticles as additives for pine oil–diesel blend. Results showed that at full load condition, CO decreases by about 27.27% more than diesel fuel. According to the authors, the oxygen content in both rice husk nanoparticles and pine oil increased the combustion rate and aided the addition of CO into CO_2. Chinnasamy et al. [88] reported that by adding 50 ppm of Al_2O_3 nanoparticles into pyrolyzed biomass oil, there was a reduction in CO emissions compared to that of neat diesel and pyrolyzed biomass oil. From the authors, the presence of the nanoparticles improved the ignition characteristics and led to high catalytic activity. This is due to the higher surface area to volume ratio of the Al_2O_3, which resulted in an enhanced air-fuel mixing in the combustion chamber which further reduced the CO emissions. According to Sheriff et al. [86], 50 ppm of cerium oxide in lemon peel oil–diesel and orange peel oil–diesel resulted in percentage values for CO emissions of 0.223% and 0.092%, respectively, at full load; in a similar manner, that of 50 ppm of carbon nanotubes in lemon peel oil–diesel and orange peel oil–diesel were both 0.225%. At the same load, the CO emissions for neat diesel was 0.251%. By blending two different nanoparticles in mahua oil fuel, Purushothaman et al. [125] showed that Al_2O_3 had better CO emission reduction than TiO_2. This was as a result of Al_2O_3 higher thermal conductivity than TiO_2. Sathiyamoorthu et al. [85] provided evidence to the fact that the addition of cerium oxide to emulsified LGO25 decreased the base fuel's CO emission by 7.14% and 4.87% when compared to LGO25. From the investigators, the presence of cerium oxide shortened ignition delay with better fuel–air mixing that could have led to the uniform burning process in the combustion chamber and promoted more complete combustion.

Hydrocarbon

In the study of Purushothaman et al. [125], HC emissions from emulsified mahua oil with Al_2O_3 and TiO_2 were significantly lowered compared to other neat test fuels. The HC emission values of 100 ppm Al_2O_3 and TiO_2 nanoparticle-blended EMO were found to be 57 ppm and 61 ppm, respectively, in contrast to 65 ppm and 91 ppm for diesel and mahua oil. The combined effect of micro explosion and high in-cylinder temperature due to the nanoparticles may have contributed to this reduction in HC emissions. Elumalai et al. [129] experimented on harmful pollution reduction technique in a low heat rejection (LHR) engine fuelled with blends of pre-heated linseed oil and TiO_2. The blends with nanoparticles had lower HC emissions than the other base fuels without nanoparticles. The HC emissions of blends PLSNP50, PLSNP100, PLSNP150, PLSNP200 are −7.35%, −22.10%, −29.41%, and −33.82%, respectively, compared with the PLS20 in LHR engine. From their work, the authors explained that the addition of nanoparticles to preheated fuel led to a rapid burning of fuel due to the oxygen influx from TiO_2 and minimized the carbon content during combustion. With cerium oxide acting as an oxidizing agent, Dhinesh and Annamalai [87] showed that NENOB (emulsion with nanoparticle) provided a 20%, 30%, and 36.3% reduction in HC emissions when compared to the other fuels without the nanoparticles (NOB, ENOB, and SFDF, respectively). Furthermore, hydrocarbons react with cerium oxide to form various products such as CO_2, water, and cerous oxide, thereby limiting HC emissions. Panithasan et al. [128] showed that rice husk nanoparticles in the blend of pine oil–diesel (B20-0.1%RH) at full load conditions decreases the HC emissions by 19.64% compared to neat diesel. The researchers explained that the excess oxygen molecules delivered by the nanoparticle prevented the hydrocarbons from escaping the combustion process, thereby reducing the HC emissions from the exhaust gases. In the work of Sheriff et al. [86], it was revealed that 50 ppm CNT nanoparticle in lemon peel oil blend showed relatively less HC emission than that of cerium oxide due to the higher surface to volume ratio of the former, which led to improved air–fuel mixing. However, the trend reverses when lemon peel oil is exchanged with orange peel oil.

Nitrogen Oxides

In the presence of iron oxide nanoparticles, NO_x emission was reduced to 50% of that of neat diesel at higher Polanga oil, according to Santhanamuthu et al. [91]. The reason was that the iron oxide acted as a catalyst for the reaction of the hydroxyl radicals present in the Polanga oil and lowered the oxidation temperature. Balasubramanian et al. [127] presented results for their investigation on a diesel engine fuelled with lemongrass oil and alumina nanoparticles. The results showed that the addition of the nanoparticle lowered the NO_x emissions of the base fuels. At medium and high load conditions, there was a 19.23% and 1.73% increase in NO_x emissions for B20 blend and neat diesel. These values corresponded to a significant decrease in NO_x emission values for the B20A20 blend of 1.53% and 2.25% at medium and higher load conditions, respectively. They gave reasons that the alumina nanoparticles acted as a reducing agent and an oxygen absorber to reduce the NO_x emissions. Ceria nanoparticles of dosage 30 ppm were blended in emulsion fuel of lemongrass oil for assessment on performance, combustion, and emission characteristics in the work of Annamalai et al. [82]. It was revealed that the NO_x emissions of LGO nanoemulsion reduced by 24.8% and 20.3% compared with LGO and diesel fuels, respectively. According to the authors, the nanoparticle acted as a reduction agent. The oxides of nitrogen are reduced to form nitrogen and oxygen as a result of the high thermal stability of cerous oxide formed from the oxidation of unburned hydrocarbon. The soot remained stable and active after enhancing the initial combustion cycle, which may have significantly reduced NO_x emission. Equation (5) represents the chemical reaction described above.

$$Ce_2O_3 + NO \rightarrow 2CeO_2 + \frac{1}{2}N_2 \tag{5}$$

Unlike the case of CO emissions, the higher thermal conductivity of Al_2O_3 became a disadvantage in terms of NO_x emissions when compared to TiO_2 as nano-additives for mahua oil [125]. The NO values of 100 ppm Al_2O_3 and TiO_2 nanoparticle-blended EMO in the work of Purushothaman et al. [125] were found to be 260 ppm and 275 ppm, respectively, whereas, for diesel and mahua oil, the respective values were 537 ppm and 289 ppm. According to Panithasan et al. [128], the addition of rice husk nanoparticles to diesel-pine oil blend had a negative impact on NO_x emissions; at full load conditions, the NO_x emission of B20-0.1%RH increased by 8.76% compared to neat diesel. The authors attributed this observation to the additional oxygen content provided by the nanoparticles, which caused an increase in the in cylinder temperature of the combustion chamber and thereby assisted in increasing the NO_x level.

4.2.6. Effect of Nano-Additives and Diesel–Vegetable Oil Blend Fuel on Engine Combustion

Table 5 summarizes the most recent experiments on CI engines fuelled with various nano-additions and vegetable-based oils. Despite the benefits associated with bio-oil/diesel blends in CI engines, the usage of bio-oil as blend fuel gives a few drawbacks, such as large variation in fuel consumption [129], less calorific value and density [130], and decrement in mileage on vitality premise by ~10% [84]. To overcome these drawbacks, it is often necessary to improve it with suitable nano-additives and by appropriate combustion management. However, distinct species of nano-additives, such as Al_2O_3, CeO_2, MgO, rice husk NPs, SiO_2, MWCNTs, and TiO_2 in bio-oils, are used to obtain better fuel properties over a long period of time [25,131,132], and may improve the engine combustion [86].

Among other research investigated, Balasubramanian et al. [127], Chinnasamy et al. [133], and Purushothaman et al. [125] examined the effect of Al_2O_3 NPs with bio-oil based fuels from the diesel engine, and a significant increase was observed with heat release rate and in-cylinder pressure leading to an increase in thermal efficiency. They attributed these results to a rise in ignition delay, and combustion duration causes the in-cylinder soot to be more highly oxidized, hence promoting the oxidation rate of soot particles which is then higher than the specific active surface rate [134–136]. Similarly, the diffusive combustion phase is shortened due to the addition of CeO_2 NPs into waste pyrolysis and orange oils, which raises the range of ignition delay that helps in accelerating the combustion [86,137].

Although other researchers investigated that increasing the concentration of nano-additives such as CeO_2, $Ce_{0.7}Zr_{0.3}O_2$, and MgO in bio-oil affects the peak heat rate, in-cylinder, and peak pressure due to the higher energy droplet aggregation during spray atomization [82,87,138,139] later resulted in high fuel consumption [30]. That means not all nano-additives and bio-oil fuels contribute to the in-cylinder chamber; assessment needs to be made for notable nanofluids selected. Besides, a rise in the concentration of TiO_2 and MWCNTs with vegetable-based fuels resulted in a higher heat release rate and cylinder pressure [140,141]. Furthermore, the presence of water molecules in emulsion fuel and vegetable-oil fuels leads to an increase in ignition delay and in-cylinder pressure, which suddenly favours heat release rate [106,142]. However, most of the literature reported that the addition of nano-additives facilitated the uniform distribution and stable suspension of fuel in the combustion chamber, resulting in an increase in the penetration length of the spray [25,27,143].

Table 5. Summary of the most recent experiments on different nanoparticles addition in vegetable oil/pure bio-oil fuels used in CI engines.

Type of NPs Used	Vegetable Based Fuel	Blends	Size of NPs/NPs Concentration	Engine Sp.	Combustion		Performance		Gaseous Emission			Observation	Reference
					HRR	ICP	BFSC	BTE	CO	HC	NO$_x$		
Al$_2$O$_3$	Mahua oil	MO; EMOA (EMOA25, EMOA50, EMOA75 and EMOA100) and DF	25 ppm, 50 ppm, 75 ppm and 100 ppm	Kirloskar AVI, 1C, 4S, WC, CI, CR16.5:1, RP3.7 kW, RS1500 rpm, IT 23° bTDC, IP 220 bars.	↑ with Al$_2$O$_3$ NPs.	↑ with Al$_2$O$_3$ NPs.	—	↑ by ~29.2% with EMOA100 compared to other fuels.	↓ by 87.4% with EMOA100 and ↓ by 24.2% with MO than DF.	↓ by 37.3% with EMOA100 compared to DF and MO.	↓ by 10% with EMOA than MO and ↓ by 51% with MO than DF.	The result of Al$_2$O$_3$ NP fuels showed better combustion reactivity due to high thermal conductivity than the latter.	[125]
	Waste plastic oil	WPO20; WPO20A10-I, WPO20A20-I, WPO20A10-II and WPO20A20-II	20 nm and 100 nm/10 ppm and 20 ppm	Kirloskar 240PE, 1C, 4S, DI, CR17.5:1, RP3.7 kW, RS1500 rpm, IT 23° bTDC, IP 200 bar.	↑ with WPO20A10-I and WPO20A20-I.	↑ with WPO20A10-I and WPO20A20-I.	↓ by 8–11% with WPO NP fuels.	↓ by 8–12.2% with WPO NP fuels.	↓ with NP fuels.	↓ with NP fuels.	↑ with an ↑ in load.	Low amount of NP had better emission and combustion performance compared to large amount.	[133]
	Lemon grass oil	B20; BA (B20A10, B20A20 and B20A30)	20 nm–30 nm/10 ppm, 20 ppm and 30 ppm	Kirloskar TV1, 1C, 4S, DI, CI, CR17.5:1, RP5.2 kW, RS1500 rpm, IT 23° bTDC, IP 220 bar	↑ by 20.4% with B20A20 compared to B20.	↑ by 4.75% with B20A20 compared to B20.	BSEC ↓ with an ↑ in loads and NPs rate.	↑ by 2.71% with B20A20 at full load.	↓ by ~15.15% with NPs fuels compared to B20.	~ by ~5.98% with NPs fuels compared to B20.	↓ by ~2.2% with NPs fuels compared to B20.	HRR was better in case of Al$_2$O$_3$ NP and ICP is better with CeO$_2$, as compared in literature.	[127]
CeO$_2$	Tyre pyrolysis oil	B5, B10, B15 and B20; and B10D85C100	50 ppm and 100 ppm	Kirloskar TV1, 1C, 4S, DI, CR17.5:1, RP3.7 kW, RS1500 rpm, IT 23° bTDC IP 200 bar	↑ with CeO$_2$ NP fuel	↓ slightly with B5 NP fuel compared to DF	—	Improved by 2.85% with B5D85C100	↓ by 13.33% with B5D85C100 compared to DF.	↓ by 3.0% with B5D85C100 compared to DF.	↑ slightly by 1.4% with B5D85C100.	Best result can be achieved with low blend rate and NP conc.	[137]
	Orange peel oil	OPO20; OPO20C50 and OPO20C100	32 nm/50 ppm and 100 ppm	Kirloskar TV1, 1C, 4S, DI, CI, CR17.5:1, RP5.2 kW, RS1500 rpm, IT 23° bTDC, IP 200 bar	↑ with an ↑ in loads and ICP.	Improved due to Ä.	↓ with an ↑ in loads and NPs rate.	↑ with an ↑ in loads and NPs rate.	↓ with and without NPs addition compared to DF due to Ä.	~ with and without NPs addition compared to DF due to Ä.	↓ with an ↑ in CeO$_2$ NPs due to Þ.	Better aromatic and good solubility of CeO$_2$ in orange oil led to good combustion.	[86]
	Lemon grass oil	LGO; LGO emulsion, LGO nano-emulsion and DF.	10 nm–20 nm/20 ppm–80 ppm	Kirloskar TV1, 1C, 4S, WC, CI, CR17.5:1, RP5.2 kW, RS1500 rpm, IT 23° bTDC, IP 200 bar.	—	—	↓ with an ↑ in EP for all the fuels.	↑ by 31.25% with LGO nano-emulsion compared to other test fuels..	↓ by 15.21% with LGO nano-emulsion compared to other fuels.	↓ by 16.12% with LGO nano-emulsion compared to other fuels.	↑ with LGO emulsion compared to other fuels.	Results shows that CeO$_2$ NPs help to reduce the CT of the A/F mixture, as the AE of N$_2$ is higher.	[144]

Table 5. *Cont.*

Type of NPs Used	Vegetable Based Fuel	Blends	Size of NPs/NPs Concentration	Engine Sp.	Combustion		Performance		Gaseous Emission				Observation	Reference
					HRR	ICP	BFSC	BTE	CO	HC	NO$_x$			
	Ginger grass oil	G10C30, G20C30, G30C30 and G40C30; and DF	30 ppm	Kirloskar, 1C, 4S, CI, CR17.5:1, RP5.2 kW, RS1500 rpm, IT 23° bTDC, IP 200 bar.	—	—	↓ slightly with an ↑ in load.	↓ with an ↑ in load.	↑ with NP fuels	↓ with G40C30 compared to DF.	↓ with DF compared to NP fuels.	•	It was found that G10C30 had better result means of emission and performance.	[145]
	Nerium oleander	SFDF, NOB, ENOB and NENOB	15.01 nm/30 ppm	Kirloskar, 1C, 4S, WC, CI, CR17.5:1, RP5.2 kW, RS1500 rpm, IT 23° bTDC, IP 200 bar.	↓ with NP fuels compared to DF.	↓ with NP fuels	BSEC ↑ higher with NENOB compared to other fuels.	↓ with NOB compared to other fuels	↓ with NP fuels but ↑ with highest EP.	↓ significantly with NENOB compared to DF.	↓ slightly with NENOB compared to DF.	•	CI engine needs modification to have thermal efficiency. comparable to diesel.	[87]
	Lemon grass oil	LGO; LGO emulsion, LGO nano-emulsion and DF.	16.27 nm/30 ppm	Kirloskar, 1C, 4S, WC, CI, CR17.5:1, RP5.2 kW, RS1500 rpm, IT 27° bTDC, IP 200 bar.	↓ with NP fuels due to §	↓ with NP fuels due to Ÿ	BSEC ↑ with LGO nano-emulsion compared to other fuels.	↑ with LGO compared to other fuels.	↓ with LGO nano-emulsion but ↑ at highest EP.	↓ with LGO nano-emulsion compared to other fuels.	↑ with an ↑ in EP across all fuels.	•	Both ICP and HRR decreases with NP fuels. However, not all NPs contributes to the enhancement of engine combustion.	[82]
Ce$_{0.7}$Zr$_{0.3}$O$_2$	Corn stalk pyrolysis bio-oil	CB10C50, CB15C50, CB20C50 and CB25C50; and DF	50 ppm	1C, 4S, WC, CI, CR17.0:1, RP13.2 kW, RS2200 rpm, IP 190 bar.	—	↓ with CBs fuels	↓ with CBs and ↑ with an ↑ in load.	↑ by CBs with an ↑ in EP.	↑ with CB25C50 and ↑ with an ↑ in load.	↑ with CB25C50 and ↑ with an ↑ in load.	↓ with CB25C50 compared to other fuels.	•	CBs exhibit lower CV compared to DF, which might consume more fuel to maintain the same EP with low comparability.	[138]
MgO	Municipal waste plastic oil	MPO20; MPO20M100 and DF	100 ppm	Kirloskar, 1C, 4S, DI, CI, CR17.5:1, RP3.5 kW, RS1500 rpm, IT 23° bTDC, IP 220 bar	↓ by 7.04% with MPO20M100 compared to DF and ↓ by 17.5% with MPO20 than DF.	↓ by 11.96% with MPO20M100 compared to DF and ↓ by 19.52% with MPO20 than DF.	↓ with MPO20M100 compared to MPO20.	↑ with MPO20M100 compared to MPO20.	↓ by 18.18% with MPO20M100 compared to MPO20.	↓ by 21.87% with MPO20M100 compared to DF.	↑ by 14.47% with MPO20M100 compared to DF.	•	Addition of NPs in plastic oil led to an increased max HRR compared to diesel fuel.	[139]
MWCNT	Lemon peel oil	LPO20; LPO20CNT50 and LPO20CNT100	10 nm/50 ppm and 100 ppm	Kirloskar TV1, 1C, 4S, DI, CI, CR17.5:1, RP5.2 kW, RS1500 rpm, IT 23° bTDC, IP 200 bar	↑ with an ↑ in loads and ICP.	Improved due to Ä.	↓ with an ↑ in loads and NPs rate.	↑ with an ↑ in loads and NPs rate.	↓ with and without NPs addition compared to DF due to Ä.	↓ with and without NPs addition compared to DF due to Ä.	↑ with an ↑ in MWCNT NPs due to ß.	•	Excess carbon led to improper mixing and thus increasing the HRR and ID period.	[86]

Table 5. *Cont.*

Type of NPs Used	Vegetable Based Fuel	Blends	Size of NPs/NPs Concentration	Engine Sp.	Combustion		Performance		Gaseous Emission				Observation	Reference
					HRR	ICP	BFSC	BTE	CO	HC	NO$_x$			
	Waste fishing net oil	WFNO; WM50 and DF	50 ppm	Kirloskar, 1C, 4S, WC, CI, CR17.5:1, RP3.5 kW, RS1500 rpm, IT 23° bTDC.	↑ with WFNO compared to other fuels.	↑ with WFNO compared to other fuels.	↓ by 3.87% with MWCNT fuel compared to WFNO.	↑ by 3.83% with MWCNT fuel compared to WFNO.	↓ by 25%	↓ by 9.09%	↓ by 5.25%		• The engine results showed high efficiency with WFNO compared to MWCNT fuel.	[141]
Rice husk	Pine oil	B10 and B20; B10RH and B20RH; and DF	<100 rm/0.1% (1 g/l)	Kirloskar TV1, 1C, 4S, DI, CI, CR17.5:1, RP5.2 kW, RS1500 rpm, IT 23° bTDC, IP 210 bar	—	—	↑ by 4.1–8.7% with BRH compared to DF.	↓ by 3.04% with RH NPs compared to DF.	↓ by 27.27% with B20RH compared to other fuels.	↓ by 19.64% with B20RH compared to other fuels.	↑ with an ↑ in RH NPs.		• Result indicated a significant change in performance and emission with RH NP fuels.	[128]
	Linseed oil	LS100; PLS20; PLS50, PLS100, PLS150 and PLS200	25 nm–150 nm/50 ppm, 100 ppm, 150 ppm and 200 ppm	Kirloskar TV1, 1C, 4S, WC, CI, CR17.5:1, RP5.2 kW, RS1500 rpm, IT 23° bTDC, IP 200 bar	—	—	↓ with an ↑ in load and TiO$_2$ NPs conc.	↑ slightly by 8.11% with PLS200 compared to LS20	↓ by 21.05% with PLS200 compared to LS20.	↓ by 33.82% with an ↑ in TiO$_2$ NPs conc.	↑ by ~6.53% with an ↑ in TiO$_2$ NPs conc.		• It was observed that the linseed oil values of viscosity and density are almost equal to the diesel when pre-heated to 100 °C.	[129]
TiO$_2$	Orange oil	OM; OMT50 and OMT100; DF	20 nm/50 ppm and 100 ppm	Kirloskar TV1, 1C, 4S, WC, CI, CR17.5:1, RP5.2 kW, RS1500 rpm, IT 23° bTDC, IP 200 bar	↑ with TiO$_2$ NPs fuels.	Improved with an ↑ in NPs.	↑ slightly with an ↑ in TiO$_2$ NPs conc.	Improved for OMT50 and OMT100 by 1.6% and 3.0%, resp. compared to DF.	↓ significantly by 22.4% with OMT100 compared to DF.	↓ by 18.7% with OMT100 compared to DF.	↑ slightly by 7.2–10.4% with an ↑ in TiO$_2$ NPs conc.		• The presence of water molecules and NPs in the fuel lead to increase in ICP and ID period, which suddenly favours HRR.	[106]
	Plastic oil	CPD 2S 5W; PDO 2S 5W; CWT, PWT and DF	40 nm–50 nm/20 ppm, 40 ppm and 60 ppm	Kirloskar TAF1, 1C, 4S, AC, CI, CR17.5:1, RP4.4 kW, RS1500 rpm, IT 26° bTDC, IP 215 bar	↑ by 4.12% PDO compared to CPD.	Improved with an ↑ in NPs.	↓ with an ↑ in load and TiO$_2$ NPs conc.	↓ with an ↑ in load.	↓ with an ↑ in load and TiO$_2$ NPs conc.	↓ with an ↑ in load and TiO$_2$ NPs conc.	↓ with an ↑ in load and TiO$_2$ NPs conc.		• With TiO$_2$ NP fuel, combustion and emission behaviour significantly improved.	[140]

Note: All engines considered are research-based, solely for testing purposes; ↑ = increase; ↓ = decrease; Þ = high heat absorbing capacity; Ã = better solubility and higher surface; Ä = fine atomization behaviour; AC = air-cooled; AE = activation energy; ß = longer ignition delay; bTDC = before top dead centre; BSEC = brake specific energy consumption; C = cylinder; CD = combustion duration; CI = compressive ignition; CR = compression ratio; Conc = concentration; CRDI = common rail direct injection; CWT = (CPD 2S 5W 20T, CPD 2S 5W 40T, CPD 2S 5W 60T); DF = diesel fuel; DI = direct injection; Eff = Efficiency; EP = engine power; HRR: heat release rate; ICP = in-cylinder pressure; ID = ignition delay; IT = ignition timing; IP = injection pressure; Max. = Maximum; Min = Minimum; MDF = methanol-diesel fuel; NPs = nanoparticles; PWT = (PDO 2S 5W 20T, PDO 2S 5W 40T, and PDO 2S 5W 60T); RS = rated speed; RP = rated power; S = stroke; Sp. = specification; WC = water cooled.

4.2.7. Engine Performance Characteristics of Nanoparticles in Biodiesel-Based Fuels

Brake Thermal Efficiency

By comparing neat biodiesel to Ag-ZnO/ZnO-biodiesel, Sam Sukumar et al. [146] showed that the BTE of the nano-based fuels increased by 24% and 19.35%, respectively. According to the authors, this observation was as a result of the higher surface area and reactive surfaces of the nanoparticles, which generated maximum chemical reactivity within the fuel. Karthikeyan et al. [99] investigated the effect of cerium oxide additive on performance and emission characteristics of a CI engine operating on rice bran biodiesel and its blends. It was revealed that the presence of CeO_2 nanoparticles enhanced proper fuel mixing and reduced fuel consumption, which consequently led to an increase in BTE compared to the neat base fuels. Baluchamy and Karuppusamy [103] investigated the combined effect of cobalt chromite nanoparticles and variable injection timing of preheated biodiesel and diesel on performance, combustion, and emission characteristics of CI engine. Their study showed that by advancing ignition timing, the BTE of blends SIT KC1 -ADV, SIT KC2 -ADV, SIT KC3 -ADV, and SIT KC4 -ADV increased by 3.2%, 3.7%, 4.5%, and 7.2%, respectively, when compared with the 23 CAD bTDC (i.e., standard injection timing) due to the presence of nanoparticles in the fuel and fuel burns completely during the combustion process. Various nano-additives (cerium oxide, zirconium oxide, and titanium oxide were blended in Garcinia gummi-gutta biodiesel–diesel blends (B20) for a comparative study conducted by Janakiraman et al. [97]. It was seen that adding metal oxide-based nano additive to B20 fuel reduced ignition delay and showed a slight rise in the net heat release rate and cylinder pressure, causing the BTE of the nano-based blends to increase compared to neat B20 fuel. Kumaran et al. [147] showed that 100 ppm of methanol-based hydroxyapatite nanorods has the ability to increase the BTE of waste cooking oil biodiesel as a result of the improved combustion atomization and rapid evaporation associated with the nanoparticles.

Brake Specific Fuel Consumption

According to Debbarma and Misra [148], the presence of iron nanoparticles in biodiesel–diesel blends reduces energy consumption at full load engine conditions compared to the other test fuels without the nanoparticles. They explained that the presence of the iron nanoparticles increased the calorific value of the base fuel to generate some intensity of power with low consumption of fuel. However, increasing the concentration of the iron additives in the modified biodiesel will increase the energy consumption due to the increase in density and viscosity of the fuel in the presence of a higher concentration of nanoparticles. Nano-copper additive was added to *Calophyllum inophyllum* by Tamilvanan et al. [149] to investigate its effect on the performance, combustion, and emission characteristics of a CI engine. According to the researchers, a reduction of 3–6% in BSFC was achieved for biodiesel blends with the copper additives compared to a neat biodiesel blend at maximum load. They explained that the presence of the metal additive may have caused an increase in combustion temperature, which led to an increase in the conversion efficiency of heat energy into mechanical work and resulted in a reduction in BSFC. Rice husk nanoparticles (0.1%) were blended into B10 and B20 neem oil biodiesel-diesel blends by Sivasaravanan et al. [150]. The addition of rice husk nanoparticles slightly reduced BSFC when added to B10 and B20, in the range of 3.8–6.9% and 2.5–6.1%, respectively. The presence of the rice husk nanoparticle improved the combustion efficiency of all test fuels, and hence lower fuel was required to produce the same amount of work as that of the test fuels without nanoparticles. In a comparative study of two different carbon-based nano-additives, Chacko and Jeyaseelan [105] used graphene oxide and graphene nanoplatelets as fuel additives in diesel and biodiesel blends. Results revealed that by adding the nano-additives, the BSFC of the base fuels was positively impacted. According to the authors, the nanoparticles reduced ignition delay and combustion duration to ensure efficient combustion of the fuel supplied. It was also reported that the graphene oxide-based blends had a better BSFC than the graphene nanoplatelets-based blends. The former blend possessed a higher in-cylinder pressure which in turn produces better power output than the combustion with the latter blends. Microalgae biodiesel produced from Botryococcus braunii algal oil were doped with a mixture of titanium dioxide and silicon dioxide nanoparticles at dosages of 50 and 100 ppm in the work of Karthikeyan and Prathima [151]. Results from the study indicated that BSFC at all BMEPs were lower for B20 + 50 ppm (TiO_2 + SiO_2) and B20 + 100 ppm (TiO_2 + SiO_2) than diesel and B20 fuels. From the researchers' point of view, the addition of the mixed nanoparticles oxidized the carbon deposits from the engine which led to an efficient operation and reduced fuel consumption.

Brake Power and Brake Torque

With the aim of investigating the effect of aqueous nanofluids on the performance and exhaust emissions of diesel engines, Khalife et al. [98] added cerium oxide into diesel–biodiesel–water blends (WBDE). Results revealed that adding cerium oxide nanoparticles in WBDE resulted in higher BP values compared with neat WBDE. The authors assigned this observation to the fact that metal-based additives could react with water at higher temperatures during the combustion process, resulting in hydrogen generation and, consequently, promoting the engine cylinder's combustion process. Hoseini et al. [108] investigated the effect of graphene oxide nanoparticles on biodiesels from three different feedstocks; Evening primrose, Tree of heaven, and Camelina. It was seen in their results that the BP of all three biodiesels increases with the addition of the nanoparticles. Graphene oxide nanoparticles have the ability to increase the heat of evaporation of fuel. They have high energy content and a high surface-to-volume ratio. These properties lead to an increase in density of fuel–air charge, better oxidation of fuel blends, and high enthalpy of combustion, which caused an increase in the BP of the base fuels. It is also worth noting that, due to the high lower heating value of Camelina, its BP was greater than the other two test biodiesels. Alumina nanoparticles (40, 80, 120, and 160 ppm) were prepared and added as an additive to waste cooking oil biodiesel–diesel blend by Ghanbari et al. [152]. The process resulted in a significant increase in BT and BP. The researchers explained that this result was due to the improvement of the surface-to-volume ratio and catalytic effect of the alumina nanoparticles in the fuel blend, which improved combustion quality. In another study, palm–sesame biodiesel was blended with oxygenated alcohols in the presence of 100 ppm CNT and TiO_2 nanoparticles [153]. Compared to B30 fuel, B30+ TiO_2 and B30 + CNT blended fuels showed a slight decrease in average BT values by 1.28% and 0.88%, respectively. This was a result of an increase in viscosity and density values of the base fuel with the addition of the nanoparticles. The trend was quite similar to BP. Compared to B30 fuel, B30 + TiO_2 and B30 + CNT blended fuels showed a slight decrease in average BP values by 1.47% and 1.04%, respectively. According to the study of Shekofteh et al. [154], functionalized MWCNTs–OH were blended into diesel–biodiesel–bioethanol blends for performance and emission characteristics. MWCNTs–OH into the base fuel improved BT and BP. In comparison to D100 and B5, adding MWCNTs–OH to B5E4 and B5E8 at 1800 rpm resulted in an increase in torque of 8.61 and 7.41 percent on average. Similarly, when MWCNTs–OH was added to B5E4 and B5E8 fuels at 2400 rpm, the torque increased by 14.19 and 11.32 percent, respectively, as compared to D100 and B5. As MWCNTs–OH was added to B5E4 and B5E8 at 1800 rpm, power increased by 7.33 and 4.35 percent, respectively, when compared to D100 and B5. Similarly, adding MWCNTs–OH to B5E4 and B5E8 at 2400 rpm increased power by 18.90 and 17.71 percent, respectively. The observed findings, according to the researchers, were attributable to the inclusion of the nanoparticles, which generated greater peak cylinder pressure and a faster heat release rate by lowering the ignition delay and combustion duration of fuel in the engine, resulting in a more complete combustion of the engine.

4.2.8. Engine Emission Characteristics of Nanoparticles in Biodiesel–Based Fuels

Carbon Monoxide

In the study by Anbarasu and Karthikeyan [101], it is seen that cerium oxide blended biodiesel emulsion fuels recorded a significant reduction in CO emissions compared to the other test fuels. The presence of the cerium oxide in the base fuel shortened ignition delay and improved combustion characteristics. As a result, there was an improvement in the quantity of fuel–air mixing and uniform burning. Srinivasan et al. [155] studied the effect of Al_2O_3 and TiO_2 of dosages 25 ppm and 50 ppm on a diesel engine fueled with rubber seed oil methyl ester. The addition of the nanoparticles resulted in a reduction in CO from 0.31% to 0.76% and 0.3% to 0.75% by adding 25 ppm and 50 ppm of Al_2O_3 to B100, respectively. On the other hand, adding 25 ppm and 50 ppm of TiO_2 to B100 led to a reduction in CO emissions from 0.29% to 0.74% and 0.28% to 0.73%, respectively. According to the researchers, the nanoparticles presented additional oxygen molecules to facilitate oxidation of reaction, leading to better combustion and decreased CO emissions. Additionally, the nanoparticles underwent a catalytic oxidation reaction, which improved the mixing rate of air with fuel. Nithya et al. [96] made an investigation into the effect of engine emission operating on canola biodiesel blends with TiO_2. Their results revealed that by adding the nanoparticles to B20 fuel, the CO emission decreases nearly 30% at full engine load due to the shorter ignition delay and provision of additional oxygen molecules by the nanoparticles, which helped achieve a more complete combustion. Orange peel oil biodiesel was converted to a nanofluid under the influence of titanium dioxide by Kumar et al. [106] for assessing its effects on the performance, emission, and combustion characteristics of a diesel engine. It was reported in this work that at peak power output, the CO emissions for diesel fuel, pure OOME, OOME–T50, and OOME–T100 were 0.58%, 0.55%, 0.51%, and 0.45%, respectively. The authors claimed that the ability of titanium dioxide to act as an oxidation catalyst offered more oxygen for the burning of the fuel inside the chamber, which resulted in complete combustion and reduced the creation of CO emissions. Cerium oxide and Gadolinium doped cerium oxide (CeO_2:Gd) nanoparticles were dispersed in blended Pongamia oil biodiesel by Dhanasekar et al. [156] to analyse the emission from a four stroke single cylinder diesel engine. Results showed that compared to pure biodiesel, the CO emission is drastically decreased in the ceria and GDC blended fuels as a result of the nanoparticles' reaction with surface oxygen which is released from cerium oxide and GDC nanoparticles. The investigators also reported that, GDC nanoparticles showed better CO emission reduction pure diesel, which may be due to the high content of surface oxygen of GDC nanoparticles as it converts CO into CO_2.

Hydrocarbons

Copper (II) oxide nanoparticles were added to Jojoba biodiesel blend (JB20) by Rastogi et al. [94] to investigate its effect on the performance and emission characteristics of a diesel engine. It was reported from this work that the CuO had a significant impact on reducing HC emissions of the base fuel. The average HC emissions for the JB20, JB20CN25, JB20CN50, and JB20CN75 were reduced by 5.18%, 9.39%, 12.17%, 7.45% with respect to diesel fuel at 5.2 kW engine load. The results were ascribed to the excellent characteristics introduced by CuO such as, increase in calorific value, decreasing fuel viscosity due to which proper fuel atomization occurred, better fuel explosion process, shortened ignition delay, and enhanced heat release rate during fuel combustion, which promoted complete combustion process inside the combustion chamber. Ramakrishnan et al. [102] presented findings to show the role of nano additive blended biodiesel on emission characteristics of a diesel engine. The authors added carbon nanotubes to neem biodiesel (NBD) and observed that the addition of CNT at 50 ppm and 100 ppm to NBD reduced HC emissions by 5.1% and 6.7%, respectively, at all loads. The authors gave the following reasons for the obtained results for HC emissions: (1) the positive effect of CNT, which acted as an oxidation catalyst, lowers the carbon combustion activation temperature and improve the oxidation of NBD, and (2) CNT reduces ignition delay and improves secondary atomization leading to enhanced combustion. Solmaz et al. [104] predicted the performance and exhaust emission characteristics of a CI engine fuelled with MWCNTs doped biodiesel–diesel blends using response surface methodology. They reported that exhaust HC concentrations of the test fuels decreased with the addition of the MWCNTs into B20 fuel. The authors argued that the improved combustion characteristics and catalyst activity of MWCNTs were responsible for such decrement in the exhaust HC concentrations. In the experiment of Gad and Jayaraj [157], carbon nanotubes, titanium dioxide, and aluminum dioxide of concentrations 25, 50, and 100 ppm were blended into Jatropha biodiesel–diesel blend. The maximum decrease in HC emissions of J20C25, J20C50, J20C100, J20T25, J20T50, JT20T100, J20A25, J20A50, and JT20A100 were 4%, 12%, 15%, 22%, 17%, 15%, 19%, 21%, 18%, respectively, compared to neat diesel. The decrease in HC emissions for the nanofuels was attributed to the higher catalytic activity and the higher surface-to-volume ratio of the nanoparticles as well as their ability to lower carbon combustion activation temperature and enhance oxidation of fuel. Further, 25 and 50 ppm of MWCNT nanoparticles were blended in Honge oil methyl ester (HOME) by Tewari et al. [109]. HC emission for HOME operation was higher compared to diesel but lower for the HOME–MWCNTs than pure HOME. The HC emissions for HOME50MWCNT, HOME25MWCNT, HOME and for diesel were 58, 70, 82, and 32 ppm at 80% load, respectively. The lower HC emissions for the MWCNT–blends were lower than pure HOME due to the catalytic activity and improved combustion characteristics of MWCNT which promoted complete combustion.

Nitrogen Oxides

El-Seesy et al. [92] revealed that the addition of MWCNTs to JB20D leads to a decrease in NO_x emissions compared to neat JB20D. The researchers attributed this trend to the catalytic effect of the MWCNTs that may have accelerated the combustion process to be completed, forming final products with a minimum thermal breakdown of the hydrocarbon compounds. To analyse the emission and performance of a direct injection engine fuelled with Mahua biodiesel blends, Rastogi et al. [95] blended Al_2O_3 nanoparticles into the base fuel. They showed that with the addition of the nanoparticles, the NO_x emissions of the blended fuels were lower than neat diesel. They accounted for this observation by explaining that the role Al_2O_3 plays in increasing surface area, reducing ignition delay, helping the active reaction of hydrocarbon with oxygen, and reducing the reaction of nitrogen with oxygen was the cause for the reduction in NO_x formation in the cylinder. The effect of 50–150 ppm of MWCNTs in the diesel–biodiesel blend was studied by Alenezi et al. [93] with a focus on the emission and combustion characteristics of diesel engines. It was shown in this work that the addition of MWCNTs to B20 and B40 base fuel resulted in a significant reduction in NO_x formation. However, the opposite trend of NO_x emissions was reported for B10 when the concentration of the MWCNTs was increased. Sulochana and Bhatti [107] added MWCNTs in 25 ppm and 50 ppm mass fractions to waste fry oil biodiesel. It was highlighted that, due to their higher premixed combustion heat release rate and complete combustion, WFOME25MWCNT produced higher NO_x emissions than WFOME50MWCNT and pure WFOME. The recorded emission of NO_x for diesel, WFOME, WFOME50MWCNT and WFOME25MWCNT were 654 ppm, 731 ppm, 764 ppm, and 884 ppm, respectively. In the work of Deepak Kumar et al. [158], biodiesel derived from cottonseed oil has been investigated along with 80 ppm of ZnO. NO_x emission of ZnO-based fuels was lower compared to pure diesel and biodiesel. In their work, the authors explained that zinc oxide raises the average temperature of the combustion chamber (due to its higher calorific value), allowing more oxygen in the mixture to react, resulting in fewer NO_x emissions. For NO_x reduction, zinc oxide absorbs oxygen. Shorter ignition delays result in better fuel–air mixing, which results in an oxygen deficit for NO_x, lowering NO_x emissions.

4.2.9. Effect of Nano-Additives and Diesel–Biodiesel Blends on Engine Combustion

This subsection critically evaluates the combustion effect of diesel fuel blended with different biodiesel and nano-additives in various CI engines. The most recent studies on diesel/biodiesel–nanoparticle blends discovered that the combustion of such mixture in diesel engines could offer many remarkable outputs such as better atomization behavior [129], good solubility, and a higher surface to volume ratio [104,106], low combustion activation temperature due to high thermal conductivity and catalytic activity [126], shorter ignition delay [157], and high burning rate due to the improved latent heat of vaporization [159].

The key factors in evaluating the combustion characteristics are in-cylinder pressure rise rate and heat release rate [160]. As recently reported, the ignition delay increases with an increase in diesel–biodiesel blend fuels [97,161]; this was accompanied by a large amount of fuels in the premixed combustion phasing [27], thereby resulting in minimal thermal efficiency [162]. This was due to the improvement in cetane number, viscosity and density of the blend fuel. This trend is in conformity with the result found with Al_2O_3, CeO_2, $Co(Al, Cr)_2O_4$, and SiO_2 [103,163–165], which reported fast burning rate during premixed combustion phase, resulting in high peak cylinder pressure with an improved heat release rate, as summarized in Table 6. This occurrence is inconsistent to the final remark found by Ranjan et al. [166] with the addition of MgO NPs and [167] with ZnO NPs, which had a low ignition delay due to lower viscosity. The result was later accompanied by a low heat release rate and a decrease in peak cylinder pressure.

Improving the conversion efficiency of the engine using nano-catalyst is a novel concept that was found suitable for a modified and unmodified engine. Janakiraman et al. [97] evaluates the effect of ZrO_2 at 25 ppm concentration into *Garcinia* biodiesel (B20) to enhance the diesel engine efficiency. It was found that adding ZrO_2 NPs, enhanced the ratio of surface area to volume, hence improving the HRR and ignition properties of the fuel blend. A similar trend in HRR was observed using biodiesel-based fuels on diesel engines with the addition of Al_2O_3 [168,169], CeO_2 [159], and TiO_2 [106]. Several researchers, such as [27,101,161,170,171] investigated and reported that the rationale behind the trend of HRR is similar to that of ICR in most cases.

Table 6. Summary of the most recent experiments on different NPs addition in biodiesel fuels used in CI engines.

Type of NPs Used	Source of Biodiesel Fuel	Blends	Size of NPs/NPs Concentration	Engine Sp.	Combustion		Performance		Gaseous Emission			Observation	Reference
					HRR	ICP	BFSC	BTE	CO	HC	NOx		
Al_2O_3	Honge oil	HOME20, HOME2020, HOME2040, and HOME2060	10 nm/20 ppm, 40 ppm and 60 ppm	Kirloskar TV1, 1C, 4S, WC, CI, CR17.5:1, RP5.5 kW, RS1500 rpm, IT 23° bTDC, IP 205 bar.	↑ with HOME2040 compared to other fuels.	↑ with HOME2040 compared to other fuels.	↓ by 11.6% HOME2040 compared to HOME20.	↑ by 10.57% with HOME2040 compared to HOME20.	↓ significantly by 47.43% HOME2040 compared to HOME20.	↓ by 37.72% HOME2040 compared to HOME20.	↑ with an ↑ in Al_2O_3 NPs conc.	• Min values of ID period and CD were achieved for HOME2040.	[165]
	Tamarind seed oil	TS20; TSA (TS20A30 and TS20A60) and DF	20 nm/30 ppm and 60 ppm	Kirloskar TVI, 1C, 4S, WC, DI, CI, CR17.5:1, RP5.2 kW, RS1500 rpm, IT 23° bTDC, IP 230 bars.	↑ with TS20 compared to TSA.	↑ with DF compared to TSA blend.	↑ with TSA fuels compared to TS20 blend.	↑ by 1.56% with TS20A60 compared to DF.	↓ by 56.6% with NPs fuels compared to DF.	↓ with an ↑ in Al_2O_3 NPs.	↑ higher with TS20A60 but lower than TS20.	• Shorter ID period and better ignition properties of the NP additives results in enhanced premixed combustion.	[168]
	Pongamia oil methyl ester	B25; B25A (B25A50 and B25A100) and DF	50 ppm and 100 ppm	Kirloskar TVI, 1C, 4S, WC, DI, CI, CR16.5:1, RP3.7 kW, RS1500 rpm, IT 23° bTDC, IP 220 bars.	↑ with an ↑ in Al_2O_3 NPs rate.	↑ with an ↑ in Al_2O_3 NPs rate.	↓ highly with B25A100 compared to other fuels.	Improved with an ↑ in Al_2O_3 NPs.	~ marginal with B25A compared to B25 and DF.	↓ marginal with B25A compared to B25 and DF.	↑ with an ↑ in Al_2O_3 NPs rate.	• The current challenge is unburnt NPs from the exhaust of the diesel engine, which need to be investigated in order to safeguard the global environment.	[169]
CeO_2	WCO	B20; B20C80-10, B20C80-30 and B20C80-80.	10 nm, 30 nm and 80 nm/80 ppm	Kirloskar TV1, 1C, 4S, WC, CI, CR17.5:1, RP5.2 kW, RS1500 rpm, IP 180 bar.	↑ slightly with B20C80-30	↑ by ~1.7% with B20C80-30 compared to other fuels.	↓ by 2.5% with B20C80-30 compared to DF.	↑ with CeO_2 NPs conc.	↓ by 56% with B20C80-30 compared to DF.	↓ by 27% with B20C80-30 compared to DF.	↓ by 17% with B20C80-30.	• 30 nm-sized CeO_2 NP was found the most effective in decreasing NOx compared to 10 nm and 80 nm-sized.	[159]
	Corn oil	CO10; CO10C25, CO10C50 and CO10C75	50 nm-70 nm/25 ppm, 50 ppm, and 75 ppm	Kirloskar TV1, 1C, 4S, WC, CI, CR17.5:1, RP5.2 kW, RS1500 rpm, IT 23° bTDC, IP 200 bar.	—	↑ with CeO_2 NPs conc.	↓ with an ↑ in CeO_2 NPs conc.	↑ with CO10C50 at max. eff. 34.8% and load.	↓ with CO10C50	↓ with CO10C50	↑ with an ↑ in load and CeO_2 NPs conc.	• High rate of ICP and temperature in combustion influenced the formation of NOx.	[164]
Co(Al, Cr)$_2$O$_4$	Kapok oil	SIT KC1-RET, SIT KC2-RET, SIT KC3-RET, SIT KC4-RET, SIT KC1-ADV, SIT KC2-ADV, SIT KC3-ADV, SIT KC4-ADV	50 ppm, 100 ppm, 150 ppm, and 200 ppm	Kirloskar SV1, 1C, 4S, WC, CR17.5:1, RP5.9 kW, RS1800 rpm, IT varies.	↑ by 5.09% with IT of 23CAD bTDC	↑ by 5.27% with IT of 23CAD bTDC	↓ by 21.23% with SIT KC4-ADV.	↑ by 7.2% with KC4-ADV than SIT.	↓ by 41.66% with SIT KC4-ADV	↓ by 37.86% with SIT KC4-ADV	↓ by 16.45% with SIT KC1-RET.	• Better result can be obtained by IT retardation of the engine.	[103]

Table 6. *Cont.*

Type of NPs Used	Source of Biodiesel Fuel	Blends	Size of NPs/NPs Concentration	Engine Sp.	Combustion		Performance		Gaseous Emission			Observation	Reference
					HRR	ICP	BFSC	BTE	CO	HC	NOx		
$FeO.Fe_2O_3$	Chicken fat oil	B10, B20 and B30; B10F50, B10F100, B10F150, B20F50, B20F100, B20F150, B30F50, B30F100 and B30F150	18.21 nm/50 ppm, 100 ppm, and 150 ppm	Kirloskar TV1, 1C, 4S, WC, CI, CR17.5:1, RP5.2 kW, RS1500 rpm, IT 23° bTDC, IP 200 bar.	—	—	↓ highly by 10.64% with B20F100 compared to B20.	Improved by 4.84% for B20F100 compared to B20.	↓ Max 56.66% with B30F100 compared to B30.	↓ by 22.72% with B30F100 compared to B30.	↓ by ~15.39% with $FeO.Fe_2O_3$ NP fuels.	• No any side effects observed related to engine efficiency after four months. • No carbon deposits were observed in the fuel tank and carburettor.	[172]
MgO	WCO	B's (B10, B20 and B100); MgO NPs fuels (B10W30A, B20W30A and B100W30A) and DF	30 ppm	Kirloskar TVI, 1C, 4S, WC, CI, CR17.5:1, RP7 kW, RS1500 rpm, IT 23° bTDC, IP 220 kgf/cm².	↓ with an ↑ in NPs.	↓ highly with MgO NPs fuels.	↑ with MgO NPs fuels compared to B's.	↑ with MgO NPs fuels compared to B's.	↓ with MgO NPs fuels compared to other fuels.	↓ with MgO NPs fuels compared to other fuels.	↑ with MgO NPs fuels and B's compared to DF.	• Result showed that MgO NPs can be used to improve the cold flow properties when used in CI engine.	[166]
MW-CNTs	WCO	B20; B20M20, B20M50, B20M75 and B20M100.	2nm–16 nm/20 ppm, 50 ppm, 75 ppm, and 100 ppm	Lombardini, 1C, 4S, WC, CI, CR17.5:1, RP5.77 kW, RS3000 rpm, IP 190 bar.	—	—	↓ by ~6.7% as the load ↑	Improved by ~7.4% with an ↑ NPs.	↓ with an ↑ in MWCNTs fuels.	↓ with an ↑ in MWCNTs fuels.	↑ significantly	• Increasing engine load from significantly enhanced the emission behaviour,	[104]
	Jatropha	J20; J20C25, J20C50 and J20C100	20 nm–25 nm/25 ppm, 50 ppm, and 100 ppm	DEUTZ F1L511, 1C, 4S, AC, CI, CR17.5:1, RP5.77 kW, RS1500 rpm, IT 24° bTDC, IP 175 bar.	—	—	↓ with an ↑ in load.	↑ with an ↑ in load and NPs conc.	↓ with an ↑ in CNTs NP conc.	↓ with an ↑ in CNTs NP conc.	↓ partially with CNTs fuels at high loads..	• J20C50 achieved improvement in engine performance and emissions reductions compared to other fuels.	[157]
NiO	Neem oil	NB25, NB25N25, NB25N50, NB25N75 and NB25N100	7 nm–10 nm/25 ppm, 50 ppm, 75 ppm, and 100 ppm	Rocket Engg, VCR, 1C, 4S, WC, CR17.5:1, RP4.8 kW, RS1500 rpm, IT 23–27° bTDC.	—	—	↓ with an ↑ in NP conc. at 27° bTDC.	↑ by 6.3% with NiO fuels.	↓ significantly by NiO fuels.	↓ significantly by NiO fuels.	↑ with NiO NP fuels by advanced fuel injection.	• Advancing fuel IT with presence of NP improves the performance and reduces the engine emissions.	[162]
SiO_2	Soybean	SB25; SB25S25, SB25S50 and SB25S75; DF	5 nm–20 nm/25 ppm, 50 ppm, and 75 ppm	Kirloskar VCR, 1C, 4S, CI, CR21.5:1, RP5HP, RS1800 rpm.	—	—	↓ with an ↑ in load.	↓ slightly with SiO_2 fuels compared to DF.	↑ significantly by SiO_2 NP fuels with an ↑ in load.	↑ slightly with SiO_2 NP fuels compared to DF.	↑ significantly with an ↑ in load.	• Not all NP and biofuels are considered as clean energy, but assessment needs to be done.	[173]

Table 6. *Cont.*

Type of NPs Used	Source of Biodiesel Fuel	Blends	Size of NPs/NPs Concentration	Engine Sp.	Combustion		Performance		Gaseous Emission				Observation	Reference
					HRR	ICP	BFSC	BTE	CO	HC	NOx			
	WCO	B20; B20SiO$_2$ and DF.	100 ppm	Lombardini 15 LD 350, 1C, 4S, WC, DI, CR20 3:1, RP7.5HP, RS3600 rpm, IP207 bar.	↑ with an ↑ in load for all fuels.	↑ with an ↑ in load for all fuels.	↓ by B10SiO$_2$ with an ↑ in loads.	↑ with an ↑ in load for all fuels.	↓ slightly with B10SiO$_2$ compared to DF.	↓ by 80.98% with B10SiO$_2$ compared to DF.	↑ significantly with B10SiO$_2$ compared to DF.	•	SiO$_2$-based NPs into biodiesel gives better results than biodiesel alone.	[163]
TiO$_2$	Cottonseed oil	CSBD; CSBD50 and CSBD100; and DF	17 nm–28 nm, 50 ppm, and 100 ppm	Kirloskar AV1, 1C, 4S, WC, CI, CR18.5:1, RP3.5 kW, RS1400 rpm IT 23° bTDC, IP 200 bar.	—	—	↓ with an ↑ in load at all tested fuels.	↓ with CSBD and TiO$_2$ NP fuels compared to DF.	↓ with an ↑ in TiO$_2$ NP.	↓ by 14.7–16.2% with CSBD100 compared to DF.	↓ with an ↑ in TiO$_2$ NP at all load conditions.	•	CSBD100 fuel exhibits 1.1–1.5%. improvement compared to CSBD.	[174]
	Palm oil	B0, B20; B20T60 and B20AOT60; DF	60 ppm	TECH-ED, 1C, 4S, WC, VRC, CR20:1, RP4 kW, RS1500 rpm.	—	—	↓ with B20AOT60 compared to B20.	↑ higher with B20AOT60 compared to other fuels.	↓ with B20AOT60	↓ with B20AOT60	↑ with B20 but much lesser with B20AOT60 compared to DF.	•	Better results were obtained with NP and AO as it acts as a deterrent in the fuel reaction.	[175]
ZnO	Grapeseed oil	GS; GSZ50 and GSZ100	36 nm / 50 ppm and 100 ppm	Kirloskar TV1, 1C, 4S, CI, CR17.5:1, RP5.2 kW, RS1500 rpm, IT 27° bTDC, IP 200 bar.	↓ with GS compared to DF.	↓ with GS compared to DF.	↓ with DF compared to other fuels.	Max. ↑ was at GSZ100.	↓ with an ↑ in EP and NPs conc.	↓ by ~13% with GSZ100 compared to other fuels.	↑ with GS compared to DF.	•	The fuel consumption increases with ZnO NPs fuel conc.	[167]
ZrO$_2$	*Garcinia gummi-gutta*	B100, B20 and B20Z25	25 ppm	Kirloskar TAF-1, 1C, 4S, AC, CR17.6:1, RP5.2 kW, RS1500 rpm, IT 23° bTDC.	↑ with B20	↑ with B20	↓ with B20Z25 compared to B100.	↓ with B20Z25 compared to DF	↓ with B20Z25 compared to DF and B100).	↓ with NP fuel compared to DF.	↓ slightly by B20z25 with an ↑ in EP.	•	B20Z25 acquired better efficiency and minimal emissions than B20 and B100 fuels.	[97]

Note: All engines considered are research-based, solely for testing purposes; ↑ = increase; ↓ = decrease; AO = antioxidant; AC = air-cooled; bTDC = before top dead centre; BSEC = brake specific energy consumption; C = cylinder; CD = combustion duration; CI = compressive ignition; CR = compression ratio; Conc = concentration; CRDI = common rail direct injection; CV = calorific value; DF = diesel fuel; DI = direct injection; Eff = Efficiency; Engg = Engineering; EP = engine power; HRR: heat release rate; ICP = in-cylinder pressure; ID = ignition delay; IT = ignition timing; IP = injection pressure; Max. = Maximum; Min = Minimum; NPs = nanoparticles; RS = rated speed; RP = rated power; S = stroke; Sp. = specification; WC = water cooled; VCR = variable compression ratio; WCO = waste cooking oil.

5. Comparative Strengths of Different Nanoparticles in Same Base Fuel

Although nanoparticles can significantly influence the behaviour of fuels during combustion, the degree of improvement varies from one nanoparticle to another. With respect to their own unique characteristics, some nanoparticles perform better than their counterparts upon their addition in the same base fuels. For example, Ağbulut et al. [70] have shown that when Al_2O_3 and TiO_2 of 100 ppm are blended in diesel–ethanol fuel, Al_2O_3 exhibits better performance and combustion characteristics than TiO_2. This conclusion has also been reached in the study of Purushothaman et al. [125] where Al_2O_3 and TiO_2 of 100 ppm were blended in Mahua oil. CNT, Al_2O_3 and TiO_2 (50 ppm) were used as fuel additives in biodiesel–diethyl ether blends by Mardi K et al. [115]. At 1700 rpm, CNT showed better performance and emission characteristics than the other two nanoparticles, except for NO_x emissions, which were positively impacted the most by Al_2O_3. However, an opposite result to Mardi K et al. [115] has been reported by Gad and Jayaraj [157]; using the same nanoparticles in Jatropha biodiesel–diesel blends, at maximum load, Al_2O_3 showed the best characteristics for BSFC and BTE compared to CNT and TiO_2, whereas CNT was best suited for CO and NO_x emissions. For HC emissions, TiO_2 was the best nanoparticle. By doping ZnO with silver (Ag) nanoparticles for Pongamia biodiesel–diesel fuel, Sam Sukumar et al. [146] showed that Ag-ZnO had a better effect on engine characteristics than neat ZnO. Finally, Chacko and Jeyaseelan [105] used graphene oxide and graphene nanoplatelets as blend components in Karanja oil biodiesel/waste cooking oil biodiesel–diesel blends. It was reported in this study that, at 2250 rpm and BMEP of 3.45 bar, graphene oxide showed a better effect on engine performance, whereas graphene nanoplatelets were more favourable for engine emission reduction.

In general, for one nanoparticle to perform better than another, it means that the superior nanoparticle has enhanced catalytic activity, higher surface to volume ratio, better oxygen buffering, higher evaporation rate, and higher thermal conductivity compared to the inferior nanoparticles. Other factors such as nanoparticle size, viscosity, and density could also vary the performance of nanoparticles. However, it is worth noting that, by varying the concentrations of the base fuels and nanoparticles or engine operating conditions, a superior nanoparticle could become relatively inferior, and vice versa.

6. Similarities and Differences in Engine Characteristics of the Same Nanoparticle in Low-Carbon Fuels

During our examination, we noticed several comparable and contrasting themes across the diverse research studies. This section describes the similarities and differences in engine characteristics for all three liquid fuels when blended with the same nanoparticle under three independent and varying experimental conditions, including engine load and speed, nanoparticle concentration and size, and base fuel concentration. The addition of iron oxide nanoparticles in alcohol [69], vegetable oil [91], and biodiesel [172] were all reported to have led to a decrease in CO, NO_x, and HC, while BTE was increased in all three cases. According to Ağbulut et al. [70], Chinnasamy et al. [88], and Anchupogu et al. [111], the addition of alumina nanoparticles to alcohol, bio-oil, and biodiesel, respectively, results in a decrease in HC and CO. However, NO_x emissions increase in the bio-oil while they decrease in the alcohol and biodiesel. Cerium oxide in alcohol [75], vegetable oil [85], and biodiesel [97] resulted in a CO, HC, NO_x, and increase in BTE. However, unlike alcohol and biodiesel, BSFC worsened in vegetable oil. Örs et al. [81], Elumalai et al. [129], Kumar et al. [106] to alcohol, vegetable oil, and biodiesel, respectively, BSFC, CO, HC reduced, but NO_x increased in the alcohol and vegetable oil systems whilst it reduced in the biodiesel fuels. It was shown in the experimental findings of Wei et al. [74] (alcohol) and Gavhane et al. [173] (biodiesel) that upon the addition of silicon dioxide nanoparticles, CO, HC, smoke reduces and BTE increases. However, the opposing trend was that, in the alcohol system, BSFC and NO_x improve while they worsen in the biodiesel system. By blending MWCNT in alcohol [68], bio-oil [141], and biodiesel [107], NO_x emissions were reported to have increased in all three experiments. Results from El-Seesy and Hassan [77] and Chacko and Jeyaseelan [105] revealed that when GO/GNP nanoparticles are blended in alcohol and biodiesel-based fuels, respectively, NO, CO, and HC emissions of the base fuels reduce significantly. In the works of Heydari-Maleney et al. [65] (alcohol), Sharma et al. [90] (bio-oil), and Ramakrishnan et al. [102] (biodiesel), all three investigations reported that CO and HC decrease upon the addition of CNT. However, unlike the bio-oil and biodiesel, NOx of the alcohol fuel increased. The analysis made in this section also reveals that, though nanoparticles can positively or negatively impact the performance, emission, and combustion characteristics of the base fuels, the type or extent of impact depends on certain inherent factors such as the type of nanoparticle, type of base fuel, concentration and size of nanoparticle, engine conditions such as load and speed, and the approach in which the nanoparticles were prepared and blended into the base fuels.

7. Summary of the Mechanism Involved with Nanoparticle's Role during Low Carbon Fuel Combustion in ICE

Evidence from literary sources, as presented in the previous sections, points to one obvious fact: nanoparticles generally produce better engine performance, combustion, and emission characteristics when blended in liquid biofuels. However, as mentioned earlier, the extent of improvement will significantly be determined by the type of nanoparticle, type of base fuel, concentration and size of the nanoparticle, engine conditions such as load and speed, and the approach in which the nanoparticles were prepared and blended into the base fuels. Various researchers have given several reasons to explain how these oxides of metal and carbon nano-additives improve the engine characteristics upon their addition into the base fuels. In this section, we only present more general and the most consistent reasons given by investigators of the reviewed literature in the previous sections. First of all, nanoparticles play a role as oxygen buffers. By doing so, additional oxygen molecules are provided in the combustion chamber, which promotes complete combustion and lowers unburnt emissions. Secondly, most of the reviewed nanoparticles have a higher surface area to volume ratio. This enhances catalytic behaviour by providing a larger surface area for the fuel particles to interact and also produce more energy inside the cylinder, which provides an efficient burning process to obtain a more complete combustion and reduced emission of pollutants. Next the nanoparticles exhibit micro-explosive properties, thereby promoting better atomization and air–fuel mixing. Furthermore, due to their high thermal conductivity, nanoparticles can act as heat sinks which helps decrease the temperature and NO_x emissions. In addition, nanoparticles show catalytic activity, which lowers combustion activation temperature and helps increase the burning rate. Moreover, there is a higher evaporation rate with nanoparticles which leads to an enhanced mixture of fuel vapour with air, reducing ignition delay and combustion duration to increase the chances of complete combustion. Nanoparticles can also act as an oxidation catalyst that promotes the oxidation of hydrocarbons to reduce HC emissions.

Some authors have also reported other factors such as surface tension and latent heat of vaporization of nanoparticles to be associated with the effective combustion of nanofluids. The wettability of the fluid improves as the surface tension is reduced. Spray parameters such as droplet size, dispersion, and spray angle are heavily influenced by surface tension in combustion applications. One of the most important factors in influencing the burning rate of liquid fuels is the latent heat of vaporization. This is an important result since a greater burning rate suggests more efficient combustion and maybe a smaller combustor, and this can be dramatically altered when the latent heat of vaporization of nanofluids is varied.

Depending on several conditions such as fuel concentration, dosage and size of nanoparticles, experimental setup, and researcher(s)' experience, each nanoparticle affects the physicochemical properties of base fuels differently. In other words, for the same nanoparticle, fuel property adjustment varies in trend from one study to another. However, a general trend could still be observed in the reported studies. For metals and their oxides, titanium and aluminum have proven to be excellent additives for enhancing the calorific value and cetane number of the base fuels. Similarly, amongst the carbon-based nanomaterials, it would appear that the multi-walled carbon nanotubes are more suited for improving the energy content and ignition qualities of the base fuels. For viscosity and density, the general trend shows slight increase in the values of the base fuels after the addition of the nanoparticles. It is advisable for researchers to carefully consider the type of base fuel and its concentration, the dosage and size of the nano-additives, and the specific targets of the study before selecting a particular nanoparticle(s) for modifying or designing any new fuel. This would ensure a more consistent and reliable trend in the fuel properties adjustment of nanofuels.

8. Exergy, Exergoeconomic, Exergoenvironmental, and Sustainability of Nano-Additives and Low Carbon Fuels in ICE

In the previous sections, the energy-based analysis such as BTE, BSFC, BT, and BP as well as the emission characteristics of diesel engines fuelled with low carbon fuels under the influence of nano-additives have been discussed. Indeed, these energy-based indicators have been extensively studied and reported in literature. Energy analysis has been criticized in the open literature for failing to consider the effect of second law's limitation on an energy conversion process [36]. On the other hand, exergy analysis fills this gap by providing relevant information on the irreversibility aspects (availability losses) of energy conversion systems [176]. Evaluating the performance of thermal systems based solely on exergy analysis is not sufficient, and the analysis could be more comprehensive when economic, environmental, and sustainability aspects are included. These aspects put together provide a complete understanding on the profitability and sustainability of an improvement achieved through exergy analysis. [35]. Figure 7 shows the nexus between exergy efficiency, environmental impact, and sustainability of a thermal system [177].

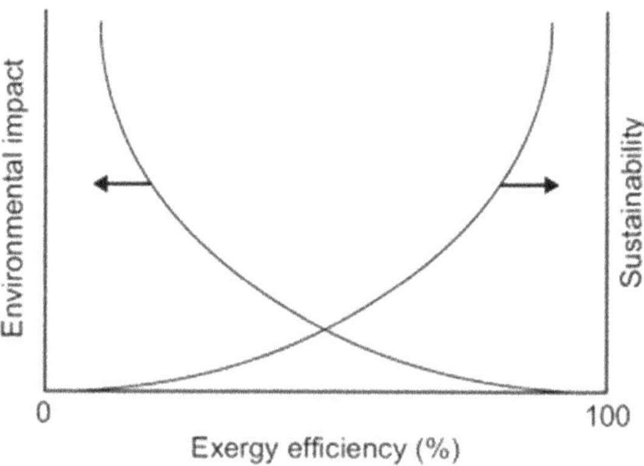

Figure 7. Intercourse between exergy efficiency, environmental impact, and sustainability of a thermal system. (Reprinted with permission from Ref. [177] with permission. Copyright © 2001 Elsevier.

From Figure 7, it is observed how, by increasing exergy efficiency of a process, there is a corresponding improvement in system sustainability and reduction in environmental impact [177]. As environmental impact approaches zero, exergy efficiency approaches 100% and, simultaneously, sustainability is promoted by virtue of the fact that the process approaches an ideal reversible process [36]. Furthermore, reduction in exergy efficiency towards 0% affects the sustainability negatively (also approaches zero). This is because the exergy-containing resources are being utilized but there is no meaningful outcome. In this same direction, environmental impact approaches infinity since for a provision of fixed service, an ever-increasing amount of resources must be consumed which leads to the creation of exergy-containing waste by the same magnitude [177]. In order to prevent wrong decisions while evaluating modification of thermal processes, it is important to consider the multidirectional analysis of the thermal systems. This will create an environment for accurately determining the quantity and quality of nanofuels used in ICE.

In the last five years, a limited number of studies have been conducted on nanofuels in diesel engines based on several versions of exergy analysis. Table 7 summarizes the main findings from these studies.

According to the reviewed literature, in general, per the second law of efficiency, results have shown that the thermal system performance and sustainability of low carbon fuels blended in diesel becomes worse compared to pure diesel. This could be attributed to relatively inferior properties of the low carbon fuels such as higher viscosity, cetane number, latent of vaporization, and lower calorific value, resulting in the engines' poor combustion characteristics. Nonetheless, low carbon fuels such as alcohols and biodiesels help reduce overdependence on fossil fuels for transport applications. Hence, their use is still ongoing and a major research hotspot. The existing studies have shown that the aforementioned situation significantly improves when the base fuel is modified with nanoparticles. The surface area, catalytic activity, and oxygen buffering of these NPs are very favourable for improving the ignition qualities of the base fuel, accelerating chemical reactions, promoting complete combustion, and enhancing the thermal properties. These events work together to enhance exergy efficiencies, reduce unaccounted thermal losses and entropy generation. Against this backdrop, other exergy indicators such as exergoeconomic, exergoenvironmental, and sustainability of the nanofuels become more optimal compared to the base fuels. Nanoparticle size and dosage also affect the performance and sustainability of thermal systems fuelled with nanofuels. The main challenge now has to do with the production of nanoparticles which is quite an expensive venture in today's market. However, as research continues to improve, the unit price of nano-additives could be greatly reduced in the near foreseeable future, further bolstering the feasibility and attractiveness of nanofuels from a technical, economical, and environmental point of view.

Table 7. Observations from exergy-based studies regarding low carbon fuels and nano-additives in diesel engines.

Base Fuel	Nanoparticles	Remarks				Ref.
		Exergy	Economic	Environmental	Sustainability	
Diesel–ethanol (D90E10)	Al_2O_3 and TiO_2 at 100 ppm	The exergy efficiency at all loads followed a decreasing trend in superiority in the order: $D90E10Al_2O_3$ > $D90E10TiO_2$ > D100 > D90E10. Clearly the addition of nanoparticles to diesel–ethanol blends improved the exergy. The presence of NPs increased the heating values of the fuels. In addition to this, the combustion efficiency and exergy efficiency of the fuels improved by virtue of the catalytic effect, micro-explosions, oxygen buffering, and large surface area-to-volume ratio of the NPs which causes chemical reactions to accelerate and provides excellent thermal properties.	The presence of NPs to D90E10 led to a reduction in fuel consumption and the specific exergy of the base fuel was increased. This led to a decrease in the fuel cost flow rate. At all loads, the cost of crankshaft work per unit energy is \$/GJ followed a decreasing trend in inferiority in the order: D90E10 > $D90E10TiO_2$ > $D90E10Al_2O_3$ > D100. The exergoeconomic analysis thus favoured the nanofuels compared to diesel–ethanol blend.	The presence of NPs ensured higher exergy efficiencies and this led to the production of nanofuels with relatively lower environmental impact. At all loads, the environmental impact rate pr unit of break power followed a decreasing trend in inferiority in the order: D90E10 > D100 > $D90E10TiO_2$ > $D90E10Al_2O_3$. Nanofuels have thus presented better exergoenvironmental feasibility compared to both pure diesel and diesel–ethanol blend.	The most sustainable test fuel according to their sustainability index was $D90E10Al_2O_3$. This is sequentially followed by $D90E10TiO_2$, D100, and D90E10.	[34]
Diesel–canola oil biodiesel (C10)	TiO_2 at 100 ppm and 3 different sizes (29 nm, 45 nm, and 200 nm)	The exergy loss and exergy destruction increase with increase in NP size. As NP size gets larger, there is a general reduction in surface area-to-volume ratio, catalytic activity while fuel consumption and exergy inlet rate increases. The aggregation of these events at larger NP sizes leads to a lower exergy efficiency. At all loads, the cumulative exergy efficiency followed a decreasing trend in superiority in the order: C10 + 28 nm TiO_2 (81.60%) > C10 + 45 nm TiO_2 (79.06%) > C10 + 200 nm TiO_2 (77.37%) > D100 (74.98%) > C10 (71.50%).	Similarly, the presence of NPs led to a superior thermoeconomic results in nanofuels compared to pure diesel and its blend with biodiesel. The NPs improve energy and exergy efficiencies and this produced optimal thermoeconomic results. The best thermoeconomic results was obtained at the smallest NP size. However, an opposite trend is observed for the unit cost and specific exergy cost. In this context, neat diesel and C10 had an economic advantage over their NP-doped counterparts. Reducing the grain size of the NPs led to the production of a worst fuel from an economic point of view. The heating value of the base fuel increases in the presence of the NPs, causing an increment in specific exergy cost for the nanofuels. Despite this trend, it is worth noting that per their advantage in exergy efficiencies, nanofuels showed beneficial and superior exergoeconomic results against the base fuel.	-	At all engine loads, the highest sustainability index of the diesel engine was recorded for C10 + 28 nm TiO_2 test fuel as a result of its superior exergy efficiencies in contrast to other test fuels. This is followed by C10 + 45 nm TiO_2 > C10 + 200 nm TiO_2 > D100 > C10 in a decreasing order of sustainability.	[39]

Table 7. Cont.

Base Fuel	Nanoparticles	Remarks				Ref.
		Exergy	Economic	Environmental	Sustainability	
Diesel–biodiesel (B5)	Al_2O_3 at 50 and 100 ppm	Averagely, the presence of Al_2O_3 increased the exergy efficiency by 7.28% compared to BF. Similarly, the Addition of the NP to the base fuel reduced unaccounted losses by 31.8% on an average. Additionally, there was a slight change in exergy loss to the cooling water when the NP was used. It is worth noting that, increase in NP dosage led to superior exergy efficiencies and entropy generation results. The high surface area of the Al_2O_3 NP led high ignition qualities–shortening the combustion time. Al_2O_3's high catalytic activity and surface area-to-volume ratio also ensure that the carbon activation temperature is lowered–leading to the promotion of fuel oxidation and complete combustion. Thermal properties thus increase and causes enhanced exergy efficiencies of the low carbon fuelled-diesel engines under the influence of the NPs.	-	-	-	[41]
Diesel–biodiesel (B5 and B10)	Hybrid nano catalysts additives comprising cerium oxide and molybdenum oxide on amide-functionalized MWCNTs at 30, 60, 90 ppm	The nano-additives provided sufficient oxygen to promote complete combustion and decrease the amount of exhaust air pollutants. The occurrence of these mechanisms in the cylinder by virtue of the inclusion of the nano-additives ensured that there is a decrease in the exergy rate of the exhaust gas and the heat transfer exergy rate of the diesel engine. Increasing the concentration of the nano-additives made this observation more obvious. Furthermore, the net exergy work rate of the diesel engine benefits from the presence of the nano-additives compared to the nano-additive-free blends. In addition to their oxygen buffering characteristics, the nano-additives exhibit nanocluster explosiveness which help the decomposition of sediments and deposits, and prevents their reformation. The absence of iron and carbon deposits reduces friction of the engine's movable parts. These factors contributed to an increase in engine power and causes the net exergy work rate of the engine to increase. The net exergy work is directly proportional to the exergy efficiency. Hence, the exergy efficiency of the diesel engine increases with increase in the amount of nano-additives.	-	-	-	[42]

Table 7. *Cont.*

Base Fuel	Nanoparticles	Remarks				Ref.
		Exergy	Economic	Environmental	Sustainability	
Diesel–biodiesel (B5) emulsified with water at concentrations of 3, 5, and 7 wt%. Tween 80 and Span 80 used as surfactants	Aqueous nano CeO_2 at 0 and 90 ppm	The presence of water decreases the exergy efficiency of pure B5, but the situation is greatly improved with the addition of Aqueous nano CeO_2. At all loads, B5 with 3 wt% water and 90 ppm of NP (B5W3$_m$) showed the best exergy efficiency amongst all test blended fuels. Similar findings are witnessed for the thermal efficiency.	Despite the excellent results for exergy efficiency, the exergoeconomic analysis revealed that pure diesel was more favourable than B5W5m.	-	Due to its high exergy efficiency, B5W3m in the test engine had the most favourable sustainability index among all test blended fuels.	[38,178]
Diesel–waste cooking oil biodiesel (D90B10)	Al_2O_3, TiO_2, SiO_2 at 100 ppm	The exergy efficiency of pure diesel degrades after the addition of biodiesel. The trend is significantly reversed with the inclusion of the NPs. D90B10Al$_2$O$_3$ recorded the highest exergy efficiency. This is followed by D90B10SiO$_2$ > D90B10TiO$_2$ > D100 > D90B10. Similarly, the lowest and highest exergy destruction was observed D90B10Al$_2$O$_3$ and D90B10, respectively. In addition, the crankshaft work followed an increasing trend of superiority in the order D90B10 < D100 < D90B10TiO$_2$ < D90B10SiO$_2$ < D90B10Al$_2$O$_3$.	Adding NPs led to a decrease in fuel consumption–hence, at all load conditions, the highest and lowest cost flow rate was recorded by D100 and D90B10SiO$_2$. In the same way, D90B10SiO$_2$ recorded the lowest exhaust cost flow rate and loss cost flow rate, closely followed by D90B10Al$_2$O$_3$. However, for cost flow rate of crankshaft work, D90B10Al$_2$O$_3$ was the most economical ahead of D90B10SiO$_2$. The exergo-economic factor for the nanofuels were superior than the base blend and pure diesel at all engine loads with D90B10SiO$_2$ being the highest of all.	-	At each engine load, the depletion number of the diesel engine followed a decreasing trend in the order D90B10 < D100 < D90B10TiO$_2$ < D90B10SiO$_2$ < D90B10Al$_2$O$_3$. In addition, the sustainability index of the nanofuels were better than the base blend and pure diesel at all engine loads with D90B10Al$_2$O$_3$ being the highest of all.	[40]

9. Toxicity and Health Impacts of Nanoparticles

One of the primary obstacles to broad commercial deployment of these nano-additives, as with many nanomaterials, is their potential toxicity and health effects. Nanotoxicology is concerned with the research and improved understanding of nanoparticle toxicity. Nanoparticles are significantly linked to toxicity, according to several in vivo and in vitro studies. Despite the advantages of nanoparticles, humans, animals, and plants have been exposed to their potential toxicity through different nanotechnology applications (see Figure 8).

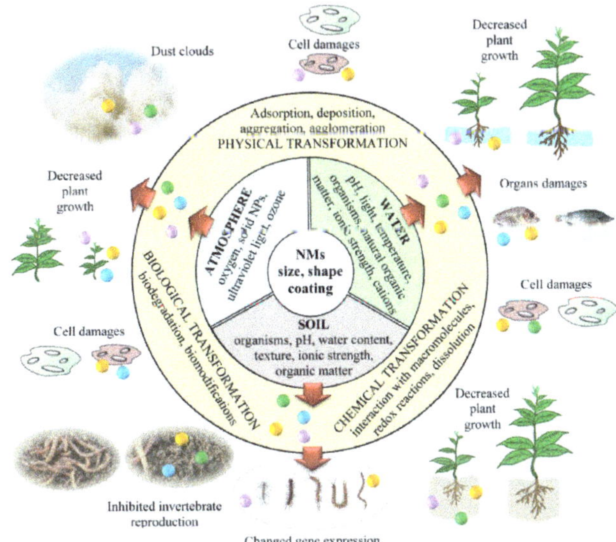

Figure 8. Toxicity and health impact of NPs. (Adapted with permission from Ref. [25]. Copyright © 2022 Elsevier.)

Inhalation, ingestion, skin absorption, injection, and implantation are among ways that nanoparticles can enter the human body [179]. Due to the small size of nanoparticles, their ease of penetration and biocompatibility, and their potential ability to breach the placental barrier, the widespread use of nanomaterials has raised concerns about their negative impact on human health, particularly on men's and women's reproductive systems, as well as fetal health. Early research on anthropogenic nanoparticles, such as diesel exhaust, shows that they aggregate and attach to human cells as a result of regular exposure, causing disruption to normal physiological processes. Moreover, nanoparticles have been linked to pulmonary injury, hepatotoxicity, immuno-nanotoxicity neurotoxicity, renal toxicity, and permanent testicular damage in animals [180]. Nanomaterials can clump together to form larger particles or longer fibre chains, altering their characteristics and potentially affecting their behaviour in both indoor and outdoor environments, as well as their potential exposure and entrance into the human body [181]. Due to large surface area, high surface activity, unique shape, tiny diameters, or decomposition into smaller particles after deposition, they might deposit in the respiratory system and exhibit nanostructure-influenced toxicity. If nanomaterial-derived particles display nanostructure-dependent biological activity, they may pose a danger. Nanoparticles have high deposition efficiency in healthy people's lungs, and much greater deposition efficiencies in those with asthma or chronic obstructive pulmonary disease [182,183]. When nanoparticles are breathed, they deposit dispersedly on the alveolar surface, causing a scattered chemo-attractant signal and lowering identification and alveolar macrophage responses. Karlsson et al. [44]

looked into the cytotoxicity and capacity to produce DNA damage and oxidative stress of various nanoparticles and nanotubes. Their research evaluated the toxicity of metal oxide nanoparticles (CuO, TiO_2, ZnO, $CuZnFe_2O_4$, Fe_3O_4, Fe_2O_3) to carbon nanoparticles and multi-walled carbon nanotubes (MWCNT). Cell viability and DNA damage were both affected by ZnO, but DNA damage was solely induced by TiO_2 particles (a combination of rutile and anatase). No or low toxicity was reported for iron oxide particles (Fe_3O_4, Fe_2O_3), while $CuZnFe_2O_4$ particles were rather effective in causing DNA damages. Finally, even at the lowest dose tested, carbon nanotubes were cytotoxic and caused DNA damage.

External nanoparticles comprising zinc and aluminum have recently been shown to have harmful effects on seedling germination and root growth in a range of plant species according to Lin and Xing [45] and Doshi et al. [43]. They found that when ZnO nanoparticles were exposed to the root surface, Ryegrass biomass was reduced substantially, root tips shrunk, and root epidermal and cortical cells were severely vacuolated or collapsed. Further, according to Soutter [48], diesel fuels enriched with cerium oxide nanoparticles have been observed to produce pulmonary consequences in exposed rats, including increased bronchial alveolar lavage fluid and lung inflammation. Long et al. [46] reported that titanium dioxide nanoparticles found in sunscreens could cause brain damage in mice. Nano size titanium dioxide stimulates reactive oxygen species in brain microglia and damages neurons in vitro [47]. Balasubramanyam et al. [184] reported that aluminum oxide nanoparticles (30–40 nm) contain genotoxic characteristics that are dosage-dependent. They used rat blood cells to test for genotoxicity using the comet assay and the micronucleus test. Another study employing a mouse lymphoma cell line found that aluminum oxide nanoparticles (50 nm) have genotoxic effects in the form of DNA damage without being mutagenic [185]. Titanium dioxide possesses some toxic health effects in experimental animals, including DNA damage as well as genotoxicity and lung inflammation [186,187]. Titanium dioxide nanoparticles (<100 nm) induce oxidative stress and form DNA adducts [188]. Titanium dioxide nanoparticles (5–200 nm) are harmful to immune function, liver, kidney, spleen, myocardium, hyperglycaemia, and lipid homeostasis in experimental animals, in addition to genotoxicity [189,190]. In vivo investigations have demonstrated that iron oxide nanoparticles stay in cell organelles (endosomes/lysosomes) after entering the cells, decompose in the cytoplasm, and contribute to cellular iron pool [191]. After inhalation, magnetic iron oxide nanoparticles were shown to collect in the liver, spleen, lungs, and brain. Murine macrophage cells, human macrophages, human hepatocellular carcinoma cells, and rat mesenchymal stem cells were all found to be at risk after exposure to the nanoparticles made from iron oxide. On murine macrophage cells, iron oxide nanoparticles were found to be lethal at concentrations of 25–200 g/mL after a 2-h exposure. Their study also reported consequences such as a reduction in cell viability [192]. On human brain microvascular endothelial cells (HBMVECs), aluminum nanoparticles in the size range of 1–10 μM were utilized for 24 h. Treatment led in a decrease in mitochondrial activity, cell viability, and an increase in oxidative stress, according to the research of Chen et al. [193]. The effect of MWCNTs was also evaluated in a study conducted on rat [194]. Rats were exposed to MWCNTs intratracheally in this study. Both histologically and biochemically, the researchers looked at inflammation, lung persistence, and fibrotic responses. The bronchial lumen was found to have pulmonary lesions, which were characterized by collagen-rich granulomas. Cha and Myung [195] tested the cytotoxicity of zinc, iron, and silicon at various doses against cell lines from the liver (Huh7), brain (A-172), stomach (MKN-1), lung (A-549), and kidney (HEK293). The decrease in DNA content, as well as mitochondrial activity, was easily detected in brain and liver cells. In a research of zebrafish embryos (*Danio rerio*), Asharani et al. [196] discovered that uncoated silver nanoparticles caused higher genotoxicity because they were able to reach the nucleus cells, causing DNA strands to break.

It is evident from the reviewed works that almost all nanoparticles are closely associated with toxicity and have shown to have detrimental health impacts. Over-exposure to nanoparticles has been proved to cause DNA and reproductive damage, cytotoxicity, and even cancer. There are currently ongoing studies aimed at providing better nanotoxicity evaluation and measures to minimize nanotoxicity levels in the environment. Other areas of research are still being conducted to mitigate the dangers posed by metal nanoparticles in the production of biodegradable and biocompatible nanoparticles. The current state-of-the art is thus on the development of nanoparticles that interact better with the environment and have less harmful effects.

It is worth mentioning that studies have also made conscious attempts to reduce the toxicity of nanoparticles. Some of the approaches include degradable nanoparticles, next generation lipids, surface coating, doping, and alteration of surface properties [197]. Doping of nanoparticles with dopants such as aluminum titanium and iron has been found to decrease nanoparticle dissolution and cause a reduction in toxic ions released, and this would cause an alteration to the reactive surfaces leading to a decrease in reactive oxygen species generation [197,198]. On the other hand, surface coating is an approach for modifying or diminishing the adverse effects associated with nanomaterials. It includes modifying properties such as stability of nanoparticles, agglomeration and arrest dissolution and discharge of noxious ions [199]. Cai et al. [200] have reported that ethylenediamine tetra coating could passivate the surface of metal oxides, thereby reducing their toxicity and pulmonary hazard effect. Methods focusing on altering properties of nanoparticles to reduce their toxicity also include alteration of surface charge, aggregation characteristics and/or hydrodynamic diameter of nanoparticles [201]. There are, however, ongoing studies to improve the efficacy of the abovementioned methods.

10. Conclusions and Future Research Direction

As automotive industries continue to look for more efficient combustion of liquid fuels coupled with the existence of stringent environmental regulations, nanoparticles as fuel additives for combustion in diesel engines have become an important research field in recent years. Several studies have been conducted to review the effect of nanoparticles on the performance, emission, and combustion characteristics of liquid fuels. However, these studies, to a large extent, have primarily focused on biodiesel, whereas those on vegetable oils and alcohols remain scarce. In our quest to bridge the existing gap in the literature, the current study was set out to simultaneously and holistically review experimental results related to all three biofuels (alcohols, biodiesels, vegetable oils) in the context of the effect nanoparticles may have on their fuel properties, performance, emission, and combustion characteristics when operating in a diesel engine. Another novelty presented in this work relates to the evolutionary trends, research hotspots, and key contributors of this research field from 2000 to 2021. Of the three biofuels reviewed, biodiesels have been the most investigated on how they perform in diesel engines under the influence of nanoparticles. Earlier research focused extensively on carbon nanotubes, but the recent trend shows a shift towards cerium dioxide, titanium dioxide, and, mainly, aluminum dioxide. The key contributors to this field originate from Asia, largely represented by India, China, and Iran. It became apparent that the key interest of this research field hinges on the effect of nanoparticles on the performance, emission, and combustion characteristics of alcohol/biodiesel/vegetable oil-based fuels.

Nanoparticles can positively impact key physical properties of the base fuels such as density, kinematic viscosity, cetane number, flash point, calorific value, etc., but the extent of the impact will greatly depend on the type and size of the nanoparticle, type of base fuel, and concentration of blends. For performance characteristics, it is evident that most of the studies carried out with the addition of nano-particles into prospective renewable additives (such as biodiesel, vegetable-based oil, and alcohol) showed a significant reduction in BSFC, while BTE tended to increase. These were attributed to the improved fuel properties, excess oxygen content, better atomization, high thermal conductivity, and good catalytic activity of nanoparticles in renewable additives. For combustion characteristics, the heat release rate and in-cylinder pressure can either decrease or increase as investigated. These inconsistencies occur due to many factors: (i) when the ignition delay increases due to higher viscosity, the peak heat release rate and in-cylinder pressure increase, leading to high fuel droplets in the premixed combustion phase, and (ii) when the flame temperature inside the combustion increases due to low viscosity and ignition delay, the in-cylinder pressure, and thermal efficiency increase, promoting soot particles' oxidation rate. However, most of the in-cylinder pressure and heat release rates significantly improved with nano-additives/renewable fuels in relation to control conditions. In most of the cases, the addition of nanoparticles showed a slight reduction in NOx due to an increase in the cooling effect of nanofluids with a substantial reduction in CO and HC emissions compared to diesel fuel. Various investigators attributed the role of nanoparticles in fuels to their oxygen buffering, the higher surface-to-volume ratio, micro-explosive property, thermal conductivity, evaporation rate, and oxidation catalysis. Based on these characteristics, there is efficient and more complete combustion of fuel to the positive impact the performance, emission, and combustion characteristics.

Beyond energy-based indicators, the exergy, economic, environmental, and sustainability aspects of the blends in diesel engines were discussed. It is observed that the performance of the diesel engine fuelled with low carbon fuels, according to the second law of efficiency, improves under the influence of the nano-additives. By virtue of their oxygen buffering, higher surface-to-volume ratio, micro-explosive property, thermal conductivity, evaporation rate, and oxidation catalysis, nanoparticles in low carbon fuels lead to high combustion efficiency and accelerated chemical reactions, which result in improved exergy efficiencies. In return, the exergoeconomic, exergoenvironmental, and sustainability aspects of these nanofuels are superior compared to the base fuels.

Some key recommendations and future perspectives are provided as follows. Contributions from some parts of Asia, South America, Africa, and Oceania are very underrepresented, and the most active researchers could attempt collaborative works with authors from these continents for more ground-breaking discoveries and development of this research field. Future studies should devote more attention to alcohols (especially > C2 alcohols) and vegetable oils. Researchers can also experiment on hybrid nanoparticles by blending multiple nanoparticles and studying how it affects engine characteristics. There is still more work that needs to be undertaken in the area of different nanoparticles in the same base fuel. The optimum concentration and size of the nanoparticles together with the base fuels for an efficient combustion and reduced emissions should be studied. Since certain nanoparticles are very surface reactive, long-term studies of the engine or engine exhaust resistivity are necessary. More studies are needed on the exergy, exergo-economic, exergo-environmental, and sustainability aspects of nanofuels in ICE to complement the highly existing energy-based studies. Furthermore, because some researchers have identified the cost of nanoparticles as an inherent problem, future research can also look at finding an optimum balance between the performance and cost of nanoparticles to increase their feasibility and wide use. Metal nanoparticles represent a threat to human health, and biodegradable and biocompatible nanoparticles may help to minimize this risk. As a result, the focus is on developing nanoparticles that interact better with the environment and have fewer negative consequences. It should be noted that nanoparticles are not directly released into the atmosphere. They are mixed as an additive in the fuel and go through a complex combustion process. The effect of the nanoparticle additives on the atmosphere after being combusted in the engine needs to be researched.

Author Contributions: Conceptualization, J.D.A. and A.A.Y.; methodology, C.J.; software, S.A.; validation, S.A., E.B.A., S.K., M.S. and M.H.; formal analysis, J.D.A. and A.A.Y.; investigation, J.D.A. and A.A.Y.; resources, E.B.A., S.K., M.S. and M.H.; data curation, E.B.A., S.K., M.S. and M.H.; writing—original draft preparation, J.D.A., A.A.Y. and I.M.R.F., writing—review and editing, C.J., H.L., M.S., M.H. and P.L.S.; visualization, S.A.; supervision, C.J., H.L. and I.M.R.F.; funding acquisition, C.J. All authors have read and agreed to the published version of the manuscript.

Funding: This research was funded by Tianjin Science Fund for Distinguished Young Scholars, grant number 20JCJQJC00160, and the Universiti Tenaga Nasional grant no. IC6-BOLDREFRESH2025 (HCR) under the BOLD2025 Program. The APC was funded by Cardiff University.

Institutional Review Board Statement: Not applicable.

Informed Consent Statement: Not applicable.

Data Availability Statement: The data used in the bibliometric section of this review are available on the Web of Science Collection Database. All data used to support the findings of the remaining sections are included within the review.

Conflicts of Interest: The authors declare no conflict of interest. The funders had no role in the design of the study; in the collection, analyses, or interpretation of data; in the writing of the manuscript, or in the decision to publish the results.

References

1. Joshi, G.; Pandey, J.K.; Rana, S.; Rawat, D.S. Challenges and opportunities for the application of biofuel. *Renew. Sustain. Energy Rev.* **2017**, *79*, 850–866. [CrossRef]
2. Atabani, A.E.; Silitonga, A.S.; Badruddin, I.A.; Mahlia, T.M.I.; Masjuki, H.H.; Mekhilef, S. A comprehensive review on biodiesel as an alternative energy resource and its characteristics. *Renew. Sustain. Energy Rev.* **2012**, *16*, 2070–2093. [CrossRef]
3. Garg, P. Energy Scenario and Vision 2020 in India. *J. Sustain. Energy Environ.* **2012**, *3*, 7–17.
4. EIA. *International Energy Outlook 2010.* Available online: http://www.eia.doe.gov/oiaf/ieo/pdf/0484%282010%29.pdf (accessed on 18 August 2021).
5. EIA. Annual Energy Review (AER), International Energy. Available online: http://www.eia.gov/emeu/aer/inter.html (accessed on 18 August 2021).
6. Kumar, N. Oxidative stability of biodiesel: Causes, effects and prevention. *Fuel* **2017**, *190*, 328–350. [CrossRef]
7. Rawat, D.S.; Joshi, G.; Lamba, B.Y.; Tiwari, A.K.; Mallick, S. Impact of additives on storage stability of *Karanja (Pongamia Pinnata)* biodiesel blends with conventional diesel sold at retail outlets. *Fuel* **2014**, *120*, 30–37. [CrossRef]
8. Peer, M.S.; Kasimani, R.; Rajamohan, S.; Ramakrishnan, P. Experimental evaluation on oxidation stability of biodiesel/diesel blends with alcohol addition by rancimat instrument and FTIR spectroscopy. *J. Mech. Sci. Technol.* **2017**, *31*, 455–463. [CrossRef]
9. Shahir, S.A.; Masjuki, H.H.; Kalam, M.A.; Imran, A.; Ashraful, A.M. Performance and emission assessment of diesel–biodiesel–ethanol/bioethanol blend as a fuel in diesel engines: A review. *Renew. Sustain. Energy Rev.* **2015**, *48*, 62–78. [CrossRef]
10. Balat, M. Modeling Vegetable Oil Viscosity. *Energy Sources Part A Recovery Util. Environ. Eff.* **2008**, *30*, 1856–1869. [CrossRef]
11. Atabani, A.E.; Silitonga, A.S.; Ong, H.C.; Mahlia, T.M.I.; Masjuki, H.H.; Badruddin, I.A.; Fayaz, H. Non-edible vegetable oils: A critical evaluation of oil extraction, fatty acid compositions, biodiesel production, characteristics, engine performance and emissions production. *Renew. Sustain. Energy Rev.* **2013**, *18*, 211–245. [CrossRef]
12. Kuszewski, H. Experimental investigation of the effect of ambient gas temperature on the autoignition properties of ethanol–diesel fuel blends. *Fuel* **2018**, *214*, 26–38. [CrossRef]
13. Abu-Qudais, M.; Haddad, O.; Qudaisat, M. The effect of alcohol fumigation on diesel engine performance and emissions. *Energy Convers. Manag.* **2000**, *41*, 389–399. [CrossRef]
14. Ampah, J.D.; Liu, X.; Sun, X.; Pan, X.; Xu, L.; Jin, C.; Sun, T.; Geng, Z.; Afrane, S.; Liu, H. Study on characteristics of marine heavy fuel oil and low carbon alcohol blended fuels at different temperatures. *Fuel* **2022**, *310*, 122307. [CrossRef]
15. Jin, C.; Liu, X.; Sun, T.; Ampah, J.D.; Geng, Z.; Ikram, M.; Ji, J.; Wang, G.; Liu, H. Preparation of ethanol and palm oil/palm kernel oil alternative biofuels based on property improvement and particle size analysis. *Fuel* **2021**, *305*, 121569. [CrossRef]
16. Jin, C.; Zhang, X.; Han, W.; Geng, Z.; Tessa Margaret Thomas, M.; Ampah, J.D.; Wang, G.; Ji, J.; Liu, H. Macro and micro solubility between low carbon alcohols and rapeseed oil using different co-solvents. *Fuel* **2020**, *270*, 117511. [CrossRef]
17. Zhang, Z.; Liu, X.; Liu, H.; Wu, Y.; Zaman, M.; Geng, Z.; Jin, C.; Zheng, Z.; Yue, Z.; Yao, M. Effect of soybean oil/PODE/ethanol blends on combustion and emissions on a heavy-duty diesel engine. *Fuel* **2021**, *288*, 119625. [CrossRef]
18. Kumar, S.; Dinesha, P.; Bran, I. Experimental investigation of the effects of nanoparticles as an additive in diesel and biodiesel fuelled engines: A review. *Biofuels* **2019**, *10*, 615–622. [CrossRef]
19. Singh, G.; Sharma, S. Performance, combustion and emission characteristics of compression ignition engine using nano-fuel: A review. *Int. J. Eng. Sci. Res. Technol.* **2015**, *4*, 1034–1039.
20. Khond, V.W.; Kriplani, V.M. Effect of nanofluid additives on performances and emissions of emulsified diesel and biodiesel fueled stationary CI engine: A comprehensive review. *Renew. Sustain. Energy Rev.* **2016**, *59*, 1338–1348. [CrossRef]
21. Saxena, V.; Kumar, N.; Saxena, V.K. A comprehensive review on combustion and stability aspects of metal nanoparticles and its additive effect on diesel and biodiesel fuelled C.I. engine. *Renew. Sustain. Energy Rev.* **2017**, *70*, 563–588. [CrossRef]
22. Razzaq, L.; Mujtaba, M.A.; Soudagar, M.E.M.; Ahmed, W.; Fayaz, H.; Bashir, S.; Fattah, I.M.R.; Ong, H.C.; Shahapurkar, K.; Afzal, A.; et al. Engine performance and emission characteristics of palm biodiesel blends with graphene oxide nanoplatelets and dimethyl carbonate additives. *J. Environ. Manag.* **2021**, *282*, 111917. [CrossRef]
23. Hussain, F.; Soudagar, M.E.M.; Afzal, A.; Mujtaba, M.A.; Fattah, I.M.R.; Naik, B.; Mulla, M.H.; Badruddin, I.A.; Khan, T.M.Y.; Raju, V.D.; et al. Enhancement in Combustion, Performance, and Emission Characteristics of a Diesel Engine Fueled with Ce-ZnO Nanoparticle Additive Added to Soybean Biodiesel Blends. *Energies* **2020**, *13*, 4578. [CrossRef]

24. Ribeiro, N.M.; Pinto, A.C.; Quintella, C.M.; da Rocha, G.O.; Teixeira, L.S.G.; Guarieiro, L.L.N.; do Carmo Rangel, M.; Veloso, M.C.C.; Rezende, M.J.C.; Serpa da Cruz, R.; et al. The Role of Additives for Diesel and Diesel Blended (Ethanol or Biodiesel) Fuels: A Review. *Energy Fuels* **2007**, *21*, 2433–2445. [CrossRef]

25. Kegl, T.; Kovač Kralj, A.; Kegl, B.; Kegl, M. Nanomaterials as fuel additives in diesel engines: A review of current state, opportunities, and challenges. *Prog. Energy Combust. Sci.* **2021**, *83*, 100897. [CrossRef]

26. Venkatesan, H.; Sivamani, S.; Sampath, S.; Gopi, V.; Kumar M, D. A Comprehensive Review on the Effect of Nano Metallic Additives on Fuel Properties, Engine Performance and Emission Characteristics. *Int. J. Renew. Energy Res.* **2017**, *7*, 825–843.

27. Hoang, A.T. Combustion behavior, performance and emission characteristics of diesel engine fuelled with biodiesel containing cerium oxide nanoparticles: A review. *Fuel Process. Technol.* **2021**, *218*, 106840. [CrossRef]

28. Shaafi, T.; Sairam, K.; Gopinath, A.; Kumaresan, G.; Velraj, R. Effect of dispersion of various nanoadditives on the performance and emission characteristics of a CI engine fuelled with diesel, biodiesel and blends—A review. *Renew. Sustain. Energy Rev.* **2015**, *49*, 563–573. [CrossRef]

29. Dewangan, A.; Mallick, A.; Yadav, A.K.; Richhariya, A.K. Effect of metal oxide nanoparticles and engine parameters on the performance of a diesel engine: A review. *Mater. Today Proc.* **2020**, *21*, 1722–1727. [CrossRef]

30. Nanthagopal, K.; Kishna, R.S.; Atabani, A.E.; Al-Muhtaseb, A.a.H.; Kumar, G.; Ashok, B. A compressive review on the effects of alcohols and nanoparticles as an oxygenated enhancer in compression ignition engine. *Energy Convers. Manag.* **2020**, *203*, 112244. [CrossRef]

31. Soudagar, M.E.M.; Nik-Ghazali, N.-N.; Abul Kalam, M.; Badruddin, I.A.; Banapurmath, N.R.; Akram, N. The effect of nano-additives in diesel-biodiesel fuel blends: A comprehensive review on stability, engine performance and emission characteristics. *Energy Convers. Manag.* **2018**, *178*, 146–177. [CrossRef]

32. Goyal, S.; Chauhan, S.; Mishra, P. Circular economy research: A bibliometric analysis (2000–2019) and future research insights. *J. Clean. Prod.* **2021**, *287*, 125011. [CrossRef]

33. Meyer, T. Decarbonizing road freight transportation–A bibliometric and network analysis. *Transp. Res. Part D Transp. Environ.* **2020**, *89*, 102619. [CrossRef]

34. Ağbulut, Ü.; Uysal, C.; Cavalcanti, E.J.C.; Carvalho, M.; Karagöz, M.; Saridemir, S. Exergy, exergoeconomic, life cycle, and exergoenvironmental assessments for an engine fueled by diesel–ethanol blends with aluminum oxide and titanium dioxide additive nanoparticles. *Fuel* **2022**, *320*, 123861. [CrossRef]

35. Aghbashlo, M.; Tabatabaei, M.; Mohammadi, P.; Khoshnevisan, B.; Rajaeifar, M.A.; Pakzad, M. Neat diesel beats waste-oriented biodiesel from the exergoeconomic and exergoenvironmental point of views. *Energy Convers. Manag.* **2017**, *148*, 1–15. [CrossRef]

36. Aghbashlo, M.; Tabatabaei, M.; Mohammadi, P.; Mirzajanzadeh, M.; Ardjmand, M.; Rashidi, A. Effect of an emission-reducing soluble hybrid nanocatalyst in diesel/biodiesel blends on exergetic performance of a DI diesel engine. *Renew. Energy* **2016**, *93*, 353–368. [CrossRef]

37. Mojarab Soufiyan, M.; Aghbashlo, M.; Mobli, H. Exergetic performance assessment of a long-life milk processing plant: A comprehensive survey. *J. Clean. Prod.* **2017**, *140*, 590–607. [CrossRef]

38. Aghbashlo, M.; Tabatabaei, M.; Khalife, E.; Najafi, B.; Mirsalim, S.M.; Gharehghani, A.; Mohammadi, P.; Dadak, A.; Roodbar Shojaei, T.; Khounani, Z. A novel emulsion fuel containing aqueous nano cerium oxide additive in diesel–biodiesel blends to improve diesel engines performance and reduce exhaust emissions: Part II—Exergetic analysis. *Fuel* **2017**, *205*, 262–271. [CrossRef]

39. Ağbulut, Ü. Understanding the role of nanoparticle size on energy, exergy, thermoeconomic, exergoeconomic, and sustainability analyses of an IC engine: A thermodynamic approach. *Fuel Process. Technol.* **2022**, *225*, 107060. [CrossRef]

40. Karagoz, M.; Uysal, C.; Agbulut, U.; Saridemir, S. Exergetic and exergoeconomic analyses of a CI engine fueled with diesel-biodiesel blends containing various metal-oxide nanoparticles. *Energy* **2021**, *214*, 118830. [CrossRef]

41. Özcan, H. Energy and exergy analyses of Al_2O_3-diesel-biodiesel blends in a diesel engine. *Int. J. Exergy* **2019**, *28*, 29–45. [CrossRef]

42. Shadidi, B.; Alizade, H.H.A.; Najafi, G. Performance and exergy analysis of a diesel engine run on petrodiesel and biodiesel blends containing mixed CeO_2 and MoO_3 nanocatalyst. *Biofuels* **2022**, *13*, 1–7. [CrossRef]

43. Doshi, R.; Braida, W.; Christodoulatos, C.; Wazne, M.; O'Connor, G. Nano-aluminum: Transport through sand columns and environmental effects on plants and soil communities. *Environ. Res.* **2008**, *106*, 296–303. [CrossRef]

44. Karlsson, H.L.; Cronholm, P.; Gustafsson, J.; Möller, L. Copper Oxide Nanoparticles Are Highly Toxic: A Comparison between Metal Oxide Nanoparticles and Carbon Nanotubes. *Chem. Res. Toxicol.* **2008**, *21*, 1726–1732. [CrossRef] [PubMed]

45. Lin, D.; Xing, B. Phytotoxicity of nanoparticles: Inhibition of seed germination and root growth. *Environ. Pollut.* **2007**, *150*, 243–250. [CrossRef]

46. Long, T.C.; Saleh, N.; Tilton, R.D.; Lowry, G.V.; Veronesi, B. Titanium dioxide (P25) produces reactive oxygen species in immortalized brain microglia (BV2): Implications for nanoparticle neurotoxicity. *Environ. Sci Technol.* **2006**, *40*, 4346–4352. [CrossRef]

47. Long, T.C.; Tajuba, J.; Sama, P.; Saleh, N.; Swartz, C.; Parker, J.; Hester, S.; Lowry, G.V.; Veronesi, B. Nanosize titanium dioxide stimulates reactive oxygen species in brain microglia and damages neurons in vitro. *Environ. Health Perspect.* **2007**, *115*, 1631–1637. [CrossRef]

48. Soutter, W. Nanoparticles as Fuel Additives. Available online: https://www.azonano.com/article.aspx?ArticleID=3085 (accessed on 25 August 2021).

49. World Bank. World Bank Open Data. Available online: https://data.worldbank.org/ (accessed on 7 May 2020).

50. British Petroleum(BP). BP Energy Outlook. Available online: https://www.bp.com/content/dam/bp/business-sites/en/global/corporate/pdfs/energy-economics/energy-outlook/bp-energy-outlook-2019.pdf (accessed on 25 September 2021).

51. International Energy Agency (IEA). Tracking Transport International Energy Agency and Organization for Economic Cooperation and Development, Paris. Available online: https://www.iea.org/reports/tracking-transport-2019 (accessed on 25 September 2021).

52. Umar, M.; Ji, X.; Kirikkaleli, D.; Xu, Q. COP21 Roadmap: Do innovation, financial development, and transportation infrastructure matter for environmental sustainability in China? *J. Environ. Manag.* **2020**, *271*, 111026. [CrossRef]

53. Umar, M.; Ji, X.; Mirza, N.; Naqvi, B. Carbon neutrality, bank lending, and credit risk: Evidence from the Eurozone. *J. Environ. Manag.* **2021**, *296*, 113156. [CrossRef]

54. Jain, A.; Sarsaiya, S.; Kumar Awasthi, M.; Singh, R.; Rajput, R.; Mishra, U.C.; Chen, J.; Shi, J. Bioenergy and bio-products from bio-waste and its associated modern circular economy: Current research trends, challenges, and future outlooks. *Fuel* **2022**, *307*, 121859. [CrossRef]

55. KAPSARC. Guide to the Circular Carbon Economy. Available online: https://www.cceguide.org/guide/ (accessed on 3 December 2021).

56. Pritchard, A. Statistical bibliography or bibliometrics. *J. Doc.* **1969**, *25*, 348–349.

57. Schneider, J.W.; Borlund, P. Introduction to bibliometrics for construction and maintenance of thesauri. *J. Doc.* **2004**, *60*, 524–549. [CrossRef]

58. Aria, M.; Cuccurullo, C. bibliometrix: An R-tool for comprehensive science mapping analysis. *J. Informetr.* **2017**, *11*, 959–975. [CrossRef]

59. Mao, G.; Liu, X.; Du, H.; Zuo, J.; Wang, L. Way forward for alternative energy research: A bibliometric analysis during 1994–2013. *Renew. Sustain. Energy Rev.* **2015**, *48*, 276–286. [CrossRef]

60. Min, Z.; Hao, M. The Scientometric Evaluation on the Research of Biofuels Based on CiteSpace. *Adv. Eng. Res.* **2019**, *184*, 6–11.

61. Jin, C.; Ampah, J.D.; Afrane, S.; Yin, Z.; Liu, X.; Sun, T.; Geng, Z.; Ikram, M.; Liu, H. Low-carbon alcohol fuels for decarbonizing the road transportation industry: A bibliometric analysis 2000–2021. *Environ. Sci. Pollut. Res.* **2021**, *29*, 5577–5604. [CrossRef]

62. Zhang, M.; Gao, Z.; Zheng, T.; Ma, Y.; Wang, Q.; Gao, M.; Sun, X.J.J.o.M.C.; Management, W. A bibliometric analysis of biodiesel research during 1991–2015. *J. Mater. Cycles Waste Manag.* **2018**, *20*, 10–18. [CrossRef]

63. Hosseini, S.H.; Taghizadeh-Alisaraei, A.; Ghobadian, B.; Abbaszadeh-Mayvan, A. Performance and emission characteristics of a CI engine fuelled with carbon nanotubes and diesel-biodiesel blends. *Renew. Energy* **2017**, *111*, 201–213. [CrossRef]

64. Sadhik Basha, J.; Anand, R.B. Performance, emission and combustion characteristics of a diesel engine using Carbon Nanotubes blended Jatropha Methyl Ester Emulsions. *Alex. Eng. J.* **2014**, *53*, 259–273. [CrossRef]

65. Heydari-Maleney, K.; Taghizadeh-Alisaraei, A.; Ghobadian, B.; Abbaszadeh-Mayvan, A. Analyzing and evaluation of carbon nanotubes additives to diesohol-B2 fuels on performance and emission of diesel engines. *Fuel* **2017**, *196*, 110–123. [CrossRef]

66. Cobo, M.J.; López-Herrera, A.G.; Herrera-Viedma, E.; Herrera, F. An approach for detecting, quantifying, and visualizing the evolution of a research field: A practical application to the Fuzzy Sets Theory field. *J. Informetr.* **2011**, *5*, 146–166. [CrossRef]

67. LCPTi. Key to a Sustainable Economy-Low Carbon Transport Fuels. Available online: http://docs.wbcsd.org/2015/11/LCTPi-LCTF-FinalReport.pdf (accessed on 20 April 2022).

68. El-Seesy, A.I.; Kosaka, H.; Hassan, H.; Sato, S. Combustion and emission characteristics of a common rail diesel engine and RCEM fueled by n-heptanol-diesel blends and carbon nanomaterial additives. *Energy Convers. Manag.* **2019**, *196*, 370–394. [CrossRef]

69. Nutakki, P.K.; Gugulothu, S.K.; Ramachander, J.; Sivasurya, M. Effect Of n-amyl alcohol/biodiesel blended nano additives on the performance, combustion and emission characteristics of CRDi diesel engine. *Environ. Sci. Pollut. Res.* **2021**, *29*, 82–97. [CrossRef]

70. Ağbulut, Ü.; Polat, F.; Sarıdemir, S. A comprehensive study on the influences of different types of nano-sized particles usage in diesel-bioethanol blends on combustion, performance, and environmental aspects. *Energy* **2021**, *229*, 120548. [CrossRef]

71. Venu, H.; Raju, V.D.; Lingesan, S.; Elahi M Soudagar, M. Influence of Al_2O_3 nano additives in ternary fuel (diesel-biodiesel-ethanol) blends operated in a single cylinder diesel engine: Performance, combustion and emission characteristics. *Energy* **2021**, *215*, 119091. [CrossRef]

72. El-Seesy, A.I.; Hassan, H. Combustion Characteristics of a Diesel Engine Fueled by Biodiesel-Diesel-N-Butanol Blend and Titanium Oxide Additives. *Energy Procedia* **2019**, *162*, 48–56. [CrossRef]

73. Prabakaran, B.; Udhoji, A. Experimental investigation into effects of addition of zinc oxide on performance, combustion and emission characteristics of diesel-biodiesel-ethanol blends in CI engine. *Alex. Eng. J.* **2016**, *55*, 3355–3362. [CrossRef]

74. Wei, J.; He, C.; Lv, G.; Zhuang, Y.; Qian, Y.; Pan, S. The combustion, performance and emissions investigation of a dual-fuel diesel engine using silicon dioxide nanoparticle additives to methanol. *Energy* **2021**, *230*, 120734. [CrossRef]

75. Pan, S.; Wei, J.; Tao, C.; Lv, G.; Qian, Y.; Liu, Q.; Han, W. Discussion on the combustion, performance and emissions of a dual fuel diesel engine fuelled with methanol-based CeO_2 nanofluids. *Fuel* **2021**, *302*, 121096. [CrossRef]

76. Ramachander, J.; Gugulothu, S.K.; Sastry, G.R.K. Performance and Emission Reduction Characteristics of Metal Based SiO_2 Nanoparticle Additives Blended with Ternary Fuel (Diesel-MME-Pentanol) on CRDI Diesel Engine. *Silicon* **2022**, *14*, 2249–2263. [CrossRef]

77. El-Seesy, A.I.; Hassan, H. Investigation of the effect of adding graphene oxide, graphene nanoplatelet, and multiwalled carbon nanotube additives with n-butanol-Jatropha methyl ester on a diesel engine performance. *Renew. Energy* **2019**, *132*, 558–574. [CrossRef]

78. Heidari-Maleni, A.; Mesri Gundoshmian, T.; Jahanbakhshi, A.; Ghobadian, B. Performance improvement and exhaust emissions reduction in diesel engine through the use of graphene quantum dot (GQD) nanoparticles and ethanol-biodiesel blends. *Fuel* **2020**, *267*, 117116. [CrossRef]
79. Shaafi, T.; Velraj, R. Influence of alumina nanoparticles, ethanol and isopropanol blend as additive with diesel–soybean biodiesel blend fuel: Combustion, engine performance and emissions. *Renew. Energy* **2015**, *80*, 655–663. [CrossRef]
80. Venu, H.; Madhavan, V. Effect of Al_2O_3 nanoparticles in biodiesel-diesel-ethanol blends at various injection strategies: Performance, combustion and emission characteristics. *Fuel* **2016**, *186*, 176–189. [CrossRef]
81. Örs, I.; Sarıkoç, S.; Atabani, A.E.; Ünalan, S.; Akansu, S.O. The effects on performance, combustion and emission characteristics of DICI engine fuelled with TiO2 nanoparticles addition in diesel/biodiesel/n-butanol blends. *Fuel* **2018**, *234*, 177–188. [CrossRef]
82. Annamalai, M.; Dhinesh, B.; Nanthagopal, K.; SivaramaKrishnan, P.; Isaac JoshuaRamesh Lalvani, J.; Parthasarathy, M.; Annamalai, K. An assessment on performance, combustion and emission behavior of a diesel engine powered by ceria nanoparticle blended emulsified biofuel. *Energy Convers. Manag.* **2016**, *123*, 372–380. [CrossRef]
83. Dhinesh, B.; Niruban Bharathi, R.; Isaac JoshuaRamesh Lalvani, J.; Parthasarathy, M.; Annamalai, K. An experimental analysis on the influence of fuel borne additives on the single cylinder diesel engine powered by Cymbopogon flexuosus biofuel. *J. Energy Inst.* **2017**, *90*, 634–645. [CrossRef]
84. Kumaravel, S.T.; Murugesan, A.; Vijayakumar, C.; Thenmozhi, M. Enhancing the fuel properties of tyre oil diesel blends by doping nano additives for green environments. *J. Clean. Prod.* **2019**, *240*, 118128. [CrossRef]
85. Sathiyamoorthi, R.; Sankaranarayanan, G.; Pitchandi, K. Combined effect of nanoemulsion and EGR on combustion and emission characteristics of neat lemongrass oil (LGO)-DEE-diesel blend fuelled diesel engine. *Appl. Therm. Eng.* **2017**, *112*, 1421–1432. [CrossRef]
86. Sheriff, S.A.; Kumar, I.K.; Mandhatha, P.S.; Jambal, S.S.; Sellappan, R.; Ashok, B.; Nanthagopal, K. Emission reduction in CI engine using biofuel reformulation strategies through nano additives for atmospheric air quality improvement. *Renew. Energy* **2020**, *147*, 2295–2308. [CrossRef]
87. Dhinesh, B.; Annamalai, M. A study on performance, combustion and emission behaviour of diesel engine powered by novel nano nerium oleander biofuel. *J. Clean. Prod.* **2018**, *196*, 74–83. [CrossRef]
88. Chinnasamy, C.; Tamilselvam, P.; Ranjith, R. Environmental effects of adding aluminum oxide nano-additives on diesel engine fueled with pyrolyzed biomass oil–diesel blends. *Energy Sources Part A Recovery Util. Environ. Eff.* **2018**, *40*, 2735–2744. [CrossRef]
89. Dobrzyńska, E.; Szewczyńska, M.; Pośniak, M.; Szczotka, A.; Puchałka, B.; Woodburn, J. Exhaust emissions from diesel engines fueled by different blends with the addition of nanomodifiers and hydrotreated vegetable oil HVO. *Environ. Pollut.* **2020**, *259*, 113772. [CrossRef]
90. Sharma, S.K.; Das, R.K.; Sharma, A. Improvement in the performance and emission characteristics of diesel engine fueled with jatropha methyl ester and tyre pyrolysis oil by addition of nano additives. *J. Braz. Soc. Mech. Sci. Eng.* **2016**, *38*, 1907–1920. [CrossRef]
91. Santhanamuthu, M.; Chittibabu, S.; Tamizharasan, T.; Mani, T.P. Evaluation of CI engine performance fuelled by Diesel-Polanga oil blends doped with iron oxide nanoparticles. *Int. J. ChemTech Res.* **2014**, *6*, 1299–1308.
92. El-Seesy, A.I.; Abdel-Rahman, A.K.; Bady, M.; Ookawara, S. Performance, combustion, and emission characteristics of a diesel engine fueled by biodiesel-diesel mixtures with multi-walled carbon nanotubes additives. *Energy Convers. Manag.* **2017**, *135*, 373–393. [CrossRef]
93. Alenezi, R.A.; Norkhizan, A.M.; Mamat, R.; Erdiwansyah; Najafi, G.; Mazlan, M. Investigating the contribution of carbon nanotubes and diesel-biodiesel blends to emission and combustion characteristics of diesel engine. *Fuel* **2021**, *285*, 119046. [CrossRef]
94. Rastogi, P.M.; Sharma, A.; Kumar, N. Effect of CuO nanoparticles concentration on the performance and emission characteristics of the diesel engine running on jojoba (Simmondsia Chinensis) biodiesel. *Fuel* **2021**, *286*, 119358. [CrossRef]
95. Rastogi, P.M.; Kumar, N.; Sharma, A.; Vyas, D.; Gajbhiye, A. Sustainability of Aluminium Oxide Nanoparticles Blended Mahua Biodiesel to the Direct Injection Diesel Engine Performance and Emission Analysis. *Pollution* **2020**, *6*, 25–33. [CrossRef]
96. Nithya, S.; Manigandan, S.; Gunasekar, P.; Devipriya, J.; Saravanan, W.S.R. The effect of engine emission on canola biodiesel blends with TiO_2. *Int. J. Ambient Energy* **2019**, *40*, 838–841. [CrossRef]
97. Janakiraman, S.; Lakshmanan, T.; Chandran, V.; Subramani, L. Comparative behavior of various nano additives in a DIESEL engine powered by novel Garcinia gummi-gutta biodiesel. *J. Clean. Prod.* **2020**, *245*, 118940. [CrossRef]
98. Khalife, E.; Tabatabaei, M.; Najafi, B.; Mirsalim, S.M.; Gharehghani, A.; Mohammadi, P.; Aghbashlo, M.; Ghaffari, A.; Khounani, Z.; Roodbar Shojaei, T.; et al. A novel emulsion fuel containing aqueous nano cerium oxide additive in diesel–biodiesel blends to improve diesel engines performance and reduce exhaust emissions: Part I–Experimental analysis. *Fuel* **2017**, *207*, 741–750. [CrossRef]
99. Karthikeyan, S.; Elango, A.; Prathima, A. The effect of cerium oxide additive on the performance and emission characteristics of a CI engine operated with rice bran biodiesel and its blends. *Int. J. Green Energy* **2016**, *13*, 267–273. [CrossRef]
100. Hajjari, M.; Ardjmand, M.; Tabatabaei, M. Experimental investigation of the effect of cerium oxide nanoparticles as a combustion-improving additive on biodiesel oxidative stability: Mechanism. *RSC Adv.* **2014**, *4*, 14352–14356. [CrossRef]
101. Anbarasu, A.; Karthikeyan, A. Performance and Emission Characteristics of a Diesel Engine Using Cerium Oxide Nanoparticle Blended Biodiesel Emulsion Fuel. *J. Energy Eng.* **2016**, *142*, 04015009. [CrossRef]

102. Ramakrishnan, G.; Krishnan, P.; Rathinam, S.; Thiyagu, R.; Devarajan, Y. Role of nano-additive blended biodiesel on emission characteristics of the research diesel engine. *Int. J. Green Energy* **2019**, *16*, 435–441. [CrossRef]

103. Baluchamy, A.; Karuppusamy, M. The combined effects of cobalt chromite nanoparticles and variable injection timing of preheated biodiesel and diesel on performance, combustion and emission characteristics of CI engine. *Heat Mass Transf.* **2021**, *57*, 1565–1582. [CrossRef]

104. Solmaz, H.; Ardebili, S.M.S.; Calam, A.; Yılmaz, E.; İpci, D. Prediction of performance and exhaust emissions of a CI engine fueled with multi-wall carbon nanotube doped biodiesel-diesel blends using response surface method. *Energy* **2021**, *227*, 120518. [CrossRef]

105. Chacko, N.; Jeyaseelan, T. Comparative evaluation of graphene oxide and graphene nanoplatelets as fuel additives on the combustion and emission characteristics of a diesel engine fuelled with diesel and biodiesel blend. *Fuel Process. Technol.* **2020**, *204*, 106406. [CrossRef]

106. Kumar, A.R.M.; Kannan, M.; Nataraj, G. A study on performance, emission and combustion characteristics of diesel engine powered by nano-emulsion of waste orange peel oil biodiesel. *Renew. Energy* **2020**, *146*, 1781–1795. [CrossRef]

107. Sulochana, G.; Bhatti, S.K. Performance, Emission and Combustion Characteristics of a Twin Cylinder 4 Stroke Diesel Engine Using Nano-tubes Blended Waste Fry Oil Methyl Ester. *Mater. Today Proc.* **2019**, *18*, 75–84. [CrossRef]

108. Hoseini, S.S.; Najafi, G.; Ghobadian, B.; Ebadi, M.T.; Mamat, R.; Yusaf, T. Biodiesels from three feedstock: The effect of graphene oxide (GO) nanoparticles diesel engine parameters fuelled with biodiesel. *Renew. Energy* **2020**, *145*, 190–201. [CrossRef]

109. Tewari, P.; Doijode, E.; Banapurmath, N.R.; Yaliwal, V.S. Experimental investigations on a diesel engine fuelled with multiwalled carbon nanotubes blended biodiesel fuels. *Int. J. Emerg. Technol. Adv. Eng.* **2013**, *3*, 72–76.

110. Venkatraman, N.; Suppandipillai, J.; Muthu, M. Experimental investigation of DI diesel engine performance with oxygenated additive and SOME Biodiesel. *J. Therm. Sci. Technol.* **2014**, *10*, JTST0014.

111. Anchupogu, P.; Rao, L.N.; Banavathu, B. Effect of alumina nano additives into biodiesel-diesel blends on the combustion performance and emission characteristics of a diesel engine with exhaust gas recirculation. *Environ. Sci. Pollut. Res.* **2018**, *25*, 23294–23306. [CrossRef]

112. Zenghui, Y.; Jing, H.; Wei, J. Study on the influence of alumina nanomethanol fluid on the performance, combustion and emission of DMDF diesel engine. *E3S Web Conf.* **2021**, *268*, 01004.

113. Mirbagheri, S.A.; Safieddin Ardebili, S.M.; Kiani Deh Kiani, M. Modeling of the engine performance and exhaust emissions characteristics of a single-cylinder diesel using nano-biochar added into ethanol-biodiesel-diesel blends. *Fuel* **2020**, *278*, 118238. [CrossRef]

114. Safieddin Ardebili, S.M.; Taghipoor, A.; Solmaz, H.; Mostafaei, M. The effect of nano-biochar on the performance and emissions of a diesel engine fueled with fusel oil-diesel fuel. *Fuel* **2020**, *268*, 117356. [CrossRef]

115. Mardi K, M.; Antoshkiv, O.; Berg, H.P. Experimental analysis of the effect of nano-metals and novel organic additives on performance and emissions of a diesel engine. *Fuel Process. Technol.* **2019**, *196*, 106166. [CrossRef]

116. Soudagar, M.E.M.; Afzal, A.; Safaei, M.R.; Manokar, A.M.; El-Seesy, A.I.; Mujtaba, M.A.; Samuel, O.D.; Badruddin, I.A.; Ahmed, W.; Shahapurkar, K.; et al. Investigation on the effect of cottonseed oil blended with different percentages of octanol and suspended MWCNT nanoparticles on diesel engine characteristics. *J. Therm. Anal. Calorim.* **2020**, *147*, 525–542. [CrossRef]

117. Khan, H.; Soudagar, M.E.M.; Kumar, R.H.; Safaei, M.R.; Farooq, M.; Khidmatgar, A.; Banapurmath, N.R.; Farade, R.A.; Abbas, M.M.; Afzal, A.; et al. Effect of Nano-Graphene Oxide and n-Butanol Fuel Additives Blended with Diesel—Nigella sativa Biodiesel Fuel Emulsion on Diesel Engine Characteristics. *Symmetry* **2020**, *12*, 961. [CrossRef]

118. Mehregan, M.; Moghiman, M. Numerical Investigation of Effect of Nano-Aluminum Addition on NOx and CO Pollutants Emission in Liquid Fuels Combustion. *Int. J. Mater. Mech. Manuf.* **2014**, *2*, 60–63. [CrossRef]

119. Nour, M.; El-Seesy, A.I.; Abdel-Rahman, A.K.; Bady, M. Influence of adding aluminum oxide nanoparticles to diesterol blends on the combustion and exhaust emission characteristics of a diesel engine. *Exp. Therm. Fluid Sci.* **2018**, *98*, 634–644. [CrossRef]

120. Yusuf, A.A.; Inambao, F.L.; Ampah, J.D. Evaluation of biodiesel on speciated PM2.5, organic compound, ultrafine particle and gaseous emissions from a low-speed EPA Tier II marine diesel engine coupled with DPF, DEP and SCR filter at various loads. *Energy* **2022**, *239*, 121837. [CrossRef]

121. Wei, J.; Yin, Z.; Wang, C.; Lv, G.; Zhuang, Y.; Li, X.; Wu, H. Impact of aluminium oxide nanoparticles as an additive in diesel-methanol blends on a modern DI diesel engine. *Appl. Therm. Eng.* **2021**, *185*, 116372. [CrossRef]

122. Yaşar, A.; Koçkin, A.; Yıldızhan, Ş.; Uludamar, E., Ocakoğlu, K. Effects of titanium-based additive with blends of butanol and diesel fuel on engine characteristics. *Int. J. Glob. Warm.* **2018**, *15*, 38. [CrossRef]

123. Wei, J.; He, C.; Fan, C.; Pan, S.; Wei, M.; Wang, C. Comparison in the effects of alumina, ceria and silica nanoparticle additives on the combustion and emission characteristics of a modern methanol-diesel dual-fuel CI engine. *Energy Convers. Manag.* **2021**, *238*, 114121. [CrossRef]

124. Somasundaram, D.; Elango, A.; Karthikeyan, S. Estimation of carbon credits of fishing boat diesel engine running on diesel-ethanol-bio-diesel blends with nano alumina doped ceria-zirconia. *Mater. Today Proc.* **2020**, *33*, 2923–2928. [CrossRef]

125. Purushothaman, P.; Masimalai, S.; Subramani, V. Effective utilization of mahua oil blended with optimum amount of Al_2O_3 and TiO_2 nanoparticles for better performance in CI engine. *Environ. Sci. Pollut. Res.* **2021**, *28*, 11893–11903. [CrossRef]

126. Ramesh, P.; Vivekanandan, S.; Sivaramakrishnan; Prakash, D. Performance optimization of an engine for canola oil blended diesel with Al_2O_3 nanoparticles through single and multi-objective optimization techniques. *Fuel* **2021**, *288*, 119617. [CrossRef]

127. Balasubramanian, D.; Venugopal, I.P.; Viswanathan, K. *Characteristics Investigation on Di Diesel Engine with Nano-Particles as an Additive in Lemon Grass Oil*; SAE Technical Paper 2 019-28-0081; SAE Mobilus: Warrendale, PL, USA, 2019.
128. Panithasan, M.S.; Gopalakichenin, D.; Venkadesan, G.; Veeraraagavan, S. Impact of rice husk nanoparticle on the performance and emission aspects of a diesel engine running on blends of pine oil-diesel. *Environ. Sci. Pollut. Res.* **2019**, *26*, 282–291. [CrossRef]
129. Elumalai, P.V.; Balasubramanian, D.; Parthasarathy, M.; Pradeepkumar, A.R.; Mohamed Iqbal, S.; Jayakar, J.; Nambiraj, M. An experimental study on harmful pollution reduction technique in low heat rejection engine fuelled with blends of pre-heated linseed oil and nano additive. *J. Clean. Prod.* **2021**, *283*, 124617. [CrossRef]
130. Wamankar, A.K.; Murugan, S. Combustion, performance and emission of a diesel engine fuelled with diesel doped with carbon black. *Energy* **2015**, *86*, 467–475. [CrossRef]
131. Ampah, J.D.; Yusuf, A.A.; Afrane, S.; Jin, C.; Liu, H. Reviewing two decades of cleaner alternative marine fuels: Towards IMO's decarbonization of the maritime transport sector. *J. Clean. Prod.* **2021**, *320*, 128871. [CrossRef]
132. Rashedul, H.K.; Masjuki, H.H.; Kalam, M.A.; Ashraful, A.M.; Ashrafur Rahman, S.M.; Shahir, S.A. The effect of additives on properties, performance and emission of biodiesel fuelled compression ignition engine. *Energy Convers. Manag.* **2014**, *88*, 348–364. [CrossRef]
133. Chinnasamy, C.; Tamilselvam, P.; Ranjith, R. Influence of aluminum oxide nanoparticle with different particle sizes on the working attributes of diesel engine fueled with blends of diesel and waste plastic oil. *Environ. Sci Pollut. Res. Int.* **2019**, *26*, 29962–29977. [CrossRef]
134. Lu, T.; Cheung, C.S.; Huang, Z. Effects of engine operating conditions on the size and nanostructure of diesel particles. *J. Aerosol. Sci.* **2012**, *47*, 27–38. [CrossRef]
135. Wei, J.; Lu, W.; Pan, M.; Liu, Y.; Cheng, X.; Wang, C. Physical properties of exhaust soot from dimethyl carbonate-diesel blends: Characterizations and impact on soot oxidation behavior. *Fuel* **2020**, *279*, 118441. [CrossRef]
136. Wei, J.; Wang, Y. Effects of biodiesels on the physicochemical properties and oxidative reactivity of diesel particulates: A review. *Sci. Total Environ.* **2021**, *788*, 147753. [CrossRef]
137. Thangavelu S, K.; Arthanarisamy, M. Experimental investigation on engine performance, emission, and combustion characteristics of a DI CI engine using tyre pyrolysis oil and diesel blends doped with nanoparticles. *Environ. Prog. Sustain. Energy* **2020**, *39*, e13321. [CrossRef]
138. Fu, P.; Bai, X.; Yi, W.; Li, Z.; Li, Y.; Wang, L. Assessment on performance, combustion and emission characteristics of diesel engine fuelled with corn stalk pyrolysis bio-oil/diesel emulsions with $Ce_{0.7}Zr_{0.3}O_2$ nanoadditive. *Fuel Process. Technol.* **2017**, *167*, 474–483. [CrossRef]
139. Jeyakumar, N.; Narayanasamy, B. Investigation of performance, emission, combustion characteristics of municipal waste plastic oil fueled diesel engine with nano fluids. *Energy Sources Part A: Recovery Util. Environ. Eff.* **2020**, 1–22. [CrossRef]
140. Karisathan Sundararajan, N.; Ammal, A.R.B. Improvement studies on emission and combustion characteristics of DICI engine fuelled with colloidal emulsion of diesel distillate of plastic oil, TiO_2 nanoparticles and water. *Environ. Sci. Pollut. Res. Int.* **2018**, *25*, 11595–11613. [CrossRef]
141. Sivathanu, N.; Valai Anantham, N. Impact of multi-walled carbon nanotubes with waste fishing net oil on performance, emission and combustion characteristics of a diesel engine. *Environ. Technol.* **2020**, *41*, 3670–3681. [CrossRef]
142. Vellaiyan, S. Enhancement in combustion, performance, and emission characteristics of a biodiesel-fueled diesel engine by using water emulsion and nanoadditive. *Renew. Energy* **2020**, *145*, 2108–2120. [CrossRef]
143. Soukht Saraee, H.; Taghavifar, H.; Jafarmadar, S. Experimental and numerical consideration of the effect of CeO_2 nanoparticles on diesel engine performance and exhaust emission with the aid of artificial neural network. *Appl. Therm. Eng.* **2017**, *113*, 663–672. [CrossRef]
144. Perumal Venkatesan, E.; Kandhasamy, A.; Sivalingam, A.; Kumar, A.S.; Ramalingam, K.; Joshua, P.j.t.; Balasubramanian, D. Performance and emission reduction characteristics of cerium oxide nanoparticle-water emulsion biofuel in diesel engine with modified coated piston. *Environ. Sci. Pollut. Res.* **2019**, *26*, 27362–27371. [CrossRef]
145. Senthil Kumar, J.; Ramesh Bapu, B.R. Cerium oxide nano additive impact of VCR diesel engine characteristics by using Ginger grass oil blended with diesel. *Int. J. Ambient Energy* **2019**, *43*, 578–583. [CrossRef]
146. Sam Sukumar, R.; Maddula, M.R.; Gopala Krishna, A. Experimental investigations with zinc oxide and silver doped zinc oxide nanoparticles for performance and emissions study of biodiesel blends in diesel engine. *Int. J. Ambient Energy* **2020**, 1–9. [CrossRef]
147. Kumaran, P.; Joel Godwin, A.; Amirthaganesan, S. Effect of microwave synthesized hydroxyapatite nanorods using Hibiscus rosa-sinensis added waste cooking oil (WCO) methyl ester biodiesel blends on the performance characteristics and emission of a diesel engine. *Mater. Today: Proc.* **2020**, *22*, 1047–1053. [CrossRef]
148. Debbarma, S.; Misra, R.D. Effects of Iron Nanoparticles Blended Biodiesel on the Performance and Emission Characteristics of a Diesel Engine. *J. Energy Resour. Technol.* **2017**, *139*, 042212. [CrossRef]
149. Tamilvanan, A.; Balamurugan, K.; Vijayakumar, M. Effects of nano-copper additive on performance, combustion and emission characteristics of Calophyllum inophyllum biodiesel in CI engine. *J. Therm. Anal. Calorim.* **2019**, *136*, 317–330. [CrossRef]
150. Sivasaravanan, S.; Booma Devi, P.; Nagaraj, M.; Jeya Jeevahan, J.; Britto Joseph, G. Influence of Rice Husk Nanoparticles on Engine Performance and Emission Characteristics of Diesel and Neem Oil Biodiesel Blends in a Single Cylinder Diesel Engine. *Energy Sources Part A Recovery Util. Environ. Eff.* **2019**, 1–16. [CrossRef]

151. Karthikeyan, S.; Prathima, A. Environmental effect of CI engine using microalgae methyl ester with doped nano additives. *Transp. Res. Part D Transp. Environ.* **2017**, *50*, 385–396. [CrossRef]
152. Ghanbari, M.; Mozafari-Vanani, L.; Dehghani-Soufi, M.; Jahanbakhshi, A. Effect of alumina nanoparticles as additive with diesel–biodiesel blends on performance and emission characteristic of a six-cylinder diesel engine using response surface methodology (RSM). *Energy Convers. Manag. X* **2021**, *11*, 100091. [CrossRef]
153. Mujtaba, M.A.; Kalam, M.A.; Masjuki, H.H.; Gul, M.; Soudagar, M.E.M.; Ong, H.C.; Ahmed, W.; Atabani, A.E.; Razzaq, L.; Yusoff, M. Comparative study of nanoparticles and alcoholic fuel additives-biodiesel-diesel blend for performance and emission improvements. *Fuel* **2020**, *279*, 118434. [CrossRef]
154. Shekofteh, M.; Gundoshmian, T.M.; Jahanbakhshi, A.; Heidari-Maleni, A. Performance and emission characteristics of a diesel engine fueled with functionalized multi-wall carbon nanotubes (MWCNTs-OH) and diesel–biodiesel–bioethanol blends. *Energy Rep.* **2020**, *6*, 1438–1447. [CrossRef]
155. Srinivasan, S.K.; Kuppusamy, R.; Krishnan, P. Effect of nanoparticle-blended biodiesel mixtures on diesel engine performance, emission, and combustion characteristics. *Environ. Sci. Pollut. Res.* **2021**, *28*, 39210–39226. [CrossRef]
156. Dhanasekar, K.; Sridaran, M.; Arivanandhan, M.; Jayavel, R. A facile preparation, performance and emission analysis of pongamia oil based novel biodiesel in diesel engine with CeO_2:Gd nanoparticles. *Fuel* **2019**, *255*, 115756. [CrossRef]
157. Gad, M.S.; Jayaraj, S. A comparative study on the effect of nano-additives on the performance and emissions of a diesel engine run on Jatropha biodiesel. *Fuel* **2020**, *267*, 117168. [CrossRef]
158. Deepak Kumar, T.; Sameer Hussain, S.; Ramesha, D.K. Effect of a zinc oxide nanoparticle fuel additive on the performance and emission characteristics of a CI engine fuelled with cotton seed biodiesel blends. *Mater. Today Proc.* **2020**, *26*, 2374–2378. [CrossRef]
159. Dinesha, P.; Kumar, S.; Rosen, M.A. Effects of particle size of cerium oxide nanoparticles on the combustion behavior and exhaust emissions of a diesel engine powered by biodiesel/diesel blend. *Biofuel Res. J.* **2021**, *8*, 1374–1383. [CrossRef]
160. Yusuf, A.A.; Inambao, F.L.; Farooq, A.A. Impact of n-butanol-gasoline-hydrogen blends on combustion reactivity, performance and tailpipe emissions using TGDI engine parameters variation. *Sustain. Energy Technol. Assess.* **2020**, *40*, 100773. [CrossRef]
161. Subramani, L.; Parthasarathy, M.; Balasubramanian, D.; Ramalingam, K. Novel Garcinia gummi-gutta methyl ester (GGME) as a potential alternative feedstock for existing unmodified DI diesel engine. *Renew. Energy* **2018**, *125*, 568–577. [CrossRef]
162. Srinidhi, C.; Madhusudhan, A.; Channapattana, S.V. Effect of NiO nanoparticles on performance and emission characteristics at various injection timings using biodiesel-diesel blends. *Fuel* **2019**, *235*, 185–193. [CrossRef]
163. Ağbulut, Ü.; Karagöz, M.; Sarıdemir, S.; Öztürk, A. Impact of various metal-oxide based nanoparticles and biodiesel blends on the combustion, performance, emission, vibration and noise characteristics of a CI engine. *Fuel* **2020**, *270*, 117521. [CrossRef]
164. Muruganantham, P.; Pandiyan, P.; Sathyamurthy, R. Analysis on performance and emission characteristics of corn oil methyl ester blended with diesel and cerium oxide nanoparticle. *Case Stud. Therm. Eng.* **2021**, *26*, 101077.
165. Soudagar, M.E.M.; Nik-Ghazali, N.-N.; Kalam, M.A.; Badruddin, I.A.; Banapurmath, N.R.; Bin Ali, M.A.; Kamangar, S.; Cho, H.M.; Akram, N. An investigation on the influence of aluminium oxide nano-additive and honge oil methyl ester on engine performance, combustion and emission characteristics. *Renew. Energy* **2020**, *146*, 2291–2307. [CrossRef]
166. Ranjan, A.; Dawn, S.S.; Jayaprabakar, J.; Nirmala, N.; Saikiran, K.; Sai Sriram, S. Experimental investigation on effect of MgO nanoparticles on cold flow properties, performance, emission and combustion characteristics of waste cooking oil biodiesel. *Fuel* **2018**, *220*, 780–791. [CrossRef]
167. Praveena, V.; Martin, M.L.J.; Geo, V.E. Experimental characterization of CI engine performance, combustion and emission parameters using various metal oxide nanoemulsion of grapeseed oil methyl ester. *J. Therm. Anal. Calorim.* **2020**, *139*, 3441–3456. [CrossRef]
168. Dhana Raju, V.; Kishore, P.S.; Nanthagopal, K.; Ashok, B. An experimental study on the effect of nanoparticles with novel tamarind seed methyl ester for diesel engine applications. *Energy Convers. Manag.* **2018**, *164*, 655–666. [CrossRef]
169. Sivakumar, M.; Shanmuga Sundaram, N.; Ramesh kumar, R.; Syed Thasthagir, M.H. Effect of aluminium oxide nanoparticles blended pongamia methyl ester on performance, combustion and emission characteristics of diesel engine. *Renew. Energy* **2018**, *116*, 518–526. [CrossRef]
170. Jaichandar, S.; Thamaraikannan, M.; Yogaraj, D.; Samuelraj, D. A comprehensive study on the effects of internal air jet piston on the performance of a JOME fueled DI diesel engine. *Energy* **2019**, *185*, 1174–1182. [CrossRef]
171. Kumar, S.; Dinesha, P.; Rosen, M.A. Effect of injection pressure on the combustion, performance and emission characteristics of a biodiesel engine with cerium oxide nanoparticle additive. *Energy* **2019**, *185*, 1163–1173. [CrossRef]
172. Suhel, A.; Abdul Rahim, N.; Abdul Rahman, M.R.; Bin Ahmad, K.A.; Teoh, Y.H.; Zainal Abidin, N. An Experimental Investigation on the Effect of Ferrous Ferric Oxide Nano-Additive and Chicken Fat Methyl Ester on Performance and Emission Characteristics of Compression Ignition Engine. *Symmetry* **2021**, *13*, 265. [CrossRef]
173. Gavhane, R.S.; Kate, A.M.; Soudagar, M.E.M.; Wakchaure, V.D.; Balgude, S.; Rizwanul Fattah, I.M.; Nik-Ghazali, N.-N.; Fayaz, H.; Khan, T.M.Y.; Mujtaba, M.A.; et al. Influence of Silica Nano-Additives on Performance and Emission Characteristics of Soybean Biodiesel Fuelled Diesel Engine. *Energies* **2021**, *14*, 1489. [CrossRef]
174. Rameshbabu, A.; Senthilkumar, G. Emission and performance investigation on the effect of nano-additive on neat biodiesel. *Energy Sources Part A Recovery Util. Environ. Eff.* **2021**, *43*, 1315–1328. [CrossRef]
175. Reddy, S.N.K.; Wani, M.M. An investigation on the performance and emission studies on diesel engine by addition of nanoparticles and antioxidants as additives in biodiesel blends. *Int. Rev. Appl. Sci. Eng.* **2021**, *12*, 111–118. [CrossRef]

176. Oyedepo, S.O.; Fagbenle, R.O.; Adefila, S.S.; Alam, M.M. Thermoeconomic and thermoenvironomic modeling and analysis of selected gas turbine power plants in Nigeria. *Energy Sci. Eng.* **2015**, *3*, 423–442. [CrossRef]
177. Rosen, M.A.; Dincer, I. Exergy as the confluence of energy, environment and sustainable development. *Exergy Int. J.* **2001**, *1*, 3–13. [CrossRef]
178. Aghbashlo, M.; Tabatabaei, M.; Khalife, E.; Roodbar Shojaei, T.; Dadak, A. Exergoeconomic analysis of a DI diesel engine fueled with diesel/biodiesel (B5) emulsions containing aqueous nano cerium oxide. *Energy* **2018**, *149*, 967–978. [CrossRef]
179. Oberdörster, G.; Oberdörster, E.; Oberdörster, J. Nanotoxicology: An Emerging Discipline Evolving from Studies of Ultrafine Particles. *Environ. Health Perspect.* **2005**, *113*, 823–839. [CrossRef] [PubMed]
180. Brohi, R.D.; Wang, L.; Talpur, H.S.; Wu, D.; Khan, F.A.; Bhattarai, D.; Rehman, Z.-U.; Farmanullah, F.; Huo, L.-J. Toxicity of Nanoparticles on the Reproductive System in Animal Models: A Review. *Front. Pharmacol.* **2017**, *8*, 606. [CrossRef] [PubMed]
181. Ray, P.C.; Yu, H.; Fu, P.P. Toxicity and environmental risks of nanomaterials: Challenges and future needs. *J. Environ. Sci. Health. Part C Environ. Carcinog. Ecotoxicol. Rev.* **2009**, *27*, 1–35. [CrossRef] [PubMed]
182. Kreyling, W.G.; Semmler-Behnke, M.; Möller, W. Ultrafine particle-lung interactions: Does size matter? *J. Aerosol Med. Off. J. Int. Soc. Aerosols Med.* **2006**, *19*, 74–83. [CrossRef] [PubMed]
183. Anderson, P.J.; Wilson, J.D.; Hiller, F.C. Respiratory tract deposition of ultrafine particles in subjects with obstructive or restrictive lung disease. *Chest* **1990**, *97*, 1115–1120. [CrossRef]
184. Balasubramanyam, A.; Sailaja, N.; Mahboob, M.; Rahman, M.F.; Hussain, S.M.; Grover, P. In vivo genotoxicity assessment of aluminium oxide nanomaterials in rat peripheral blood cells using the comet assay and micronucleus test. *Mutagenesis* **2009**, *24*, 245–251. [CrossRef]
185. Kim, Y.; Choi, H.; Song, M.; Youk, D.; Kim, J.; Ryu, J. Genotoxicity of Aluminum Oxide Nanoparticle in Mammalian Cell Lines. *Mol. Cell. Toxicol.* **2009**, *5*, 172–178.
186. Trouiller, B.; Reliene, R.; Westbrook, A.; Solaimani, P.; Schiestl, R.H. Titanium dioxide nanoparticles induce DNA damage and genetic instability in vivo in mice. *Cancer Res.* **2009**, *69*, 8784–8789. [CrossRef]
187. Liu, R.; Yin, L.; Pu, Y.; Liang, G.; Zhang, J.; Su, Y.; Xiao, Z.; Ye, B. Pulmonary toxicity induced by three forms of titanium dioxide nanoparticles via intra-tracheal instillation in rats. *Prog. Nat. Sci.* **2009**, *19*, 573–579. [CrossRef]
188. Bhattacharya, K.; Davoren, M.; Boertz, J.; Schins, R.P.; Hoffmann, E.; Dopp, E. Titanium dioxide nanoparticles induce oxidative stress and DNA-adduct formation but not DNA-breakage in human lung cells. *Part. Fibre Toxicol.* **2009**, *6*, 17. [CrossRef]
189. Liu, R.; Zhang, X.; Pu, Y.; Yin, L.; Li, Y.; Zhang, X.; Liang, G.; Li, X.; Zhang, J. Small-sized titanium dioxide nanoparticles mediate immune toxicity in rat pulmonary alveolar macrophages in vivo. *J. Nanosci. Nanotechnol.* **2010**, *10*, 5161–5169. [CrossRef]
190. Liu, H.; Ma, L.; Zhao, J.; Liu, J.; Yan, J.; Ruan, J.; Hong, F. Biochemical toxicity of nano-anatase TiO_2 particles in mice. *Biol. Trace Elem. Res.* **2009**, *129*, 170–180. [CrossRef]
191. Liu, G.; Gao, J.; Ai, H.; Chen, X. Applications and potential toxicity of magnetic iron oxide nanoparticles. *Small* **2013**, *9*, 1533–1545. [CrossRef]
192. Naqvi, S.; Samim, M.; Abdin, M.; Ahmed, F.J.; Maitra, A.; Prashant, C.; Dinda, A.K. Concentration-dependent toxicity of iron oxide nanoparticles mediated by increased oxidative stress. *Int. J. Nanomed.* **2010**, *5*, 983–989. [CrossRef]
193. Chen, L.; Yokel, R.A.; Hennig, B.; Toborek, M. Manufactured Aluminum Oxide Nanoparticles Decrease Expression of Tight Junction Proteins in Brain Vasculature. *J. Neuroimmune Pharmacol.* **2008**, *3*, 286–295. [CrossRef]
194. Muller, J.; Huaux, F.; Moreau, N.; Misson, P.; Heilier, J.-F.; Delos, M.; Arras, M.; Fonseca, A.; Nagy, J.B.; Lison, D. Respiratory toxicity of multi-wall carbon nanotubes. *Toxicol. Appl. Pharmacol.* **2005**, *207*, 221–231. [CrossRef]
195. Cha, K.E.; Myung, H. Cytotoxic effects of nanoparticles assessed in vitro and in vivo. *J. Microbiol. Biotechnol.* **2007**, *17*, 1573–1578.
196. Asharani, P.V.; Lian Wu, Y.; Gong, Z.; Valiyaveettil, S. Toxicity of silver nanoparticles in zebrafish models. *Nanotechnology* **2008**, *19*, 255102. [CrossRef]
197. Najahi-Missaoui, W.; Arnold, R.D.; Cummings, B.S. Safe Nanoparticles: Are We There Yet? *Int. J. Mol. Sci.* **2021**, *22*, 385. [CrossRef]
198. George, S.; Pokhrel, S.; Xia, T.; Gilbert, B.; Ji, Z.; Schowalter, M.; Rosenauer, A.; Damoiseaux, R.; Bradley, K.A.; Mädler, L.; et al. Use of a Rapid Cytotoxicity Screening Approach To Engineer a Safer Zinc Oxide Nanoparticle through Iron Doping. *ACS Nano* **2010**, *4*, 15–29. [CrossRef]
199. Khare, V.; Saxena, A.K.; Gupta, P.N. *Chapter 15-Toxicology Considerations in Nanomedicine*; William Andrew Publishing, Oxford, Institution: Oxford, UK, 2015; pp. 239–261.
200. Cai, X.; Lee, A.; Ji, Z.; Huang, C.; Chang, C.H.; Wang, X.; Liao, Y.-P.; Xia, T.; Li, R. Reduction of pulmonary toxicity of metal oxide nanoparticles by phosphonate-based surface passivation. *Part. Fibre Toxicol.* **2017**, *14*, 13. [CrossRef]
201. Sukhanova, A.; Bozrova, S.; Sokolov, P.; Berestovoy, M.; Karaulov, A.; Nabiev, I. Dependence of Nanoparticle Toxicity on Their Physical and Chemical Properties. *Nanoscale Res. Lett.* **2018**, *13*, 44. [CrossRef]

 *nanomaterials*

Article

Rheological Modeling of Metallic Oxide Nanoparticles Containing Non-Newtonian Nanofluids and Potential Investigation of Heat and Mass Flow Characteristics

Muhammad Rizwan [1], Mohsan Hassan [1], Oluwole Daniel Makinde [2], Muhammad Mubashir Bhatti [3,*] and Marin Marin [4]

1   Department of Mathematics, COMSATS University Islamabad, Lahore Campus, Lahore 54000, Pakistan; m.rizwan7571@gmail.com (M.R.); mohsan.hassan@cuilahore.edu.pk (M.H.)
2   Faculty of Military Science, Stellenbosch University, Private Bag X2, Saldanha 7395, South Africa; makinded@gmail.com
3   College of Mathematics and Systems Science, Shandong University of Science and Technology, Qingdao 266590, China
4   Department of Mathematics and Computer Science, Transilvania University of Brasov, 500036 Brasov, Romania; m.marin@unitbv.ro
*   Correspondence: mmbhatti@sdust.edu.cn or mubashirme@hotmail.com

Abstract: Nanofluids have great potential due to their improved properties that make them useful for addressing various industrial and engineering problems. In order to use nanofluids on an industrial scale, it is first important to discuss their rheological behavior in relation to heat transfer aspects. In the current study, the flow characteristics of nanofluids are discussed using a mathematical model that is developed by fundamental laws and experimental data. The data are collected in the form of viscosity versus shear rate for different homogeneous ethylene glycol- (EG) based nanofluids, which are synthesized by dispersing 5–20% nanoparticle concentrations of SiO_2, MgO, and TiO_2 with diameters of (20–30 nm, 60–70 nm), (20 nm, 40 nm), and (30 nm, 50 nm), respectively. The data are fitted into a rheological power-law model and further used to govern equations of a physical problem. The problem is simplified into ordinary differential equations by using a boundary layer and similarity transformations and then solved through the numerical Runge–Kutta (RK) method. The obtained results in the form of velocity and temperature profiles at different nanoparticle concentrations and diameters are displayed graphically for discussion. Furthermore, displacement and momentum thicknesses are computed numerically to explain boundary-layer growth. The results show that the velocity profile is reduced and the temperature profile is raised by increasing the nanoparticle concentration. Conversely, the velocity profile is increased and the temperature profile is decreased by increasing the nanoparticle diameter. The results of the present investigation regarding heat and mass flow behavior will help engineers design equipment and improve the efficacy and economy of the overall process in the industry.

Keywords: non-Newtonian nanofluids; mathematical modeling based on experimental data; heat-flow characteristics; power-law fluid model

Citation: Rizwan, M.; Hassan, M.; Makinde, O.D.; Bhatti, M.M.; Marin, M. Rheological Modeling of Metallic Oxide Nanoparticles Containing Non-Newtonian Nanofluids and Potential Investigation of Heat and Mass Flow Characteristics. *Nanomaterials* **2022**, *12*, 1237. https://doi.org/10.3390/nano12071237

Academic Editor: Mikhail Sheremet

Received: 10 March 2022
Accepted: 4 April 2022
Published: 6 April 2022

Publisher's Note: MDPI stays neutral with regard to jurisdictional claims in published maps and institutional affiliations.

1. Introduction

Nanofluids are advanced classes of heat-transfer fluids designed by diffusing nanometer-scale particles of metal, metal oxide, carbon nanotubes, nitride, carbide, and compound materials in traditional base fluids such as ethylene glycol (EG), water, oils, etc. [1–4]. As a consequence of the improved properties related to heat transfer and chemical stability [5–8], nanofluids have become immensely attractive and demonstrate several potential applications in many fields relating to solar collection, transportation, the energy industry, refrigeration, the cooling process, chemistry, biomedicine, and the environment [8–15].

Many studies are reported in the literature on the different properties of nanofluids, such as density, thermal conductivity, viscosity, specific heat, etc. [16–20]. However, it is observed that the most critical properties of nanofluids are their rheological properties. For example, based on the different types of base fluids, the nanofluids containing MWCNTs exhibit both Newtonian and non-Newtonian behavior. The composition of MWCNTs with water, resin, and oil displays non-Newtonian behavior [21–30]. Normally, nanofluids consisting of MWCNTs reveal Newtonian behavior at low volume fractions and non-Newtonian behavior at high volume fractions [21,22]. In the case of TiO_2-water nanofluids, the fluids mostly exhibit non-Newtonian behavior [31–34]. Contrarily, Bobbo et al. [35] and Penkavova et al. [36] reported that, for all compositions, TiO_2/water nanofluids displayed Newtonian behavior. In the case of SiO_2, the nanofluids with different base fluids showed behavior that was close to Newtonian behavior [37–42]. In short, the nanofluids exhibited non-Newtonian behavior [43–47] in many cases, whereas few showed Newtonian behavior [35,36,48].

Non-Newtonian nanofluids that were synthesized by water or EG revealed shear-thinning behavior [43–47]. Shear-thinning fluid is a kind of non-Newtonian fluid wherein viscosity declines by the rise of the shear rate. There are several mathematical models in the literature used to investigate the rheological behavior of such fluids. In the list of models, the power-law model demonstrates the relation between viscosity and shear rate. It is very popular in various disciplines, such as the biosciences and food and processing reservoir engineering [6–14]. Specifically, it is widely used in fluid flow problems under different conditions, and it is even used as a working principle in different kinds of rheometers.

In previous numerical studies, a deficiency was found in the theoretical models that can guess the exact behavior of nanofluids. The researchers used the Newtonian model for homogeneous nanofluids, which does not apply to the experimental behavior of all cases. No one study is available where homogeneous nanofluids deal with the non-Newtonian model. In the current study, we used a non-Newtonian power-law fluid model according to the trend of an experimental study. The parameters of the model are expressed as the function of nanoparticles, and we developed new mathematical relations to these parameters. By using these relations that govern equations, the physical problem became complex because every parameter is a function of the volume fraction, and solving this problem is not easy.

In this study, our goal is to investigate the flow behavior of non-Newtonian nanofluids, which are synthesized by the dispersion of metallic oxides in EG. For the flow, wedge shape is adopted as the geometry for our problem, which is favorable for accelerating or decelerating the fluids. The mathematical problem for flow is developed by the fundamental equations of fluid mechanics and modifies its parameters in view of the experimental evidence. The results from these equations are obtained in the form of velocity and temperature profiles and displayed in graphical form for discussion.

2. Nanofluid Modeling

In this section, the mathematical models of physical properties are developed by using experimental data for three homogenous nanofluids: SiO_2-EG, MgO-EG, and TiO_2-EG. The experimental data for nanofluids, which contain 5%, 10%, 15%, and 20% nanoparticle concentrations and (20–30 nm, 60–70 nm), (30 nm, 50 nm), and (20 nm, 40 nm) nanoparticle diameters, respectively, are collected for the study [1]. We have chosen the nanoparticles of materials SiO_2, MgO, and TiO_2 for the nanofluid because these materials are used in manufacturing on a large scale at industry level. EG is used as a base fluid because it can be utilized within sufficiently large temperature ranges. For rheological behavior, the power-law equation is applied for nanofluid modeling in the formation of the viscosity–shear rate relationship as follows [49]:

$$\mu\left(\dot{\gamma}\right) = \mu_{nf}\dot{\gamma}^{n-1} \tag{1}$$

where μ_{nf} is named as the consistency coefficient and n is the power index, which is justifiable according to the Newtonian or non-Newtonian behavior of fluids. When $n = 1$, the fluid exhibits Newtonian behavior, and the fluid shows shear-thinning behavior when $n < 1$. The rheological behavior of said nanofluids at different nanoparticle concentrations is displayed in Figures 1–3.

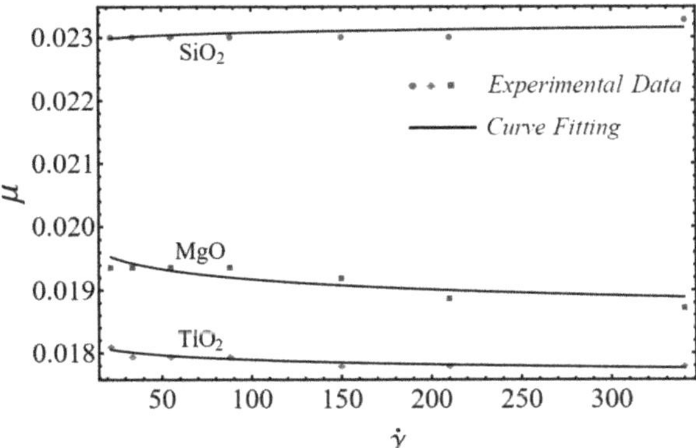

Figure 1. Experimental and mathematical rheological behavior of SiO_2-EG, MgO-EG, and TiO_2-EG nanofluids at 5% particle concentration.

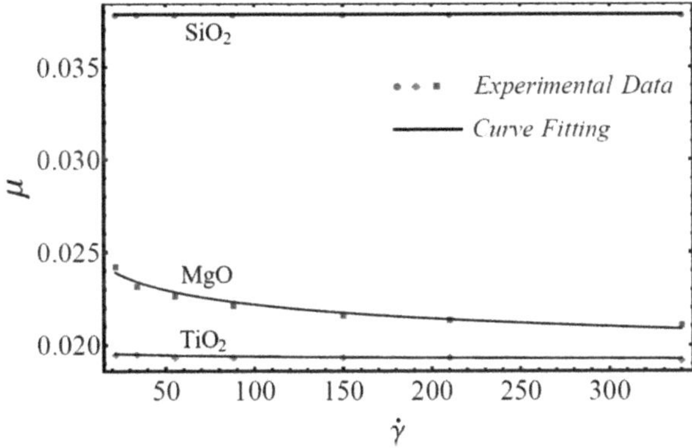

Figure 2. Experimental and mathematical rheological behavior of SiO_2-EG, MgO-EG, and TiO_2-EG nanofluids at 10% particle concentration.

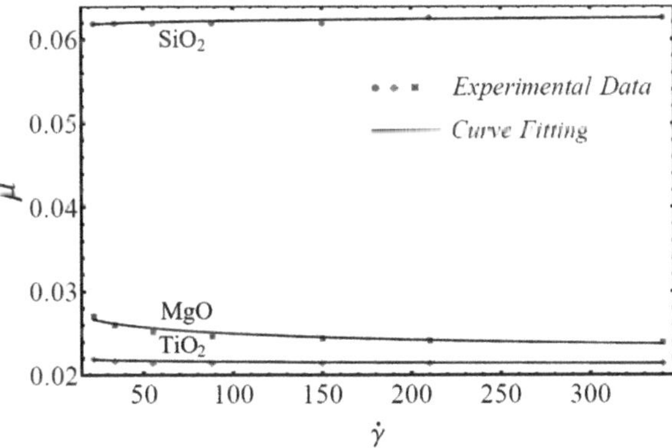

Figure 3. Experimental and mathematical rheological behavior of SiO_2-EG, MgO-EG, and TiO_2-EG nanofluids at 15% particle concentration.

The curve-fitting technique is used to calculate the values of the two empirical parameters $\left(\mu_{nf}\right)$ and (n) of Equation (1) for different nanofluids at different nanoparticle concentrations. For the curve fitting, we used the FindFit package from Mathematica, which is used for fitting in non-linear models. Afterward, these parameters are expressed in a second-order polynomial of the nanoparticle concentration ϕ. The equation of the power-law index and consistency index is defined as follows:

$$\left.\begin{array}{c} n = 1 + A_1\phi + A_2\phi^2 \\ \mu_{nf} = \left(1 + B_1\phi + B_2\phi^2\right)\mu_{bf} \end{array}\right\} \tag{2}$$

where A_1, A_2, B_1, and B_2 are constants calculated by using the curve-fitting technique in Table 1.

Table 1. The values of A_1, A_2, B_1, and B_2 for different nanofluids at different particle diameters.

	SiO$_2$-EG Nanofluid		MgO-EG Nanofluid		TiO$_2$-EG Nanofluid	
$d(nm) \rightarrow$	20–30	60–70	20	40	30	50
A_1	0.025	−0.04	−0.617	−0.0037	1.08	−0.013
A_2	−0.042	0.104	2.19	−0.112	12.63	−0.17
B_1	−4.8	−4.2	6.9	1.53	1.08	0.49
B_2	172.59	160.6	5.02	2.9	12.6	12.2

A good agreement is found between the calculated values and the polynomial equations, as seen in Figures 4 and 5.

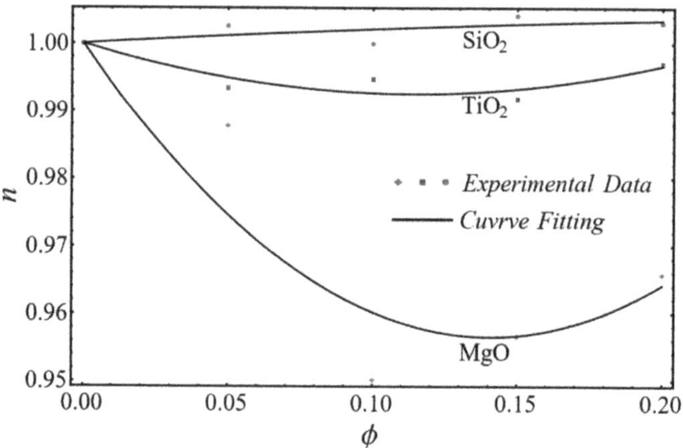

Figure 4. Curve fitting with experimental data for power-law index by using Equation (2).

Therefore, the viscosity model of Equation (1) is written as

$$\frac{\mu}{\mu_{bf}} = \left(1 + B_1\phi + B_2\phi^2\right)\dot{\gamma}^{1 + A_1\phi + A_2\phi^2} \tag{3}$$

The co-relation models for other physical properties, such as heat capacitance $\left(\rho C_p\right)_{nf}$, effective density $\left(\rho_{nf}\right)$, and thermal conductivity $\left(k_{nf}\right)$, are defined by [50].

$$\rho_{nf} = \phi\,\rho_{np} + (1 - \phi)\rho_{bf} \tag{4}$$

$$\left(\rho C_p\right)_{nf} = \phi\left(\rho C_p\right)_{np} + (1 - \phi)\rho_{bf} \tag{5}$$

$$k_{nf} = \frac{k_{pe} + 2k_{bf} + 2\left(k_{pe} - k_{bf}\right)\left(1 + \beta^*\right)^3 \phi}{k_{pe} + 2k_{bf} - \left(k_{pe} - k_{bf}\right)\left(1 + \beta^*\right)^3 \phi} k_{bf} \qquad (6)$$

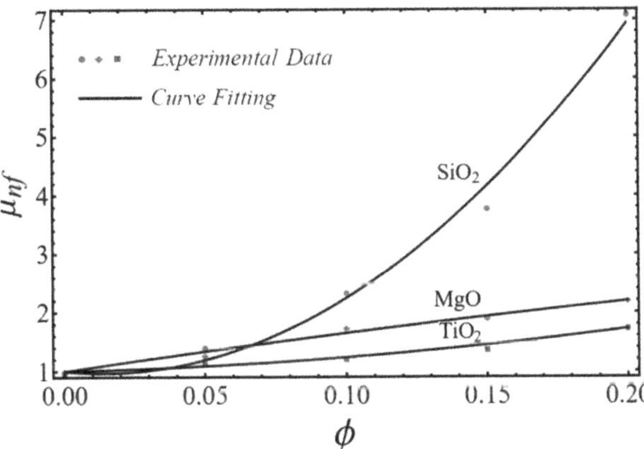

Figure 5. Curve fitting with experimental data for consistency index by using Equation (2).

In the above equations, the thermal conductivity of nanoparticle k_{npl} with a layer around the particle is defined by

$$k_{pe} = \frac{\left[2(1 - \gamma) + (1 + \beta^*)^3(1 + 2\gamma^*)\right]\gamma^*}{-(1 - \gamma^*) + (1 + \beta^*)^3(1 + 2\gamma^*)} k_{np} \qquad (7)$$

where $\gamma^* = k_{layer}/k_{np}$ is the ratio of the layers of the nanoparticle to the thermal conductivities of the nanoparticle and $\beta^* = h_{layer}/R_p$ is the ratio of the layer's height to the radius of the nanoparticle.

3. Flow Modeling

We considered the boundary-layer fluid flow of nanofluids over a moving wedge, which has incompressible and steady-state properties. The wedge is moved with the velocity $u_w(x) = cx^m$, whereas $u_\infty(x) = bx^m$ is the velocity of the nanofluid over an inviscid region, as shown in Figure 6.

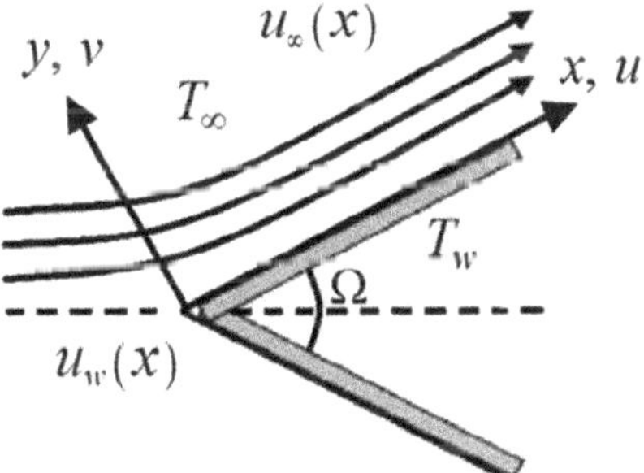

Figure 6. Structure of the flow.

On the basis of the results in Figures 1–3, the power-law model is used in momentum Equation (9). Under the boundary-layer approximation, the continuity, momentum, and energy equations are written as [51,52]

$$u_x + v_y = 0 \tag{8}$$

$$\rho_{nf}\left(u\,u_x + v\,u_y\right) = -\partial_x(p_e) + \mu_{nf}\,\partial_y\left(\left|u_y\right|^{n-1} u_y\right) \tag{9}$$

$$\left(\rho C_p\right)_{nf}\left(u\,T_x + v\,T_y\right) = k_{nf}\,\partial_y(T_y) \tag{10}$$

subjected to the boundary conditions

$$\left.\begin{array}{l} u(x,0) = -u_w(x),\ v(x,0) = 0,\ T(x,0) = T_w \\ u(x,\infty) = u_\infty(x),\ T(x,\infty) = T_\infty \end{array}\right\} \tag{11}$$

In the above equations, u and v are the velocity component parts in x and y directions. T_w is the temperature on the wedge's surface, and T_∞ is the temperature away from the surface $(T_w > T_\infty)$.

The relationship between the Falkner–Skan power-law parameter (m) and the wedge's angle $\beta = \Omega/\pi$ is stated as

$$m = \frac{\beta}{2 - \beta} \tag{12}$$

For simplicity, we introduced the similarity transformations as follows:

$$\left.\begin{array}{l} \eta = \frac{y}{x}(\mathrm{Re}_x)^{\frac{1}{1+n}},\ \psi = u_\infty x(\mathrm{Re}_x)^{\frac{-1}{1+n}} f(\eta), \\ \theta(\eta) = \frac{T-T_\infty}{T_w-T_\infty},\ u = \frac{\partial\psi}{\partial x}, v = -\frac{\partial\psi}{\partial y} \end{array}\right\} \tag{13}$$

Substituting Equation (14) into Equations (9)–(12), we obtain the following equations:

$$n\frac{\mu_{nf}}{\mu_{bf}}\left(\left|f''\right|^{n-1} f'''\right) + \frac{\rho_{nf}}{\rho_{bf}}\left[\left(\frac{m(2n-1)+1}{(n+1)(m+1)}\right)ff'' - \frac{m}{m+1}f'^2\right] + \frac{m}{m+1} = 0 \tag{14}$$

$$\frac{k_{nf}}{k_{bf}}\theta'' + \Pr\frac{\left(\rho C_p\right)_{nf}}{\left(\rho C_p\right)_{bf}}\left(\frac{m(2n-1)+1}{(n+1)(m+1)}\right)f\theta' = 0 \tag{15}$$

with the boundary conditions

$$\left.\begin{array}{l} f(0) = 0,\ f'(0) = -\lambda,\ \theta(0) = 1, \\ \theta(\infty) = 0,\ f'(\infty) = 1, \end{array}\right\} \tag{16}$$

Here, the modified Prandtl number is $\Pr_x = \frac{\rho_{bf} C_{pbf} u_\infty x}{k_{bf}(\mathrm{Re}_x)^{\frac{2}{n-1}}}$, and the Reynold number is $\mathrm{Re}_x = (m+1)\frac{u_\infty^{2-n} x^n}{\nu_{bf}}$ and $\lambda = \frac{u_w}{u_\infty}$ velocities ratio.

4. Physical Parameters

4.1. Displacement Thickness

Displacement thickness is recognized as the vertical distance that is produced by the absent mass flow rate due to the boundary-layer phenomena. The expression for displacement thickness is written as

$$\delta^* = \int_0^\infty \left(1 - \frac{u}{u_\infty}\right) dy \tag{17}$$

Dimensionless displacement thickness is composed as

$$\delta^* = x(\mathrm{Re}_x)^{\frac{-1}{1+n}} \int_0^\infty \left(1 - f'\right) d\eta \tag{18}$$

4.2. Momentum Thickness

Momentum thickness is the height of an imaginary stream, which is transmitted by the loss of the momentum flow rate due to the boundary-layer phenomena and is described mathematically as

$$\delta^{**} = \int_0^\infty \frac{u}{u_\infty} \left(1 - \frac{u}{u_\infty}\right) dy \tag{19}$$

Dimensionless momentum thickness is illustrated as

$$\delta^{**} = x(\mathrm{Re}_x)^{\frac{-1}{1+n}} \int_0^\infty f'\left(1 - f'\right) d\eta \tag{20}$$

4.3. Skin Friction

The skin-friction coefficient is a dimensionless parameter that represents the shear stress at the wall. It is expressed as

$$C_f = \frac{\tau_w}{\frac{1}{2}\rho_{bf} u_\infty{}^2} \tag{21}$$

The wall shear stress τ_w can be written as

$$\tau_w = \mu\, u_y\big|_{y=0} \tag{22}$$

After applying transformation Equation (13), we receive

$$C_f = 2(m+1)(\mathrm{Re}_x)^{\frac{-1}{1+n}} \frac{\mu_{nf}}{\mu_{bf}} |f''(0)|^n \tag{23}$$

4.4. Nusselt Number

The Nusselt number is a dimensionless parameter that represents the convective heat-transfer rate at the wall. It is written in the following form:

$$Nu_x = \frac{hx}{k_{bf}} \tag{24}$$

Here, h is a convective heat-transfer coefficient that can be written as

$$h = -\frac{k_{nf}\, \partial_y(T - T_\infty)\big|_{y=0}}{(T_w - T_\infty)} \tag{25}$$

After applying transformation Equation (13), it is written as

$$Nu_x = -(\mathrm{Re}_x)^{\frac{1}{n+1}} \frac{k_{nf}}{k_{bf}} \theta'(0), \tag{26}$$

5. Solution Technique

The solution of Equations (14) and (15) with respect to Equation (16) is obtained by using the RK method. The method is executed in the following manner:

Let $f = F_1$, $\theta = G_1$, and convert Equations (14) and (15) into a system of first-order differential equations as

$$
\left.
\begin{aligned}
F_1{}' &= F_2 \\
F_2{}' &= F_3 \\
F_3{}' &= \frac{\frac{\rho_{nf}}{\rho_{bf}}\left[\frac{m}{m+1}F_2{}^2 - \left(\frac{m(2n-1)+1}{(n+1)(m+1)}\right)F_1 F_3\right] - \frac{m}{m+1}}{n\frac{\mu_{nf}}{\mu_{bf}}|F_3|^{n-1}} \\
G_1{}' &= G_2 \\
G_2{}' &= -\Pr\frac{k_{bf}}{k_{nf}}\frac{(\rho C_p)_{nf}}{(\rho C_p)_{bf}}\left(\frac{m(2n-1)+1}{(n+1)(m+1)}\right)F_1 G_2
\end{aligned}
\right\}
\tag{27}
$$

along the initial conditions

$$
\left.
\begin{aligned}
F_1(0) &= 0, \\
F_2(0) &= -\lambda, \\
F_3(0) &= \Omega_1 \\
G_1(0) &= 1 \\
G_2(0) &= \Omega_2
\end{aligned}
\right\}
\tag{28}
$$

Here Ω_1 and Ω_2 represent unknown boundary conditions.

To evaluate the accuracy of the results, the values of $f''(0)$ and $-\theta'(0)$ against the parameters of β and $\Pr$ are compared with the existing limited results [53,54] in Tables 2 and 3.

Table 2. Comparison of results for $f''(0)$ with numerical results in [53] when $\phi = 0$ and $n = 1$.

β	Present	[53]
0	0.46961	0.4696
1	0.92773	0.9277
2	1.23262	1.2326

Table 3. Comparison of results for $-\theta'(0)$ with numerical results in [54] when $\phi = 0$ and $\beta = 1$.

$\Pr$	Present	[54]
1	0.57052	0.5705
2	0.74370	0.7437
6	1.11471	1.1147

6. Result and Discussion

In this segment, the results under the influence of the governing parameters are displayed for discussion. The first set of results is displayed graphically in the form of velocity and temperature profiles related to three homogenous metallic oxide nanofluids under the impacts of different nanoparticle concentrations and diameters. To see the effects of these parameters, the values of the geometry's parameters, for example, wedge angle $\Omega = \pi/6$, wedge speed $u_w = 0.01$, and free-stream velocity $u_\infty = 0.04$, are taken as fixed, and other parameters, such as the Reynold number and the modified Prandtl number, are varied according to different nanoparticle concentrations and diameters, which are listed in Tables 4 and 5. It is seen that the Reynold number is varied due to the variation in inertial force. The inertial force is a function of the power-law index, which becomes less than unity when the nanoparticle concentration is increased in the case of the MgO -EG and TiO$_2$-EG nanofluids. The modified Prandtl number is a function of the Reynold number and is varied by it.

Table 4. The values of Pr_x and Re_x numbers at different nanoparticle concentrations.

		SiO$_2$-EG Nanofluid $D = 20-30$ nm			MgO-EG Nanofluid $D = 20$ nm			TiO$_2$-EG Nanofluid $D = 30$ nm		
$\phi \rightarrow$ / $x \downarrow$		5%	10%	15%	5%	10%	15%	5%	10%	15%
Pr_x	2	139.5	139.6	139.6	137.9	137.0	136.8	69.7	69.7	138.2
	4	139.5	139.5	139.5	139.1	138.9	138.8	139.4	139.0	278.1
Re_x	2	6045	6067	6085	5448	5149	5079	5898	5867	5851
	4	12,098	12,153	12,195	10,705	10,017	9858	11755	11,635	11,667

Table 5. The values of Pr_x and Re_x numbers at different nanoparticle diameters.

		SiO$_2$-EG Nanofluid		MgO -EG Nanofluid		TiO$_2$-EG Nanofluid	
$D(nm) \rightarrow$ / $x \downarrow$		20–30	60–70	30	50	20	40
Pr_x	2	139.6	139.2	137.0	139.4	69.7	139.3
	4	139.5	278.5	138.9	278.7	139.0	278.5
Re_x	2	6067	5941	5149	5982	5847	5946
	4	12,153	11,855	10,017	11,951	11,635	11,866

6.1. Velocity Profiles

The results of the velocity profiles for the SiO$_2$-EG, MgO-EG, and TiO$_2$-EG nanofluids under the impact of distinct nanoparticle concentrations are illustrated in Figures 7–9. It is seen that the velocity profile for all specified nanofluids is reduced with the increase in nanoparticle concentration. In the current situation, the viscosity of present nanofluids is enhanced due to the increase in nanoparticle concentration, which causes the reduction in velocity. The trend of viscosity and velocity profile for the current study is found to be similar in existing studies [1,55–60]. These graphs assert that the dominant effects are found in the velocity profile of the SiO$_2$-EG nanofluid due to maximum viscosity as compared with the MgO-EG and TiO$_2$-EG nanofluids.

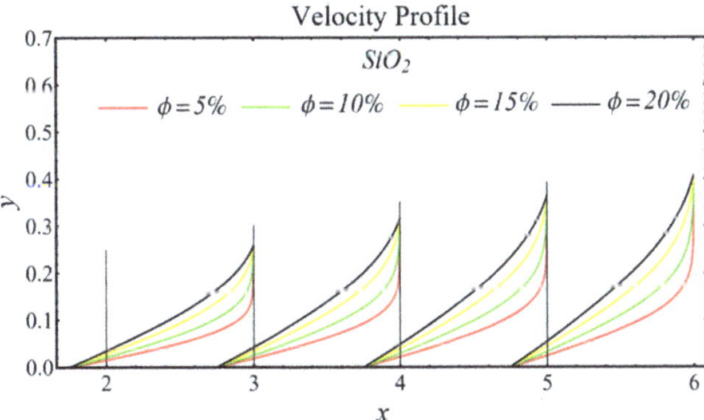

Figure 7. Velocity profile of SiO$_2$-EG nanofluid under influence of nanoparticle concentration.

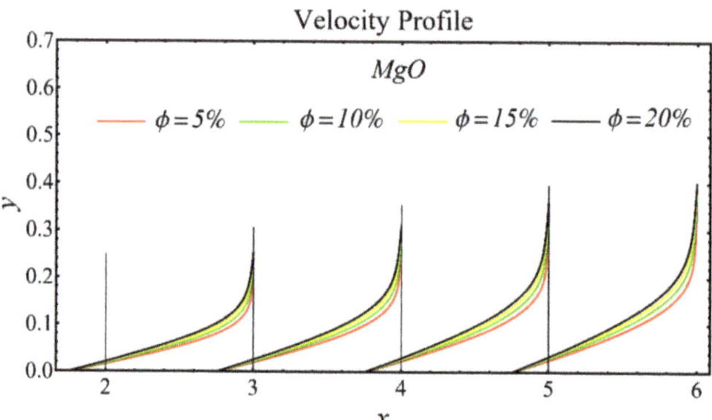

Figure 8. Velocity profile of MgO-EG nanofluid under influence of nanoparticle concentration.

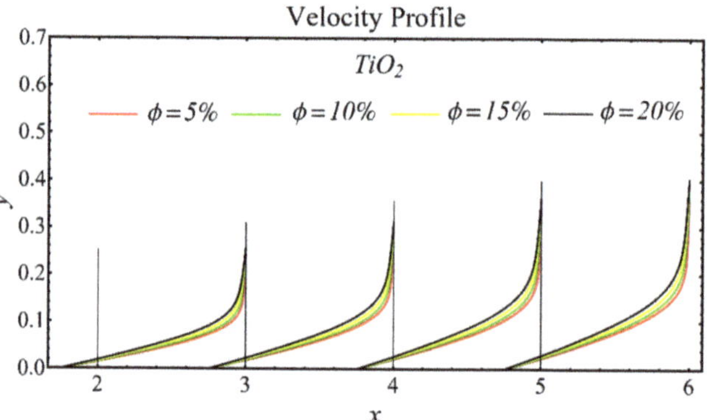

Figure 9. Velocity profile of TiO$_2$-EG nanofluid under influence of nanoparticle concentration.

The velocity profiles of the schematic nanofluids under the effect of two distinct nanoparticle diameters are shown in Figures 10–12. It is clear that the velocity profile is increased by increasing the nanoparticle diameter due to the decline in the viscosity of the nanofluids. In addition, the dominant impact of this parameter on the profile is found to be more pronounced in the MgO-EG nanofluid than in the other nanofluids.

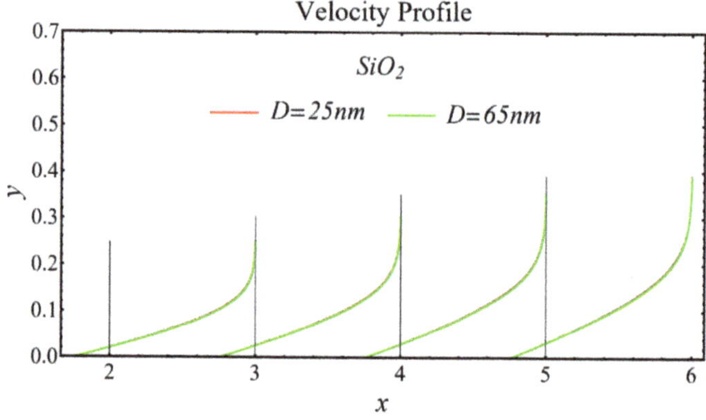

Figure 10. Velocity profile of SiO$_2$-EG nanofluid under influence of nanoparticle diameter.

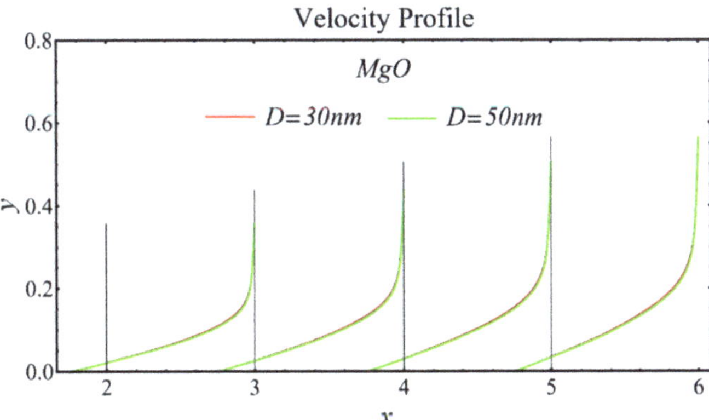

Figure 11. Velocity profile of MgO-EG nanofluid under influence of nanoparticle diameter.

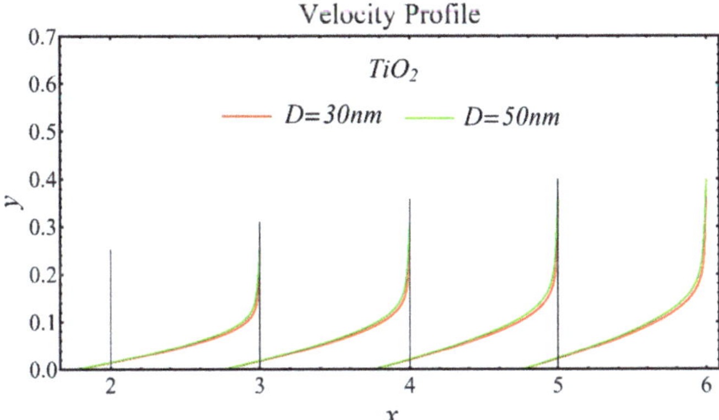

Figure 12. Velocity profile of TiO$_2$-EG nanofluid under influence of nanoparticle diameter.

6.2. Temperature Profiles

The graphs of the temperature profiles for the nanofluids with respect to various nanoparticle concentrations are displayed in Figures 13–15. The figures show that the temperature is raised with the increase in nanoparticle concentration. The fact is that the temperature distribution is enhanced due to the increase in thermal conductivity and the decline in specific heat. The trend of this profile is also matched with the trend demonstrated in existing studies [55–60]. Moreover, the graphs disclose that the temperature of the SiO$_2$-EG nanofluid is extensively affected by the enhancement of nanoparticle concentrations as compared with the other nanofluids.

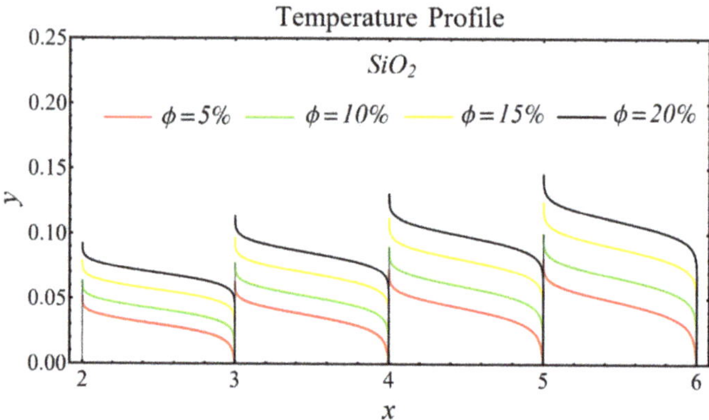

Figure 13. Temperature profile of SiO$_2$-EG nanofluid under influence of nanoparticle concentration.

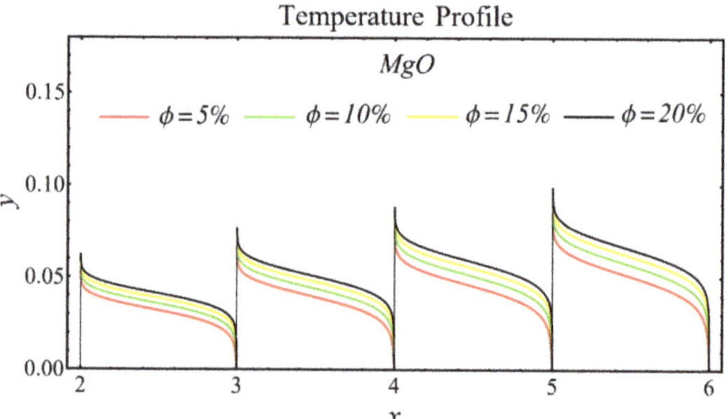

Figure 14. Temperature profile of MgO-EG nanofluid under influence of nanoparticle concentration.

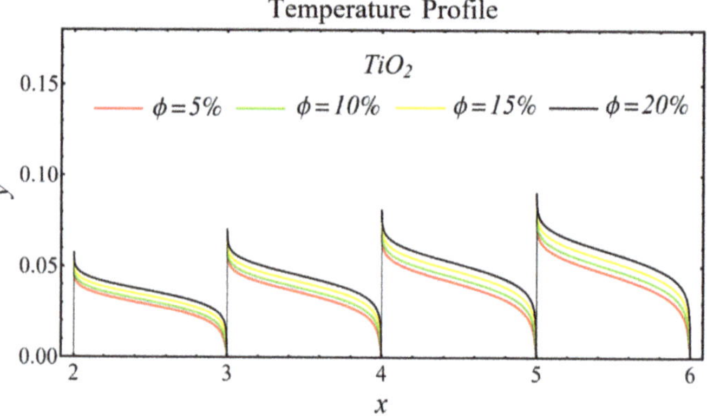

Figure 15. Temperature profile of TiO$_2$-EG nanofluid under influence of nanoparticle concentration.

The results of the temperature profiles for the schematic nanofluids under the influence of the nanoparticle diameter are illustrated in Figures 16–18. It is revealed that the temperature is reduced by enhancing the nanoparticle's diameter, as shown in all graphs. This is because of the increase in the Prandtl number that occurs by decreasing thermal diffusion.

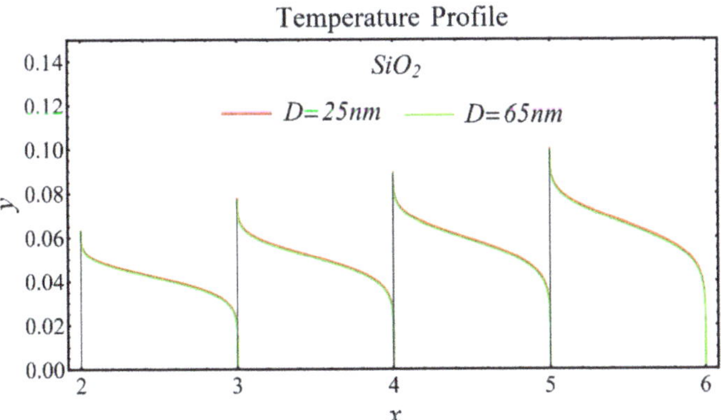

Figure 16. Temperature profile of SiO$_2$-EG nanofluid under influence of nanoparticle diameter.

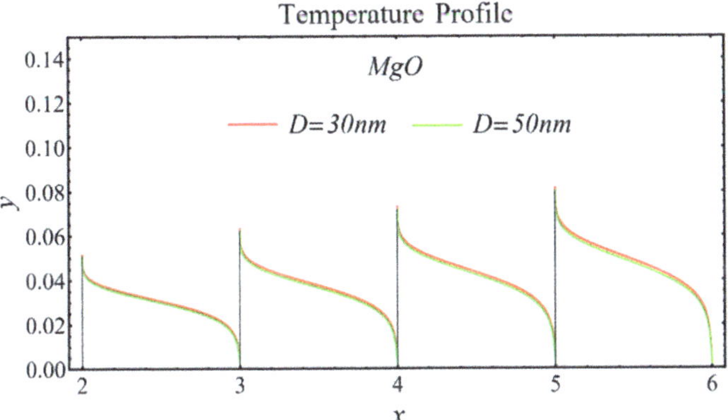

Figure 17. Temperature profile of MgO-EG nanofluid under influence of nanoparticle diameter.

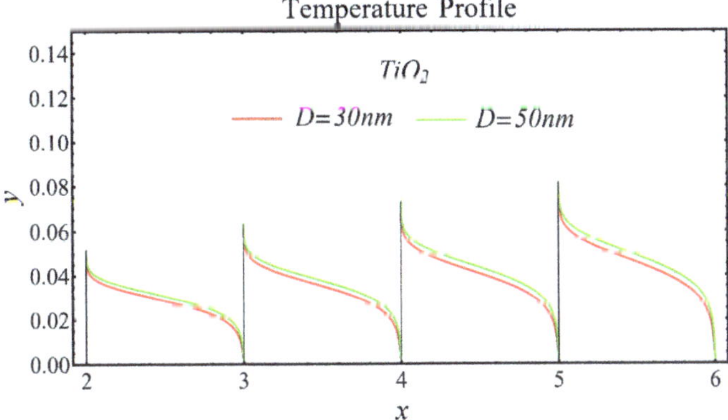

Figure 18. Temperature profile of TiO$_2$-EG nanofluid under influence of nanoparticle diameter.

6.3. Physical Parameters

The results of the boundary-layer parameters, such as the velocity and temperature boundary-region thicknesses, displacement thicknesses, and momentum thicknesses, are presented in Figures 19–22, whereas the results of the coefficient of skin friction and the Nusselt number are displayed in Figures 23 and 24.

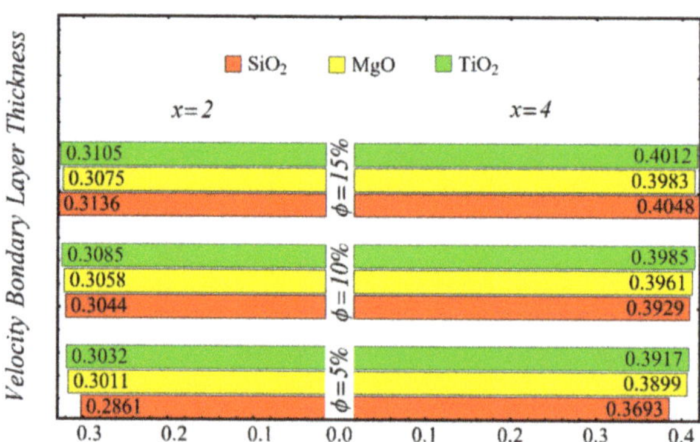

Figure 19. Velocity boundary-layer thickness at different nanoparticle concentrations.

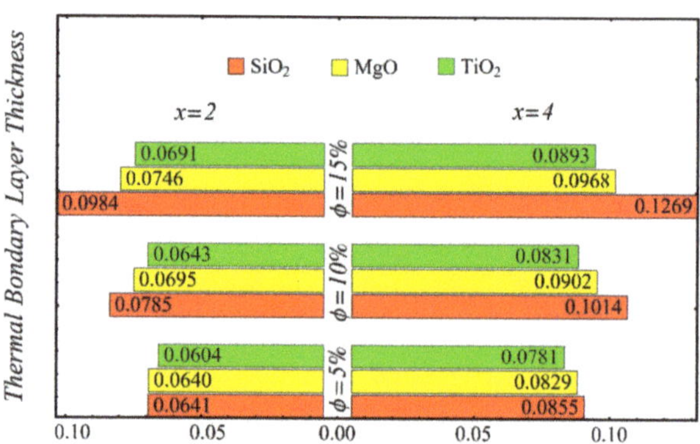

Figure 20. Temperature boundary-layer thickness at different nanoparticle concentrations.

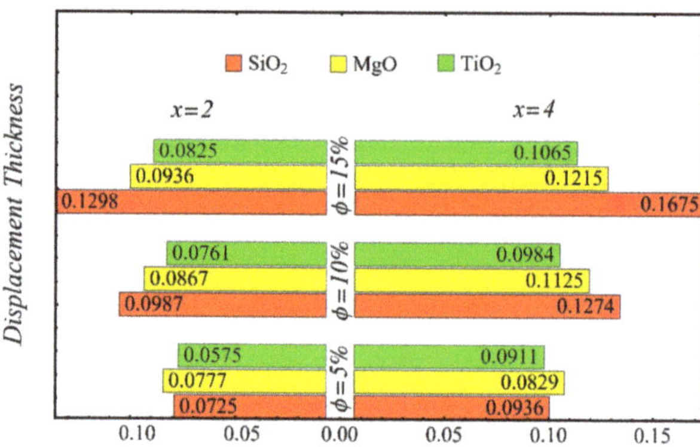

Figure 21. Displacement thickness at different nanoparticle concentrations.

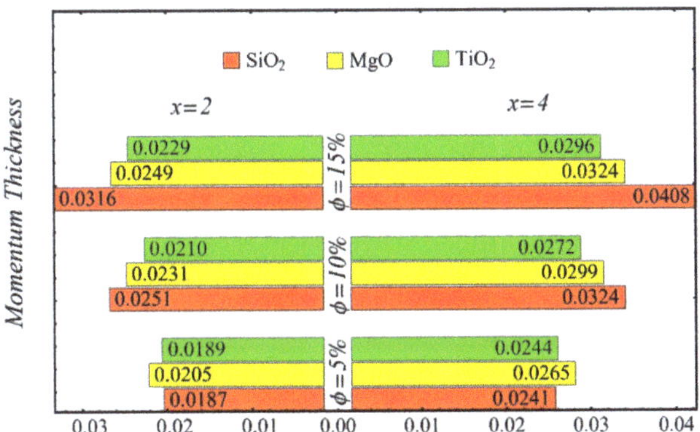

Figure 22. Momentum thickness at different nanoparticle concentrations.

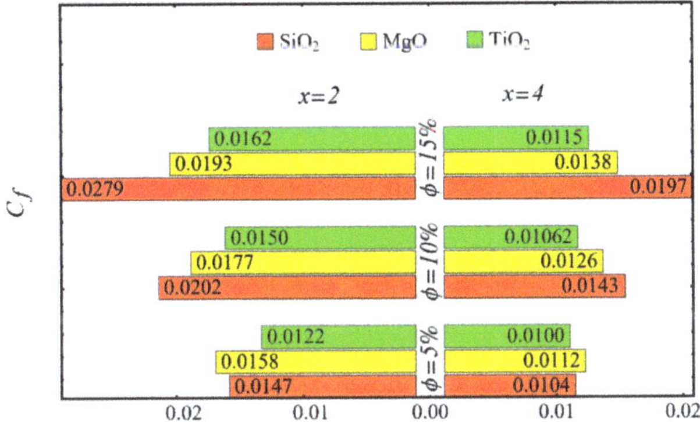

Figure 23. Coefficient of skin friction at different nanoparticle concentrations.

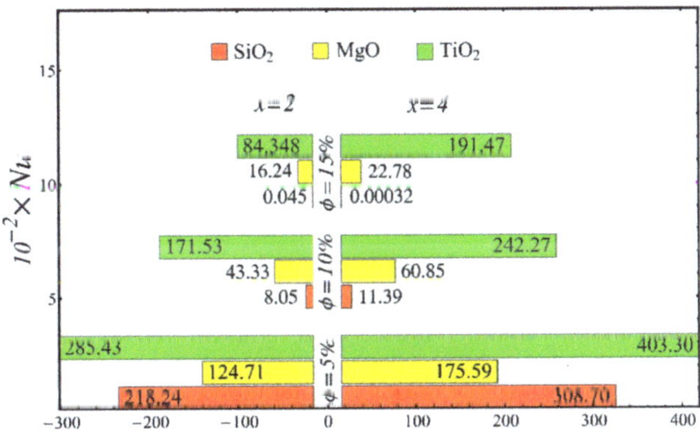

Figure 24. Nusselt number at different nanoparticle concentration.

The velocity and temperature boundary regions' thicknesses are computed at distinct positions on the surface of the wedge under the impact of nanoparticle concentrations in Figures 19 and 20. It is noted that the thickness of the velocity boundary-layer region is enlarged with the enhancement of the nanoparticle concentration and increases at a position far from the origin. It is also observed that the velocity thickness of the SiO$_2$-EG nanofluid is lowest at low concentrations but highest at high concentrations compared with both the MgO-EG and TiO$_2$-EG nanofluids. For temperature boundary-layer thickness, it is perceived that the thickness is increased by enhancing the nanoparticle concentration. In addition, the temperature boundary-layer thickness is found to be smaller as compared with the thickness of the velocity boundary layer due to the dominant effects of mass diffusion as compared with thermal diffusion. It is further observed that the thermal boundary-layer thickness is found to be the greatest in the SiO_2-EG nanofluid as compared with both the TiO_2-EG and MgO-EG nanofluids.

Figures 21 and 22 show the results of displacement and momentum thicknesses against different nanoparticle concentrations at distinct locations on the wedge's surface. The value of the displacement thickness is raised by enhancing the nanoparticle concentration and also increases along the distance of the wedge. It is seen that the maximum displacement is found in the case of the SiO_2-EG nanofluid, and the lowest displacement is found in the TiO_2-EG nanofluid. The momentum thickness indicates a reduction in the momentum of the nanofluid, and it is observed that the value is increased by raising the nanoparticle concentration.

The values of the coefficient of skin friction for the nanofluids against the nanoparticle concentration are displayed in Figure 23. It is to be noted that the value of the coefficient of skin friction is increased when the nanoparticle concentration is enhanced, and maximum enhancement is found in the case of the SiO_2-EG nanofluid. The values of the Nusselt number with respect to the nanoparticle concentration are displayed in Figure 24. It is seen that the Nusselt number is decreased with the increase in nanoparticle concentration, and a maximum decline is found in the SiO_2-EG nanofluid as well. In addition, the trend of the results for the coefficient of skin friction and the Nusselt number agrees with the trend of published studies [55–60].

The results of the coefficient of skin friction and the Nusselt number at different values of Re_x are shown in Table 6. To calculate the results, the values of Re_x in Table 4 are used. The results show the same trend as seen in Figures 23 and 24 because Re_x is dependent on the value of the volume fraction.

Table 6. The values of C_f and Nu_x at different values of Re_x at fixed $x = 2$.

	SiO$_2$-EG Nanofluid $D = 20-30$ nm			MgO -EG Nanofluid $D = 20$ nm			TiO$_2$-EG Nanofluid $D = 30$ nm		
Re_x	6045	6067	6085	5448	5149	5079	5898	5847	5861
C_f	0.0147	0.0202	0.0279	0.0158	0.0177	0.0193	0.0122	0.0150	0.0162
Nu_x	218.24	8.05	0.045	124.71	43.33	16.24	285.43	171.53	84.348

7. Conclusions

In the present analysis, the boundary-layer fluid flow of three homogenous non-Newtonian nanofluids over a moving wedge is investigated. The mathematical results are presented graphically in the form of velocity and temperature profiles and further used to obtain the values of boundary-layer parameters. The main conclusions from the results are as follows:

- The profile of velocity is decreased and increased by raising the values of nanoparticle concentration and diameter, respectively.
- The profile of temperature is increased and decreased by enhancing the values of nanoparticle concentration and diameter, respectively.
- The velocity and temperature boundary-layer regions are increased by increasing the nanoparticle concentration.
- The displacement and momentum thicknesses are increased by the rise of nanoparticle concentration.
- The skin-friction coefficient is enhanced whereas the Nusselt number is decreased with the increase in nanoparticle concentration.

Author Contributions: Conceptualization, M.R., M.H. and O.D.M.; methodology, M.M.B. and M.M.; software, M.R., M.H. and O.D.M.; validation, M.M.B. and M.M.; formal analysis, M.R., M.H. and O.D.M.; investigation, M.M.B. and M.M.; resources, M.M.B. and M.M.; data curation, M.R. and M.H.; writing—original draft preparation, M.R. and M.H.; writing—review and editing, M.M.B. and M.M.; visualization, O.D.M.; supervision, M.M.B. and M.M. All authors have read and agreed to the published version of the manuscript.

Funding: This research received no external funding.

Institutional Review Board Statement: Not applicable.

Informed Consent Statement: Not applicable.

Data Availability Statement: Not applicable.

Conflicts of Interest: The authors declare no conflict of interest.

Nomenclature

u, v	Velocity components
T	Temperature of fluid
T_∞	Temperatures of inviscid region
u_∞	Velocity of fluid in inviscid region
k	Thermal conductivity
C_p	Specific heat
Ω	Wedge angle
τ	Shear stress
$\dot{\gamma}$	Shear rate
ϕ	Volume fraction
x, y	Rectangular coordinates
T_w	Temperatures of wall
ψ	Stream function
u_w	Wall velocity
ρ	Density
μ_{nf}	Consistency index
η	Similarity variable
$\mathrm{Pr}_x = \dfrac{\rho_{bf} c_{pbf} u_\infty x}{k_{bf}(\mathrm{Re}_x)^{\frac{2}{n-1}}}$	Local Prandtl number
$\mathrm{Re}_x = (m+1)\dfrac{\rho x^n}{\mu u_\infty^{n-2}}$	Local Reynold number

Subscripts

nf	Nanofluid
np	Nanoparticle
bf	Base fluid

References

1. Yapici, K.; Osturk, O.; Uludag, Y. Dependency of nanofluid rheology on particle size and concentration of various metal oxide nanoparticles. *Braz. J. Chem. Eng.* **2018**, *35*, 575–586. [CrossRef]
2. Das, S.K.; Choi, S.U.; Yu, W.; Pradeep, T. *Nanofluids. Science and Technology*; John and Wiley and Sons: Hoboken, NJ, USA, 2007.
3. Wang, X.Q.; Mujumdar, A.S. Heat transfer characteristics of nanofluids: A review. *Int. J. Therm. Sci.* **2007**, *46*, 1–19. [CrossRef]
4. Chen, H.; Ding, Y. Heat transfer and rheological behaviour of nanofluids—A review. *Adv. Transp. Phenom.* **2009**, *1*, 135–177.
5. Nikkam, N.; Haghighi, E.B.; Saleemi, M.; Behi, M.; Khodabandeh, R.; Muhammed, M.; Palm, B.; Toprak, M.S. Experimental study on preparation and base liquid effect on thermo-physical and heat transport characteristics of α-SiC nanofluids. *Int. Commun. Heat Mass Transf.* **2014**, *55*, 38–44. [CrossRef]
6. Teja, A.S.; Beck, M.P.; Yuan, Y.; Warrier, P. The limiting behavior of the thermal conductivity of nanoparticles and nanofluids. *J. Appl. Phys.* **2010**, *107*, 114319. [CrossRef]
7. Sudeep, P.M.; Taha-Tijerina, J.; Ajayan, P.M.; Narayanan, T.N.; Anantharaman, M.R. Nanofluids based on fluorinated graphene oxide for efficient thermal management. *RSC Adv.* **2014**, *4*, 24887–24892. [CrossRef]
8. Baglioni, M.; Raudino, M.; Berti, D.; Keiderling, U.; Bordes, R.; Holmberg, K.; Baglioni, P. Nanostructured fluids from degradable nonionic surfactants for the cleaning of works of art from polymer contaminants. *Soft Matter* **2014**, *10*, 6798–6809. [CrossRef]
9. Sreeremya, T.S.; Krishnan, A.; Satapathy, L.N.; Ghosh, S. Facile synthetic strategy of oleophilic zirconia nanoparticles allows preparation of highly stable thermo-conductive coolant. *RSC Adv.* **2014**, *4*, 28020–28028. [CrossRef]

10. Cabaleiro, D.; Pastoriza-Gallego, M.J.; Gracia-Fernández, C.; Piñeiro, M.M.; Lugo, L. Rheological and volumetric properties of TiO2-ethylene glycol nanofluids. *Nanoscale Res. Lett.* **2013**, *8*, 286. [CrossRef]

11. Shima, P.D.; Philip, J. Role of thermal conductivity of dispersed nanoparticles on heat transfer properties of nanofluid. *Ind. Eng. Chem. Res.* **2014**, *53*, 980–988. [CrossRef]

12. Kole, M.; Dey, T.K. Viscosity of alumina nanoparticles dispersed in car engine coolant. *Exp. Therm. Fluid Sci.* **2010**, *34*, 677–683. [CrossRef]

13. Wen, D.; Lin, G.; Vafaei, S.; Zhang, K. Review of nanofluids for heat transfer applications. *Particuology* **2009**, *7*, 141–150. [CrossRef]

14. Rao, Y. Nanofluids: Stability, phase diagram, rheology and applications. *Particuology* **2010**, *8*, 549–555. [CrossRef]

15. Cheng, L.; Cao, D. Designing a thermo-switchable channel for nanofluidic controllable transportation. *ACS Nano* **2011**, *5*, 1102–1108. [CrossRef] [PubMed]

16. Kristiawan, B.; Rifa'i, A.I.; Enoki, K.; Wijayanta, A.T.; Miyazaki, T. Enhancing the thermal performance of TiO2/water nanofluids flowing in a helical microfin tube. *Powder Technol.* **2020**, *376*, 254–262. [CrossRef]

17. Kristiawan, B.; Wijayanta, A.T.; Enoki, K.; Miyazaki, T.; Aziz, M. Heat transfer enhancement of TiO2/water nanofluids flowing inside a square minichannel with a microfin structure: A numerical investigation. *Energies* **2019**, *12*, 3041. [CrossRef]

18. Jamshed, W.; Prakash, M.; Devi, S.; Ibrahim, R.W.; Shahzad, F.; Nisar, K.S.; Eid, M.R.; Abdel-Aty, A.H.; Khashan, M.M.; Yahia, I.S. A brief comparative examination of tangent hyperbolic hybrid nanofluid through a extending surface: Numerical Keller–Box scheme. *Sci. Rep.* **2021**, *11*, 24032. [CrossRef]

19. Reddy, M.; Rao, V.V.; Reddy, B.; Sarada, S.N.; Ramesh, L. Thermal conductivity measurements of ethylene glycol water based TiO2 nanofluids. *Nanosci. Nanotechnol. Lett.* **2012**, *4*, 105–109. [CrossRef]

20. Mariano, A.; Pastoriza-Gallego, M.J.; Lugo, L.; Camacho, A.; Canzonieri, S.; Piñeiro, M.M. Thermal conductivity, rheological behaviour and density of non-Newtonian ethylene glycol-based SnO2 nanofluids. *Fluid Phase Equilibria* **2013**, *337*, 119–124. [CrossRef]

21. Pötschke, P.; Fornes, T.D.; Paul, D.R. Rheological behavior of multiwalled carbon nanotube/polycarbonate composites. *Polymer* **2002**, *43*, 3247–3255. [CrossRef]

22. Yang, Y.; Grulke, E.A.; Zhang, Z.G.; Wu, G. Thermal and rheological properties of carbon nanotube-in-oil dispersions. *J. Appl. Phys.* **2006**, *99*, 114307. [CrossRef]

23. Seyhan, A.T.; Gojny, F.H.; Tanoğlu, M.; Schulte, K. Rheological and dynamic-mechanical behavior of carbon nanotube/vinyl ester–polyester suspensions and their nanocomposites. *Eur. Polym. J.* **2007**, *43*, 2836–2847. [CrossRef]

24. Yang, Y.; Grulke, E.A.; Zhang, Z.G.; Wu, G. Temperature effects on the rheological properties of carbon nanotube-in-oil dispersions. *Colloids Surf. A Physicochem. Eng. Asp.* **2007**, *298*, 216–224. [CrossRef]

25. Ko, G.H.; Heo, K.; Lee, K.; Kim, D.S.; Kim, C.; Sohn, Y.; Choi, M. An experimental study on the pressure drop of nanofluids containing carbon nanotubes in a horizontal tube. *Int. J. Heat Mass Transf.* **2007**, *50*, 4749–4753. [CrossRef]

26. Wang, B.; Wang, X.; Lou, W.; Hao, J. Rheological and tribological properties of ionic liquid-based nanofluids containing functionalized multi-walled carbon nanotubes. *J. Phys. Chem. C* **2010**, *114*, 8749–8754. [CrossRef]

27. Phuoc, T.X.; Massoudi, M.; Chen, R.H. Viscosity and thermal conductivity of nanofluids containing multi-walled carbon nanotubes stabilized by chitosan. *Int. J. Therm. Sci.* **2011**, *50*, 12–18. [CrossRef]

28. Ruan, B.; Jacobi, A.M. Ultrasonication effects on thermal and rheological properties of carbon nanotube suspensions. *Nanoscale Res. Lett.* **2012**, *7*, 127. [CrossRef]

29. Wang, J.; Zhu, J.; Zhang, X.; Chen, Y. Heat transfer and pressure drop of nanofluids containing carbon nanotubes in laminar flows. *Exp. Therm. Fluid Sci.* **2013**, *44*, 716–721. [CrossRef]

30. Shahid, A.; Bhatti, M.M.; Ellahi, R.; Mekheimer, K.S. Numerical experiment to examine activation energy and bi-convection Carreau nanofluid flow on an upper paraboloid porous surface: Application in solar energy. *Sustain. Energy Technol. Assess.* **2022**, *52*, 102029. [CrossRef]

31. Tseng, W.J.; Lin, K.C. Rheology and colloidal structure of aqueous TiO2 nanoparticle suspensions. *Mater. Sci. Eng. A* **2003**, *355*, 186–192. [CrossRef]

32. He, Y.; Jin, Y.; Chen, H.; Ding, Y.; Cang, D.; Lu, H. Heat transfer and flow behaviour of aqueous suspensions of TiO2 nanoparticles (nanofluids) flowing upward through a vertical pipe. *Int. J. Heat Mass Transf.* **2007**, *50*, 2272–2281. [CrossRef]

33. Chen, H.; Yang, W.; He, Y.; Ding, Y.; Zhang, L.; Tan, C.; Lapkin, A.A.; Bavykin, D.V. Heat transfer and flow behaviour of aqueous suspensions of titanate nanotubes (nanofluids). *Powder Technol.* **2008**, *183*, 63–72. [CrossRef]

34. Chen, H.; Witharana, S.; Jin, Y.; Kim, C.; Ding, Y. Predicting thermal conductivity of liquid suspensions of nanoparticles (nanofluids) based on rheology. *Particuology* **2009**, *7*, 151–157. [CrossRef]

35. Bobbo, S.; Fedele, L.; Benetti, A.; Colla, L.; Fabrizio, M.; Pagura, C.; Barison, S. Viscosity of water based SWCNH and TiO2 nanofluids. *Exp. Therm. Fluid Sci.* **2012**, *36*, 65–71. [CrossRef]

36. Penkavova, V.; Tihon, J.; Wein, O. Stability and rheology of dilute TiO2-water nanofluids. *Nanoscale Res. Lett.* **2011**, *6*, 273. [CrossRef] [PubMed]

37. Chevalier, J.; Tillement, O.; Ayela, F. Rheological properties of nanofluids flowing through microchannels. *Appl. Phys. Lett.* **2007**, *91*, 233103. [CrossRef]

38. Tavman, I.; Turgut, A.; Chirtoc, M.; Schuchmann, H.P.; Tavman, S. Experimental investigation of viscosity and thermal conductivity of suspensions containing nanosized ceramic particles. *Arch. Mater. Sci.* **2008**, *100*, 99–104.

39. Chevalier, J.; Tillement, O.; Ayela, F. Structure and rheology of SiO 2 nanoparticle suspensions under very high shear rates. *Phys. Rev. E* **2009**, *80*, 051403. [CrossRef]
40. Jung, Y.; Son, Y.H.; Lee, J.K.; Phuoc, T.X.; Soong, Y.; Chyu, M.K. Rheological behavior of clay–nanoparticle hybrid-added bentonite suspensions: Specific role of hybrid additives on the gelation of clay-based fluids. *ACS Appl. Mater. Interfaces* **2011**, *3*, 3515–3522. [CrossRef]
41. Mondragon, R.; Julia, J.E.; Barba, A.; Jarque, J.C. Determination of the packing fraction of silica nanoparticles from the rheological and viscoelastic measurements of nanofluids. *Chem. Eng. Sci.* **2012**, *80*, 119–127. [CrossRef]
42. Anoop, K.; Sadr, R.; Al-Jubouri, M.; Amani, M. Rheology of mineral oil-SiO2 nanofluids at high pressure and high temperatures. *Int. J. Therm. Sci.* **2014**, *77*, 108–115. [CrossRef]
43. Żyła, G.; Cholewa, M.; Witek, A. Rheological properties of diethylene glycol-based MgAl 2 O 4 nanofluids. *RSC Adv.* **2013**, *3*, 6429–6434. [CrossRef]
44. Mostafizur, R.M.; Aziz, A.A.; Saidur, R.; Bhuiyan, M.H.U.; Mahbubul, I.M. Effect of temperature and volume fraction on rheology of methanol based nanofluids. *Int. J. Heat Mass Transf.* **2014**, *77*, 765–769. [CrossRef]
45. Hojjat, M.; Etemad, S.G.; Bagheri, R.; Thibault, J. Rheological characteristics of non-Newtonian nanofluids: Experimental investigation. *Int. Commun. Heat Mass Transf.* **2011**, *38*, 144–148. [CrossRef]
46. Chen, H.; Ding, Y.; Lapkin, A. Rheological behaviour of nanofluids containing tube/rod-like nanoparticles. *Powder Technol.* **2009**, *194*, 132–141. [CrossRef]
47. Lu, K. Rheological behavior of carbon nanotube–alumina nanoparticle dispersion systems. *Powder Technol.* **2007**, *177*, 154–161. [CrossRef]
48. Chen, H.; Ding, Y.; Tan, C. Rheological behaviour of nanofluids. *New J. Phys.* **2007**, *9*, 367. [CrossRef]
49. Chhabra, R.P.; Richardson, J.F. *Non-Newtonian Flow and Applied Rheology: Engineering Applications*; Butterworth-Heinemann: Oxford, UK, 2011.
50. Mahian, O.; Kolsi, L.; Amani, M.; Estellé, P.; Ahmadi, G.; Kleinstreuer, C.; Marshall, J.S.; Siavashi, M.; Taylor, R.A.; Niazmand, H.; et al. Recent advances in modeling and simulation of nanofluid flows-Part I: Fundamentals and theory. *Phys. Rep.* **2019**, *790*, 1–48. [CrossRef]
51. Bhatti, M.M.; Bég, O.A.; Abdelsalam, S.I. Computational Framework of Magnetized MgO–Ni/Water-Based Stagnation Nanoflow Past an Elastic Stretching Surface: Application in Solar Energy Coatings. *Nanomaterials* **2022**, *12*, 1049. [CrossRef]
52. Zhang, L.; Bhatti, M.M.; Michaelides, E.E.; Marin, M.; Ellahi, R. Hybrid nanofluid flow towards an elastic surface with tantalum and nickel nanoparticles, under the influence of an induced magnetic field. *Eur. Phys. J. Spec. Top.* **2021**, 1–13. [CrossRef]
53. Yacob, N.A.; Ishak, A.; Pop, I. Falkner–Skan problem for a static or moving wedge in nanofluids. *Int. J. Therm. Sci.* **2011**, *50*, 133–139. [CrossRef]
54. Kuo, B.L. Heat transfer analysis for the Falkner–Skan wedge flow by the differential transformation method. *Int. J. Heat Mass Transf.* **2005**, *48*, 5036–5046. [CrossRef]
55. Reddy, P.S.; Sreedevi, P. Effect of thermal radiation and volume fraction on carbon nanotubes based nanofluid flow inside a square chamber. *Alex. Eng. J.* **2021**, *60*, 1807–1817. [CrossRef]
56. Marzougui, S.; Mebarek-Oudina, F.; Assia, A.; Magherbi, M.; Shah, Z.; Ramesh, K. Entropy generation on magneto-convective flow of copper–water nanofluid in a cavity with chamfers. *J. Therm. Anal. Calorim.* **2021**, *143*, 2203–2214. [CrossRef]
57. Gul, T.; Bilal, M.; Alghamdi, W.; Asjad, M.I.; Abdeljawad, T. Hybrid nanofluid flow within the conical gap between the cone and the surface of a rotating disk. *Sci. Rep.* **2021**, *11*, 1180. [CrossRef]
58. Khan, U.; Ahmed, N.; Mohyud-Din, S.T.; Hamadneh, N.N.; Khan, I.; Andualem, M. The dynamics of H2O suspended by multiple Shaped Cu nanoadditives in rotating system. *J. Nanomater.* **2021**, *2021*, 7299143. [CrossRef]
59. Jamshed, W.; Eid, M.R.; Aissa, A.; Mourad, A.; Nisar, K.S.; Shahzad, F.; Saleel, C.A.; Vijayakumar, V. Partial velocity slip effect on working magneto non-Newtonian nanofluids flow in solar collectors subject to change viscosity and thermal conductivity with temperature. *PLoS ONE* **2021**, *16*, e0259881. [CrossRef]
60. Ali Zaidi, S.Z.; Khan, U.; Abdeljawad, T.; Ahmed, N.; Mohyud-Din, S.T.; Khan, I.; Nisar, K.S. Investigation of thermal transport in multi-Shaped Cu nanomaterial-based nanofluids. *Materials* **2020**, *13*, 2737.

nanomaterials

Article

Analytical Investigation of the Time-Dependent Stagnation Point Flow of a CNT Nanofluid over a Stretching Surface

Ali Rehman [1], Anwar Saeed [2,*], Zabidin Salleh [1], Rashid Jan [3] and Poom Kumam [2,4,*]

[1] Department of Mathematics, Faculty of Ocean Engineering Technology and Informatics, Universiti Malaysia Terengganu, Kuala Nerus 21030, Terengganu, Malaysia; alirehmanchd8@gmail.com (A.R.); zabidin@umt.edu.my (Z.S.)

[2] Center of Excellence in Theoretical and Computational Science (TaCS-CoE), Science Laboratory Building, Faculty of Science, King Mongkut's University of Technology Thonburi (KMUTT), 126 Pracha-Uthit Road, Bang Mod, Thung Khru, Bangkok 10140, Thailand

[3] Department of Mathematics, Univesity of Swabi, Swabi 94640, Pakistan; rashid_ash2000@yahoo.com

[4] Department of Medical Research, China Medical University Hospital, China Medical University, Taichung 40402, Taiwan

* Correspondence: anwarsaeed769@gmail.com (A.S.); poom.kum@kmutt.ac.th (P.K.)

Abstract: The heat transfer ratio has an important role in industry and the engineering sector; the heat transfer ratios of CNT nanofluids are high compared to other nanofluids. This paper examines the analytical investigation of the time-dependent stagnation point flow of a CNT nanofluid over a stretching surface. For the investigation of the various physical restrictions, single and multi-walled carbon nanotubes (SWCNTs, MWCNTs) were used and compared. The defined similarity transformation was used, to reduce the given nonlinear partial differential equations (PDEs) to nonlinear ordinary differential equations (ODEs). The model nonlinear ordinary differential equations were solved, with an approximate analytical (OHAM) optimal homotopy asymptotic method being used for the model problem. The impact of different parameters such as magnetic field parameter, unsteady parameter, dimensionless nanoparticles volume friction, Prandtl number, and Eckert number are interpreted using graphs, in the form of the velocity and temperature profile.

Keywords: CNTs; nanofluid; stretching surface; heat transfer; stagnation point

check for
updates

Citation: Rehman, A.; Saeed, A.; Salleh, Z.; Jan, R.; Kumam, P. Analytical Investigation of the Time-Dependent Stagnation Point Flow of a CNT Nanofluid over a Stretching Surface. *Nanomaterials* **2022**, *12*, 1108. https://doi.org/10.3390/nano12071108

Academic Editors: M. M. Bhatti, Kambiz Vafai, Sara I. Abdelsalam and Yang-Tse Cheng

Received: 1 February 2022
Accepted: 21 March 2022
Published: 28 March 2022

Publisher's Note: MDPI stays neutral with regard to jurisdictional claims in published maps and institutional affiliations.

1. Introduction

The study of Magneto–Marangoni convection, which is caused by surface tension, has been a popular research topic for scientists and engineers in recent years. This is due to the numerous applications, including thin liquid layer scattering, atomic reactors, semiconductor processing, dynamic uses in the welding process, crystal development, material science, varnishes, silicon melting, and many more. In addition, the Marangoni phenomena are commonly exploited in fine art mechanisms, such as pigment on the ground. The colorant or dye is suspended on the outside surface of the needed medium, such as water or another thickness fluid, in this technique. To make a print, the medium is encased in rag or paper. Pop et al. [1] investigated the various characteristics of the Marangoni convection technique for thermo-solutal boundary films. Marangoni convection was researched by Al-Mudhaf and Chamkha [2], utilizing a temperature and solute gradient moving through a porous medium. Wang [3] used perturbation solutions to implement the Marangoni convection concept in a thin film spray. By examining the power-law model under Marangoni, Chen [4] extended Wang's idea. Magyari and Chamkha [5] used the flow assumption of a high Reynolds number to study the influence of Marangoni convection. Lin et al. [6,7] investigated the architectures of a MHD Marangoni-convective exchange of power-law nanoliquid with temperature gradients in a MHD Marangoni-convective exchange of power-law nanoliquid. Aly and Ebaid [8] used the Laplace transformation on a permeable surface and convective boundary film conditions, to compute the exact solution of Marangoni flow of a viscid nanoliquid. Ellahi et al. [9] studied the effects of various forms of nanoscale materials in an ethylene glycol-based aqueous nanofluid solution. The analytical results of Marangoni convective heat conversation of power-law liquids in permeable media were computed by Xu and Chen [10]. Sheikholeslami [11] proposed an ideal model of two-phase Marangoni convective MHD nanoliquid hydrothermal flow. Various types of research have recently focused on the Marangoni convection flow in various flow features [12–16].

The energy crisis has become a major topic in recent years, with many academics attempting to discover advanced resources for restoring energy and controlling the consumption of heat transfer devices. Solids, gases, and liquids are all part of these resources. The addition of tiny (1–100 nm) solid metal particles into common liquids can improve their thermal efficiency. Working liquids with low heat transfer rates, such as bioliquids, polymeric solutions, various oils, greases, water, toluene, refrigerants, ethylene glycol, and others, are widely utilized in a variety of engineering and scientific applications. The well-known metallic entities and their oxides, $TiO_2 Al_2O_3$, are placed in operating liquids to improve the thermal efficiency of these liquids. Choi [17] established the optimal theory of nanoscale metallic object dispersal in molecular fluids from this perspective. He conducted an experiment on nanoliquids and came to the conclusion that the contribution of these metallic items is a fantastic means for improving thermal proficiency. After combining nanoparticles of Cu at a rate of 0.3 percent volume fraction, Eastman et al. [18] observed a 40% improvement in the thermal performance of $C_2H_6O_2$. Khamis et al. [19] investigated the heat transport of $Cu_2Al_3 + H_2O$ and nanoliquids through a porous pipe. Using Al_2O_3 water nanoliquids, Malvandi and Ganji [20] investigated the effect of nanoparticle volume fraction on heat transfer rate improvement. Using the $Al_2OAg_3 + H_2O$ nanoliquid flow through an upright cone, Reddy and Chamkha [21] discovered a significant increase in the rate of heat exchange. The movement of Al_2O_3-based non-Newtonian nanoliquids over a circular region was studied by Barnoon and Toghraie [22]. A study of nanofluid flow in a microchannel was completed by Hossenine et al. [23]. Carbon nanotubes (CNTs) are carbon tube-shaped particles of nanometer dimensions. The strong connection between carbon atoms distinguishes CNTs, and the tubes can have a high aspect ratio. CNTs are classed as single-walled or multi-walled nanotubes (SWNTs or MWNTs), depending on their structure. Different properties of CNT nanoparticles, such as physical, electrical, optical, and thermal, have boosted the use of nanofluids in engineering processes such as nano- and microelectronics, biosensors, ultra-capacitors, atomic reactors, gas storage, textile engineering, flat-plate display, and medicinal tools.

Nonporous cleansers, solar collectors, and a variety of coatings are among the many other applications for CNTs. The flow of a nanoliquid containing nanoparticles was discussed by Xie and colleagues [24] (MWCNTs). The thermal characteristics of nanoliquids were improved using these nanoparticles. The motion and heat transport properties of CNT nanofluids over a tube were studied by Ding et al. [25]. Haq et al. [26] used a stretching plate to calculate the numerical results of three distinct molecular liquid flows incorporating CNT nanoparticles. When the volume fraction of particles was 15%, Ueki et al. [27] found that the improvement in thermal properties of carbon blocks increased by 7% and nanopowder increases by 19%. The 3D flow of both nanoparticles (SWCNTs/water and MWCNTs/water) and motor oil was studied by Rehman et al. [28]. The numerical research of CNT–nanoliquids flow and heat transport through a perpendicular cone was developed by Sreedevi et al. [29]. The flow of SWCNT- and MWCNT-nanofluid with a revolving and extending disc was studied by Jyothi et al. [30]. In this respect, a number of researchers [31,32] have recently looked at the flow of CNT nanoparticles from various physical perspectives. Due to their presence in a variety of technical and engineering applications, such as cylinder surface patterning [33], the cooling of optical fibers [34], greasing films on the inner walls of tubes, and condensing vapor on heat pipes [35] have been thoroughly investigated. Frenkel [36] pioneered modelling research on liquid film flow on a cylinder surface. Different models have been investigated to analyze the falling of a liquid film on a perpendicular cylinder, including the thin liquid model [37] and the thick film model [38]. The experimental examination of thin fluid film motion on a vertical cylinder by Duprat et al. [39] revealed that four different forms of flow can occur in the scheme. The 3D flow of a thin liquid sheet on a vertical cylinder was studied by Ding and Wong [40]. Alshomrani and Gul [41] investigated a thin film spray of an Al_2OCu_3 nanoliquid over a slick cylinder. Ali et al. [42] investigated the CNTs nanofluid across a stretched surface utilizing blood as a base fluid. The nonlinear dynamics of thin fluid flows through a hot cylinder were studied by Chao et al. [43].

Inspired by the literature, we present an analytical investigation of time-dependent stagnation point flow of a CNT nanofluid over a stretching surface. The key objective of this analysis was to improve the heat transfer ratio, because heat transfer ratio has some important applications in the engineering and industrial sectors. In our research paper, we used a CNT nanofluid, which has a higher heat transfer ratio compared to other nanofluids; we also discuss the convergence of the approximate analytical method and the effects of different parameters on fluid particle motion, as well as a comparison of the present research work with the previously published work. The flow analysis is considered over a stretching surface; the similarity transformation is used to convert the nondimensionless form of the differential equation to the dimensionless form. The following characteristics indicate the study's novelty: A uniform magnetic field was used to study the approximate analytical solution of the thin layer flow of a nanofluid. This was the first investigation of the analytic solution of this model. Liao et al. [44–46] found an approximate solution technique that does not depend upon the small parameters used to discover the problem's series solution. Using the BVP 2.0 program, the residual sum of the acquired results was calculated up to the 25th order approximation [47–51].

2. Mathematical Formulation

Consider the time-dependent 2D stagnation point flow of a CNT nanofluid along with heat transfer in the presence of a magnetic field on a stretching surface. The continuity, momentum, and temperature equation for the given flow problem is given below:

$$\frac{\partial u}{\partial x} + \frac{\partial v}{\partial y} = 0,$$ (1)

$$\rho_{nf}\left(\frac{\partial u}{\partial t} + u\frac{\partial u}{\partial x} + v\frac{\partial u}{\partial y}\right) = \left(U_s\frac{dU_s}{dx} + \mu_{nf}\frac{\partial^2 u}{\partial y^2} - \sigma_{nf}B_0^2(u - U_s)\right),$$ (2)

$$\frac{\partial T}{\partial t} + u\frac{\partial T}{\partial t} + v\frac{\partial T}{\partial t} = \frac{k}{(\rho C_p)_{nf}}\frac{\partial^2 T}{\partial y^2} + \frac{\mu_{nf}}{(\rho C_p)_{nf}}\left(\frac{\partial u}{\partial y}\right)^2$$ (3)

T represents temperature of the surface, k represents the thermal conductivity, ρ represents the density of the fluid, and c_p represents the specific heat. u and v are the velocity component along the x and y directions, the distance along the sheet is denoted by x and the distance perpendicular to the sheet is denoted by y, the velocity of the stagnation point is denoted by $U_w = ax$ with a > 0, v represents kinematic viscosity of the fluid.

The boundary condition for velocity and temperature are

$$u = U_w,\ v = 0\ at\ y = 0;\ u \rightarrow U_s\ as\ y \rightarrow \infty,$$
$$T = T_w\ at\ y = 0;\ T \rightarrow T_\infty\ as\ y \rightarrow \infty.$$ (4)

T_w represents the temperature of the sheet, and T_∞ represents free stream temperature The stream functions for the given flow problem are

$$u = \frac{\partial \psi'}{\partial y}\ and\ v = -\frac{\partial \psi'}{\partial x}.$$ (5)

Similarity transformations of the defined flow problem are

$$\psi = \sqrt{\frac{a v_f}{1 - \alpha t}}\, x f(\eta) \ \text{And}\ \eta = y\sqrt{\frac{a}{v_f(1 - \alpha t)}},\ \theta(\eta) = \frac{T - T_\infty}{T_W - T_\infty}. \tag{6}$$

Putting Equations (5) and (6) into Equations (1)–(4), gives

$$f''' + (1 - \phi)^{2.5}\left((1 - \phi) + \phi\frac{\rho_s}{\rho_s}\right)\left[ff'' - f'^2 - S\left(f' + \frac{\eta}{2}f''\right)\right] - (1 - \phi)^{2.5} M f' = 0 \tag{7}$$

$$\frac{k_{nf}}{k_f}\theta'' - \left((1 - \phi) + \phi\frac{(\rho C_p)_{CNT}}{(\rho C_p)_f}\right)\left(\Pr(\theta' f) + \frac{S}{2}Ec(\eta\theta')\right) = 0 \tag{8}$$

With boundary conditions

$$\begin{aligned}
&f(\eta) = 0, f'(\eta) = \frac{c}{a}\ at\ \eta = 0, f'(\eta) \to 1 as \eta \to \infty,\\
&\theta(\eta) = 1 at\ \eta = 0; \theta(\eta) \to 0 as \eta \to \infty
\end{aligned} \tag{9}$$

$$\frac{\rho_{nf}}{\rho_f} = (1 - \phi) + \phi\left(\frac{\rho_{CNT}}{\rho_f}\right),\ \frac{\mu_{nf}}{\mu_f} = (1 - \phi)^{-2.5},\ \frac{(\rho\,Cp)_{nf}}{(\rho\,Cp)_f} = \left[(1 - \phi) + \phi\left(\frac{(\rho\,cp)_{CNT}}{(\rho\,cp)_f}\right)\right],$$

$$\frac{k_{nf}}{k_f} = \frac{1 - \phi + 2\phi\left(\frac{k_{CNT}}{k_{CNT} - k_f}\right) ln\left(\frac{k_{CNT} + k_f}{2\,k_f}\right)}{1 - \phi + 2\phi\left(\frac{k_f}{k_{CNT} - k_f}\right) ln\left(\frac{k_{CNT} + k_f}{2\,k_f}\right)}. \tag{10}$$

where prime denotes differentiation with respect to η and the constant $a > 0, b > 0$, $\Pr = \frac{v}{\alpha}$ is the Prandtl number. α is thermal diffusivity, $Ec = \frac{u_w^2}{C_p(T_w - T_\infty)}$ represents the Eckert number, $S = \frac{\gamma}{b}$, represents the unsteady parameter, ϕ represents a dimensionless nanoparticle volume friction, and $M = \frac{\sigma^* B_0^2}{\rho_f b}$ represents the magnetic field

The Skin Friction and Nusselt Number

In this analysis, the two key important quantities are skin friction and Nusselt number, given as

$$\mathrm{Re}_x^{\frac{1}{2}} C_{nf} = -\frac{(1 - \phi)^{-2.5}}{(1 - \phi) + \phi\left(\frac{\rho_1}{\rho_f}\right)} f''(0), \mathrm{Re}_x^{-\frac{1}{2}} Nu_x = -\frac{k_s}{k_{nf}}\theta'(0), \tag{11}$$

where $\mathrm{Re}_x = U_w L / v_f$.

3. Method of Solution

We apply the basic idea of OHAM to the given problem given below.

$$f''' - ff'' - f'^2 - f'' - f' = 0,\ f(0) = 0,\ f'(\infty) = 1,\ f'(1) = \frac{c}{a} \tag{12}$$

where η is a similarity variable, $f(\eta)$ is related to the stream function and prime denote derivative with respect to η. Let $\lambda > 0$ denote a kind of spatial scale parameter by means of transformation

$$f(\eta) = \lambda^{-1} u(\xi),\ \xi = \lambda\eta \tag{13}$$

The original Equation (12) becomes

$$\begin{aligned}
&u'''(\xi) - uu''(\xi) - (u'(\xi))^2 - u''(\xi) - u'(\xi) = 0\\
&u(0) = 0,\ u'(\infty) = 1,\ u(1) = \frac{c}{a}
\end{aligned} \tag{14}$$

The boundary condition $u'(\infty) = 1$, the asymptotic property $u \to \xi$ *as* $\xi \to \infty$. According to the theses information, $u(\xi)$ can be stated in the form

$$u(\xi) = A_{0,0} + \xi + \sum_{m=1}^{\infty} \sum_{n=0}^{\infty} A_{m,n} \xi^n \exp(-m\xi) \tag{15}$$

where $A_{m,n}$ represent constants. The so called solution expression of $u(\xi)$ has an important role in OHAM. Subsequently, we have four boundary conditions, rendering to the solution (15) the simplest four terms of $u(\xi)$ are $A_{0,0}, \xi, A_{1,0}\exp(-\xi), A_{2,0}\exp(-\xi)$, we chose the initial solution in the form of

$$u_0(\xi) = A_{0,0} + \xi + A_{1,0}\exp(-\xi) + A_{2,0}\exp(-\xi) \tag{16}$$

where $A_{0,0}, A_{1,0}, A_{2,0}$ are unknown constants. Enforcing $u_0(\xi)$ to satisfy the three boundary conditions, we have $A_{0,0} = 0, A_{1,0} = -1, A_{2,0} = 0$ so we have

$$u_0(\xi) = \xi - 1 + e^{-\xi} \tag{17}$$

According to the solution (17), we should chose the linear operator L, in a manner that $L(u) = 0$

$$L\left(C_0 + C_1\xi + C_2 e^{-\xi}\right) \tag{18}$$

The estimation for velocity and for temperature are

$$f_0(\eta) = \frac{c}{a} + e^{-\eta} \tag{19}$$

$$\theta_0(\eta) = e^{-\eta} + 1 \tag{20}$$

which is obtained from the linear operator

$$L_f = f''' - ff'' = 0 \ , L_\theta = \ \theta'' = 0, \tag{21}$$

4. Results and Discussion

In this section, we discuss the influence of the different parameters, such as $\{M, S, \phi, \text{Pr}, \text{Ec}\}$ (magnetic field parameter, unsteady parameter, dimensionless nanoparticles volume friction, Prandtl number and Eckert number) on both $f'(\eta)$ *and* $\theta(\eta)$. The CNT nanofluid is used for the enhancement of heat improvement applications. The flow investigation has studied on a movable surface, along with the magnetic field. To transform the nonlinear partial differential equation (PDE) to a nonlinear ordinary differential equation (ODE), we used the defined similarity transformation. The nonlinear differential equations were solved with the help of the approximate analytical method, named the optimal homotopy asymptotic method (OHAM). The convergence control parameter for the particular problems is also discussed in table form. With the help of the analytical method, a solution of the nonlinear differential equations for velocity and temperature profiles were obtained. The results of the particular problem are emphasized in Figures 1–9; Figures 1–3 represent the effect of dissimilar parameters on velocity distribution, Figures 4 and 5 represent the effect of dissimilar parameters on temperature profile, Figures 6 and 7 represent the effect of dissimilar parameters on C_f (skin friction), and Figures 8 and 9 represent the effect of dissimilar parameters on the Nusselt number. From Figures 6 and 7 we can observe that skin friction is the function increasing the magnetic field parameter, unsteady parameter, and dimensionless nanoparticle volume friction; by increasing the magnitude of this parameter, resistance forces are produced which oppose the fluids particle motion, and due to this effect the skin friction increases. From Figures 8 and 9, we can see that the Nusselt number is the increasing function of the Eckert number, Prandtl number, and magnetic field parameter; physically, by enhancing the magnitude of these parameters the fluid particles motion is reduced, and, as a result, friction forces are produced, due to this Nusselt number increasing. The convergence control parameter of the given problem was obtained up to the 25th iteration for both velocity and temperature distribution and presented in Tables 1 and 2. From Tables 1 and 2 we can see that as we increased the number of iterations, the residual error decreased, and a strong convergence was obtained on the approximate analytical solution. A comparison of the approximate analytical method to the numerical method is presented in Tables 3 and 4. Tables 5 and 6 represent the thermo-physical properties and thermo-conductivity of the nanofluids. The variation in magnetic field parameter M on the velocity profile is captured in Figure 1, for both MWCNT and SWCNT on $f'(\eta)$; from Figure 1 we see that velocity is the decreasing function of the magnetic field parameter. Such a state happens due to the creation of a resistive force known as the Lorentz force. The magnitude of these forces increases with the increase of the magnetic field parameter, which oppose the fluid particle motion in the opposite direction; the motion of the fluid particles decreases, and as a result the velocity profile decreases. The variation in the dimensionless nanoparticle volume fraction parameter ϕ on velocity profile is captured in Figure 2, for both the MWCNT and SWCNT velocity profile; from Figure 2 we can see that velocity is the decreasing function of the dimensionless nanoparticle volume fraction parameter ϕ. Such a state happens due to the production of a resistive type force, known as viscose force. The strength of such a force increase with the increase in strength of the dimensionless nanoparticle volume fraction parameter ϕ, which counteracts the motion of the fluid within the boundary layer and reduces the thickness of the boundary layer. The variation in the time-dependent parameter S in velocity profile is captured in Figure 3, and for both MWCNT and SWCNT we can see that velocity is the decreasing function of the time-dependent parameter S. Figure 4 shows the relation between Eckert number Ec and temperature profile; this relation is in direct relation, or the temperature profile is the increasing function of, Eckert number Ec. By increasing the Eckert number, this will improve the kinetic energy due to the intermolecular collision increasing, and, as a result, the temperature profile increases. Figure 5 shows the Pr for both MWCNT and SWCNT and the temperature distribution. It is noticeable from Figure 5 that the relation between $\theta(\eta)$ and Pr is an inverse relation, where the largest value of Pr reduces the temperature distribution. Actually the thickness of the momentum boundary layer is superior than that of the thermal boundary layer, or the viscous diffusion is larger than the thermal diffusion, and, therefore, a larger Prandtl number reduces the thermal boundary layer; therefore, the Prandtl number was used as the Colling agent.

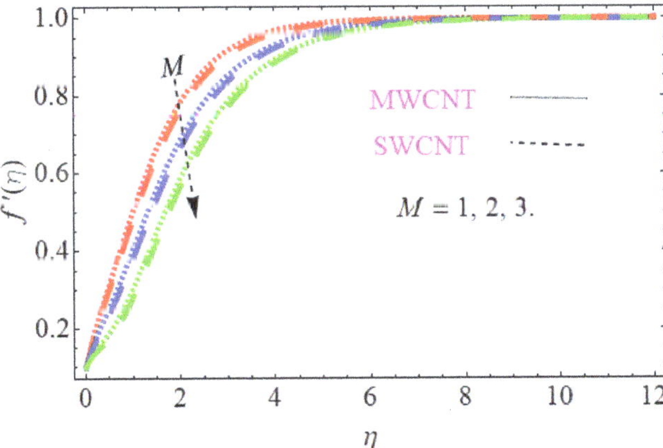

Figure 1. Consequence of M (magnetic field parameter) for velocity distribution.

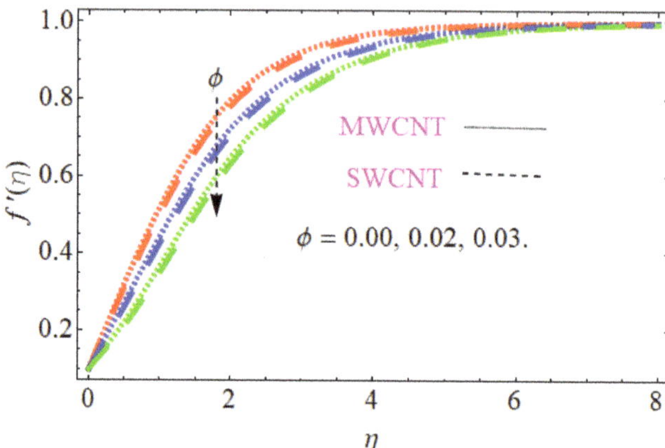

Figure 2. Consequence of dimensionless nanoparticle volume friction for velocity distribution.

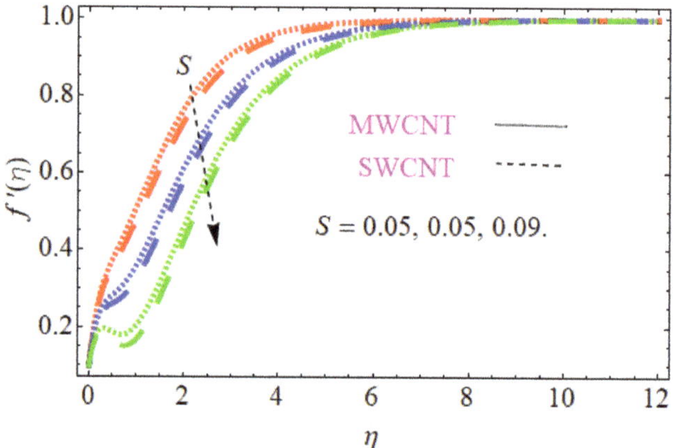

Figure 3. Consequence of unsteady parameter *S* for velocity distribution.

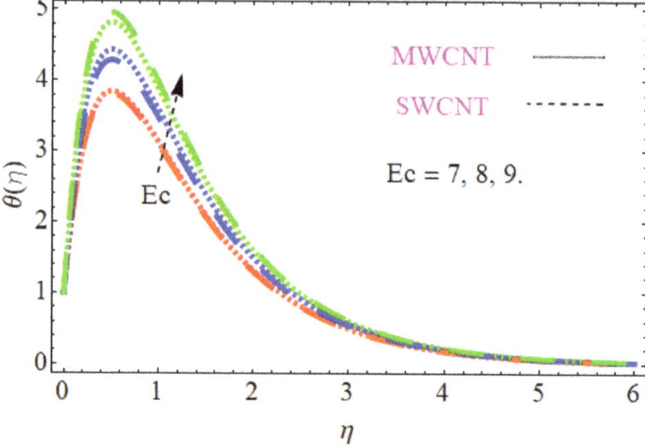

Figure 4. Consequence of the Eckert number for temperature distribution.

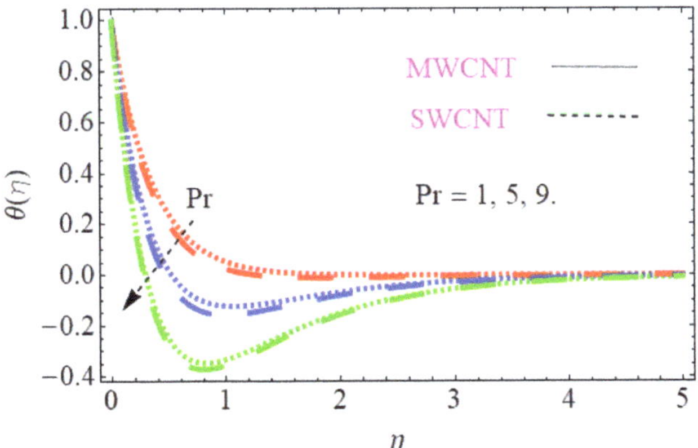

Figure 5. Consequence of Prandtl number for temperature distribution.

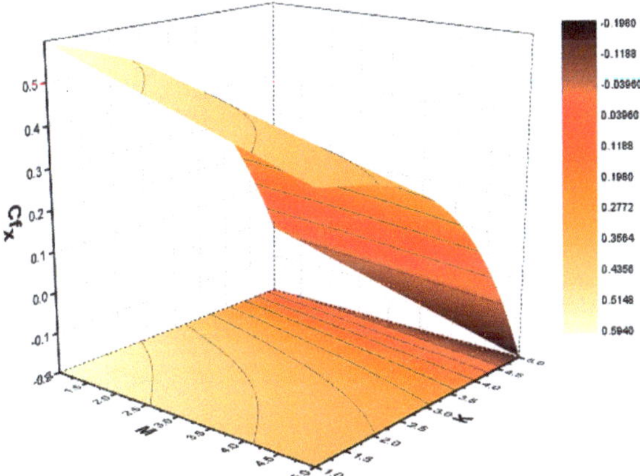

Figure 6. Effect of skin friction on M and S.

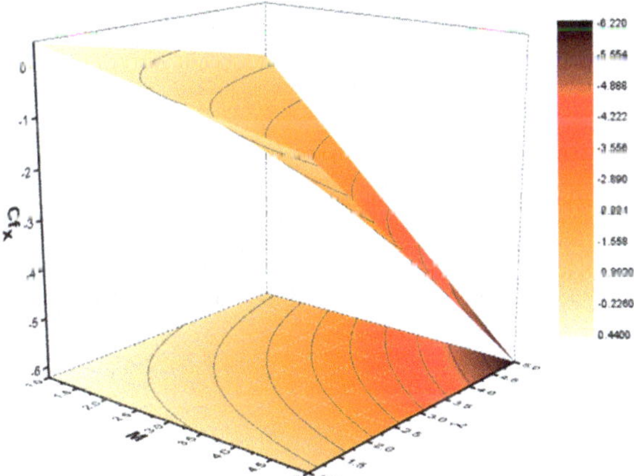

Figure 7. Effect on skin friction on M and ϕ.

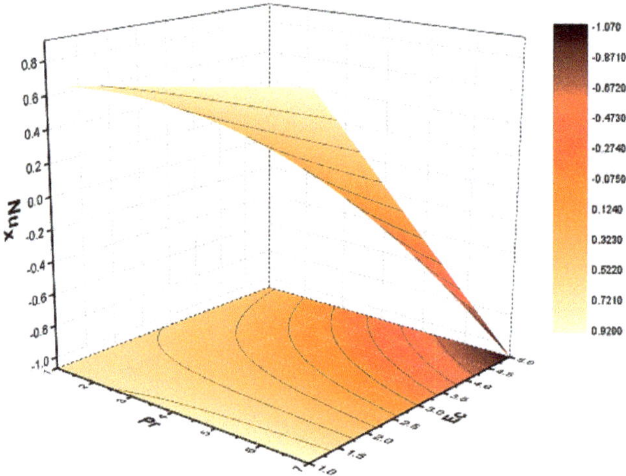

Figure 8. Effect of Nusselt number on Pr and Ec.

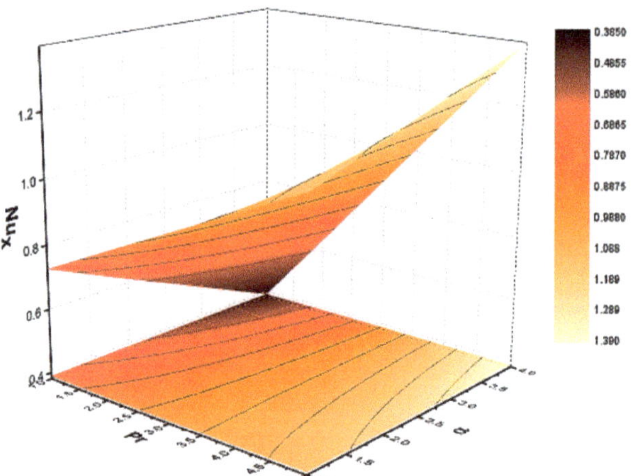

Figure 9. Effect of Nusselt number on Pr and M.

Table 1. Convergence of the method for MWCNT.

m	MWCNT	MWCNT
5	0.1431×10^{-1}	0.4775×10^{-3}
10	0.3014×10^{-2}	0.7873×10^{-5}
15	0.3443×10^{-3}	0.4729×10^{-7}
20	0.9298×10^{-5}	0.5453×10^{-8}
25	0.3787×10^{-7}	0.2412×10^{-9}

Table 2. Show the convergence of method for SWCNT.

m	SWCNT	SWCNT
5	0.4391×10^{-1}	0.4574×10^{-1}
10	0.5166×10^{-3}	0.7359×10^{-2}
15	0.8238×10^{-5}	0.7749×10^{-5}
20	0.4626×10^{-6}	0.3121×10^{-7}
25	0.9433×10^{-9}	0.5132×10^{-9}

Table 3. OHAM and numerical comparison for $f(\eta)$.

m	Numerical	OHAM	Absolute Error
1	$1.00\ldots$	$1.00\ldots$	$0.0\ldots$
2	$1.72\ldots$	$1.70\ldots$	$2 \times 10^{-2}\ldots$
3	$1.71\ldots$	$1.69\ldots$	$2 \times 10^{-2}\ldots$
4	$1.96\ldots$	$1.94\ldots$	$2 \times 10^{-2}\ldots$
5	$0.87\ldots$	$0.83\ldots$	$4 \times 10^{-2}\ldots$
6	$0.23\ldots$	$0.21\ldots$	$2 \times 10^{-2}\ldots$
7	$0.29\ldots$	$0.27\ldots$	$2 \times 10^{-2}\ldots$
8	$0.39\ldots$	$0.38\ldots$	$1 \times 10^{-2}\ldots$
9	$0.42\ldots$	$0.37\ldots$	$5 \times 10^{-2}\ldots$
10	$0.92\ldots$	$0.90\ldots$	$2 \times 10^{-2}\ldots$

Table 4. OHAM and numerical comparison for $\theta(\eta)$.

η	Numerical	OHAM	Absolute Error
1	$1.00\ldots$	$1.00\ldots$	$0.0\ldots$
2	$1.31\ldots$	$1.29\ldots$	$2 \times 10^{-2}\ldots$
3	$1.12\ldots$	$1.10\ldots$	$2 \times 10^{-2}\ldots$
4	$1.70\ldots$	$1.65\ldots$	$5 \times 10^{-2}\ldots$
5	$1.44\ldots$	$1.41\ldots$	$3 \times 10^{-2}\ldots$
6	$1.34\ldots$	$1.30\ldots$	$4 \times 10^{-2}\ldots$
7	$1.95\ldots$	$1.90\ldots$	$5 \times 10^{-2}\ldots$
8	$1.90\ldots$	$1.80\ldots$	$1 \times 10^{-2}\ldots$
9	$1.54\ldots$	$1.50\ldots$	$4 \times 10^{-2}\ldots$
10	$1.35\ldots$	$1.30\ldots$	$5 \times 10^{-2}\ldots$

Table 5. The thermo-physical properties.

Physical Properties		Thermal Conduct $K(W/mk)$	Specific Heat $C_p(J/kgK)$	Density $\rho(kg/m^3)$
Solid particles	SWCNTs	6600	2600	2600
	MWCNTs	3000	1600	1600

Table 6. The thermal conductivity values at different volume fractions.

Volume Fraction ϕ	0.0	0.01	0.02	0.03	0.04
k_{nf}(SWCNTs)	0.145	0.147	0.204	0.235	0.266
k_{nf}(MWCNTs)	0.145	0.172	0.2	0.228	0.2257

5. Conclusions

This paper examined the analytical investigation of the stagnation point flow of a CNT nanofluid over a stretching surface. In this research article, we used an approximate analytical method for the solution of the nonlinear differential equation to find a series solution for velocity and another for temperature profile. To transform the nonlinear partial differential equation (PDE) into a nonlinear ordinary differential equation (ODE), we used the defined similarity transformation. By using the approximate analytical (OHAM) optimal homotopy asymptotic method to solve the obtained nonlinear ordinary differential equations. The impact of different parameters, such as magnetic field parameter, dimensionless nanoparticle volume fraction parameter, unsteady parameter, Prandtl number, and Eckert number were interpreted through graphs. The skin friction coefficient and Nusselt number were explained in the form of graphs. The present work was found to be in very good agreement with those published previously. The main outcomes of the present analysis are the following:

1. Increasing the value of the dimensionless nanoparticle volume fraction parameter reduces velocity.
2. Increasing the value of the magnetic field parameter reduces velocity
3. Increasing the value of the unsteady parameter reduces velocity
4. Increasing the value of the Prandtl number reduces the temperature profile
5. Increasing the value of the Eckert number increases the temperature profile

Of the numerous future physical world problems to be modeled in science and engineering, most of them will be nonlinear, so we cannot find the exact solution of the nonlinear differential equation for both the partial differential equation and ordinary differential equation; therefore, this method could be used in the future to find the approximate analytical series solution of the nonlinear differential equation, which has the best convergence to the exact solution. In the future, heat transfer phenomena will play a key role in engineering and industry. In our paper, we used a CNT nanofluid and discussed the heat transfer ratio, which plays an important role.

Author Contributions: Conceptualization, A.R.; methodology, A.R., A.S., Z.S., R.J. and P.K.; software, R.J.; validation, Z.S.; formal analysis, A.R. and A.S.; investigation, A.S. and R.J., resources, Z.S.; writing—original draft preparation, A.R.; writing—review and editing, A.S. and P.K.; visualization, Z.S. and R.J.; supervision, Z.S.; project administration, P.K.; funding acquisition, P.K. All authors have read and agreed to the published version of the manuscript.

Funding: The financial support provided by the Center of Excellence in Theoretical and Computational Science (TaCS-CoE), KMUTT. Moreover, this research project was supported by Thailand Science Research and Innovation (TSRI) Basic Research Fund: Fiscal year 2022 (FF65).

Institutional Review Board Statement: Not applicable.

Informed Consent Statement: Not applicable.

Data Availability Statement: All data used in this manuscript have been presented within the article.

Acknowledgments: The authors acknowledge the financial support provided by the Center of Excellence in Theoretical and Computational Science (TaCS-CoE), KMUTT. Moreover, this research project was supported by Thailand Science Research and Innovation (TSRI) Basic Research Fund: Fiscal year 2022 (FF65).

Conflicts of Interest: The authors have no conflict of interest.

Nomenclature

$x, y,$	Cartesian coordinates
$u, v,$	Velocity components
U_w, V_w	Velocities of the stretching sheet
S	Time dependent parameter
T	Local Temperature
M	Magnetic field
Nu_x	Local Nusselt number
pr	Prandtl number
T_w	Surface temperature
B_0	Constant magnetic field

T_∞	Ambient temperature-
Ec	Eckert number
C_{fx}	Skin friction coefficient in $x-$ direction-
θ	Dimensionless temperature
ρ	Density of the fluid
ν	Fluid viscosity
η	Independent variable
λ	Spatial scale parameter
k	Thermal conductivity
c_p	Specific heat

References

1. Pop, I.; Postelnicu, A.; Grosan, T. Thermo-solutal Marangoni forced convection boundarylayers. *Meccanica* **2001**, *36*, 555–571.
2. Al-Mudhaf, A.; Chamkha, A.J. Similarity solutions for MHD thermo-solutal Marangoni convection over a flat surface in the presence of heat generation or absorption effects. *Heat Mass Transf.* **2005**, *42*, 112–121.
3. Wang, C.Y. Liquid film sprayed on a stretching surface. *Chem. Eng. Commun.* **2006**, *193*, 869–878.
4. Chen, C.-H. Marangoni effects on forced convection of power-law liquids in a thin film over a stretching surface. *Phys. Lett. A* **2007**, *370*, 51–57. [CrossRef]
5. Magyari, E.; Chamkha, A.J. Exact analytical solutions for thermo-solutal Marangoni convection in the presence of heat and mass generation or consumption. *Heat Mass Transf.* **2007**, *43*, 965–974.
6. Lin, Y.; Zheng, L.; Zhang, X. Magneto-hydrodynamics thermo-capillary Marangoni convection heat transfer of power-law fluids driven by temperature gradient. *J. Heat Transf.* **2013**, *135*, 051702.
7. Lin, Y.; Zheng, L.; Zhang, X. Radiation effects on Marangoni convection flow and heat transfer in pseudo-plastic non-Newtonian nanofluids with variable thermal conductivity. *Int. J. Heat Mass Transf.* **2014**, *77*, 708–716. [CrossRef]
8. Aly, E.; Ebaid, A. Exact analysis for the effect of heat transfer on MHD and radiation Marangoni boundary layer nanofluid flow past a surface embedded in a porous medium. *J. Mol. Liq.* **2016**, *215*, 625–639. [CrossRef]
9. Ellahi, R.; Zeeshan, A.; Hassan, M. Particle shape effects on Marangoni convection boundary layer flow of a nanofluid. *Int. J. Numer. Methods Heat Fluid Flow* **2016**, *26*, 2160–2174. [CrossRef]
10. Jiao, C.; Zheng, L.; Lin, Y.; Ma, L.; Chen, G. Marangoni abnormal convection heat transfer of power-law fluid driven by temperature gradient in porous medium with heat generation. *Int. J. Heat Mass Transf.* **2016**, *92*, 700–707. [CrossRef]
11. Sheikholeslami, M.; Chamkha, A.J. Influence of Lorentz forces on nanofluid forced convection considering Marangoni convection. *J. Mol. Liq.* **2017**, *225*, 750–757. [CrossRef]
12. Sheikholeslami, M.; Ganji, D.D. Influence of magnetic field on $CuOeH_2O$ nanofluid flow considering Marangoni boundary layer. *Int. J. Hydrog. Energy* **2017**, *42*, 2748–2755.
13. Mahanthesh, B.; Gireesha, B.J.; PrasannaKumara, B.C.; Shashikumar, N.S. Marangoni convection radiative flow of dusty nanoliquid with exponential space dependent heat source. *Nucl. Eng. Technol.* **2017**, *49*, 1660–1668. [CrossRef]
14. Sheikholeslami, M.; Ganji, D.D. Analytical investigation for Lorentz forces effect on nanofluid Marangoni boundary layer hydrothermal behavior using HAM. *Indian J. Phys.* **2017**, *91*, 1581–1587. [CrossRef]
15. Mahanthesh, B.; Gireesha, B. Thermal Marangoni convection in two-phase flow of dusty Casson fluid. *Results Phys.* **2018**, *8*, 537–544. [CrossRef]
16. Mahanthesh, B.; Gireesha, B.J. Scrutinization of thermal radiation, viscous dissipation and Joule heating effects on Ma-rangoni convective two-phase flow of Casson fluid with fluid particle suspension. *Results Phys.* **2018**, *8*, 869–878.
17. Choi, S.U.S. Enhancing thermal conductivity of fluids with nanoparticles. *ASME-Publ. Fed* **1995**, *231*, 99–106.
18. Eastman, J.A.; Choi, S.U.S.; Li, S.; Yu, W.L.; Thompson, J. Anomalously increased effective thermal conductivities of ethylene glycol-based nano-fluids containing copper nanoparticles. *Appl. Phys.* **2001**, *78*, 718–720.
19. Khamis, S.; Makinde, O.D.; Nkansah-Gyekye, Y. Unsteady flow of variable viscosity Cu-water and Al_2O_3 water nanofluids in a porous pipe with buoyancy force. *Int. J. Numer. Methods Heat Fluid Flow* **2015**, *25*, 1638–1657. [CrossRef]
20. Malvandi, A.; Ganji, D.D. Mixed Convection of Alumina/Water Nanofluid in Microchannels using Modified Buongiorno's Model in Presence of Heat Source/Sink. *J. Appl. Fluid Mech.* **2016**, *9*, 2277–2289. [CrossRef]
21. Reddy, S.; Chamkha, A.J. Heat and mass transfer characteristics of Al_2O_3-water and Ag-water nanofluid through porous media over a vertical cone with heat generation/absorption. *J. Porous Media* **2017**, *20*, 1–17. [CrossRef]
22. Barnoon, P.; Toghraie, D. Numerical investigation of laminar flow and heat transfer of non-Newtonian nanofluid within a porous medium. *Powder Technol.* **2018**, *325*, 78–91. [CrossRef]
23. Hosseini, S.; Sheikholeslami, M.; Ghasemian, M.; Ganji, D. Nanofluid heat transfer analysis in a microchannel heat sink (MCHS) under the effect of magnetic field by means of KKL model. *Powder Technol.* **2018**, *324*, 36–47. [CrossRef]
24. Xie, H.; Lee, H.; Youn, W.; Choi, M. Nanofluids containing multiwalled carbon nanotubes and their enhanced thermal conductivities. *J. Appl. Phys.* **2003**, *94*, 4967. [CrossRef]
25. Ding, Y.; Alias, H.; Wen, D.; Williams, R.A. Heat transfer of aqueous suspensions of carbon nanotubes (CNT nanofluids). *Int. J. Heat Mass Transf.* **2006**, *49*, 240–250. [CrossRef]

26. Haq, R.U.; Khan, Z.H.; Khan, W.A. Thermo-physical effects of carbon nanotubes on MHD flow over a stretching surface. *Phys. E Low-Dimens. Syst. Nanostructu.* **2014**, *63*, 215–222.

27. Ueki, Y.; Aoki, T.; Ueda, K.; Shibahara, M. Thermophysical properties of carbon-based material nanofluid. *Int. J. Heat Mass Transf.* **2017**, *113*, 1130–1134. [CrossRef]

28. Rehman, A.; Mehmood, R.; Nadeem, S.; Akbar, N.S.; Motsa, S.S. Effects of single and multi-walled carbon nanotubes on water and engine oil based rotating fluids with internal heating. *Adv. Powder Technol.* **2017**, *28*, 1991–2002.

29. Sreedevi, P.; Reddy, P.S.; Chamkha, A.J. Magneto-hydrodynamics heat and mass transfer analysis of single and multi-wall carbon nanotubes over vertical cone with convective boundary condition. *Int. J. Mech. Sci.* **2018**, *135*, 646–655. [CrossRef]

30. Jyothi, K.; Reddy, P.S.; Reddy, M.S. Influence of magnetic field and thermal radiation on convective flow of SWCNTs-water and MWCNTs-water nanofluid between rotating stretchable disks with convective boundary conditions. *Powder Technol.* **2018**, *331*, 326–337. [CrossRef]

31. Rahmati, A.; Reiszadeh, M. An experimental study on the effects of the use of multi-walled carbon nanotubes in ethylene glycol/water-based fluid with indirect heaters in gas pressure reducing stations. *Appl. Therm. Eng.* **2018**, *134*, 107–117. [CrossRef]

32. Muhammad, T.; Lu, D.-C.; Mahanthesh, B.; Eid, M.R.; Ramzan, M.; Dar, A. Significance of Darcy-Forchheimer Porous Medium in Nanofluid Through Carbon Nanotubes. *Commun. Theor. Phys.* **2018**, *70*, 361. [CrossRef]

33. Gau, H.; Herminghaus, S.; Lenz, P.; Lipowsky, R. Liquid morphologies on structured surfaces: From microchannels to mi-crochips. *Science* **1999**, *283*, 46–49. [PubMed]

34. Binda, M.; Natali, D.; Iacchetti, A.; Sampietro, M. Integration of an organic photo-detector onto a plastic optical fiber by means of spray coating technique. *Adv. Mater.* **2013**, *25*, 4335–4339.

35. Kundan, A.; Nguyen, T.T.T.; Plawsky, J.L.; Wayner, J.P.C.; Chao, D.F.; Sicker, R.J. Condensation on Highly Superheated Surfaces: Unstable Thin Films in a Wickless Heat Pipe. *Phys. Rev. Lett.* **2017**, *118*, 094501. [CrossRef]

36. Frenkel, A.L. Nonlinear Theory of Strongly Undulating Thin Films Flowing Down Vertical Cylinders. *Eur. Lett.* **1992**, *18*, 583–588. [CrossRef]

37. Chang, H.-C.; Demekhin, E.A. Mechanism for drop formation on a coated vertical fibre. *J. Fluid Mech.* **1999**, *380*, 233–255. [CrossRef]

38. Kliakhandler, I.L.; Davis, S.H.; Bankoff, S.G. Viscous beads on vertical fibre. *J. Fluid Mech.* **2001**, *429*, 381–390. [CrossRef]

39. Duprat, C.; Giorgiutti-Dauphiné, F.; Tseluiko, D.; Saprykin, S.; Kalliadasis, S. Liquid film coating a fiber as a model system for the formation of bound states in active dispersive-dissipative noninear media. *Phys. Rev. Lett.* **2009**, *103*, 234501.

40. Ding, Z.; Wong, T.N. Three-dimensional dynamics of thin liquid films on vertical cylinders with Marangoni effect. *Phys. Fluids* **2017**, *29*, 011701. [CrossRef]

41. Alshomrani, A.S.; Gul, T. A convective study of Al2O3-H2O and Cu-H2Onano-liquid films sprayed over a stretching cyl-inder with viscous dissipation. *Eur. Phys. J. Plus* **2017**, *132*, 1–16.

42. AlSagri, A.S.; Nasir, S.; Gul, T.; Islam, S.; Nisar, K.; Shah, Z.; Khan, I. MHD Thin Film Flow and Thermal Analysis of Blood with CNTs Nanofluid. *Coatings* **2019**, *9*, 175. [CrossRef]

43. Chao, Y.; Ding, Z.; Liu, R. Dynamics of thin liquid films flowing down the non-isothermal cylinder with wall slippage. *Chem. Eng. Sci.* **2018**, *175*, 354–364.

44. Liao, S.-J. A kind of approximate solution technique which does not depend upon small parameters—II. An application in fluid mechanics. *Int. J. Non-Linear Mech.* **1997**, *32*, 815–822. [CrossRef]

45. Liao, S. On the homotopy analysis method for nonlinear problems. *Appl. Math. Comput.* **2004**, *147*, 499–513. [CrossRef]

46. Liao, S.J. *Homotopy Analysis Method in Non-Linear Differential Equations*; Springer: Berlin/Heidelberg, Germany, 2012.

47. Gul, T.; Ferdous, K. The experimental study to examine the stable dispersion of the graphene nanoparticles and to look at the GO–H2O nanofluid flow between two rotating disks. *Appl. Nanosci.* **2018**, *8*, 1711–1728.

48. Gul, T.; Nasir, S.; Islam, S.; Shah, Z.; Khan, M.A. Effective Prandtl Number Model Influences on the γAl_2O_3–AH_2O and $\gamma Al2O3$–$C_2H_6O_2$ Nanofluids Spray Along a Stretching Cylinder. *Arab. J. Sci. Eng.* **2018**, *44*, 1–16.

49. Gohar, T.; Gul, W.; Khan, M.; Shuaib, M.; Khan, M.; Bonyah, E. MWCNTs/SWCNTs Nanofluid Thin Film Flow over a Non-linear Extending Disc: OHAM Solution. *J. Therm. Sci.* **2019**, *28*, 115–122.

50. Gul, T.; Haleem, I.; Ullah, I.; Khan, M.A.; Bonyah, E.; Khan, I.; Shuaib, M. The study of the entropy generation in a thin film flow with variable fluid properties past over a stretching sheet. *Adv. Mech. Eng.* **2018**, *10*, 1687814018789522. [CrossRef]

51. Khan, W.; Idress, M.; Gul, T.; Khan, A.M.; Bonyah, E. Three non-Newtonian fluids flow considering thin film over an un-steady stretching surface with variable fluid properties. *Adv. Mech. Eng.* **2018**, *10*, 1687814018807361.

Article

Computational Framework of Magnetized MgO–Ni/Water-Based Stagnation Nanoflow Past an Elastic Stretching Surface: Application in Solar Energy Coatings

Muhammad Mubashir Bhatti [1,*], Osman Anwar Bég [2] and Sara I. Abdelsalam [3]

[1] College of Mathematics and Systems Science, Shandong University of Science and Technology, Qingdao 266590, China
[2] Multi-Physical Engineering Sciences Group, Mechanical Engineering, School of Science, Engineering and Environment (SEE), Salford University, Manchester M5 4WT, UK; o.a.beg@salford.ac.uk
[3] Basic Science, Faculty of Engineering, The British University in Egypt, Al-Shorouk City 11837, Egypt; sara.abdelsalam@bue.edu.eg
* Correspondence: mmbhatti@sdust.edu.cn

Abstract: In this article, motivated by novel nanofluid solar energy coating systems, a mathematical model of hybrid magnesium oxide (MgO) and nickel (Ni) nanofluid magnetohydrodynamic (MHD) stagnation point flow impinging on a porous elastic stretching surface in a porous medium is developed. The hybrid nanofluid is electrically conducted, and a magnetic Reynolds number is sufficiently large enough to invoke an induced magnetic field. A Darcy model is adopted for the isotropic, homogenous porous medium. The boundary conditions account for the impacts of the velocity slip and thermal slip. Heat generation (source)/absorption (sink) and also viscous dissipation effects are included. The mathematical formulation has been performed with the help of similarity variables, and the resulting coupled nonlinear dimensionless ordinary differential equations have been solved numerically with the help of the shooting method. In order to test the validity of the current results and the convergence of the solutions, a numerical comparison with previously published results is included. Numerical results are plotted for the effect of emerging parameters on velocity, temperature, magnetic induction, skin friction, and Nusselt number. With an increment in nanoparticle volume fraction of both MgO and Ni nanoparticles, the temperature and thermal boundary layer thickness of the nanofluid are elevated. An increase in the porous medium parameter (Darcy number), velocity slip, and thermal Grashof number all enhance the induced magnetic field. Initial increments in the nanoparticle volume fraction for both MgO and Ni suppress the magnetic induction near the wall, although, subsequently, when further from the wall, this effect is reversed. Temperature is enhanced with heat generation, whereas it is depleted with heat absorption and thermal slip effects. Overall, excellent thermal enhancement is achieved by the hybrid nanofluid.

Keywords: MgO–Ni nanoparticles; magnetic hybrid nanofluids; porous medium; thermal and velocity slip; solar coatings; stagnation flow

Citation: Bhatti, M.M.; Bég, O.A.; Abdelsalam, S.I. Computational Framework of Magnetized MgO–Ni/ Water-Based Stagnation Nanoflow Past an Elastic Stretching Surface: Application in Solar Energy Coatings. *Nanomaterials* **2022**, *12*, 1049. https:// doi.org/10.3390/nano12071049

Academic Editor: Mikhail Sheremet

Received: 25 January 2022
Accepted: 21 March 2022
Published: 23 March 2022

Publisher's Note: MDPI stays neutral with regard to jurisdictional claims in published maps and institutional affiliations.

1. Introduction

Renewable energy sources have become increasingly important in recent years, owing to the depletion of fossil fuels and the rise in the price of electricity. Furthermore, fossil fuels emit CO_2, whereas renewable energy sources do not pose this problem and are sustainable and ecologically desirable. Environmentalists believe that the adoption of renewable energy sources will help to reduce global warming and greenhouse emissions [1]. Solar energy has emerged as a viable alternative source of renewable energy in recent years since it is easily accessible, free of pollutants, and causes the least amount of harm to the environment [2]. Solar power plants are capable of supplying thermal energy for use in household applications [3]. The twenty-first century accounts for 40% of the world's fuel market and 60% of the world's energy production. According to a recent study [4], worldwide CO_2 emissions will drop by 75 percent by 2050, compared to 1985 levels. As a result, solar energy will play an important role and serve as a natural substitute. Following the process of solar energy conversion, electricity can be converted into thermal energy by using a steam turbine. However, conventional heat transfer demonstrates lower thermal conductivity, and as a result, small solid particles should be added to the base fluid in order to increase thermal conductivity. As a result, nanofluids play an important role in this application [5]. Choi and Eastman [6] were the first to use the term nanofluid, which he coined in 1995 in order to fill the gap. Nanofluids have gained significant prominence in recent years as a result of their remarkable thermal performance capabilities. The use of nanoparticles in renewable energy sources has the potential to significantly improve the heat transfer characteristics of existing devices [7]. Several authors have recently investigated the use of nanofluids in a variety of thermal engineering and renewables applications [8,9].

An entirely new class of nanofluid [10,11], referred to as hybrid nanofluid, has recently been discovered. A hybrid nanofluid can be manufactured by dispersing two or more types of nanoparticles in a base fluid. A hybrid material is a substance that combines the chemical and physical properties of several materials at the same time, resulting in a homogeneous phase. It is possible to achieve remarkable physicochemical properties in synthetic hybrid nanomaterials that are not possible with their individual components. A substantial number of studies have therefore been communicated to explore the properties of these composite [12] and hybrid nanomaterials made up of carbon nanotubes, which can be used in a variety of applications such as nanocatalysts, biosensors, solar collectors, coatings, and electrochemical sensors [13]. Devi and Devi [14] computed the hybrid nanofluid (containing aluminum oxide and copper nanoparticles suspended in water) flow from an extending surface. Ghadikolaei et al. [15] evaluated the impact of different nanoparticle shapes in the stagnation flow of copper–titanium oxide/water hybrid nanofluid. Hassan et al. [16] investigated the heat transport properties of a hybrid nanofluid, including copper and silver nanoparticles. The natural convection of hybrid nanofluid flow over a porous medium with a non-uniform magnetic and circular heater was investigated by Izadi et al. [17]. Using an Eyring–Powell fluid model, Riaz et al. [18] investigated the thermal performance of non-Newtonian hybrid nanofluids (with a selection of metallic and carbon nanoparticles) in a wavy channel. Puneeth et al. [19] used a Casson fluid to analyze a three-dimensional hybrid nanofluid flow with a modified Buongiorno's model along a nonlinear stretching surface, considering blood as the base fluid and titania metallic oxide nanoparticles.

Since MgO and Ni nanoparticles (NPs) play an important role in real-world applications, they are specifically considered in the current investigation. MgO NPs are economically feasible, environmentally friendly, and of significant industrial significance due to their unique physicochemical behaviors, which include an outstanding refractive index, higher thermal conductivity, excellent corrosion resistance, physical strength, extraordinary optical transparency, stability, flame resistance, and mechanical strength [20–22]. These features make MgO NPs useful as catalysts in organic transformation, semiconducting materials, photocatalysts, sorbents for inorganic and organic pollutants from wastewater, refractory materials, solar coatings, and electrochemical biosensors [23,24]. It is well-known that the depletion of fossil fuels contributes to global warming by increasing the level of air pollution, which, in turn, causes the sea level to rise. The advent of batteries, solar cells, and fuel cells to replace fossil fuels as an alternative energy source has helped to alleviate this problem. They release water as a bi-product, and hydrogen is produced, which is a spectacular source of fuel and an amazing alternative form of carbon-based bi-products. Magnesium, similar to other metals, is essential for hydrogen storage and plays an important part in this process. When compared to other hydrides, magnesium NPs have a distinct advantage due to their abundance in large quantities in the earth's crust, their ability to store more hydrogen, ecologically beneficial properties, and low cost. However, nickel nanoparticles (NiNPs) are also useful in a variety of disciplines, including magnetic materials [25], biomedicine [26], energy technology [27], catalytic systems [28], catalysts for CO_2 hydrogenation [29], magnetic biocatalysts [30] and electronics. Due to the high reactivity, environmentally friendly characteristics, and operational simplicity of nickel nanoparticles, they have been found to also be useful in a variety of organic reactions. These include the reduction of ketones and aldehydes [31], chemo-selective oxidizing coupling of thiols [32], $\alpha-$alkylation of methyl ketone [33], synthesis of stilbenes from alcohol via Wittig-type olefination [34], and also hydrogenation of olefins [35].

In light of the numerous applications of MgO–Ni NPs, the purpose of this study is to investigate theoretically the magnetohydrodynamic water-based hybrid nanofluid stagnation flow impinging on a porous elastic stretching surface in a porous medium, as a model of solar collector coating manufacture. Heat generation/absorption and thermal/velocity slip effects are included. As revealed by the literature study, it is well-known that hybrid nanofluids provide promising results when compared to unitary (single nanoparticle) nanofluids in many applications. The mathematical model developed comprises the mass, momentum, energy, and induced magnetic field equations with appropriate boundary conditions. In addition, the Darcy law is deployed for porous medium effects, and viscous dissipation is incorporated. Numerical solutions of the transformed, dimensionless nonlinear ordinary differential boundary value problem are obtained with Matlab via a shooting method. Extensive visualizations of velocity, temperature, magnetic induction, skin friction, and the Nusselt number are presented graphically and with tables for the impact of emerging parameters. Validation of Matlab solutions with previously reported results (special cases) is included. Detailed interpretation of the results is provided.

2. Magnetic Hybrid Nanofluid Stagnation Flow Model

The physical regime under consideration comprises a steady orthogonal stagnation point flow of an incompressible hybrid magnetized nanofluid over an elastic surface, as shown in Figure 1. The fluid contains two types of nanoparticles, i.e., nickel (Ni) and magnesium oxide (MgO).

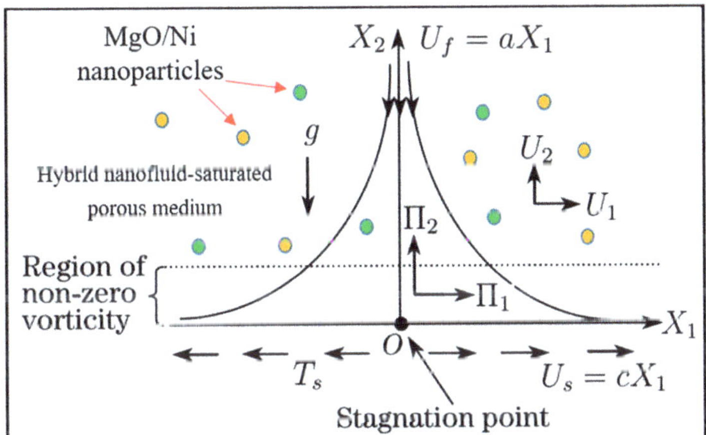

Figure 1. Two-dimensional geometrical configuration for magnetized hybrid nanofluid stagnation point flow over an elastic surface in a porous medium.

The water is a base fluid that is electrically conducting, irrotational, and impinges on an elastic surface adjacent to a Darcian porous medium. The presence of an induced magnetic field is also considered since the magnetic Reynolds number is sufficiently large. Hall current and electrical polarization effects are neglected. The elastic surface is assumed to be porous, i.e., suction/injection are present. The velocity at the free stream is considered as $U_f = aX_1$ where $a > 0$. The velocity at the surface is defined as $U_s = cX_1$ where $c > 0$

corresponds to a stretching elastic surface, $c < 0$ represents a contracting (shrinking) elastic surface, and $c = 0$ indicates a stationary surface. The surface temperature is denoted by T_s while the ambient temperature is denoted by T_{inf}. The nanoparticles are of spherical shape, having zero agglomeration. In view of the above approximations, the governing equations in vectorial form can be defined as [36–38]:

$$\nabla \cdot \Pi = 0, \ \nabla \cdot \mathbf{U} = 0, \tag{1}$$

$$\mathbf{U} \cdot \nabla \mathbf{U} - \frac{\mu_e}{4\pi\rho_{hnf}}(\Pi \cdot \nabla)\Pi = v_{hnf}\nabla^2\mathbf{U} - \frac{1}{\rho_{hnf}}\nabla P - \frac{v_{hnf}}{k}\mathbf{U} + \frac{g(T - T_{\text{inf}})(\rho\beta)_{hnf}}{\rho_{hnf}}, \tag{2}$$

$$\nabla \times (\mathbf{U} \times \Pi) = -\zeta\nabla^2\Pi, \tag{3}$$

$$(\mathbf{U} \cdot \nabla)\, T = \frac{\kappa_{hnf}}{(\rho C_p)_{hnf}}\nabla^2 T + \frac{\mathbf{S}}{(\rho C_p)_{hnf}}\nabla\mathbf{U} + \frac{\mu_{hnf}}{k(\rho C_p)_{hnf}}\mathbf{U}^2 + Q(T - T_{\text{inf}}), \tag{4}$$

Here velocity field vector is designated by $\mathbf{U}$, kinematic viscosity of hybrid nanofluid is denoted by v_{hnf}, induced magnetic field vector is represented by Π, thermal expansion coefficient is represented by β_{hnf}, the density of the hybrid nanofluid is denoted by ρ_{hnf}, the permeability of the porous medium is represented by k, the magnetic diffusivity is denoted by $\zeta = 1/4\pi\mu_e\sigma_{hnf}$ (in which σ_{hnf} denotes the electrical conductivity, the magnetic permeability parameter is represented by μ_e), $(C_p)_{hnf}$ represents the specific heat capacity, κ_{hnf} denotes the thermal conductivity, T is the temperature, pressure is represented by $P = \tilde{p} + \mu_e|\Pi|^2/8\pi$, and $\mathbf{S}$ is the viscous fluid stress tensor.

When using the boundary layer approximations, the governing Equations (1)–(4) are reduced to the following form in a two-dimensional system (X_1, X_2):

$$\left. \begin{aligned} \frac{\partial U_2}{\partial X_2} + \frac{\partial U_1}{\partial X_1} &= 0, \\ \frac{\partial \Pi_2}{\partial X_2} + \frac{\partial \Pi_1}{\partial X_1} &= 0, \end{aligned} \right\} \tag{5}$$

$$\begin{aligned} U_1\frac{\partial U_1}{\partial X_1} + U_2\frac{\partial U_1}{\partial X_2} &= v_{hnf}\frac{\partial^2 U_1}{\partial X_2^2} + \frac{\mu_e}{4\rho_{hnf}\pi}\left(\Pi_1\frac{\partial \Pi_1}{\partial X_1} + \Pi_2\frac{\partial \Pi_1}{\partial X_2}\right) \\ &\quad + \left(U_f\frac{dU_f}{dX_1} - \frac{\mu_e\Pi_f}{4\pi\rho_{hnf}}\frac{d\Pi_f}{dX_1}\right) - \frac{v_{hnf}}{k}\left(U_1 - U_f\right) + \frac{g(\rho\beta)_{hnf}}{\rho_{hnf}}(T - T_{\text{inf}}), \end{aligned} \tag{6}$$

$$U_1\frac{\partial \Pi_1}{\partial X_1} + U_2\frac{\partial \Pi_1}{\partial X_2} - \Pi_1\frac{\partial U_1}{\partial X_1} - \Pi_2\frac{\partial U_1}{\partial X_2} = \zeta\frac{\partial^2 \Pi_1}{\partial X_2^2}, \tag{7}$$

$$U_1\frac{\partial T}{\partial X_1} + U_2\frac{\partial T}{\partial X_2} = \frac{\kappa_{hnf}}{(\rho C_p)_{hnf}}\frac{\partial^2 T}{\partial X_2^2} + \frac{\mu_{hnf}}{(\rho C_p)_{hnf}}\left(\frac{\partial U_1}{\partial X_2}\right)^2 + \frac{\mu_{hnf}}{(\rho C_p)_{hnf}k}\left(U_1 - U_f\right)^2 + Q(T - T_{\text{inf}}), \tag{8}$$

In the above equations, U_1, U_2 are the velocity components in the X_1, X_2 directions (Cartesian coordinate system) where X_1 is orientated along the elastic surface stretching direction and X_2 is normal to it, Π_1, Π_2 are the magnetic induction components and Π_f represents the X_1-magnetic field towards the extremity of the boundary layer. The associated boundary conditions are defined at the wall and in the free stream as [39]:

$$\left. \begin{aligned} U_1 &= U_s + A_u\frac{\partial U_1}{\partial X_2}, \ U_2 = 0, \ \Pi_2 = \frac{\partial \Pi_1}{\partial X_2} = 0, \ T = T_s + T_u\frac{\partial T}{\partial X_2} \text{ at } X_2 \to 0, \\ U_1 &= U_f = aX_1, \ U_2 = 0, \ \Pi_1 = \Pi_f(X_1) = \Pi_0 X_1, \ T = T_{\text{inf}} \text{ at } X_2 \to \infty, \end{aligned} \right\} \tag{9}$$

Here A_u represents the velocity slip, T_u represents the thermal slip, Π_0 represents the magnetic field at infinity upstream.

The physical and thermodynamics properties of hybrid nanofluids are defined with the following relations:

a. Density:

$$\rho_{hnf} = \rho_1^{np}\psi_1 + \rho_2^{np}\psi_2 + \rho_f(1 - \psi_{hnf}), \, \psi_{hnf} = \psi_1 + \psi_2, \tag{10}$$

b. Heat capacity:

$$(\rho C_p)_{hnf} = (\rho C_p)_1^{np}\psi_1 + (\rho C_p)_2^{np}\psi_2 + (\rho C_p)_f(1 - \psi_{hnf}), \tag{11}$$

c. Dynamic viscosity:

$$\mu_{hnf} = \frac{\mu_f}{\left(1 - \psi_{hnf}\right)^{2.5}}, \tag{12}$$

d. Thermal conductivity:

$$\kappa_{hnf} = \kappa_f \times \left[\frac{2\kappa_f + \left(\frac{\psi_1\kappa_1^{np} + \psi_2\kappa_2^{np}}{\psi_{hnf}}\right) + 2\left(\psi_1\kappa_1^{np} + \psi_2\kappa_2^{np}\right) - 2\psi_{hnf}\kappa_f}{2\kappa_f + \left(\frac{\psi_1\kappa_1^{np} + \psi_2\kappa_2^{np}}{\psi_{hnf}}\right) - \left(\psi_1\kappa_1^{np} + \psi_2\kappa_2^{np}\right) + \psi_{hnf}\kappa_f} \right], \tag{13}$$

e. Thermal expansion coefficient:

$$(\rho\beta)_{hnf} = (\rho\beta)_f(1 - \psi_{hnf}) + (\rho\beta)_1^{np}\psi_1 + (\rho\beta)_2^{np}\psi_2, \tag{14}$$

f. Electric conductivity:

$$\sigma_{hnf} = \sigma_f \times \left[\frac{1 + 3\psi_{hnf}\left(\frac{\sigma_{hnp}}{\sigma_f} - 1\right)}{\left(\frac{\sigma_{hnp}}{\sigma_f} + 2\right) - \left(\frac{\sigma_{hnp}}{\sigma_f} - 1\right)\psi_{hnf}} \right], \, \sigma_{hnp} = \frac{\psi_1\sigma_1^{np} + \psi_2\sigma_2^{np}}{\psi_{hnf}}, \tag{15}$$

Here ψ_1 denotes the nanoparticle volume fraction of MgO nanoparticles, and ψ_2 denotes the nanoparticles volume fraction of Ni nanoparticles. The properties of the nanoparticles and base fluid (water) are given in Table 1.

Table 1. Thermophysical properties of water, magnesium oxide and nickel nanoparticles.

Physical Properties	Water	Nickel (Ni)	Magnesium Oxide (MgO)
$\rho\,(\mathrm{kg/m^3})$	997.1	8908	3580
$\kappa\,(\mathrm{W/mK})$	0.613	91	45
$C_p\,(\mathrm{J/kgK})$	4179	445	955
$\beta\,(1/\mathrm{K})$	0.00021	0.0000134	0.0000336
$\sigma\,(\mathrm{S/m})$	0.05	1.7×10^7	2.6×10^{-6}

3. Transformation of Mathematical Model

Let us introduce the following transformations to the mathematical model:

$$U_1 = cX_1 f'(\eta), \, \Pi_1 = \Pi_0 X_1 g'(\eta),$$
$$U_2 = -\left(v_f c\right)^{1/2} f(\eta),$$
$$\Pi_2 = -\left(v_f c\right)^{1/2} \Pi_0 g(\eta), \tag{16}$$
$$\eta = X_2\left(c v_f^{-1}\right)^{1/2}, \, \vartheta = \frac{T - T_{\inf}}{T_s - T_{\inf}},$$

Substitution of Equation (16) into Equations (5)–(9) yields the following set of dimensionless coupled nonlinear ordinary differential equations:

$$\frac{\Upsilon_1}{\Upsilon_2} f''' + ff'' + \tau^2 - f'^2 - \frac{M}{\Upsilon_2}\left(1 - g'^2 + gg''\right) - \frac{\Upsilon_1}{\Upsilon_2} D_a\left(f' - \tau\right) + G_r \frac{\Upsilon_3}{\Upsilon_2}\vartheta = 0, \tag{17}$$

$$\delta g''' - \Upsilon_4 g f'' + \Upsilon_4 f g'' = 0, \tag{18}$$

$$\frac{\Upsilon_5}{\Upsilon_6}\vartheta'' + \frac{\Upsilon_1}{\Upsilon_6} E_c P_r f''^2 + P_r \vartheta' f + \frac{\Upsilon_1}{\Upsilon_6} E_c P_r D_a\left(f' - \tau\right)^2 + P_r q\vartheta = 0. \tag{19}$$

The emerging reduced dimensionless boundary conditions are:

$$\begin{aligned}
&\text{at } \eta \to 0: \\
&f = 0, \ f' = 1 + \lambda f''(0), \ g'' = 0, \ g = 0, \ \vartheta = 1 + \chi\vartheta'(0), \\
&\text{at } \eta \to \infty: \\
&f' = \tau, \ g' = 1, \ \vartheta = 0,
\end{aligned} \tag{20}$$

Here $\tau = a/c$, $M\left(= \mu_e \Pi_0^2/4\pi\rho_f c^2\right)$ represents the magnetic parameter, $\lambda\left(= A_u\left(cv_f^{-1}\right)^{1/2}\right)$ represents the velocity slip parameter, $\chi\left(= T_u\left(cv_f^{-1}\right)^{1/2}\right)$ represents the thermal slip parameter, $G_r\left(= Gr/Re_{X_1}^2\right)$ the mixed convection parameter, $Gr\left(= (\rho\beta)_f(T_s - T_{\text{inf}})X_1^3 g/\rho_f v_f^2\right)$ is the thermal Grashof number, $E_c\left(= U_s^2/C_p(T_s - T_{\text{inf}})\right)$ is the Eckert number, $D_a\left(= v_f/ck\right)$ is the Darcy number for the porous medium, $\delta\left(= \zeta/v_f\right)$ is the reciprocal of magnetic Prandtl number, $Re_{X_1}\left(= X_1 U_s/v_f\right)$ is the local Reynolds number [40], $P_r\left(= (\rho C_p)_f v_f/\kappa_f\right)$ denotes the Prandtl number, and $q\left(= Q/(\rho C_p)_f\right)$ is the heat source/sink parameter.

The remaining parameters are as follows:

$$\Upsilon_1 = \frac{\mu_{hnf}}{\mu_f}, \ \Upsilon_2 = \frac{\rho_{hnf}}{\rho_f}, \ \Upsilon_3 = \frac{(\rho\beta)_{hnf}}{(\rho\beta)_f}, \ \Upsilon_4 = \frac{\sigma_{hnf}}{\sigma_f}, \ \Upsilon_5 = \frac{\kappa_{hnf}}{\kappa_f}, \ \Upsilon_6 = \frac{(\rho C_p)_{hnf}}{(\rho C_p)_f}. \tag{21}$$

The magnetic parameter, M, represents the ratio of kinetic to magnetic energy per unit volume, is related to the Hartmann number H_a. The magnetic parameter, M, Hartmann number, H_a, Reynolds number, Re, and magnetic Reynolds number, Re_m, are defined as:

$$M = \frac{H_a^2}{ReRe_m}, \ H_a = \mu_e\Pi_0\Gamma\sqrt{\frac{\sigma_f}{\mu_e}}, \ Re = \frac{\Gamma(c\Gamma)}{v_f}, \ Re_m = 4\mu_e\sigma_f\Gamma(c\Gamma)\pi, \tag{22}$$

In the above equation, Γ specifies the characteristic length of the elastic surface. For the magnetohydrodynamic boundary layer flows, $M \leq 1$, $\delta \geq 1$—see Kumari et al. [41]. However, for electrically non-conducting flows, in the absence of a magnetic field $M = 0$ and therefore Equation (18) is no longer required.

4. Engineering Quantities

The following are important physical quantities in engineering design for thermal coating flows, namely, the *skin friction coefficient* and the *Nusselt number*:

$$C_f = \frac{S_w}{U_s^2\rho_{hnf}}, \ N_u = \frac{Q_w X_1}{(T_{\text{inf}} - T_s)\kappa_f}, \tag{23}$$

Here S_w represents the wall shear stress, and Q_w represents the wall heat flux, which are defined as:

$$S_w = \mu_{hnf} \left.\frac{\partial U_1}{\partial X_2}\right|_{X_2=0}, \quad Q_w = -\kappa_{hnf} \left.\frac{\partial T}{\partial X_2}\right|_{X_2=0}. \tag{24}$$

Using the variables in Equation (16), the required expressions are:

$$C_f = \frac{\Upsilon_1}{\Upsilon_2 \sqrt{Re_{X_1}}} f''(0), \quad Nu = -\Upsilon_5 \sqrt{Re_{X_1}}\, \vartheta'(0), \tag{25}$$

5. Solution with MATLAB Numerical Shooting Method

A numerical shooting approach is utilized to solve the nonlinear differential Equations (17)–(19) with boundary conditions (20). MATLAB software has been used to generate all of the numerical solutions. The governing Equations (17)–(19) are first reduced to an initial value problem, which is expressed as follows:

$$\left.\begin{aligned}
f &= \hbar_1 \\
f' &= \hbar'_1 = \hbar_2 \\
f'' &= \hbar'_2 = \hbar_3 \\
f''' &= \hbar'_3 = -\frac{\Upsilon_2}{\Upsilon_1}\hbar_1\hbar_3 - \frac{\Upsilon_2}{\Upsilon_1}\tau^2 + \frac{\Upsilon_2}{\Upsilon_1}\hbar_2^2 - \frac{M}{\Upsilon_1}(\hbar_5^2 - \hbar_4\hbar_6 - 1) + D_a(\hbar_2 - \tau) - G_r\frac{\Upsilon_3}{\Upsilon_1}\hbar_7
\end{aligned}\right\} \tag{26}$$

$$\left.\begin{aligned}
g &= \hbar_4 \\
g' &= \hbar'_4 = \hbar_5 \\
g'' &= \hbar'_5 = \hbar_6 \\
g''' &= \hbar'_6 = \frac{\Upsilon_4}{\delta}\hbar_3\hbar_4 - \frac{\Upsilon_4}{\delta}\hbar_6\hbar_1
\end{aligned}\right\} \tag{27}$$

$$\left.\begin{aligned}
\vartheta &= \hbar_7 \\
\vartheta' &= \hbar'_7 = \hbar_8 \\
\vartheta'' &= \hbar'_8 = -\frac{\eta_6}{\eta_5}P_r\hbar_8\hbar_1 - \frac{\eta_1}{\eta_5}E_cP_r\hbar_3^2 - \frac{\Upsilon_1}{\Upsilon_6}E_cP_rD_a(\hbar_2 - \tau)^2P_rq\hbar_7
\end{aligned}\right\} \tag{28}$$

The boundary conditions (20) are formulated as:

$$\begin{aligned}
&\text{at } \eta \to 0: \\
&\hbar_1 = 0, \ \hbar_2 = 1 + \lambda\hbar_3, \ \hbar_4 = 0, \ \hbar_6 = 0, \ \hbar_7 = 1 + \chi\hbar_8, \\
&\text{at } \eta \to \infty: \\
&\hbar_2 = \tau, \ \hbar_5 = 1, \ \hbar_7 = 0,
\end{aligned} \tag{29}$$

In order to verify the MATLAB solutions, a comparison with previous studies is given in Tables 2–4. Excellent correlation is achieved for skin friction and the Nusselt number between the MATLAB solutions and Ali et al. [42] in Table 2. Further comparisons are given with Hassanien et al. [43], Salleh and Nazar [44] in Table 3, and Ali et al. [45] in Table 4. In all cases, very good agreement is demonstrated up to four decimal places, ensuring not only the validity of the current results but also the accuracy of the hybrid nanofluid results. Confidence in the MATLAB solutions is therefore justifiably high.

Table 2. A numerical comparison of skin friction and Nusselt number by considering the following values: $D_a = 0$; $\lambda = 0$; $\chi = 0$; $\tau = 0$; $E_c = 0$; $G_r = 0$; $\psi_1 = 0$; $\psi_2 = 0$; $q = 0$.

	C_f	$N_u(P_r = 0.72, 10)$	C_f	$N_u(P_r = 0.72, 10)$
δ	Ali et al. [42]		Present Results	
10^2	0.9914	(0.4713, 2.30818)	0.999654056	(0.463845643, 2.308119015)
10^3	0.9993	(0.4639, 2.30809)	0.999975012	(0.463616023, 2.308031227)
10^4	0.9999	(0.4632, 2.30808)	1.000004856	(0.463594549, 2.308022703)
10^5	1.0000	(0.4632, 2.30808)	1.000007819	(0.463592416, 2.308021857)
M			$\delta = 10^4$	
0.01	0.9993	(0.4639, 2.30809)	0.999975236	(0.463616264, 2.308024896)
0.05	0.9970	(0.4663, 2.30812)	0.999889007	(0.465078584, 2.308044206)
0.10	0.9945	(0.4687, 2.30817)	0.996099015	(0.467058435, 2.308027518)
0.15	0.9922	(0.4707, 2.30822)	0.992392113	(0.472321992, 2.307876202)

Table 3. A numerical comparison of Nusselt number by considering the following values: $D_a = 0$; $\lambda = 0$; $\chi = 0$; $\tau = 0$; $E_c = 0$; $G_r = 0$; $\psi_1 = 0$; $\psi_2 = 0$; $q = 0$.

P_r		$\tau = 0, M = 0$		
	Ali et al. [42]	Hassanien et al. [43]	Salleh and Nazar [44]	Present Results
0.72	0.4632	0.46325	0.46317	0.463592073
10	2.3081	2.30801	2.30821	2.308006492

Table 4. A numerical comparison of skin friction and Nusselt number by considering the following values: $D_a = 0$; $\lambda = 0$; $\chi = 0$; $\tau = 0$; $E_c = 0$; $G_r = 0$; $\psi_1 = 0$; $\psi_2 = 0$; $q = 0$.

	$\tau = 3, M = 1$			
	Skin Friction		Nusselt Number	
P_r	Ali et al. [45]	Present Results	Ali et al. [45]	Present Results
0.07			0.33814	0.338140112
0.72	4.52158	4.521582090	0.97240	0.972402233
0.5			0.82748	0.827484508
2.0			1.52147	1.521468063
6.8			2.59780	2.597812407
10			3.07902	3.079050771
M		$\tau = 3, P_r = 0.72$		
0.1	4.70928	4.709283867	0.97902	0.979021615
0.5	4.62764	4.627634204	0.97617	0.976172800
1.0	4.52158	4.521582090	0.97240	0.972402233
2.0	4.29431	4.294313656	0.96405	0.964045689
		$\tau = 0.5, P_r = 0.72$		
0.10	−0.57595	−0.575949681	0.59171	0.591705825
0.15	−0.50938	−0.509403808	0.60207	0.602061052
0.20	−0.40717	−0.407547217	0.61811	0.618079173

6. MATLAB Computational Results and Discussion

In this section, graphical and tabulated results are presented for the effects of all key parameters. The following parametric values have been chosen in order to carry out the computational formulation: $D_a = 0.5$; $\lambda = 0.2$; $\chi = 0.3$; $\tau = 2$; $\delta = 2$; $E_c = 0.5$; $G_r = 0.3$; $\psi_1 = 0.2$; $\psi_2 = 0.2$; $P_r = 6.96$; $M = 0.2$. The thermophysical properties of water, nickel (Ni), and magnesium oxide (MgO) nanoparticles have been given earlier in Table 1. Table 5 also contains the numerical data for the skin friction coefficient and the Nusselt number. It is evident that greater values of velocity slip parameter, the nanoparticle volume fraction of

MgO, the heat source/sink parameter, and the Prandtl number all reduce the skin friction coefficient, whereas an increment in the Darcy parameter and the thermal Grashof number decreases skin friction. The Nusselt number decreases as the Darcy number, Eckert number, thermal Grashof number, thermal slip, the nanoparticle volume fraction of MgO and Ni, and Prandtl number increase, whereas it increases as the heat source/sink parameter and velocity slip increase.

Table 5. Numerical results of skin friction and Nusselt number against all the leading parameters.

δ	M	P_r	λ	D_a	E_c	G_r	χ	ψ_1	ψ_2	q	C_f	N_u
1	0.2	6.96	0.2	0.5	0.5	0.3	0.3	0.2	0.2	0.5	1.386342185	1.046697391
2											1.387113826	1.046037055
3											1.387580951	1.045443513
	0.01										1.391446326	1.037575829
	0.1										1.389402333	1.041569479
	0.2										1.387113826	1.046037055
		5									1.387916242	1.093391191
		6.96									1.387113826	1.046037055
		8									1.386806321	1.015829923
			0								1.983946596	−1.847797617
			0.1								1.636922168	−0.007837483
			0.2								1.387113826	1.046037055
				0							1.343075838	1.240588255
				1							1.427934896	0.874197208
				2							1.501620476	0.584125598
					0.1						1.381818954	2.951411753
					0.2						1.383143556	2.475622258
					0.3						1.384467393	1.999464353
						0					1.374362440	1.050348820
						1					1.416662936	1.035561579
						2					1.458421488	1.019592211
							0				1.388693182	1.931909381
							0.1				1.387935006	1.506589469
							0.2				1.387450377	1.234763174
								0			1.178305552	1.329401404
								0.1			1.387113826	1.046037055
								0.2			1.699319889	0.329337847
									0		1.698735776	1.423814942
									0.1		1.599004165	1.103502632
									0.2		1.699319889	0.329337847
										0.2	1.706475625	−1.563796173
										0	1.704165102	−1.016305551
										0.2	1.702063147	−0.467389070

The variation of the velocity profile versus numerous values of various controlling parameters is shown in Figures 2–7. Higher numerical values of the Darcy number D_a enhance the velocity profile in the regime, as shown in Figure 2. The porous medium is assumed to be very sparsely packed, and therefore very high Darcy numbers are considered. The increase in permeability associated with a larger Darcy number reduces the Darcian impedance force, i.e., the resistance of solid matrix fibers to the percolating magnetic nanofluid, and this accelerates the flow leading to a depletion in momentum boundary layer thickness. With a higher Darcy number, the medium features lesser solid fibers. This assists in momentum development. A lower Darcy number, however, implies lower permeability, which results in a higher Darcian drag force and deceleration. Asymptotically smooth profiles are computed in the free stream, confirming the prescription of an adequately large infinity boundary condition in the MATLAB computations.

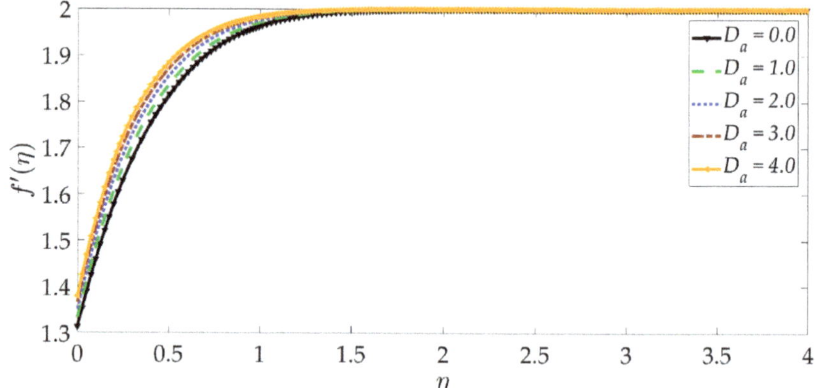

Figure 2. Consequences of Darcy number on velocity profile.

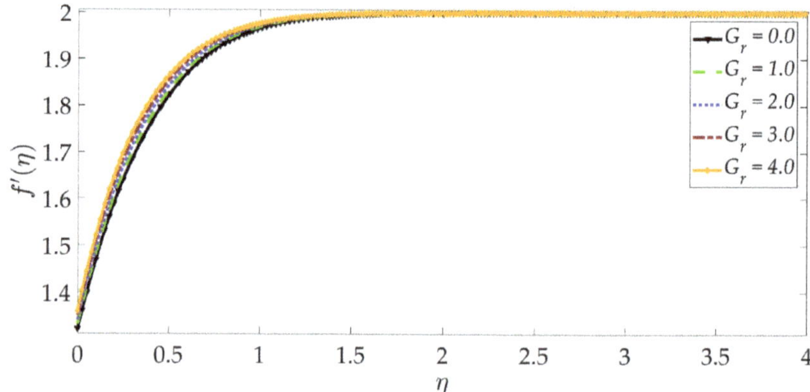

Figure 3. Consequences of thermal Grashof number on velocity profile.

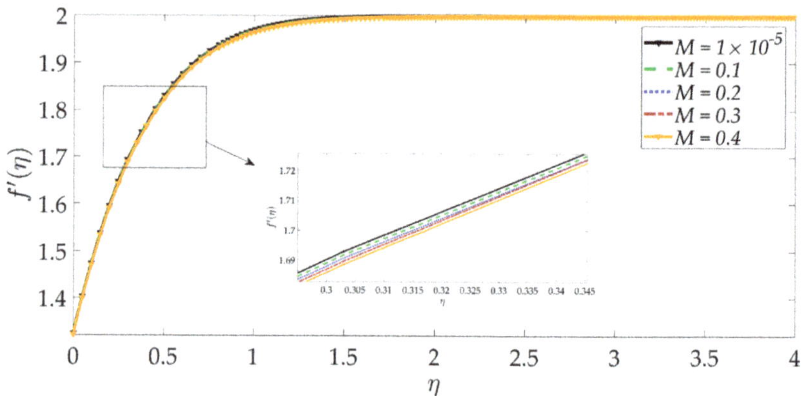

Figure 4. Consequences of magnetic parameter on velocity profile.

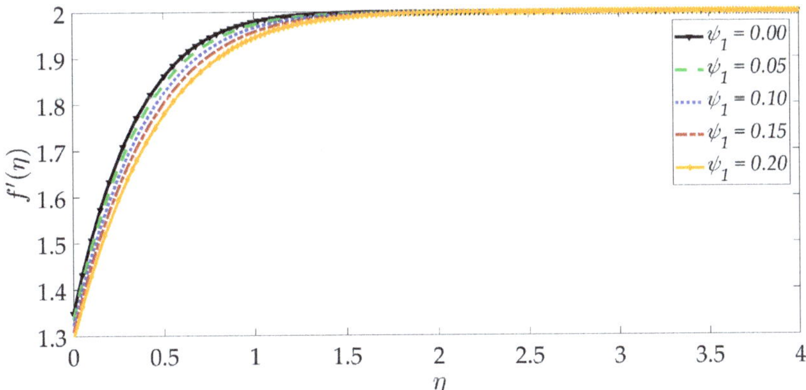

Figure 5. Consequences of nanoparticle volume fraction of MgO on velocity profile.

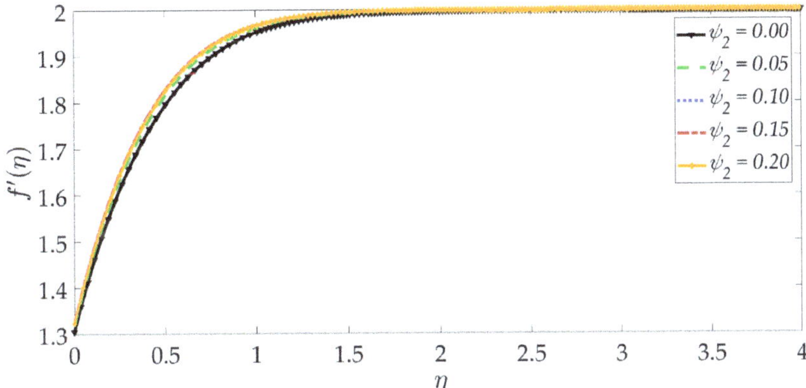

Figure 6. Consequences of nanoparticle volume fraction Ni on velocity profile.

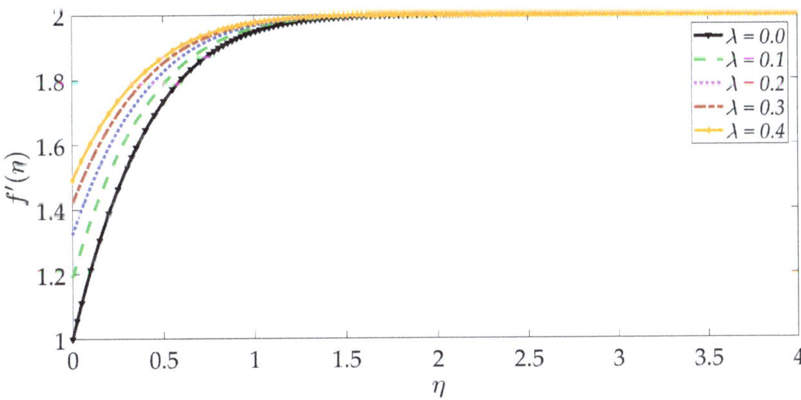

Figure 7. Consequences of velocity slip on velocity profile.

Figure 3 shows that elevation in the thermal Grashof number G_r accentuates the velocity profile. The ratio of buoyancy to viscous forces is represented by the thermal Grashof number. Higher thermal Grashof numbers indicate that buoyancy force has a stronger role relative to the inhibitive viscous force. The thermal buoyancy force, $+G_r \frac{\Upsilon_3}{\Upsilon_2} \vartheta$ in Equation (17) is therefore amplified in magnitude, and this leads to acceleration. The impact of the magnetic parameter M on the velocity profile is shown in Figure 4. We can observe that the magnetic parameter slightly reduces the velocity at an intermediate distance from the wall. The magnetic force term $-\frac{M}{\Upsilon_2}(1 - g'^2 + gg'')$ in Equation (17) inhibits the boundary layer flow and increases the momentum boundary layer thickness slightly. Figure 5 shows that when the volume fraction of MgO nanoparticles ψ_1 increases, the velocity profile decreases dramatically. Greater doping of the nanofluid with MgO nanoparticles, therefore, decelerates the flow and increases momentum boundary layer thickness. However, the behavior of a unitary nanofluid, i.e., the profile $\psi_1 = 0$ (where only Ni nanoparticles are present), achieves the maximum velocity and minimal momentum boundary layer thickness. On the other hand, with increasing volume percentage of Ni nanoparticles ψ_2, the velocity is enhanced, as seen in Figure 6. The unitary nanofluid $\psi_2 = 0$ (for which MgO nanoparticles are absent) in this case produces the minimal velocity and maximum momentum boundary layer thickness. Figure 7 indicates that increasing the values of the velocity slip parameter enhances the fluid motion since greater momentum is generated at the wall, which assists the flow; the momentum boundary layer thickness is reduced with greater velocity slip. For the case $\lambda = 0$ when hydrodynamic slip is absent, the velocity is minimal, and momentum boundary layer thickness is maximum.

The influence of selected parameters on the induced magnetic field is depicted in Figures 8–14. As seen in Figure 8, raising the numerical values of the Darcy number D_a increases the induced magnetic field g' magntiudes dramatically. Although the Darcian body force does not appear explicitly in the magnetic induction Equation (18), via coupling with the momentum Equation (17), i.e., the terms $-\Upsilon_4 g f'' + \Upsilon_4 f g''$, the Darcy number influences indirectly the magnetic induction field. A sigmoidal topology is observed from the wall (elastic surface) to the free stream. Magnetic boundary layer thickness is increased significantly with increment in Darcy number (i.e., higher permeability of the porous medium). In Figure 9, we can see that increasing the reciprocal of magnetic Prandtl number δ, i.e., decreasing magnetic Prandtl number, boosts the induced magnetic field at first; however, further from the elastic surface after $\eta > 1$, the tendency reverses, and there is a depletion in magnetic induction. Magnetic Prandtl number expresses the relative rate of momentum (viscous) diffusion to magnetic diffusion in the regime. This parameter significantly modifies the induced magnetic field distribution in the regime, but the response is dependent on location from the elastic surface. The thermal Grashof number G_r strongly enhances the induced magnetic field profile, as shown in Figure 10. For the case of forced convection, $G_r = 0$ and thermal buoyancy effects vanish. The magnetic induction is minimized for this case, as is the magnetic boundary layer thickness. Overall thermal buoyancy is assistive to the induced magnetic field and increases magnetic boundary layer thickness on the elastic surface. In Figure 11, we can see that the magnetic parameter M has a depleting effect on the induced magnetic field profile. Larger M values suppress the magnetic boundary layer thickness. In Figures 12 and 13, we can see that the volume percentage of MgO and Ni (ψ_1, ψ_2) nanoparticles initially reduces the induced magnetic field values closer to the elastic surface (wall); however, further into the boundary layer regime, transverse to the wall, the trend is opposite, and there is a clear enhancement in magnitudes of the induced magnetic field, which is sustained into the free stream. Furthermore, in these plots, the special cases of unitary (single nanoparticle) nanofluids correspond to $\psi_1 = 0$ or $\psi_2 = 0$. The effects of the hydrodynamic (velocity) slip parameter on the induced magnetic field profile are depicted in Figure 14. It is apparent that the induced magnetic profile is greatly boosted with an increment in the slip parameter. The coupling of the momentum and induced magnetic field Equations (17) and (18) again enables the momentum slip to influence the magnetic induction via the boundary condition, $f' = 1 + \lambda f''(0)$ in Equation (20). For the case $\lambda = 0$ for which wall momentum slip is absent, the induced magnetic field is minimized, and the magnetic boundary layer thickness is also minimal.

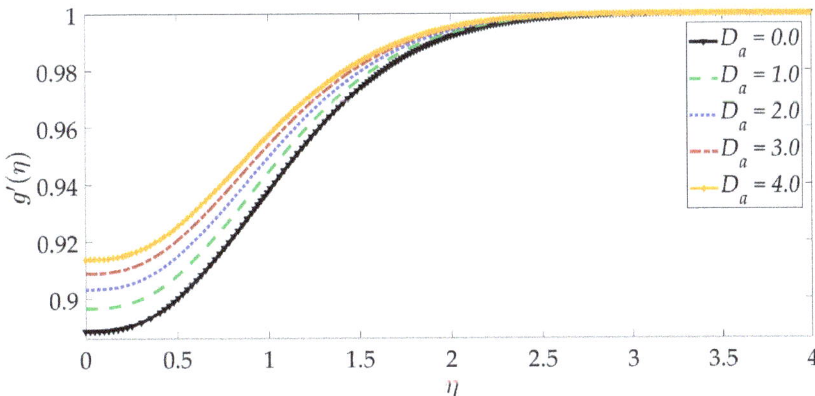

Figure 8. Consequences of Darcy number on induced magnetic field profile.

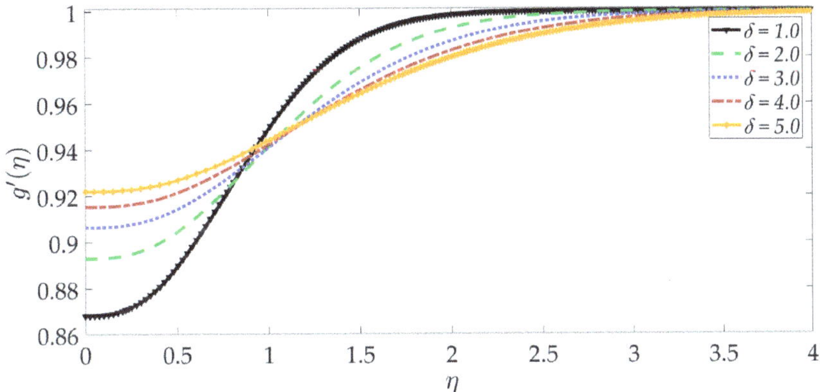

Figure 9. Consequences of reciprocal magnetic Prandtl number on induced magnetic field profile.

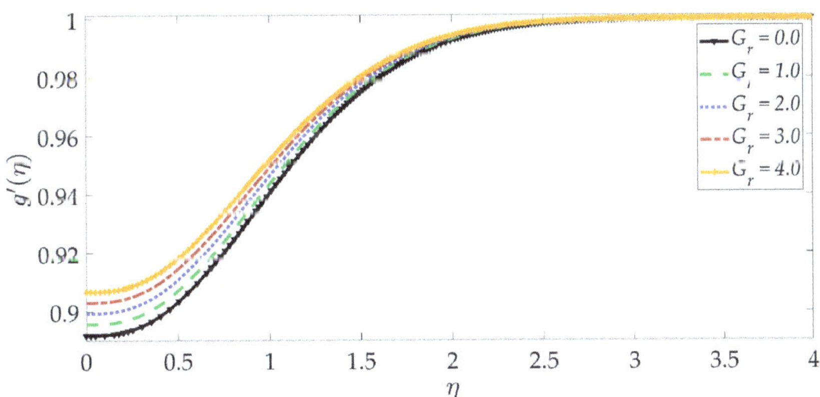

Figure 10. Consequences of thermal Grashof number on induced magnetic field profile.

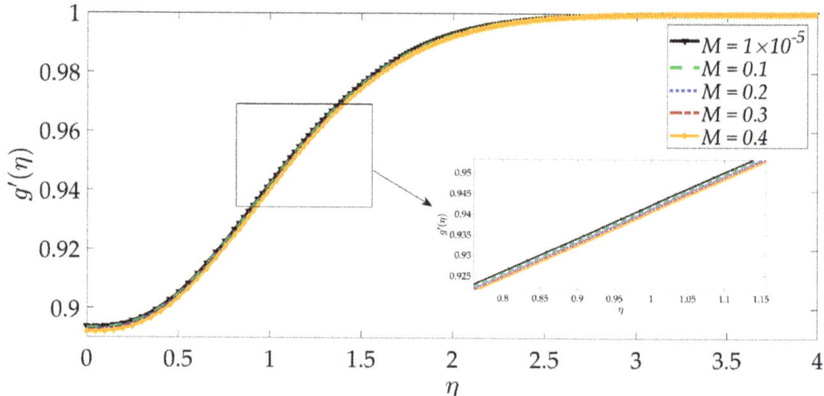

Figure 11. Consequences of magnetic parameter on induced magnetic field profile.

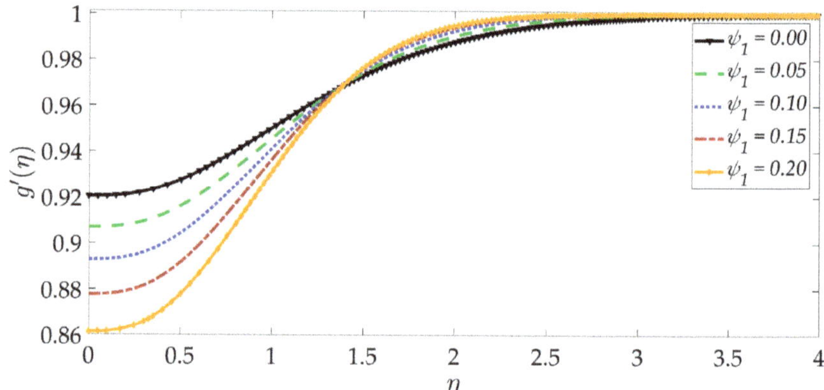

Figure 12. Consequences of nanoparticle volume fraction MgO on induced magnetic field profile.

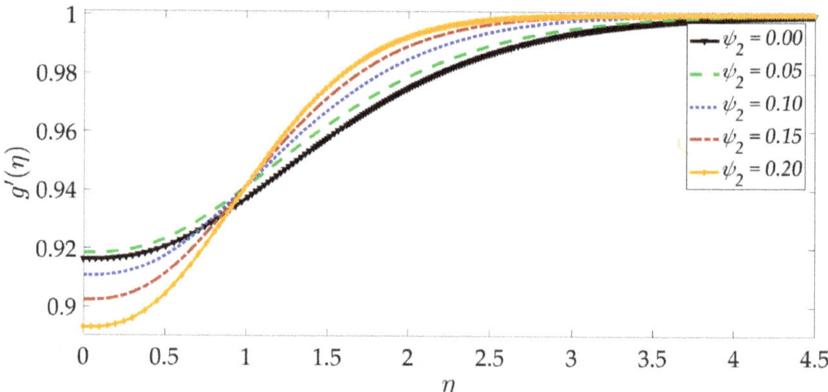

Figure 13. Consequences of nanoparticle volume fraction Ni on induced magnetic field profile.

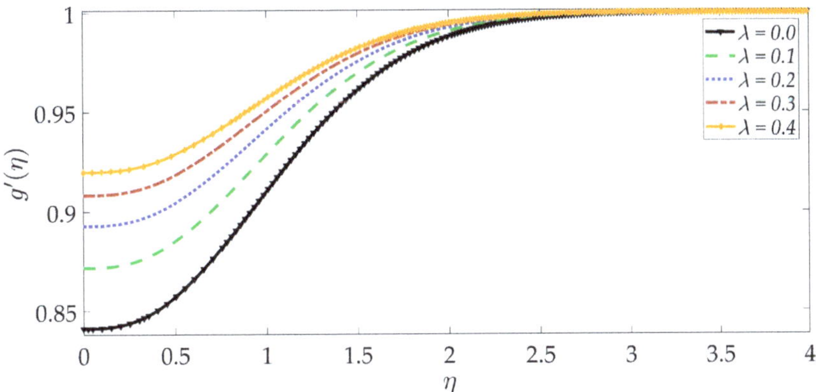

Figure 14. Consequences of velocity slip on induced magnetic field profile.

Figures 15–20 illustrate the evolution in temperature profiles for selected parameters. It can be shown in Figure 15 that with elevation in Darcy number, there is a strong increment in the temperature profile and thermal boundary layer thickness. The prominent modification, i.e., heating, is near the elastic surface (wall), and further, the effect decays when $\eta > 0.5$, which is sustained into the freestream. It can be deduced from Figures 16 and 17 that increasing the nanoparticle volume percentage of MgO and Ni nanoparticles (ψ_1, ψ_2) results in a significant and uniform enhancement in the temperature profile and thermal boundary layer thickness. Greater doping with MgO and Ni nanoparticles, therefore, achieves a desirable thermal elevation in particular close to the wall (elastic surface). Figure 18 shows that temperatures are suppressed and, therefore, also, the thermal boundary layer thickness is reduced with increment in Prandtl number P_r. Higher values of the Prandtl number indicate that momentum diffusivity is becoming more dominant over thermal diffusivity, and the net effect is a suppression of thermal diffusion. This cools the regime and depletes the thickness of the thermal boundary layer. The impact of the heat source/sink parameter (q) on the temperature profile is depicted in Figure 19. It should be noted that increasing heat generation results in a boost in temperatures, whereas increasing heat sink induces the opposite effect. Thermal boundary layer thickness is therefore also modified with either heat generation (source) or absorption (sink). The case $q = 0$ corresponds to the absence of the heat source/sink parameter and produces intermediate values of temperature between the heat source and sink cases. Finally, Figure 20 demonstrates that an increment in thermal slip has the effect of decreasing the temperature magnitudes and also the thickness of the thermal boundary layer. The thermal slip parameter, i.e., χ, arises in the modified wall thermal boundary condition $\vartheta = 1 + \chi \vartheta'(0)$, in Equation (20). It creates a thermal jump effect which delays the heat transfer from the wall to the boundary layer regime on the elastic surface. This leads to a reduction in temperatures, i.e., the cooling effect.

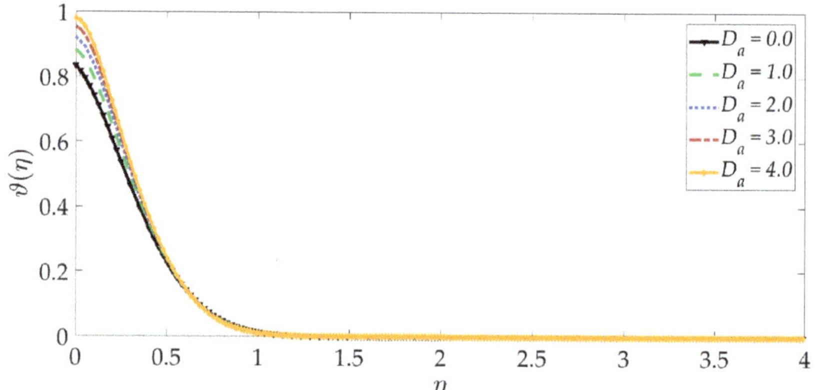

Figure 15. Consequences of Darcy number on temperature profile.

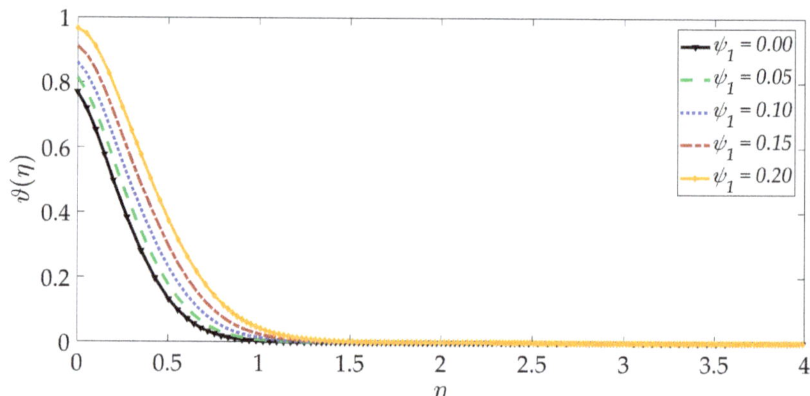

Figure 16. Consequences of nanoparticle volume fraction MgO on temperature profile.

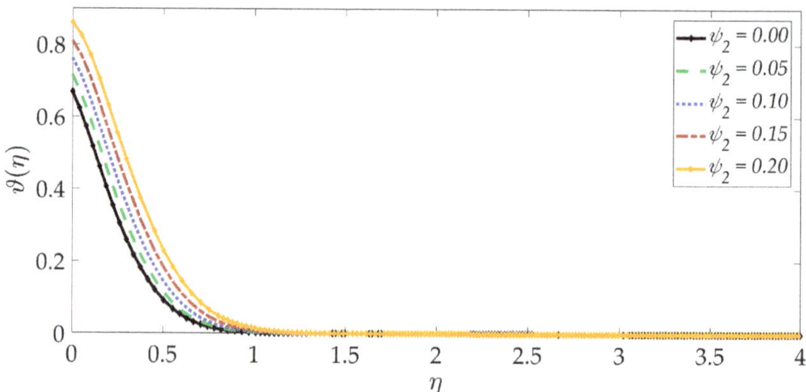

Figure 17. Consequences of nanoparticle volume fraction Ni on temperature profile.

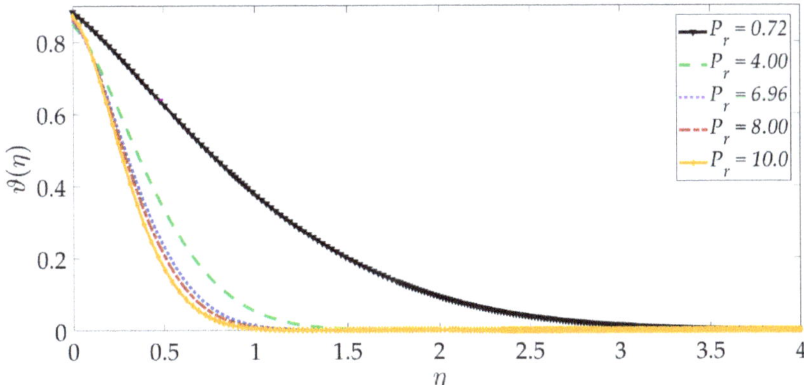

Figure 18. Consequences of Prandtl number on temperature profile.

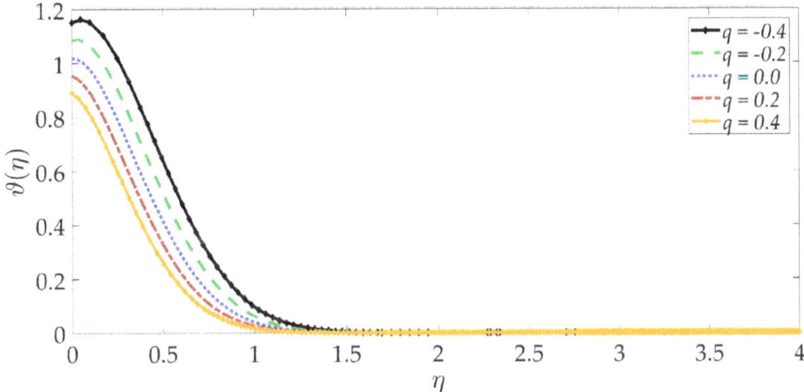

Figure 19. Consequences of heat source/sink parameter on temperature profile.

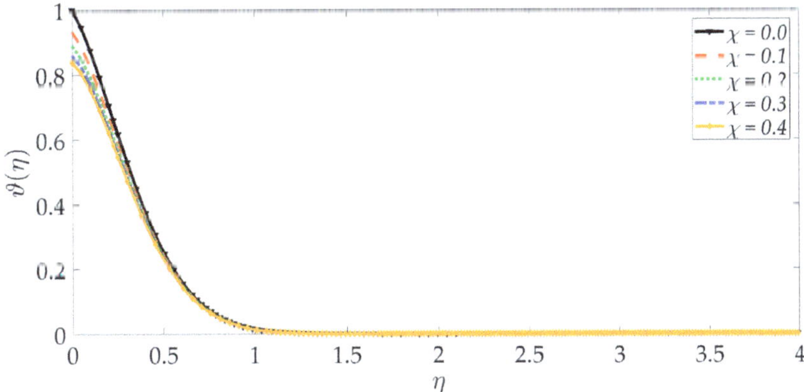

Figure 20. Consequences of thermal slip parameter on temperature profile.

7. Conclusions

A mathematical model has been derived for the steady, incompressible, magnetohydro-dynamic (MHD) stagnation flow of an electrically conducting hybrid nanofluid (comprising magnesium oxide and nickel nanoparticles in water base fluid) along an elastic surface embedded in a porous medium. The magnetic Reynolds number is sufficiently large to invoke magnetic induction effects. Darcy's law is adopted for the porous medium. Wall velocity slip, thermal slip, heat source/sink, and viscous dissipation are also incorporated in the model. The transformed dimensionless nonlinear ordinary differential boundary layer equations for momentum, induced magnetic field, and energy with appropriate boundary conditions are solved with Matlab software using a shooting method. A detailed comparison has been presented with previous studies from the literature for skin friction and Nusselt number in order to verify the accuracy and convergence of the present Matlab solutions. The primary findings of the current investigation may be summarized as follows:

a. With increasing nanoparticle volume fraction of nickel, velocity slip, thermal Grashof number, and the Darcy number, there is a strong acceleration in the flow and reduction in momentum boundary layer thickness.

b. With the elevation in the magnetic parameter and the nanoparticle volume fraction of magnesium oxide, there is a deceleration in the flow and an increase in momentum boundary layer thickness.

c. An increment in nanoparticle volume fraction of both MgO and Ni boosts the temperature and thermal boundary layer thickness.

d. Elevation in Darcy number, velocity slip, and thermal Grashof number generally enhance the induced magnetic field and increase magnetic boundary layer thickness.

e. An increment in both MgO and Ni-nanoparticle volume fractions increases magnetic induction primarily near the elastic surface (wall).

f. With an increase in the reciprocal of magnetic Prandtl number (i.e., decrease in magnetic diffusion rate relative to viscous diffusion rate), initially, there is a suppression in the induced magnetic field near the wall; however, further from the wall, this trend is reversed.

g. The temperature profile and thermal boundary layer thickness are boosted with increment in Darcy number and both MgO and Ni nanoparticle volume fractions.

h. Increasing the Prandtl number, the heat sink parameter, and the thermal slip all have the effect of decreasing the thickness of the thermal boundary layer and the temperature magnitudes.

i. Matlab shooting quadrature is found to be very accurate for solving magnetized hybrid nanofluid coating flows and achieves very good convergence characteristics.

The present study has considered *Newtonian* flow behavior. Future investigations may address the *rheology* of hybrid nanofluids (e.g., viscoelastic models [46,47]) and also consider alternative nanoparticles, e.g., combinations of metallic (gold, silver, Zinc oxide, etc.) and carbon-based (diamond, graphite, etc.) [48,49]. Efforts in these directions are also beneficial to solar energy coating manufacturing fluid dynamics and will be reported imminently.

Author Contributions: Conceptualization, M.M.B. and S.I.A.; methodology, O.A.B.; software, M.M.B. and S.I.A.; validation, M.M.B. and O.A.B.; formal analysis, S.I.A. and O.A.B.; investigation, M.M.B. and O.A.B.; resources, S.I.A. and O.A.B.; data curation, S.I.A. and O.A.B.; writing—original draft preparation, M.M.B. and O.A.B.; writing—review and editing, M.M.B. and O.A.B.; visualization, S.I.A. and O.A.B.; supervision, M.M.B. All authors have read and agreed to the published version of the manuscript.

Funding: This research received no external funding.

Institutional Review Board Statement: Not applicable.

Informed Consent Statement: Not applicable.

Data Availability Statement: Not applicable.

Conflicts of Interest: The authors declare no conflict of interest.

Nomenclature

Symbols	Names
U_f	Velocity at the free stream
U_s	Velocity at the surface
T_{inf}	Ambient temperature
T_s	Surface temperature
$\mathbf{U}$	Velocity field vector
P	Pressure
$(C_p)_{hnf}$	Specific heat
$\mathbf{S}$	Stress tensor
K	Permeability of the porous medium
X_1, X_2	Cartesian coordinate system
U_1, U_2	Velocity components
A_u	Velocity slip
T_u	Thermal slip
M	Magnetic parameter
G_r	Mixed convection parameter
Gr	Thermal Grashof number
E_c	Eckert number
D_a	Darcy number
Re_{X_1}	Local Reynolds number
P_r	Prandtl number
q	Heat source/sink parameter
H_a	Hartmann number
Re_m	Magnetic Reynolds number
S_w	Wall shear stress
Q_w	Wall heat flux

Greek Symbols	
v_{hnf}	Kinematic viscosity
Π	Induced magnetic field vector
β_{hnf}	Thermal expansion coefficient
ρ_{hnf}	Density of the hybrid nanofluid
ζ	Magnetic diffusivity
σ_{hnf}	Electrical conductivity
μ_e	Magnetic permeability parameter
Π_1, Π_2	Magnetic induction components
Π_0	Magnetic field at infinity upstream
ψ_1	Nanoparticle volume fraction of MgO nanoparticles
ψ_2	Nanoparticle's volume fraction of Ni
λ	Velocity slip parameter
χ	Thermal slip parameter
δ	Reciprocal of magnetic Prandtl number
Γ	Characteristic length of the elastic surface

Subscripts and Abbreviations	
MgO	Magnesium oxide
Ni	Nickel
hnf	Hybrid nanofluid
nf	Nanofluid
NPs	Nanoparticles

References

1. Kizilkan, O.; Kabul, A.; Dincer, I. Development and performance assessment of a parabolic trough solar collector-based integrated system for an ice-cream factory. *Energy* **2016**, *100*, 167–176. [CrossRef]
2. Verma, S.K.; Tiwari, A.K. Progress of nanofluid application in solar collectors: A review. *Energy Convers. Manag.* **2015**, *100*, 324–346. [CrossRef]

3. Marefati, M.; Mehrpooya, M.; Shafii, M.B. Optical and thermal analysis of a parabolic trough solar collector for production of thermal energy in different climates in Iran with comparison between the conventional nanofluids. *J. Clean. Prod.* **2018**, *175*, 294–313. [CrossRef]

4. Harvey, L.D. Solar-hydrogen electricity generation and global CO_2 emission reduction. *Int. J. Hydrogen Energy* **1996**, *21*, 583–595. [CrossRef]

5. Minea, A.A.; El-Maghlany, W.M. Influence of hybrid nanofluids on the performance of parabolic trough collectors in solar thermal systems: Recent findings and numerical comparison. *Renew. Energy* **2018**, *120*, 350–364. [CrossRef]

6. Choi, S.U.; Eastman, J.A. *Enhancing Thermal Conductivity of Fluids with Nanoparticles*; No. ANL/MSD/CP-84938; CONF-951135–29; Argonne National Lab.: Argonne, IL, USA, 1995.

7. Prasher, R.; Bhattacharya, P.; Phelan, P.E. Thermal conductivity of nanoscale colloidal solutions (nanofluids). *Phys. Rev. Lett.* **2005**, *94*, 025901. [CrossRef] [PubMed]

8. Azmi, W.H.; Hamid, K.A.; Usri, N.A.; Mamat, R.; Sharma, K.V. Heat transfer augmentation of ethylene glycol: Water nanofluids and applications—A review. *Int. Commun. Heat Mass Transf.* **2016**, *75*, 13–23. [CrossRef]

9. Olabi, A.G.; Elsaid, K.; Sayed, E.T.; Mahmoud, M.S.; Wilberforce, T.; Hassiba, R.J.; Abdelkareem, M.A. Application of nanofluids for enhanced waste heat recovery: A review. *Nano Energy* **2021**, *84*, 105871. [CrossRef]

10. Reese, J.M.; Zheng, Y.; Lockerby, D.A. Computing the near-wall region in gas micro-and nanofluidics: Critical Knudsen layer phenomena. *J. Comput. Theor. Nanosci.* **2007**, *4*, 807–813. [CrossRef]

11. Badur, J.; Ziółkowski, P.J.; Ziółkowski, P. On the angular velocity slip in nano-flows. *Microfluid. Nanofluid.* **2015**, *19*, 191–198. [CrossRef]

12. Li, H.; Ha, C.S.; Kim, I. Fabrication of carbon nanotube/SiO_2 and carbon nanotube/SiO_2/Ag nanoparticles hybrids by using plasma treatment. *Nanoscale Res. Lett.* **2009**, *4*, 1384–1388. [CrossRef]

13. Guo, S.; Dong, S.; Wang, E. Gold/platinum hybrid nanoparticles supported on multiwalled carbon nanotube/silica coaxial nanocables: Preparation and application as electrocatalysts for oxygen reduction. *J. Phys. Chem. C* **2008**, *112*, 2389–2393. [CrossRef]

14. Devi, S.A.; Devi, S.S.U. Numerical investigation of hydromagnetic hybrid Cu–Al2O3/water nanofluid flow over a permeable stretching sheet with suction. *Int. J. Nonlinear Sci. Numer. Simul.* **2016**, *17*, 249–257. [CrossRef]

15. Ghadikolaei, S.S.; Yassari, M.; Sadeghi, H.; Hosseinzadeh, K.; Ganji, D.D. Investigation on thermophysical properties of TiO_2–Cu/H_2O hybrid nanofluid transport dependent on shape factor in MHD stagnation point flow. *Powder Technol.* **2017**, *322*, 428–438. [CrossRef]

16. Hassan, M.; Marin, M.; Ellahi, R.; Alamri, S.Z. Exploration of convective heat transfer and flow characteristics synthesis by Cu–Ag/water hybrid-nanofluids. *Heat Transf. Res.* **2018**, *49*, 1837–1848. [CrossRef]

17. Izadi, M.; Maleki, N.M.; Pop, I.; Mehryan, S.A.M. Natural convection of a hybrid nanofluid subjected to non-uniform magnetic field within porous medium including circular heater. *Int. J. Numer. Methods Heat Fluid Flow* **2019**, *29*, 1211–1231. [CrossRef]

18. Riaz, A.; Ellahi, R.; Sait, S.M. Role of hybrid nanoparticles in thermal performance of peristaltic flow of Eyring–Powell fluid model. *J. Therm. Anal. Calorim.* **2021**, *143*, 1021–1035. [CrossRef]

19. Puneeth, V.; Manjunatha, S.; Madhukesh, J.K.; Ramesh, G.K. Three dimensional mixed convection flow of hybrid casson nanofluid past a non-linear stretching surface: A modified Buongiorno's model aspects. *Chaos Solitons Fractals* **2021**, *152*, 111428. [CrossRef]

20. Pilarska, A.A.; Klapiszewski, Ł.; Jesionowski, T. Recent development in the synthesis, modification and application of $Mg(OH)_2$ and MgO: A review. *Powder Technol.* **2017**, *319*, 373–407. [CrossRef]

21. El-Moslamy, S.H. Bioprocessing strategies for cost-effective large-scale biogenic synthesis of nano-MgO from endophytic Streptomyces coelicolor strain E72 as an anti-multidrug-resistant pathogens agent. *Sci. Rep.* **2018**, *8*, 3820. [CrossRef]

22. Shen, Y.; He, L.; Yang, Z.; Xiong, Y. Corrosion behavior of different coatings prepared on the surface of AZ80 magnesium alloy in simulated body fluid. *J. Mater. Eng. Perform.* **2020**, *29*, 1609–1621. [CrossRef]

23. Tang, H.; Zhou, X.B.; Liu, X.L. Effect of magnesium hydroxide on the flame retardant properties of unsaturated polyester resin. *Procedia Eng.* **2013**, *52*, 336–341. [CrossRef]

24. Khalil, K.D.; Bashal, A.H.; Khalafalla, M.; Zaki, A.A. Synthesis, structural, dielectric and optical properties of chitosan-MgO nanocomposite. *J. Taibah Univ. Sci.* **2020**, *14*, 975–983. [CrossRef]

25. Hyeon, T. Chemical synthesis of magnetic nanoparticles. *Chem. Commun.* **2003**, *8*, 927–934. [CrossRef]

26. Mariam, A.A.; Kashif, M.; Arokiyaraj, S.; Bououdina, M.; Sankaracharyulu, M.; Jayachandran, M.; Hashim, U. Bio-synthesis of NiO and Ni nanoparticles and their characterization. *Dig. J. Nanomater. Biostruct.* **2014**, *9*, 1007–1019.

27. Karmhag, R.; Tesfamichael, T.; Wäckelgård, E.; Niklasson, G.A.; Nygren, M. Oxidation kinetics of nickel particles: Comparison between free particles and particles in an oxide matrix. *Sol. Energy* **2000**, *68*, 329–333. [CrossRef]

28. Heilmann, M.; Kulla, H.; Prinz, C.; Bienert, R.; Reinholz, U.; Guilherme Buzanich, A.; Emmerling, F. Advances in nickel nanoparticle synthesis via oleylamine route. *Nanomaterials* **2020**, *10*, 713. [CrossRef]

29. Ochirkhuyag, A.; Sápi, A.; Szamosvölgyi, Á.; Kozma, G.; Kukovecz, Á.; Kónya, Z. One-pot mechanochemical ball milling synthesis of the MnO_x nanostructures as efficient catalysts for CO_2 hydrogenation reactions. *Phys. Chem. Chem. Phys.* **2020**, *22*, 13999–14012. [CrossRef]

30. Bussamara, R.; Eberhardt, D.; Feil, A.F.; Migowski, P.; Wender, H.; de Moraes, D.P.; Dupont, J. Sputtering deposition of magnetic Ni nanoparticles directly onto an enzyme surface: A novel method to obtain a magnetic biocatalyst. *Chem. Commun.* **2013**, *49*, 1273–1275. [CrossRef]

31. Alonso, F.; Riente, P.; Yus, M. Hydrogen-transfer reduction of carbonyl compounds promoted by nickel nanoparticles. *Tetrahedron* **2008**, *64*, 1847–1852. [CrossRef]

32. Saxena, A.; Kumar, A.; Mozumdar, S. Ni-nanoparticles: An efficient green catalyst for chemo-selective oxidative coupling of thiols. *J. Mol. Catal. A Chem.* **2007**, *269*, 35–40. [CrossRef]

33. Alonso, F.; Riente, P.; Yus, M. Alcohols for the α-Alkylation of Methyl Ketones and Indirect Aza-Wittig Reaction Promoted by Nickel Nanoparticles. *Eur. J. Org. Chem.* **2008**, *29*, 4908–4914. [CrossRef]

34. Alonso, F.; Riente Paiva, P.; Yus, M. Wittig-type olefination of alcohols promoted by nickel nanoparticles: Synthesis of polymethoxylated and polyhydroxylated stilbenes. *Eur. J. Org. Chem.* **2009**, *34*, 6034–6042. [CrossRef]

35. Dhakshinamoorthy, A.; Pitchumani, K. Clay entrapped nickel nanoparticles as efficient and recyclable catalysts for hydrogenation of olefins. *Tetrahedron Lett.* **2008**, *49*, 1818–1823. [CrossRef]

36. Muhammad, N.; Nadeem, S.; Khan, U.; Sherif, E.S.M.; Issakhov, A. Insight into the significance of Richardson number on two-phase flow of ethylene glycol-silver nanofluid due to Cattaneo-Christov heat flux. *Waves Random Complex Media* **2021**, 1–19. [CrossRef]

37. Anwar, M.I.; Firdous, H.; Zubaidi, A.A.; Abbas, N.; Nadeem, S. Computational analysis of induced magnetohydrodynamic non-Newtonian nanofluid flow over nonlinear stretching sheet. *Prog. React. Kinet. Mech.* **2022**, *47*, 14686783211072712. [CrossRef]

38. Khan, A.U.; Ullah, N.; Al-Zubaidi, A.; Nadeem, S. Finite element analysis for CuO/water nanofluid in a partially adiabatic enclosure: Inclined Lorentz forces and porous medium resistance. *Alex. Eng. J.* **2022**, *61*, 6477–6488. [CrossRef]

39. Zhang, L.; Bhatti, M.M.; Michaelides, E.E.; Marin, M.; Ellahi, R. Hybrid nanofluid flow towards an elastic surface with tantalum and nickel nanoparticles, under the influence of an induced magnetic field. *Eur. Phys. J. Spec. Top.* **2021**, 1–13. [CrossRef]

40. Ziółkowski, P.A.W.E.Ł.; Badur, J. On Navier slip and Reynolds transpiration numbers. *Arch. Mech.* **2018**, *70*, 269–300.

41. Kumari, M.; Takhar, H.S.; Nath, G. MHD flow and heat transfer over a stretching surface with prescribed wall temperature or heat flux. *Wärme-Und Stoffübertragung* **1990**, *25*, 331–336. [CrossRef]

42. Ali, F.M.; Nazar, R.; Arifin, N.M.; Pop, I. MHD boundary layer flow and heat transfer over a stretching sheet with induced magnetic field. *Heat Mass Transf.* **2011**, *47*, 155–162. [CrossRef]

43. Hassanien, I.A.; Abdullah, A.A.; Gorla, R.S.R. Flow and heat transfer in a power-law fluid over a nonisothermal stretching sheet. *Math. Comput. Model.* **1998**, *28*, 105–116. [CrossRef]

44. Salleh, M.Z.; Nazar, R. Numerical solutions of the boundary layer flow and heat transfer over a stretching sheet with constant wall temperature and heat flux. In Proceedings of the Third International Conference on Mathematical Sciences ICM 2008, Al Ain, United Arab Emirates, 3–6 March 2008; pp. 1260–1267.

45. Ali, F.M.; Nazar, R.; Arifin, N.M.; Pop, I. MHD stagnation-point flow and heat transfer towards stretching sheet with induced magnetic field. *Appl. Math. Mech.* **2011**, *32*, 409–418. [CrossRef]

46. Kuharat, S.; Anwar Bég, O.; Kadir, A.; Vasu, B.; Bég, T.A.; Jouri, W.S. Computation of gold-water nanofluid natural convection in a three-dimensional tilted prismatic solar enclosure with aspect ratio and volume fraction effects. *Nanosci. Technol. Int. J.* **2020**, *11*, 141–167. [CrossRef]

47. Bhatti, M.M.; Abdelsalam, S.I. Bio-inspired peristaltic propulsion of hybrid nanofluid flow with Tantalum (Ta) and Gold (Au) nanoparticles under magnetic effects. *Waves Random Complex Media* **2021**, 1–26. [CrossRef]

48. Ferdows, M.; Alsenafi, A.; Anwar Bég, O.; Bég, T.A.; Kadir, A. Numerical study of nano-biofilm stagnation flow from a nonlinear stretching/shrinking surface with variable nanofluid and bio-convection transport properties. *Sci. Rep.* **2021**, *11*, 9877.

49. Anwar Bég, O.; Kuharat, S.; Ferdows, M.; Dao, M.; Kadir, A.; Shamshuddin, M. Magnetic nano polymer flow with magnetic induction and nanoparticle solid volume fraction effects: Solar magnetic nano-polymer fabrication simulation. *Proc. IMechE-Part N J. Nanoeng. Nanomater. Nano-Syst.* **2019**, *233*, 27–45. [CrossRef]

 *nanomaterials*

Article

Computational Analysis of the Morphological Aspects of Triadic Hybridized Magnetic Nanoparticles Suspended in Liquid Streamed in Coaxially Swirled Disks

Zubair Akbar Qureshi [1,*], Sardar Bilal [2,*], Imtiaz Ali Shah [2], Ali Akgül [3], Rabab Jarrar [4], Hussein Shanak [4] and Jihad Asad [4]

1 Department of Mathematics, Multan Campus, AIR University, Multan 49501, Pakistan
2 Department of Mathematics, AIR University, Islamabad 44000, Pakistan; imtiazsha91@gmail.com
3 Department of Mathematics, Art and Science Faculty, Siirt University, Siirt 56100, Turkey; aliakgul00727@gmail.com
4 Department of Physics, Faculty of Applied Sciences, Palestine Technical University-Kadoorie, Tulkarm 305, West Bank, Palestine; r.jarrar@ptuk.edu.ps (R.J.); h.shanak@ptuk.edu.ps (H.S.); j.asad@ptuk.edu.ps (J.A.)
* Correspondence: zubair@aumc.edu.pk (Z.A.Q.); sardarbilal@mail.au.edu.pk (S.B.)

Abstract: Currently, pagination clearly explains the increase in the thermophysical attributes of viscous hybrid nanofluid flow by varying morphological aspects of inducted triadic magnetic nanoparticles between two coaxially rotating disks. Copper metallic nanoparticles are inserted with three different types of metallic oxide nanoparticles: Al_2O_3, Ti_2O, and Fe_3O_4. Single-phase simulation has been designed for the triadic hybrid nanofluids flow. The achieved expressions are transmuted by the obliging transformation technique because of dimensionless ordinary differential equations (ODEs). Runge–Kutta in collaboration with shooting procedure are implemented to achieve the solution of ODEs. The consequences of pertinent variables on associated distributions and related quantities of physical interest are elaborated in detail. It is inferred from the analysis that Cu-Al_2O_3 metallic type hybrid nanofluids flow shows significant results as compared with the other hybrid nanoparticles. The injection phenomenon on hybrid nanofluids gives remarkable results regarding shear stress and heat flux with the induction of hybridized metallic nanoparticles. Shape and size factors have also been applied to physical quantities. The morphology of any hybrid nanoparticles is directly proportional to the thermal conductance of nanofluids. Peclet number has a significant effect on the temperature profile.

Keywords: triadic hybridize nanofluid model; heat and mass flux; MHD; morphology effect; computational analysis (shooting technique)

Citation: Qureshi, Z.A.; Bilal, S.; Shah, I.A.; Akgül, A.; Jarrar, R.; Shanak, H.; Asad, J. Computational Analysis of the Morphological Aspects of Triadic Hybridized Magnetic Nanoparticles Suspended in Liquid Streamed in Coaxially Swirled Disks. *Nanomaterials* **2022**, *12*, 671. https://doi.org/10.3390/nano12040671

Academic Editors: M.M. Bhatti, Kambiz Vafai and Sara I. Abdelsalam

Received: 1 October 2021
Accepted: 27 January 2022
Published: 17 February 2022

Publisher's Note: MDPI stays neutral with regard to jurisdictional claims in published maps and institutional affiliations.

1. Introduction

Copper is a transitional metal and possesses the ability to accept and donate electrons along with the characteristic to carry out oxidization and reduction. Because of this tendency, it is used as an essential micronutrient for functionality of living organisms and development of enzymes involved in metabolic processes [1,2]. Moreover, copper contributes to the cure of heart disease, diabetes, and obesity by the formation of drugs [3–7]. In recent years, advancement in nanotechnology has generated intent toward findings of innovative procedures for the production of copper particles on the nano scale (1–100 nm). Especially, metallic nanoparticles including copper particles are widely used in medicinal instruments, investigative imaging, drug supply procedure, therapeutics, and cancers cells. In spite of excellent utilizations of copper nanoparticles, some serious dangers of such nanoparticles are continuously elevated as a result of tissue damaging [8–10]. For this purpose, hybridization of copper with non-metallic oxides is performed, which makes a new class of working liquids. Extensive experimental research on hybrid nanocomposites can be found, e.g., Turcu et al. [11] and Jana et al. [12]. Furthermore, Devi and Devi [13,14] examined the consequence of hybridized nanoliquid over stretchable configuration and concluded on enhancement in the thermal rate with inclusion of hybridized nanoparticles in base fluid. Farooq et al. [15] analyzed the flow of bio convective cross nanofluid with motile microorganisms in the attendance of radiative energy and melting phenomenon. Anitha et al. [16] evaluated the performance of a double heat exchanger by considering two different hybrid nanofluids composed of water and ethylene as base liquids and TiO_2-γ-AlOOH nanoparticles. Ebrahimi et al. [17] showed modelling of laminarly convective heat transfer of nanoliquid in an enclosure by implementing the finite element approach.

Rotating flows is one of the basic frameworks in the dynamics of liquids. The pioneering work in this direction was performed by Karman [18], who developed the formulation of such a problem by forming the Navier–Stokes equation in curvilinear coordinates. Following the work presented by Karman, extensive works on rotational flows have been performed, such as Griffiths [19], who executed the flow mechanism of non-Newtonian liquid over a spinning disk. Some valuable and old literature related to rotational flow problems is encapsulated in [20–24]. Waini et al. [25] examined the influence of hybrid copper and aluminum oxide nanoparticles in heat transfer elevation of water on a rotating disk. Turkyilmazoglu [26] inspected the single-phase flow of nanofluid over a rotated disk by determining Brownian diffusion aspects. Turkyilmazoglu [27] analyzed 3D laminar flow of electrically conducting viscous liquid flowing over a rotating disk. Some recent acquisitions on rotatory flow problems are divulged in [28–31].

Synthetic characteristics of hybrid nanoparticles along with superior fluidity and stability properties and practical utilization of these composite particles have been raised in different technological applications such as electronic cooling devices, thermal control of vehicles, welding, power systems, lubrication, hydroelectric manufacturing, production of paper and biomedicine, nuclear production, manufacture of spacecraft devices, and many other areas [32–34]. Xu et al. [35] explained the unsteady mixed convective flow of a hybrid nanoliquid between spinning disks. The influence of Hall current and magnetic field on hybrid nanoliquid flow between coaxially rotated disks was encapsulated by Nilankush et al. [36]. Dinarvand et al. [37] computationally scrutinized the flow behavior of a hybrid nanofluid over a porous rotating disk with induction of metallic-oxide (ZuO-Au). Khan et al. [38] probed impression of the Hall effect on a hybrid nanoliquid flowing on a spherical surface. Izadi et al. [39] delineated convective heat transfer in a water hybrid nanofluid with induction of multi-wall carbon nanotubes inside an enclosure. Arani et al. [40] depicted the heat and flow characteristics of a laminar water-based nanoliquid in a novel design of a double-layered microchannel heat sink. Safaei et al. [41] presented work on the thermal aspects of functionalized multi-walled carbon nanotubes in nanoliquid flow over a flat plate by performing numerical simulations. Coshayochi et al. [42] determined the influence of the shape and size of nanoparticles in elevation in the heat transfer rate of a pulsating heat pipe under the influence of magnetic field. An overview about work conducted by a researcher regarding a hybrid nanoliquid over rotating disks is accumulated in [43,44].

Combined evaluation of heat and mass transfer phenomenon has superb applications in chemical and food processing, hydrometallurgy, ceramics manufacturing, polymerization, and so forth. Vajravelu et al. [45] analyzed magnetically effected 3D squeezed flow of the nanoliquid between rotating discs with velocity slip. Das et al. [46] presented mathematical modelling of magnetically influenced squeezed nanoliquid flow between coaxially rotating disks. Heat and mass transfer aspects in MHD squeezed flow with dispersion of nanoparticles by providing slip effects on surface of disk were deliberated by Din et al. [47]. Qayyum et al. [48] studied heat and mass change in nanofluid thermal flux across a spinning disk with a uniform thin layer. Aziz et al. [49] adumbrated heat and mass transport in dissipated and magnetized flow of viscous fluid over a spinning disk. Reddy et al. [50] presented an analysis on enhancement in convective heat and mass transfer with the addition of metallic hybridized metallic nanoparticles in fluid flow over a rotating disk.

MHD is the study of the fluid flow mechanism under the influence of magnetic field. Magnetized fluid possesses significant applications in different fields, especially in biomedical science like laser beam scanning, drug delivery targeting, manipulation of nanoparticles, MHD base micropump, magnetic rays imaging, and many others. Muhammad et al. [51] probed the flow of viscoelastic liquid under the impact of magnetic field. Uddin et al. [52] discussed the impact of magnetization on nano viscous liquid over a rotating permeable disk. The effects of magnetic field on viscously dissipated hybrid nanofluids by performing numerical simulations were reported by Imran et al. [53]. Khan et al. [54] evaluated the magnetic field effect on flow features of viscous liquid between coaxially rotated disks. Some recent literature surveys regarding the influence of magnetic field on fluid flow problem in multiple computational domains and under the consideration of various physical variables are accumulated like Krishna et al. [55,56] discussed the consequence of magnetic field along with hall and ion slip on second grade rotating fluid on a semi-infinite vertically moving surface. 3D convective heat transfer in micro concentrated annulus generated by non-uniform heat flux at wall in water base nanofluid with induction of Al_2O_3 nanoparticles was determined by Davood et al. [57]. Some recent developments on MHD fluid in different computation domains are gathered in [58–67].

Examination of fluid flow phenomenon in porous orthogonal disks has numerous dedicated utilizations in many advanced technologies, such as lubricants bearing technology, mass and heat exchanger, viscometers, crystal growth, biomechanics, oceanography, and computer storage system. In this geometry, the foremost physical aspect is the injection/suction along with consideration of hybrid nanofluids, which makes this problem more remarkable in view of practical essence. So, the main objective here is to investigate the enhancement in the thermophysical characteristics of water by inserting triadic hybridized nanoparticles in flow between two orthogonally moving permeable disks. After reviewing the aforementioned literature, consideration of triadic nano particles inside fluid domains is not scrutinized yet. Therefore, the prime concern of this pagination is to inspect the behavior of flow concerning profiles like shear stress, velocity, temperature, and mass profile for injection/suction cases with the addition of triadic particles. A solution to the problem at hand is heeded by implementing the shooting technique and a comparison of computed data with published literature is revealed.

2. Problem Formulation

Here, we assume unsteady, laminarly, and 3D viscous liquid over rotating permeable disks in the attendance of externally produced magnetic field. Here, we ignore the Hall current effect and body forces, i.e., induced magnetic field due to the presence of external pressure by providing low magnitude of permeable Reynold number. Single-phase simulation was developed for the problem at hand in the presence of different types of nanoparticles. Permeable disks are located at equal distances from the center and moving up and down with distance 2s(t), along with velocity s'(t). Here, the base fluid is water to support our single-phase simulation of the hybrid nanofluids flow. The triadic type of hybrid composite material was introduced here, in which metallic particles of copper are commuted with different metallic-oxide nanomaterials. It is noted that the temperature and concentration of the lower disk are strictly greater than those of the upper disk, as exhibited in Figure 1.

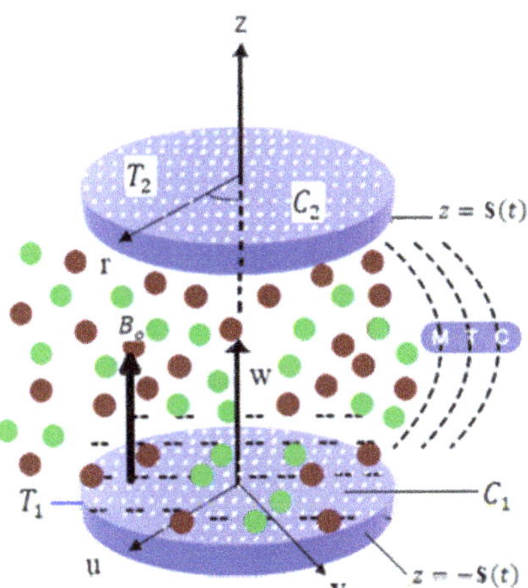

Figure 1. Physical model.

The constitutive expressions can be written as follows [60]:

$$\frac{\partial u}{\partial r} + \frac{u}{r} + \frac{\partial w}{\partial z} = 0,\tag{1}$$

$$\frac{\partial u}{\partial t} + u\frac{\partial u}{\partial r} + w\frac{\partial u}{\partial z} - \frac{v^2}{r} = -\frac{1}{\rho_{hnf}}\frac{\partial p}{\partial r} + v_{hnf}\left(\frac{\partial^2 u}{\partial r^2} + \frac{1}{r}\frac{\partial u}{\partial r} - \frac{u}{r^2} + \frac{\partial^2 u}{\partial z^2}\right) - \frac{\sigma_e B_0^2}{\rho_{hnf}}u,\tag{2}$$

$$\frac{\partial v}{\partial t} + u\frac{\partial v}{\partial r} + w\frac{\partial v}{\partial z} + \frac{uv}{r} = v_{hnf}\left(\frac{\partial^2 v}{\partial r^2} + \frac{1}{r}\frac{\partial v}{\partial r} - \frac{v}{r^2} + \frac{\partial^2 v}{\partial z^2}\right) - \frac{\sigma_e B_0^2}{\rho_{hnf}}v,\tag{3}$$

$$\frac{\partial w}{\partial t} + u\frac{\partial w}{\partial r} + w\frac{\partial w}{\partial z} = -\frac{1}{\rho_{hnf}}\frac{\partial p}{\partial z} + v_{hnf}\left(\frac{\partial^2 w}{\partial r^2} + \frac{1}{r}\frac{\partial w}{\partial r} + \frac{\partial^2 w}{\partial z^2}\right)\tag{4}$$

$$\frac{\partial T}{\partial t} + u\frac{\partial T}{\partial r} + w\frac{\partial T}{\partial z} = \alpha_{hnf}\frac{\partial^2 T}{\partial z^2} + \frac{\mu_{hnf}}{(\rho c)_{p_{hnf}}}\left(\frac{\partial u}{\partial z}\right)^2 + \frac{\sigma_e B_0^2 u^2}{(\rho c_p)_{hnf}}\tag{5}$$

$$\frac{\partial C}{\partial t} + u\frac{\partial C}{\partial r} + w\frac{\partial C}{\partial z} = \frac{\partial^2 C}{\partial z^2}\tag{6}$$

$$v_{hnf} = \frac{\mu_{hnf}}{\rho_{hnf}} \quad \text{and} \quad \alpha_{hnf} = \frac{k_{hnf}}{(\rho c_p)_{hnf}}, Pr = \frac{(\mu C_p)_{bf}}{k_{bf}}\tag{7}$$

where ρ_s and ρ_f represent the densities of solid particles, particles and fluid; $(c_p)_{hnf}$ shows the heat capacitance of hybrid nanofluids; and k_{hnf} is the thermal conductivity of the hybrid nanofluids.

Uniform mixing of nanoparticles and hosting liquid with negligible slip is presumed in the single-phase approach, which produces thermophysical characteristics that are found by experimental calculation. The accuracy of this approach is predominantly dependent on the accurate prediction of the thermophysical properties of the nanofluids, which are thus presently estimated using the same property correlations utilized in the respective experimental study [51].

Boundary Condition

The associated boundary conditions are as follows:

$$z_1 = -s(t)\ u = 0\ v = -\frac{rA_1 s'(t)}{2s}\ w = -A_1 s'(t)\ T = T_1\ C = C_1$$

and

$$z_1 = s(t)\ u = 0\ v = \frac{rA_1 s'(t)}{2s}\ w = A_1 s'(t)\ T = T_2\ C = C_2 \tag{8}$$

For elimination of the pressure term, the following similarity variables are utilized:

$$\eta = \tfrac{z_1}{s}\ u = -\tfrac{rv_f}{s^2} F_\eta(\eta, t)\ v = \tfrac{rv_f}{s^2} G(\eta, t)\ w = \tfrac{2v_f}{s} F(\eta, t) \qquad \theta = \tfrac{T-T_2}{T_1-T_2}\ ,\ \chi(\eta) = \tfrac{C-C_2}{C_1-C_2},$$

and

$$\frac{v_{hnf}}{v_f} F_{\eta\eta\eta\eta} + \alpha(3F_{\eta\eta} + \eta F_{\eta\eta\eta}) - 2FF_{\eta\eta\eta} - \frac{s^2}{v_f} F_{\eta\eta t} + 2GG_\eta - \frac{\rho_f}{\rho_{hnf}} MF_{\eta\eta} = 0, \tag{9}$$

$$\frac{v_{hnf}}{v_f} G_{\eta\eta} + \alpha(2G + \eta F_\eta) + 2GF_\eta - \frac{s^2}{v_f} G_t - 2FG_\eta - \frac{\rho_f}{\rho_{hnf}} MG = 0, \tag{10}$$

$$\theta_{\eta\eta} + \frac{v_f}{\alpha_{hnf}}(\alpha\eta - 2F)\theta_\eta + [(1 - (\varphi_1 + \varphi_2))^{-2.5} F_{\eta\eta}{}^2 + MF_\eta{}^2]E_c P_r \frac{\kappa_f}{\kappa_{hnf}} - \frac{k^2}{\alpha_{nf}}\theta_t = 0, \tag{11}$$

$$D\chi'' + v_f\left(\alpha\eta - 2F\right)\chi' - k^2 \chi_t = 0, \tag{12}$$

Boundary conditions in dimensionless form are as follows:

$$\eta = -1,\ F = -Re, F_\eta = 0, \theta = 1, \chi = 1,$$

and

$$\eta = 1,\ F = Re, F_\eta = 0, \theta = 0,\ \chi = 0$$

where $\alpha = \frac{ss'(t)}{v_f}$ stands for wall expansion ratio, $Re = \frac{A_1 ss'(t)}{2v_f}$ stands for permeable Reynold number, $Pr = \frac{(\mu c_p)_f}{k_f}$ stands for the Prandtl number, $Ec = \frac{u^2}{(T_1 - T_2)(c_p)_f}$ stands for Eckert number, $P_e = Re * P_r$ is the Peclet number, $Sc = \frac{v_f}{D}$ stands for Schmidt number, and $M = \frac{\sigma_e B_0^2 s^2}{\mu_f}$ stands for magnetic parameter referred to [60,61].

Finally, we set $F = f\ Re$, $G = g\ Re$ by following Majdalani et al. [61] when α is a constant.

$f = f(\eta)$ and $\theta = \theta(\eta)$, which leads to $\theta_t = 0$, $g_{\eta t} = 0$, $f_{\eta\eta t} = 0$, and $\chi_t = 0$. Thus, we have the following equations:

$$\frac{v_{hnf}}{v_f} f_{\eta\eta\eta\eta} + \alpha(3f_{\eta\eta} + \eta f_{\eta\eta\eta}) - 2ff_{\eta\eta\eta} + 2Regg_\eta - \frac{\rho_f}{\rho_{hnf}} Mf_{\eta\eta} = 0 \tag{13}$$

$$\frac{v_{hnf}}{v_f} g_{\eta\eta} + \alpha(2g + \eta g_\eta) + 2Re(gf_\eta - fg_\eta) - \frac{\rho_f}{\rho_{hnf}} Mg = 0 \tag{14}$$

$$\theta_{\eta\eta} + \frac{v_f}{\alpha_{hnf}} Pr(\alpha\eta - 2Ref)\theta_\eta + [(1 - (\varphi_1 + \varphi_2))^{-2.5} f_{\eta\eta}{}^2 + Mf_\eta{}^2]ReE_c P_e \frac{\kappa_f}{\kappa_{hnf}} = 0 \tag{15}$$

$$\chi'' + Sc\left(\alpha\eta - 2fRe\right)\chi' = 0 \tag{16}$$

$$\eta = -1;\ f = -1,\ f_\eta = 0, \theta = 1,\ \chi = 1$$

and

$$\eta = 1;\ f = 1,\ f_\eta = 0, \theta = 0\ \chi = 0. \tag{17}$$

3. Practical and Engineering Interest

3.1. Skin Friction Coefficients

The C_{f1} and C_{f-1} are expressed as follows:

$$
\begin{aligned}
C_{f-1} &= \frac{\xi_w|_{\eta\,=-1}}{\rho_{bf}(s'A_1)^2} \\
&= \frac{\frac{(1+0.1008((\varphi_1)^{0.69574}\left(dp_1\right)^{0.44708}+(\varphi_2)^{0.69574}\left(dp_2\right)^{0.44708})}{Re_r}}{}\sqrt{(f''(-1))^2+(g'(-1))^2} \\
C_{f1} &= \frac{\xi_w|_{\eta\,=1}}{\rho_{bf}(s'A_1)^2} = \frac{(1+0.1008((\varphi_1)^{0.69574}\left(dp_1\right)^{0.44708}+(\varphi_2)^{0.69574}\left(dp_2\right)^{0.44708})}{Re_r}\sqrt{(f''(1))^2+(g'(1))^2}
\end{aligned}
\tag{18}
$$

where ξ_w stand for total shear stress and $Re_r = 4\left(\frac{s}{r}\right)\left(\frac{1}{(Re)^2}\right)$ stands for local the Reynold number.

$$
\xi_{zr} = \mu_{hnf}\left(\frac{\partial u}{\partial z}\right)\Big|_{\eta\,=-1} = \mu_{bf}(1+0.1008((\varphi_1)^{0.69574}\left(dp_1\right)^{0.44708}+(\varphi_2)^{0.69574}\left(dp_2\right)^{0.44708})\left(\frac{rv_f}{s^3}\right)f''(-1)
$$

$$
\xi_{0z} = \mu_{hnf}\left(\frac{\partial v}{\partial z}\right)\Big|_{\eta\,--1} = \mu_{bf}(1+0.1008((\varphi_1)^{0.69574}\left(dp_1\right)^{0.44708}+(\varphi_2)^{0.69574}\left(dp_2\right)^{0.44708})\left(\frac{rv_f}{s^3}\right)g'(-1)
$$

3.2. Nusselt Numbers

Nu_{z-1} and Nu_{z1} are given as

$$
\begin{aligned}
Nu_{z-1} &= \frac{se_z}{\kappa_f(T_1-T_2)}\Big|_{\eta\,=-1} = -\frac{k_{hnf}}{k_f}\theta'(-1) \\
Nu_{z1} &= \frac{se_z}{\kappa_f(T_1-T_2)}\Big|_{\eta\,=1} = -\frac{k_{hnf}}{k_f}\theta'(1)
\end{aligned}
\tag{19}
$$

Here, heat flux is denoted as s_z, which is as follows:

$$
e_z|_{\eta\,=-1} = -k_{hnf}\left(\frac{\partial T}{\partial z}\right)\Big|_{\eta\,=-1} = -\frac{(T_1-T_2)}{s}k_{hnf}\theta'(-1)
$$

$$
e_z|_{\eta\,=1} = -k_{hnf}\left(\frac{\partial T}{\partial z}\right)\Big|_{\eta\,=1} = -\frac{(T_1-T_2)}{s}k_{hnf}\theta'(1)
$$

where $Re = \frac{A_1 ss'(t)}{2v_f}$.

3.3. Sherwood Number

Sherwood number is the ratio of convectional mass transfer and diffusion mass transfer. The mass transfer rate (Sherwood number) $Sh|_{\eta\,=-1}$ and $Sh|_{\eta\,=1}$ at the lower and upper disk have the following mathematical expression:

$$
\begin{aligned}
Sh|_{\eta\,=-1} &= \frac{k_{yz}}{D_{hnf}(C_1-C_2)}\Big|_{\eta\,=-1} = -\chi'(-1) \\
Sh|_{\eta\,=1} &- \frac{kq_z}{D_{hnf}(C_1-C_2)}\Big|_{\eta\,=1} - -\chi'(1)
\end{aligned}
\tag{20}
$$

where

$$
q_z|_{\eta\,=-1} = -D_{hnf}\left(\frac{\partial C}{\partial z}\right)\Big|_{\eta\,=-1} = -D_{hnf}\frac{(C_1-C_2)}{k}\chi'(-1)
$$

$$
q_z|_{\eta\,=1} = -D_{hnf}\left(\frac{\partial C}{\partial z}\right)\Big|_{\eta\,=1} = -D_{hnf}\frac{(C_1-C_2)}{k}\chi'(1)
$$

where $Re = \frac{A_1 ss'(t)}{2v_f}$.

3.4. Thermophysical Properties

PDE Equations (13)–(16) have appropriate thermophysical properties:

$$\left(\frac{(1+0.1008((\varphi_1)^{0.69574}\left(dp_1\right)^{0.44708}+(\varphi_2)^{0.69574}\left(dp_2\right)^{0.44708})}{\left((1-(\varphi_1+\varphi_2))+(\varphi_1)\left(\frac{\rho_{s_1}}{\rho_{bf}}\right)+(\varphi_2)\left(\frac{\rho_{s_2}}{\rho_{bf}}\right)\right)} \right) f''''[\eta] - \alpha(3f''[\eta]+\eta f'''[\eta]) -$$

$$2Ref[\eta]f'''[\eta] - \left(\frac{1}{\left((1-\varphi_1-\varphi_2)+\varphi_1\left(\frac{\rho_{s_1}}{\rho_{bf}}\right)+\varphi_2\left(\frac{\rho_{s_2}}{\rho_{bf}}\right)\right)} \right) M f''[\eta] = 0 \tag{21}$$

$$\left(\frac{(1+0.1008((\varphi_1)^{0.69574}\left(dp_1\right)^{0.44708}+(\varphi_2)^{0.69574}\left(dp_2\right)^{0.44708})}{\left((1-(\varphi_1+\varphi_2))+(\varphi_1)\left(\frac{\rho_{s_1}}{\rho_{bf}}\right)+(\varphi_2)\left(\frac{\rho_{s_2}}{\rho_{bf}}\right)\right)} \right) g''[\eta] + \alpha(2g[\eta]+\eta g'[\eta]) +$$

$$2Re(g[\eta]f'[\eta] - f[\eta]g'[\eta]) - \left(\frac{1}{\left((1-\varphi_1-\varphi_2)+\varphi_1\left(\frac{\rho_{s_1}}{\rho_{bf}}\right)+\varphi_2\left(\frac{\rho_{s_2}}{\rho_{bf}}\right)\right)} \right) Mg[\eta] = 0 \tag{22}$$

$$\theta''[\eta]$$
$$+\left((1-(\varphi_1+\varphi_2))+(\varphi_1)\left(\frac{\rho_{cps_1}}{\rho_{cpbf}}\right)\right.$$
$$+\left.(\varphi_2)\left(\frac{\rho_{cps_2}}{\rho_{cpbf}}\right)\right)\left(\frac{k_{s2}+(N-1)k_{mbf}+\varphi_2\left(k_{mbf}-k_{s2}\right)}{k_{s2}+(N-1)k_{mbf}-(N-1)\varphi_2\left(k_{mbf}-k_{s2}\right)}\right)\left(\frac{k_{s1}+(N-1)k_{bf}+\varphi_1\left(k_{bf}-k_{s1}\right)}{k_{s1}+(N-1)k_{bf}-(N-1)\varphi_1\left(k_{bf}-k_{s1}\right)}\right)$$

$$Pr\left(\alpha\eta - 2Ref[\eta]\right)\theta'[\eta] + [(1-(\varphi_1+\varphi_2))^{-2.5}f_{\eta\eta}^2+$$
$$Mf_\eta^2]ReE_cP_e\left(\frac{k_{s2}+(N-1)k_{mbf}+\varphi_2\left(k_{mbf}-k_{s2}\right)}{k_{s2}+(N-1)k_{mbf}-(N-1)\varphi_2\left(k_{mbf}-k_{s2}\right)}\right)\left(\frac{k_{s1}+(N-1)k_{bf}+\varphi_1\left(k_{bf}-k_{s1}\right)}{k_{s1}+(N-1)k_{bf}-(N-1)\varphi_1\left(k_{bf}-k_{s1}\right)}\right) = 0 \tag{23}$$

$$\chi''[\eta]+Sc(\alpha\eta - 2Ref[\eta])\chi'[\eta] = 0 \tag{24}$$

$$H_1 = \left(\frac{(1 + 0.1008((\varphi_1)^{0.69574}\left(dp_1\right)^{0.44708} + (\varphi_2)^{0.69574}\left(dp_2\right)^{0.44708})}{\left((1 - (\varphi_1+\varphi_2)) + (\varphi_1)\left(\frac{\rho_{s_1}}{\rho_{bf}}\right) + (\varphi_2)\left(\frac{\rho_{s_2}}{\rho_{bf}}\right)\right)} \right) \tag{25}$$

$$H_2 = \left(\frac{1}{\left((1 - \varphi_1 - \varphi_2) + \varphi_1\left(\frac{\rho_{s_1}}{\rho_{bf}}\right) + \varphi_2\left(\frac{\rho_{s_2}}{\rho_{bf}}\right)\right)} \right) \tag{26}$$

$$H_3 = \left((1 - (\varphi_1+\varphi_2)) + (\varphi_1)\left(\frac{\rho_{cps_1}}{\rho_{cpbf}}\right) + (\varphi_2)\left(\frac{\rho_{cps_2}}{\rho_{cpbf}}\right) \right) \tag{27}$$

$$D_1 = \left(\frac{k_{s2} + (N-1)k_{mbf} + \varphi_2\left(k_{mbf} - k_{s2}\right)}{k_{s2} + (N-1)k_{mbf} - (N-1)\varphi_2\left(k_{mbf} - k_{s2}\right)} \right) \tag{28}$$

$$D_2 = \left(\frac{k_{s1} + (N-1)k_{bf} + \varphi_1\left(k_{bf} - k_{s1}\right)}{k_{s1} + (N-1)k_{bf} - (N-1)\varphi_1\left(k_{bf} - k_{s1}\right)} \right) \tag{29}$$

$$\omega = D_1 D_2. \tag{30}$$

Putting values of (25)–(30) in Equations (21)–(24), the final result is

$$H_1 f''''[\eta] - \alpha(3f''[\eta]+\eta f'''[\eta]) - 2Ref[\eta]f''[\eta] - H_2 M f''[\eta] = 0 \tag{31}$$

$$H_1 g''[\eta] + \alpha(2g[\eta]+\eta g'[\eta]) + 2Re\left(g[\eta]f'[\eta] - f[\eta]g'[\eta]\right) - H_2 Mg[\eta] = 0 \tag{32}$$

$$\theta''[\eta] + H_3\omega Pr\left(\alpha\eta - 2Ref[\eta]\right)\theta'[\eta] + \omega[(1-(\varphi_1+\varphi_2))^{-2.5}f_{\eta\eta}^2 + Mf_\eta^2]ReE_cP_e = 0 \tag{33}$$

$$\chi''[\eta] + Sc(\alpha\eta - 2Ref[\eta])\chi'[\eta] = 0 \tag{34}$$

3.5. Solution Procedure

This segment is presented for the discussion of the implemented numerical scheme and steps involved during the simulations. For this purpose, firstly, Equations (13)–(16) with boundary conditions along with effective thermophysical properties are solved numerically by implementing the numerical scheme renowned as RK 4th order in conjunction with the shooting method. To achieve the solution from these procedures, initially, numerical values are choose carefully to accomplish the desired level of accuracy. Owing to low computation cost and memory loss and provision of accurate and consistent results in less time, Runge–Kutta and shooting methods are applied.

4. Result and Discussion

In the present section, results on both the graphical and tabular form against different parameters such as the expansion/contraction ratio parameter (α), permeable Reynold parameter (Re), magnetic parameter (M), Prandtl number (Pr), diameter/size of the nanoparticles (dp_1) and (dp_2), shape factor of the nanoparticles (N), Peclet number (Pe), and Eckert number (Ec), as well as nanoparticle volume fraction (φ) on velocity, temperature and mass distributions, shear stress, and heat and mass transfer rate, are examined thoroughly.

Tables 1 and 2 present the thermophysical properties of (HNFDs) and base fluid with different types of nanoparticles (NPs). Table 3 shows the effect of diameter, volume fraction, and Reynold number on shear stresses by considering three different types of compositions for hybridize NPs, as well as shear stress showing an increasing pattern as compared with tensional stress with an injection factor. Similarly, behavior is observed for the variation in the size of metallic oxides for all types of nanoparticles (Al_2O_3, TiO_2, and Fe_3O_4) compared with copper. Nanoparticles have unique features as compared with bulk material of the same structure. The most common properties of the nanoparticles, for example, can be easily rehabilitated by varying their size and shape. Copper nanomaterials have high thermal conductivity as well as electrical conductivity. The most common shape of copper is round visibility, such as black powder. On the other hand, metal oxide nanoparticles are a very important technological material and have many industrial applications.

Table 1. Thermophysical properties of HNFDs [62–66].

Properties	**(HNFDs)**
Density (ρ)	$\rho_{hnf} = \varphi_1\rho_{s_1} + \varphi_2\rho_{s\,2} + (1 - \varphi_1 - \varphi_2)\,\rho_{bf}$
Viscosity (μ)	$\mu_{hnf} = \mu_{bf}(1 + 0.1008\,((\varphi_1)^{0.69574}(dp_1)^{0.44708} + (\varphi_2)^{0.69574}(dp_2)^{0.44708}))$
Heat Capacity (ρC_p)	$(\rho c_p)_{hnf} = \varphi_1(\rho c_p)_{s_1} + \varphi_2(\rho c_p)_{s_2} + (1 - \varphi_1 - \varphi_2)\,(\rho c_p)_{bf}$
Thermal Conductivity (K)	$k_{hnf} = \left(\dfrac{k_{s2}+(N-1)k_{mbf}+\varphi_2\left(k_{mbf}-k_{s2}\right)}{k_{s2}+(N-1)k_{mbf}-(N-1)\varphi_2\left(k_{mbf}-k_{s2}\right)}\right)k_{bf}$ where $k_{bf} = \left(\dfrac{k_{s1}+(N-1)k_{bf}+\varphi_1\left(k_{bf}-k_{s1}\right)}{k_{s1}+(N-1)k_{bf}-(N-1)\varphi_1\left(k_{bf}-k_{s1}\right)}\right)k_f$

Table 2. Properties of base fluids and NPs [67]

Base Fluid/NP's	$\rho(\mathbf{kgm}^{-3})$	C_p (J kg^{-1}k^{-1})	$\kappa\left(\mathbf{wm}^{-1}\mathbf{k}^{-1}\right)$
H_2O	997.1	4179	0.613
Cu	8933	385	401
Al_2O_3	3970	765	40
Fe_3O_4	5180	670	9.7
TiO_2	4250	686.2	8.9538

Table 3. Different parameters of the effect in shear stress and tensional stress.

					Cu-Al$_2$O$_3$/H$_2$O		Cu-TiO$_2$/H$_2$O		Cu-Fe$_3$O$_4$/H$_2$O	
dp_1	dp_2	φ_1	φ_2	Re	$f''(-1)$	$g'(-1)$	$f''(-1)$	$g'(-1)$	$f''(-1)$	$g'(-1)$
1	1	0.01	0.01	0.2	2.0587	−1.7628	2.0593	−1.7629	2.0615	−1.7633
2					2.0604	−1.7545	2.0611	−1.7547	2.0632	−1.7551
3					2.0616	−1.7484	2.0623	−1.7485	2.0644	−1.7489
4					2.0627	−1.7434	2.0634	−1.7435	2.0654	−1.7439
1	2				2.0604	−1.7546	2.0611	−1.7547	2.0632	−1.7551
	3				2.0616	−1.7484	2.0623	−1.7485	2.0645	−1.7489
	4				2.0627	−1.7434	2.0633	−1.7435	2.0656	−1.7439
	1	0.02			2.0799	−1.7519	2.0806	−1.7521	2.0828	−1.7524
		0.03			2.1009	−1.7429	2.1016	−1.7432	2.1038	−1.7433
		0.04			2.1218	−1.7349	2.1225	−1.7351	2.1247	−1.7353
			0.02		2.0684	−1.7499	2.0697	−1.7502	2.0741	−1.7509
			0.03		2.0777	−1.7392	2.0797	−1.7395	2.0862	−1.7405
			0.04		2.0868	−1.7296	2.0894	−1.7300	2.0981	−1.7313
				0.4	2.2258	−2.1202	2.2271	−2.1215	2.2312	−2.1259
				0.6	2.4441	−2.6358	2.4463	−2.6394	2.4536	−2.6511
				0.8	2.7414	−3.4364	2.7452	−3.4452	2.758	−3.4747

If we increase the nanoparticle level fraction of copper from 1% to 4%, then shear stress is an increasing function; a similar trend is also observed for metallic oxide nanoparticle fraction. The injection phenomenon is very important for biomedical sciences [46]. When increasing the numerical values of injection number in Table 3, both shear and tensional stresses show an increasing pattern. Table 4 calculates the Nusselt number with different shape factors for hybridizing nanofluids (Cu-TiO$_2$/H$_2$O, Cu-Fe$_3$O$_4$/H$_2$O, Cu-Al$_2$O$_3$/H$_2$O). If we increase the values of nanoparticles volume fraction, φ_1 and φ_2, then the shape factor of (Cu-Al$_2$O$_3$/H$_2$O) shows better performance than the others. Table 5 demonstrates the effect of the permeable Reynold number Re, expansion ratio α, and Sc on Sherwood number of Cu-Al$_2$O$_3$/H$_2$O. Sherwood number has significant results for the suction case, as compared with injection, when $\alpha > 0$. An opposite trend is observed for contracting and expanding cases when Re < 0. If we increase the numerical values of S, then there is a very significant effect on the Sherwood number. Table 6 displays the effect of different parameters on the heat transfer rate at the lower porous disk. Increases in the Pr and Re numbers significantly enhance the heat transfer rate at the lower porous disk, while, on the other hand, N(shape factor) and M (MHD) have a small effect on the heat transfer rate. The Prandtl number Pr, in our problem, the relative importance of the fluid's viscosity and thermal conductivity, appears to raise the actual fluid temperature. Table 7 represents the influence of the magnetic parameter M > 0 on the shear stress, tensional stress, and heat transfer rate, and they are all gradually enhanced in the presence of 2% hybrid nanoparticles with the injection case too. In Figure 2, we examine the effect of four types of shape factors on hybrid nanofluid flow associated with a different numerical range of φ_1 and φ_2. All shape factors lie on the x-axis and thermal conductivity varies on the y-axis. A high thermal conductivity value is achieved at 5.7 when the nanoparticle volume fraction is at 1. Thermal conductivity is an increasing function of the shape factor with the nanoparticle volume fraction. We can say that morphology is directly proportional to the thermal conductivity of any nanofluids. By taking equal numerical values of φ_1 and φ_2, in Figure 3, with a viscosity of base fluids and diameter of all nanoparticles along with different numerical values of morphology (spherical, bricks, cylindrical, and platelets), the X-axis denotes the size factor and the y-axis represents the effective viscosity. Furthermore, the effect of viscosity is very high for a size factor of 12.73 when the nanoparticle volume fraction level is 1%. We also observed that platelet-shaped nanocomposites show a better performance on heat and mass transferability as compared with the other shapes of nanoparticles. Figures 4 and 5 show the effect of a magnetic parameter in the tangent velocity profile and the temperature profile influence of fixed value $\alpha = 1, Re = 1, Pr = 6.2, Ec = 0.00068$. By increasing the numerical values of the magnetic parameter M, the momentum boundary layer thickness decreases from both porous walls in the presence of $\varphi_1 = \varphi_2 = 1\%$. The physically generated Lorentz force by amplification of the magnetic field generates a resistance to flow and decreases the momentum boundary layer thickness, which is why heat is the main source of heat production. The conclusion is suitable for the fact that the magnetic field implemented is a resistance force that plays a crucial role in decelerating and directing fluid flow. Figure 5 shows the increasing behavior of the temperature profile from the center and covers the whole domain. Figure 6 is drawn to show the nature of the Schmidt number concerning the concentration profiles $\chi(\eta)$. The Schmidt number is, theoretically, the conceptual interaction of momentum and mass diffusivity. Owing to the value of the Schmidt number, the diffusivity increases as a function of the decline in fluid concentration. The value of the Schmidt number is inversely proportional to the diffusivity of the Brownian movement. The greater diffusivity of Brownian corresponds to lower concentration profiles of $\chi(\eta)$. With the increase in the value of the Schmidt number, concentration boundary later thickness is an increasing function of Sc in the presence of $\varphi_1 = \varphi_2 = 3\%$. Figure 7 represents the temperature profile with various values of the Peclet number. With the increase in the value of P_e, the flow of heat transfer enhancement is significantly increased from both porous disks in the presence of $\varphi_1 = \varphi_2 - 5\%$. Physically, the product of Reynold number and Prandtl number is equal to Peclet number which tends to reduce flow velocity in downstream directions and the current factors tend to be one-way properties. In the common perception, it is expected that particles made from high thermal conductivity material should impose high thermal conductivity on nanofluids, but this is not necessary; on the other hand, a famous scientist, Lee et al. [52], conducted experiments with AL_2O_3 and CuO as hybrid nanofluids and reported that, though AL_2O_3 material had a higher thermal conductivity than CuO, CuO as a nanofluid possesses higher thermal conductivity. The investigator claimed reason that AL_2O_3 nanoparticles form a larger nanocluster than CuO nanoparticles in the base fluid water. Hence, some other factor may also be involved.

Table 4. The effect of different hybridized nanoparticles on Nusselt number (Nu).

	Cu-TiO$_2$/H$_2$O				Cu-Fe$_3$O$_4$/H$_2$O				Cu-Al$_2$O$_3$/H$_2$O			
$\varphi_1 = \varphi_2$	$\lvert Nu(3)\rvert$	$\lvert Nu(3.7)\rvert$	$\lvert Nu(4.8)\rvert$	$\lvert Nu(5.7)\rvert$	$\lvert Nu(3)\rvert$	$\lvert Nu(3.7)\rvert$	$\lvert Nu(4.8)\rvert$	$\lvert Nu(5.7)\rvert$	$\lvert Nu(3)\rvert$	$\lvert Nu(3.7)\rvert$	$\lvert Nu(4.8)\rvert$	$\lvert Nu(5.7)\rvert$
1%	1.8586	2.4807	3.2553	3.7454	1.8801	2.4908	3.2586	3.7565.	2.1023	2.7721	3.6048	4.1293
2%	4.4101	4.9099	5.4021	5.6378	4.4251	4.9128	5.4092	5.6484	4.6134	5.1103	5.5781	5.7749
3%	5.5102	5.7584	5.9028	5.8985	5.5122	5.7651	5.914	5.909	5.616	5.8326	5.9358	5.8346
4%	5.9063	5.9308	6.2162	6.3539	5.8959	5.9459	6.4308	6.8651	5.9212	5.9652	6.5901	6.6571

Table 5. Different parameter Re (Reynold number), α (wall expansion parameter) Sc (Schmidt number) effects on Sherwood number.

			Cu-Al$_2$O$_3$/H$_2$O
Re	α	Sc	$\lvert Sh\rvert_{\eta=-1}$
-1.5	1	1	0.19442
-1			0.30838
1			1.30681
1.5			1.69612
-1	-2		0.62786
	-1		0.30838
	1		0.23170
	2		0.05776
	-1	2	2.58310
		4	5.6498
		6	8.7222
		8	11.754

Table 6. The effect of different physical nondimensional parameters Re, α (wall expansion), M (magneitc parameter), N (shape factor), Ec (Eckert number), Pr (Prandtl number), Pe (Peclet number) θ (temperature) on heat transfer.

Re	α	M	N	Ec	Pr	Pe	$\lvert\theta'(-1)\rvert$
0.3	0.1	1	3	0.000068	6.2	6.8	1.3807
0.6							2.9588
0.9							4.6361
1.2							5.5035
0.3	0.2						1.1974
	0.3						1.0301
	0.4						0.8791
	0.1	4					1.3661
		7					1.3539
		11					1.3405
		1	3.7				1.3660
			4.8				1.3438
			5.7				1.3265
			3	0.00078			1.3805
				0.00088			1.3803
				0.00098			1.3801
					5.7		1.2891
					6.7		1.4756
					7.7		1.6741
						2.8	1.3817
						5.8	1.3810
						8.8	1.3803

Table 7. Magnetic-filed effect in $f''(\eta)$, $g'(\eta)$ and $\theta'(\eta)$ at lower wall $= -1$, $Re = 1$, $\varphi_1 = \varphi_2 = 0.02$, $Ec = 0.00068$, $Pr = 6.2$.

	Cu-Al$_2$O$_3$/H$_2$O		
M	$\lvert f'(-1)\rvert$	$\lvert g'(-1)\rvert$	$\lvert\theta'(-1)\rvert$
3	5.5298	3.8675	0.4361
5	5.838	4.2893	1.6694
7	6.1241	4.6544	3.4804
9	6.3914	4.9794	6.0671
11	6.6429	5.2744	8.4787
13	6.8808	5.546	10.7512
15	7.1069	5.7994	12.9101

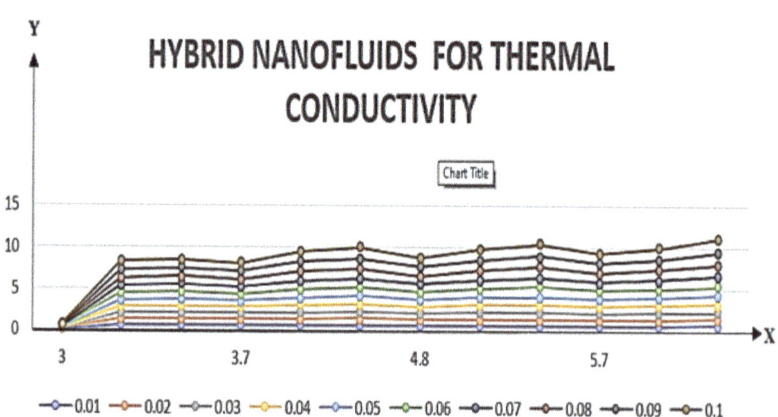

Figure 2. The influence of the scale of hybrid NPs under the thermal conductivity coloring umbrella. Reprinted from [68].

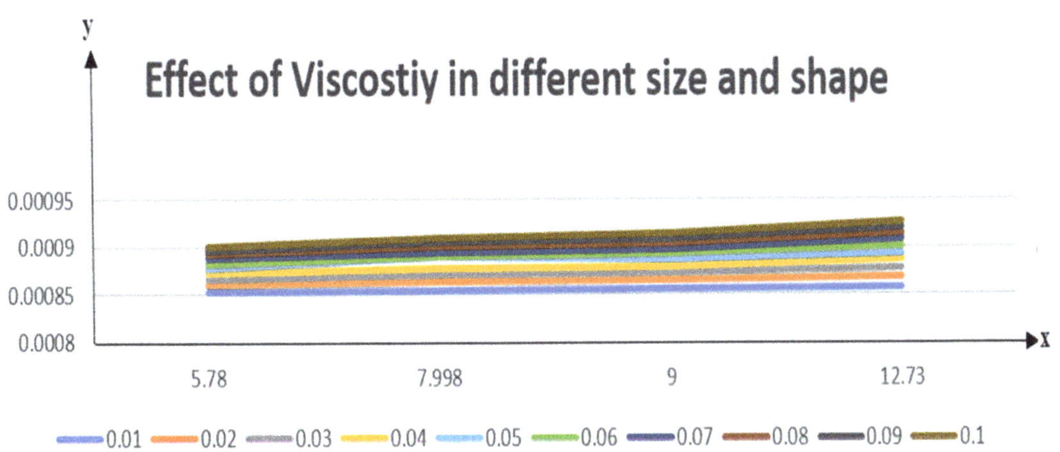

Figure 3. The influence of the scale of hybrid NPs under the viscosity coloring umbrella. Reprinted from [68].

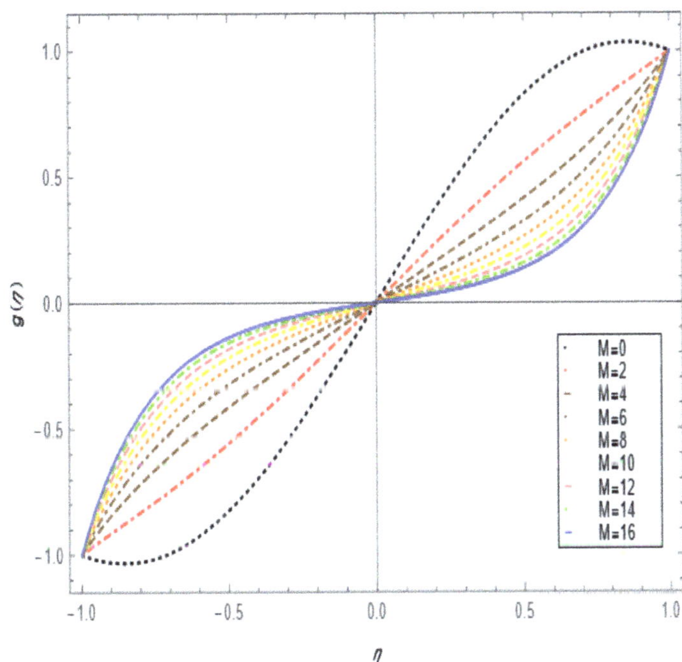

Figure 4. Tangential velocity profile effect for magnetic parameter for $= 1$, $Re = 1$, $\varphi_1 = \varphi_2 = 0.01$, $Ec = 0.00068$, $Pr = 6.2$.

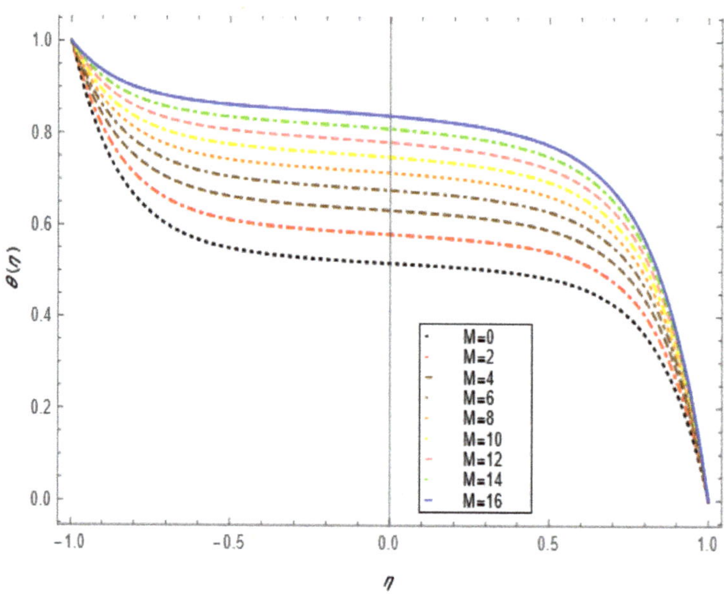

Figure 5. Temperature profile effect for magnetic parameter for $= 1$, $Re = 1$, $\varphi_1 = \varphi_2 = 0.01$, $Ec = 0.00068$, $Pr = 6.2$.

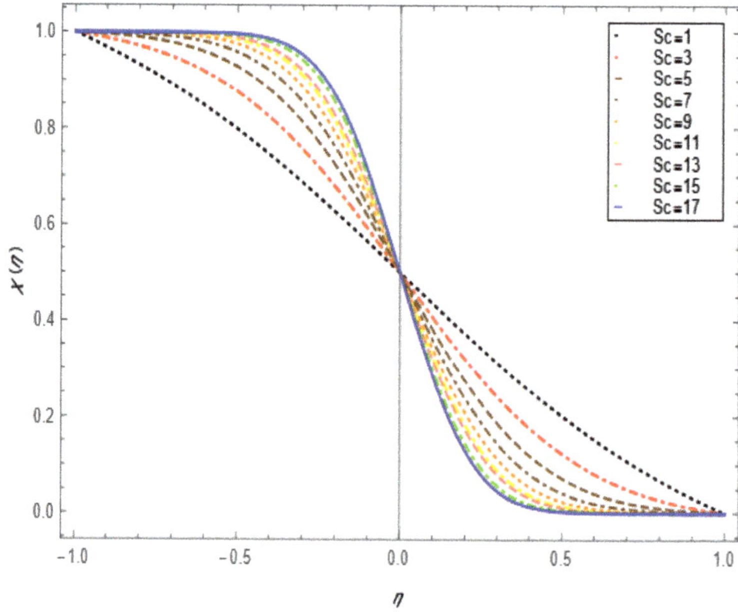

Figure 6. Concentration profile effect for Schmidt number for $= -1$, $Re = -1$, $\varphi_1 = \varphi_2 = 0.03$, $Ec = 0.00068$, $Pr = 6.2$, $M = 1$.

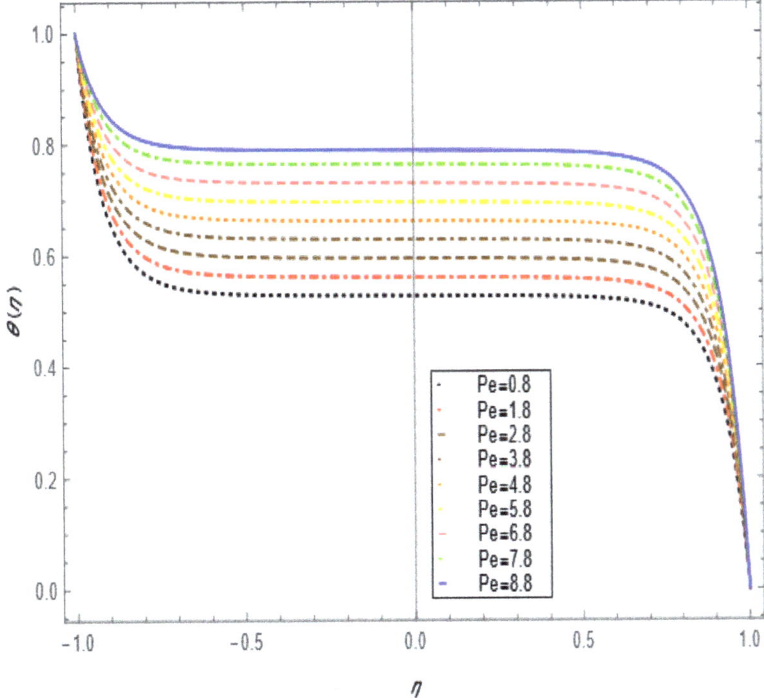

Figure 7. Temperature profile effect size for $= 0.1, Re = 1.5, M = 5, \varphi_1 = \varphi_2 = 0.05, Ec = 0.00068, Pr = 6.2$.

5. Conclusions

We explore the three-dimensional triadic hybrid nanofluid flow behavior with heat and mass transfer aspects in this manuscript, as well as the Newtonian fluid flow through orthogonal porous disks with MHD effects. Metallic and metallic-oxide nanoparticles with morphology effect are considered here. The most important findings are as follows.

Thermal conductivity and viscosity intensity are the highest in HNFD3 with platelet nanoparticles, followed by spherical, brick, and cylindrical nanoparticles, respectively.

By increasing the values magnetic field M, expansion ratio α, and Peclet Pe, there is an improvement in the rate of heat transfer on the porous disk.

- With an increase in the values of volume fraction, the Nusselt number has the largest effect on HNF3 with platelet nanoparticles.
- The momentum boundary layer is gradually increased if we increase the values of permeable Reynold number Re, the diameter of nanoparticles dp_1 and dp_2, and magnetic parameter M.
- If the volume fraction φ_1 and φ_2, the diameter of nanoparticles dp_1 and dp_2, and chemical reaction of Reynold number Re values are greater than zero, then the rate of shear stress and tensional stress is enhanced.
- By raising the Schmidt number Sc and chemical reaction Reynold number Re, the rate of mass transfer on the porous disk is enhanced.

Author Contributions: Conceptualization, Writing-original draft Z.A.Q.; Methodology, Results and discussion S.B.; Results finding, Writing—original draft I.A.S.; Code Validation A.A.; Incorporation of review comments, funding R.J.; Correction of Language, Funding H.S.; Proofreading, Incorporation of Reviewer Comments, Funding J.A. All authors have read and agreed to the published version of the manuscript.

Funding: The authors R.J., H.S.S. and J.A. would like to thank Palestine Technical university—Kadoorie for funding this work. Scientific research funding 2022.

Institutional Review Board Statement: Not applicable.

Informed Consent Statement: Not applicable.

Data Availability Statement: All data is available in the manuscript.

Conflicts of Interest: The authors declare no conflict of interest.

Nomenclature

B_o	Uniform magnetic field [T]	F_η	Dimensionless radial velocity profile
G_η	Dimensionless tangential velocity profile		
C_f	Total skin friction coefficient	θ_η	Dimensionless temperature profile
C_p	Specific heat at constant pressure	σ	Electrical conductivity $[(m^3 A^2)/\text{kg}]$
s	Time depended coefficient	υ	Kinematic viscosity $[m^2/s]$
M	Magnetic parameter	μ	Dynamic viscosity $[P_a.\text{s}]$
P_r	Prandtl number	ρ	Density $[\text{kg}/m^3]$
r z	Cylindrical coordinates system	ρC_p	Volumetric heat capacity [J/(m3 K)]
Re	Reynold number	T	(HN_{fd}) temperature [K]
u, v, w	velocity component along z axis	A_1	Dimensionless parameter coefficient
$Nu(HN_{fd})$	Nusselt number	$(\rho c_p)_{hnf}$	specific heat capacity for
ρ_{hnf}	Density for (HN_{fd})	ρ_{s1}	Density for first solid NP's
ρ_{s2}	Density for second solid NP's	k_{hnf}	Thermal conductivity for (HN_{fd})
k_{s1}	Thermal conductivity for first solid fraction	k_{s2}	Thermal conductivity for second solid fraction
k_{mbf}	Thermal conductivity for shape base fluid	P	pressure
k_{bf}	Thermal conductivity for base fluid	μ_{eff}	viscosity for effect
μ_{bf}	viscosity base fluid	dp	Dimeter of nanoparticles
dp_1	Dimeter of first nanoparticles	dp_2	dimeter of 2nd nanoparticles
υ_{hnf}	Kinematics viscosity for (HN_{fd})	Nu_z	Nusselt number
Sc	Schmidt number	Pe	Pectlet number
Ec	Eckert number	N	Size factor
Subscripts			
(B_{fd})	Base fluid	(NFs)	Nanofluids
(HNFDs)	Hybrid nanofluids		
HNFD1	Cu-TiO$_2$ /H$_2$O	HNFD2	Ag-Fe$_3$O$_4$/H$_2$O
HNFD3	Cu-Al$_2$O$_3$ /H$_2$O		
Greek Symbols			
α	Thermal diffusivity $[m^2/s]$	η	Independent similarity variable
φ	Equivalent nanoparticles volume fraction		
φ_1	Equivalent first nanoparticles volume fraction	φ_2	Equivalent first nanoparticles volume fraction

References

1. Festa, R.A.; Thiele, D.J. Copper: An essential metal in biology. *Curr. Biol.* **2011**, *21*, R877–R883. [CrossRef]
2. Lutsenko, S.; Barnes, N.L.; Bartee, M.Y.; Dmitriev, O.Y. Function and Regulation of Human Copper-Transporting ATPases. *Physiol. Rev.* **2007**, *87*, 1011–1046. [CrossRef]
3. De Bie, P.; Muller, P.; Wijmenga, C.; Klomp, L.W.J. Molecular pathogenesis of Wilson and Menkes disease: Correlation of mutations with molecular defects and disease phenotypes. *J. Med Genet.* **2007**, *44*, 673–688. [CrossRef]
4. Gupta, A.; Lutsenko, S. Human copper transporters: Mechanism, role in human diseases and therapeutic potential. *Future Med. Chem.* **2009**, *1*, 1125–1142. [CrossRef] [PubMed]
5. Majewski, M.; Ognik, K.; Zdunczyk, P.; Juskiewicz, J. Effect of dietary copper nanoparticles versus one copper (II) salt: Analysis of vasoreactivity in a rat model. *Pharmacol. Rep.* **2017**, *69*, 1282–1288. [CrossRef] [PubMed]
6. Tümer, Z.; Møller, L.B. Menkes disease. *Eur. J. Hum. Genet.* **2010**, *18*, 511–518. [CrossRef] [PubMed]
7. Lorincz, M.T. Wilson disease and related copper disorders. *Handb. Clin. Neurol.* **2018**, *147*, 279–292. [CrossRef]

8. Glod, D.; Adamczak, M.; Bednarski, W. Selected aspects of nanotechnology applications in food production. *Żywność Nauka Technol. Jakość* **2014**, *5*, 36–52.

9. Assadi, Z.; Emtiazi, G.; Zarrabi, A. Hyperbranched polyglycerol coated on copper oxide nanoparticles as a novel core-shell nano-carrier hydrophilic drug delivery model. *J. Mol. Liq.* **2018**, *250*, 375–380. [CrossRef]

10. Cholewińska, E.; Juśkiewicz, J.; Ognik, K. Comparison of the effect of dietary copper nanoparticles and one copper (II) salt on the metabolic and immune status in a rat model. *J. Trace Elem. Med. Biol.* **2018**, *48*, 111–117. [CrossRef]

11. Turcu, R.; Darabont, A.L.; Nan, A.; Aldea, N.; Macovei, D.; Bica, D.; Vekas, L.; Pana, O.; Soran, M.L.; Koos, A.A.; et al. New polypyrrole-multiwall carbon nanotubes hybrid materials. *J. Optoelectron. Adv. Mater.* **2006**, *8*, 643–647.

12. Jana, S.; Salehi-Khojin, A.; Zhong, W.-H. Enhancement of fluid thermal conductivity by the addition of single and hybrid nano-additives. *Thermochim. Acta* **2007**, *462*, 45–55. [CrossRef]

13. Devi, S.A.; Devi, S.S.U. Numerical investigation of hydromagnetic hybrid Cu–Al2O3/water nanofluid flow over a permeable stretching sheet with suction. *Int. J. Nonlinear Sci. Numer. Simul.* **2016**, *17*, 249–257. [CrossRef]

14. Devi, S.S.U.; Devi, S.A. Numerical investigation of three-dimensional hybrid Cu–Al2O3/water nanofluid flow over a stretching sheet with effecting Lorentz force subject to Newtonian heating. *Can. J. Phys.* **2016**, *94*, 490–496. [CrossRef]

15. Imran, M.; Farooq, U.; Waqas, H.; Anqi, A.E.; Safaei, M.R. Numerical performance of thermal conductivity in Bioconvection flow of cross nanofluid containing swimming microorganisms over a cylinder with melting phenomenon. *Case Stud. Therm. Eng.* **2021**, *26*, 101181. [CrossRef]

16. Anitha, S.; Safaei, M.R.; Rajeswari, S.; Pichumani, M. Thermal and energy management prospects of γ-AlOOH hybrid nanofluids for the application of sustainable heat exchanger systems. *J. Therm. Anal. Calorim.* **2021**, *27*, 123–142. [CrossRef]

17. Ebrahimi, D.; Yousefzadeh, S.; Akbari, O.A.; Montazerifar, F.; Rozati, S.A.; Nakhjavani, S.; Safaei, M.R. Mixed convection heat transfer of a nanofluid in a closed elbow-shaped cavity (CESC). *J. Therm. Anal. Calorim.* **2021**, *144*, 2295–2316. [CrossRef]

18. Karman, T.V. Über laminare und turbulente reibung. *Z. Angew. Math. Mech.* **1921**, *1*, 1233–1252. [CrossRef]

19. Griffiths, P.T. Flow of a generalized Newtonian fluid due to a rotating disk. *J. Non-Newton. Fluid Mech.* **2015**, *221*, 9–17. [CrossRef]

20. Cochran, W.G. The flow due to a rotating disk. In *Mathematical Proceedings of the Cambridge Philosophical Society*; Cambridge University Press: Cambridge, UK, 1934; Volume 30, pp. 365–375.

21. Stewartson, K. On the flow between two rotating coaxial disks. In *Mathematical Proceedings of the Cambridge Philosophical Society*; Cambridge University Press: Cambridge, UK, 1953; Volume 49, pp. 333–341.

22. Mellor, G.L.; Chapple, P.J.; Stokes, V.K. On the flow between a rotating and a stationary disk. *J. Fluid Mech.* **1968**, *31*, 95–112. [CrossRef]

23. Arora, R.C.; Stokes, V.K. On the heat transfer between two rotating disks. *Int. J. Heat Mass Transf.* **1972**, *15*, 2119–2132. [CrossRef]

24. Kumar, S.K.; Tacher, W.I.; Watson, L.T. Magnetohydrodynamic flow between a solid rotating disk and a porous stationary disk. *Appl. Math. Model.* **1989**, *13*, 494–500. [CrossRef]

25. Waini, I.; Ishak, A.; Pop, I. Agrawal flow of a hybrid nanofluid over a shrinking disk. *Case Stud. Therm. Eng.* **2021**, *25*, 100950. [CrossRef]

26. Turkyilmazoglu, M. Nanofluid flow and heat transfer due to a rotating disk. *Comput. Fluids* **2014**, *94*, 139–146. [CrossRef]

27. Turkyilmazoglu, M. Three dimensional MHD stagnation flow due to a stretchable rotating disk. *Int. J. Heat Mass Transf.* **2012**, *55*, 6959–6965. [CrossRef]

28. Hayat, T.; Qayyum, S.; Imtiaz, M.; Alsaedi, A. Flow between two stretchable rotating disks with Cattaneo-Christov heat flux model. *Results Phys.* **2016**, *7*, 126–133. [CrossRef]

29. Hayat, T.; Nasir, T.; Khan, M.I.; Alsaedi, A. Non-Darcy flow of water-based single (SWCNTs) and multiple (MWCNTs) walls carbon nanotubes with multiple slipconditions due to rotating disk. *Results Phys.* **2018**, *9*, 390–399. [CrossRef]

30. Ghadikolaei, S.S.; Hosseinzadeh, K.; Ganji, D.D. Investigation on three dimensional squeezing flow of mixture base fluid (ethylene glycol-water) suspended by hybrid nanoparticle (Fe3O4-Ag) dependent on shape factor. *J. Mol. Liq.* **2018**, *262*, 376–388. [CrossRef]

31. Chamkha, A.J.; Dogonchi, A.S.; Ganji, D.D. Magneto-hydrodynamic flow and heat transfer of a hybrid nanofluid in a rotating system among two surfaces in the presence of thermal radiation and Joule heating. *AIP Adv.* **2019**, *9*, 025103. [CrossRef]

32. Gholinia, M.; Armin, M.; Ranjbar, A.; Ganji, D. Numerical thermal study on CNTs/ $C_2H_6O_2$– H_2O hybrid base nanofluid upon a porous stretching cylinder under impact of magnetic source. *Case Stud. Therm. Eng.* **2019**, *14*, 100490. [CrossRef]

33. Ghachem, K.; Aich, W.; Kolsi, L. Computational analysis of hybrid nanofluid enhanced heat transfer in cross flow micro heat exchanger with rectangular wavy channels. *Case Stud. Therm. Eng.* **2020**, *24*, 100822. [CrossRef]

34. Abedalh, S.A.; Shaalan, A.Z.H.; Yassien, N.S. Mixed convective of hybrid nanofluids flow in a backward-facing step. *Case Stud. Therm. Eng.* **2021**, *25*, 100868. [CrossRef]

35. Xu, H. Modelling unsteady mixed convection of a nanofluid suspended with multiple kinds of nanoparticles between two rotating disks by generalized hybrid model. *Int. Commun. Heat Mass Transf.* **2019**, *108*, 104275. [CrossRef]

36. Acharya, N.; Bag, R.; Kundu, P.K. Influence of Hall current on radiative nanofluid flow over a spinning disk: A hybrid approach. *Phys. E Low-Dimens. Syst. Nanostruct.* **2019**, *111*, 103–112. [CrossRef]

37. Dinarvand, S.; Rostami, M.N. An innovative mass-based model of aqueous zinc oxide–gold hybrid nanofluids for von Karman's swirling flow. *J. Therm. Anal. Calorim.* **2019**, *138*, 845–855. [CrossRef]

38. Khan, M.; Ali, W.; Ahmed, J. A hybrid approach to study the influence of Hall current in radiative nanofluid flow over a rotating disk. *Appl. Nanosci.* **2020**, *10*, 5167–5177. [CrossRef]

39. Izadi, M.; Mohebbi, R.; Karimi, D.; Sheremet, M.A. Numerical simulation of natural convection heat transfer of nanofluid with Cu, MWCNT, and Al2O3 nanoparticles in a cavity with different aspect ratios. *Chem. Eng. Process. Process Intensif.* **2018**, *125*, 56–66. [CrossRef]

40. Arani, A.A.A.; Akbari, O.A.; Safaei, M.R.; Marzban, A.; Alrashed, A.A.; Ahmadi, G.R.; Nguyen, T.K. Heat transfer improvement of water/single-wall carbon nanotubes (SWCNT) nanofluid in a novel design of a truncated double-layered microchannel heat sink. *Int. J. Heat Mass Transf.* **2017**, *113*, 780–795. [CrossRef]

41. Safaei, M.R.; Ahmadi, G.; Goodarzi, M.S.; Kamyar, A.; Kazi, S.N. Boundary Layer Flow and Heat Transfer of FMWCNT/Water Nanofluids over a Flat Plate. *Fluids* **2016**, *1*, 31. [CrossRef]

42. Goshayeshi, H.R.; Safaei, M.R.; Goodarzi, M.; Dahari, M. Particle size and type effects on heat transfer enhancement of Ferro-nanofluids in a pulsating heat pipe. *Powder Technol.* **2016**, *301*, 1218–1226. [CrossRef]

43. Akbari, O.A.; Safaei, M.R.; Goodarzi, M.; Akbar, N.S.; Zarringhalam, M.; Shabani, G.A.S.; Dahari, M. A modified two-phase mixture model of nanofluid flow and heat transfer in a 3-D curved microtube. *Adv. Powder Technol.* **2016**, *27*, 2175–2185. [CrossRef]

44. Safaei, M.R.; Shadloo, M.S.; Goodarzi, M.; Hadjadj, A.; Goshayeshi, H.R.; Afrand, M.; Kazi, S.N. A survey on experimental and numerical studies of convection heat transfer of nanofluids inside closed conduits. *Adv. Mech. Eng.* **2016**, *8*. [CrossRef]

45. Vajravelu, K.; Prasad, K.V.; Ng, C.-O.; Vaidya, H. MHD squeeze flow and heat transfer of a nanofluid between parallel disks with variable fluid properties and transpiration. *Int. J. Mech. Mater. Eng.* **2017**, *12*, 9. [CrossRef]

46. Das, K.; Jana, S.; Acharya, N. Slip effects on squeezing flow of nanofluid between two parallel disks. *Int. J. Appl. Mech. Eng.* **2016**, *21*, 5–20. [CrossRef]

47. Mohyud-Din, S.T.; Khan, S.I.; Bin-Mohsin, B. Velocity and temperature slip effects on squeezing flow of nanofluid between parallel disks in the presence of mixed convection. *Neural Comput. Appl.* **2017**, *28*, 169–182. [CrossRef]

48. Qayyum, S.; Imtiaz, M.; Alsaedi, A.; Hayat, T. Analysis of radiation in a suspension of nanoparticles and gyrotactic microorganism for rotating disk of variable thickness. *Chin. J. Phys.* **2018**, *56*, 2404–2423. [CrossRef]

49. Aziz, A.; Alsaedi, A.; Muhammad, T.; Hayat, T. Numerical study for heat generation/absorption in flow of nanofluid by a rotating disk. *Results Phys.* **2018**, *8*, 785–792. [CrossRef]

50. Reddy, P.S.; Sreedevi, P.; Chamkha, A.J. MHD boundary layer flow, heat and mass transfer analysis over a rotating disk through porous medium saturated by Cu-water and Ag-water nanofluid with chemical reaction. *Powder Technol.* **2017**, *307*, 46–55. [CrossRef]

51. Muhammad, S.; Ali, G.; Shah, Z.; Islam, S.; Hussain, S.A. The Rotating Flow of Magneto Hydrodynamic Carbon Nanotubes over a Stretching Sheet with the Impact of Non-Linear Thermal Radiation and Heat Generation/Absorption. *Appl. Sci.* **2018**, *8*, 482. [CrossRef]

52. Uddin, I.; Akhtar, R.; Khan, M.A.R.; Zhiyu, Z.; Islam, S.; Shoaib, M.; Raja, M.A.Z. Numerical treatment for fluidic system of activation energy with non-linear mixed convective and radiative flow of magneto nanomaterials with Navier's velocity slip. *AIP Adv.* **2019**, *9*, 055210. [CrossRef]

53. Khan, M.I.; Hayat, T.; Alsaedi, A. A modified homogeneous-heterogeneous reactions for MHD stagnation flow with viscous dissipation and Joule heating. *Int. J. Heat Mass Transf.* **2017**, *113*, 310–317. [CrossRef]

54. Khan, N.; Sajid, M.; Mahmood, T. Heat transfer analysis for magnetohydrodynamics axisymmetric flow between stretching disks in the presence of viscous dissipation and Joule heating. *AIP Adv.* **2015**, *5*, 057115. [CrossRef]

55. Krishna, M.V.; Ahamad, N.A.; Chamkha, A.J. Hall and ion slip impacts on unsteady MHD convective rotating flow of heat generating/absorbing second grade fluid. *Alex. Eng. J.* **2020**, *60*, 845–858. [CrossRef]

56. Krishna, M.V.; Chamkha, A.J. Hall and ion slip effects on MHD rotating boundary layer flow of nanofluid past an infinite vertical plate embedded in a porous medium. *Results Phys.* **2019**, *15*, 102652. [CrossRef]

57. Toghraie, D.; Mashayekhi, R.; Arasteh, H.; Sheykhi, S.; Niknejadi, M.; Chamkha, A.J. Two-phase investigation of water-Al2O3 nanofluid in a micro concentric annulus under non-uniform heat flux boundary conditions. *Int. J. Numer. Methods Heat Fluid Flow* **2019**, *30*, 1795–1814. [CrossRef]

58. Krishna, M.V.; Ahamad, N.A.; Chamkha, A.J. Hall and ion slip effects on unsteady MHD free convective rotating flow through a saturated porous medium over an exponential accelerated plate. *Alex. Eng. J.* **2020**, *59*, 565–577. [CrossRef]

59. Krishna, M.V.; Chamkha, A.J. Hall and ion slip effects on MHD rotating flow of elastico-viscous fluid through porous medium. *Int. Commun. Heat Mass Transf.* **2020**, *113*, 104494. [CrossRef]

60. Ali, K.; Akbar, M.Z.; Iqbal, M.F.; Ashraf, M. Numerical simulation of heat and mass transfer in unsteady nanofluid between two orthogonally moving porous coaxial disks. *AIP Adv.* **2014**, *4*, 107113. [CrossRef]

61. Majdalani, J.; Zhou, C.; Dawson, C.A. Two-dimensional viscous flow between slowly expanding or contracting walls with weak permeability. *J. Biomech.* **2002**, *35*, 1399–1403. [CrossRef]

62. Chamkha, A.J.; Selimefendigil, F. Forced Convection of Pulsating Nanofluid Flow over a Backward Facing Step with Various Particle Shapes. *Energies* **2018**, *11*, 3068. [CrossRef]

63. Ghadikolaei, S.; Hosseinzadeh, K.; Hatami, M.; Ganji, D. MHD boundary layer analysis for micropolar dusty fluid containing Hybrid nanoparticles (Cu-Al$_2$O$_3$) over a porous medium. *J. Mol. Liq.* **2018**, *268*, 813–823. [CrossRef]

64. Rostami, M.N.; Dinarvand, S.; Pop, I. Dual solutions for mixed convective stagnation-point flow of an aqueous silica–alumina hybrid nanofluid. *Chin. J. Phys.* **2018**, *56*, 2465–2478. [CrossRef]

65. Sundar, S.; Sharma, K.V.; Singh, M.K.; Sousa, A.C.M. Hybrid nanofluids preparation, thermal properties, heat transfer and friction factor—A review. *Renew. Sustain. Energy Rev.* **2017**, *68*, 185–198. [CrossRef]
66. Selimefendigil, F.; Öztop, H.F.; Chamkha, A.J. Mixed Convection of Pulsating Ferrofluid Flow Over a Backward-Facing Step. *IJST-T Mech. Eng.* **2021**, *43*, 593–612. [CrossRef]
67. Sheikholeslami, M.; Ganji, D.D. Nanofluid flow and heat transfer between parallel plates considering Brownian motion using DTM. *Comput. Methods Appl. Mech. Eng.* **2014**, *283*, 651–663. [CrossRef]
68. Abdelmalek, Z.; Akbar Qureshi, M.Z.; Bilal, S.; Raza, Q.; Sherif, E.S.M. A case study on morphological aspects of distinct magnetized 3D hybrid nanoparticles on fluid flow between two orthogonal rotating disks: An application of thermal energy systems. *Case Stud. Therm. Eng.* **2021**, *23*, 100744. [CrossRef]

Article

Higher-Dimensional Fractional Order Modelling for Plasma Particles with Partial Slip Boundaries: A Numerical Study

Tamour Zubair [1], Muhammad Imran Asjad [2], Muhammad Usman [3] and Jan Awrejcewicz [4],*

1   School of Mathematical Sciences, Peking University, Beijing 100871, China; tamourzubair@pku.edu.cn
2   Department of Mathematics, University of Management and Technology, Lahore 40050, Pakistan; Imran.asjad@umt.edu.pk
3   Department of Mathematics, National University of Modern Languages (NUML), Islamabad 44000, Pakistan; dr.usman@numl.edu.pk
4   Department of Automation, Biomechanics, and Mechatronics, Faculty of Mechanical Engineering, Lodz University of Technology, 90-924 Lodz, Poland
*   Correspondence: jan.awrejcewicz@p.lodz.pl

Abstract: We integrate fractional calculus and plasma modelling concepts with specific geometry in this article, and further formulate a higher dimensional time-fractional Vlasov Maxwell system. Additionally, we develop a quick, efficient, robust, and accurate numerical approach for temporal variables and filtered Gegenbauer polynomials based on finite difference and spectral approximations, respectively. To analyze the numerical findings, two types of boundary conditions are used: Dirichlet and partial slip. Particular methodology is used to demonstrate the proposed scheme's numerical convergence. A detailed analysis of the proposed model with plotted figures is also included in the paper.

Keywords: partial slip boundary conditions; polynomial theory; linear polarization; fraction plasma modelling

Citation: Zubair, T.; Asjad, M.I.; Usman, M.; Awrejcewicz, J. Higher-Dimensional Fractional Order Modelling for Plasma Particles with Partial Slip Boundaries: A Numerical Study. *Nanomaterials* **2021**, *11*, 2884. https://doi.org/10.3390/nano11112884

Academic Editors: M. M. Bhatti, Kambiz Vafai, Sara I. Abdelsalam and Uroš Cvelbar

Received: 11 September 2021
Accepted: 25 October 2021
Published: 28 October 2021

Publisher's Note: MDPI stays neutral with regard to jurisdictional claims in published maps and institutional affiliations.

1. Introduction

In the study of plasma [1–7] particles, there is a ground breaking tool available in the literature named the "Vlasov Maxwell system", which is the amalgamation of the Vlasov equation and Maxwell equations. The Vlasov and Maxwell equations (MEs) are actually the mathematical formulation of performance of plasma particles and electro-magnetic field, formulated by well-known scientists Anatoly Vlasov (1938) and James Clerk Maxwell (1862), respectively. This system provides us different types of information in three-dimensional velocity $\mathbf{v}$ and position $\mathbf{r}$ coordinates under the impacts of electromagnetism. The mathematical formulation of this system is given below [8–10]:

$$\left. \begin{array}{c} D_t^\alpha f + \mathbf{v} \cdot \nabla_\mathbf{r} f + \frac{q}{m} \underbrace{(\mathbf{E} + \mathbf{v} \times \mathbf{B})}_{*} \cdot \nabla_\mathbf{v} f = 0, \\ * \left\{ \begin{array}{c} \frac{1}{c^2} D_t^\alpha \mathbf{E} - \nabla \times \mathbf{B} + \mu_0 \mathbf{J} = 0, D_t^\alpha \mathbf{B} + \nabla \times \mathbf{E} = 0, \\ \nabla \cdot \mathbf{E} = \frac{n}{\varepsilon_0}, \nabla \cdot \mathbf{B} = 0, \end{array} \right. \end{array} \right\} \tag{1}$$

In the above system defined in Equation (1), f represents distribution function and also $f = f(t, \mathbf{r}, \mathbf{v}), \mathbf{r}, \mathbf{v} \in \mathbb{R}^\Upsilon, \Upsilon = 1, 2, 3$. ME's are included due to (*) part of system (1). The importance of VMS is also due to the frequent applications which are given in [9,11–13]. D_t^α is the time-fractional operator [8], which can be seen in the system (1) to belong to the fractional calculus (FC). Fractional calculus is another innovative idea to explore the concealed procedures of the physical nature. With the help of literature study, it can be easily noticed that fractional calculus was the topic of pure-mathematics, but in a very short time, this theory has become famous due to the large number of applications [14–19]. FC turns the directions of research and generates new ideas to study the existing models in a different manner. The applications of VMS and FC are explained in Figure 1.

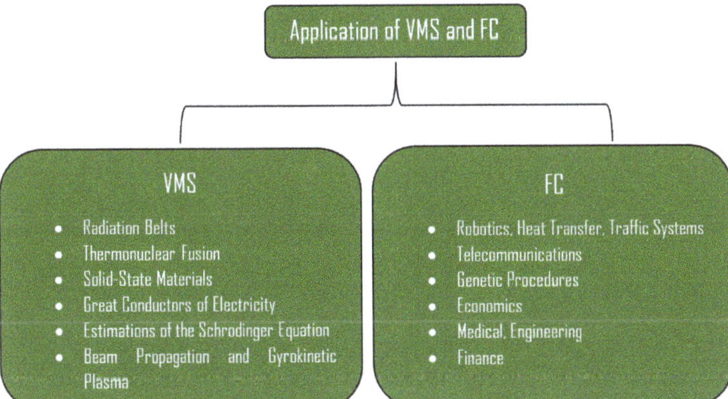

Figure 1. Applications of VMS and FC.

The first goal of this study is to transform the defined physical geometry into a mathematical model. For this purpose, we defined our geometry in Figure 2. The simple description of this geometry is in the following points as [8,20,21]:

- Collisionless plasma particles are available in a higher-dimensional computational domain, and LASER light is settled in such a way that it creates light with linear polarization properties. Particles are actually charged particles, and they generate a self-consistent electro-magnetic field.
- As light generates electromagnetism, it has its own fields in the computational domain.
- According to the geometry, we have the following components of electric **E**, vector potential **A** and magnetic **B** are given:

$$\mathbf{A} = \mathbf{A}(t, x, y) = 0, 0, A_z \hat{k}, \mathbf{B} = \mathbf{B}(t, x, y) = B_x \hat{i} + B_y \hat{j}, \mathbf{E} = \mathbf{E}(t, x, y) = E_x \hat{i} + E_y \hat{j} + E_z \hat{k},$$

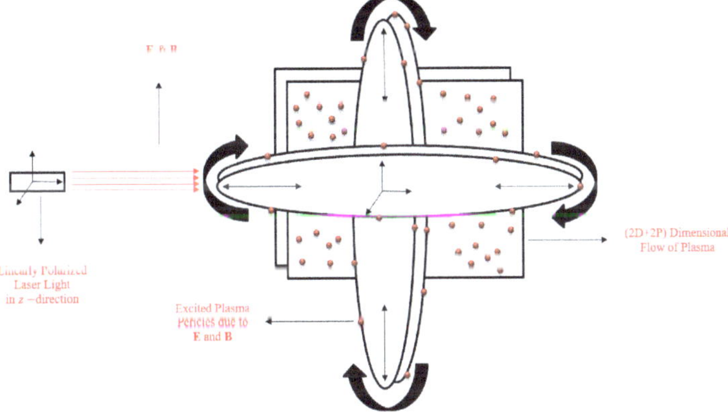

Figure 2. Geometry of the problem.

According to the assumptions, we consider the distribution function as:

$$f(t, \mathbf{r}, \mathbf{P}) = f(t, x, y, p_x, p_y)\delta(p_z - p_0(t, x, y)), \delta : \mathbb{R} \times \mathbb{R} \to \mathbb{R}, \tag{2}$$

In above Equation (2), $p_0(t,x,y)$ and δ are the momentum component and "Dirac measure", respectively. Further, we use the concepts of Hamiltonian, Coulomb gauge and canonical conjugate momentum to obtain the dimensionless form of the system as [9,20,21]:

$$\left.\begin{aligned}
&f^\alpha + \frac{p_x}{\gamma_1}\frac{\partial f}{\partial x} + \frac{p_y}{\gamma_1}\frac{\partial f}{\partial y} - \left(E_x + \frac{A_z}{\gamma_2}\frac{\partial A_z}{\partial x}\right)\frac{\partial f}{\partial p_x} - \left(E_y + \frac{A_z}{\gamma_2}\frac{\partial A_z}{\partial y}\right)\frac{\partial f}{\partial p_y} = 0, \\
&A_z^{2\alpha} - \left(\frac{\partial^2}{\partial x^2} + \frac{\partial^2}{\partial y^2}\right)A_z + n_\gamma A_z = 0, \\
&E_x^\alpha - j_x = 0, \\
&E_y^\alpha - j_y = 0, \\
&n_\gamma = \int_{\mathbb{R}} \frac{1}{\gamma_2} f\, dp_x dp_y,\ j_x = \int_{\mathbb{R}} \frac{p_x}{\gamma_1} f\, dp_x dp_y, \\
&j_y = \int_{\mathbb{R}} \frac{p_y}{\gamma_1} f\, dp_x dp_y,\ n = \int_{\mathbb{R}} f\, dp_x dp_y.
\end{aligned}\right\} \tag{3}$$

The suitable generalized partial slip boundary conditions for Equation (3), are given below [22–24];

$$\left.\begin{aligned}
&\left(f + \Lambda_f \frac{\partial f}{\partial x}\right)\Big|_{x=0,L_x} = 0,\ \left(f + \Lambda_f \frac{\partial f}{\partial y}\right)\Big|_{y=0,L_y} = 0,\ \left(f + \Lambda_f \frac{\partial f}{\partial p_x}\right)\Big|_{p_x=0,L_{p_x}} = 0,\ \left(f + \Lambda_f \frac{\partial f}{\partial p_y}\right)\Big|_{p_y=0,L_{p_y}} = 0, \\
&\left(A_z + \Lambda_{A_z} \frac{\partial A_z}{\partial x}\right)\Big|_{x=0,L_x} = 0,\ \left(A_z + \Lambda_{A_z} \frac{\partial A_z}{\partial y}\right)\Big|_{y=0,L_y} = 0,\ \left(A_z + \Lambda_{A_z} \frac{\partial A_z}{\partial p_x}\right)\Big|_{p_x=0,L_{p_x}} = 0,\ \left(A_z + \Lambda_{A_z} \frac{\partial A_z}{\partial p_y}\right)\Big|_{p_y=0,L_{p_y}} = 0, \\
&\left(E_x + \Lambda_{E_x} \frac{\partial E_x}{\partial x}\right)\Big|_{x=0,L_x} = 0,\ \left(E_x + \Lambda_{E_x} \frac{\partial E_x}{\partial y}\right)\Big|_{y=0,L_y} = 0,\ \left(E_x + \Lambda_{E_x} \frac{\partial E_x}{\partial p_x}\right)\Big|_{p_x=0,L_{p_x}} = 0,\ \left(E_x + \Lambda_{E_x} \frac{\partial E_x}{\partial p_y}\right)\Big|_{p_y=0,L_{p_y}} = 0, \\
&\left(E_y + \Lambda_{E_y} \frac{\partial E_y}{\partial x}\right)\Big|_{x=0,L_x} = 0,\ \left(E_y + \Lambda_{E_y} \frac{\partial E_y}{\partial y}\right)\Big|_{y=0,L_y} = 0,\ \left(E_y + \Lambda_{E_y} \frac{\partial E_y}{\partial p_x}\right)\Big|_{p_x=0,L_{p_x}} = 0,\ \left(E_y + \Lambda_{E_y} \frac{\partial E_y}{\partial p_y}\right)\Big|_{p_y=0,L_{p_y}} = 0.
\end{aligned}\right\}$$

The above relations change the boundaries into

- Partial slip boundary conditions (SPSBCs) using $\Lambda_f = \Lambda_{A_z} = 1, \Lambda_{E_x} = \Lambda_{E_y} = 0$.

- Dirichlet boundary conditions (DBCs) using $\Lambda_f = \Lambda_{A_z} = \Lambda_{E_x} = \Lambda_{E_y} = 0$.

The most frequently and broadly used methods for VMS are particle-in-method (PIM) [12,25,26]. Grid-dependent methods such as finite volume [27], finite element [28–33] and finite difference [11] and Galerkin [28,34,35] methods can also be inspected in the literature for VMS. D. Nunn [36] suggested an algorithm, which is a combination of spline and Fourier concepts. As we have mentioned that our modelled problem is a fractional order VMS system, so we can say it is a system of fractional differential equations (FDEs). FDEs are also treated numerically, which can be found in the literature. Some recent concepts of VMS, FDEs and the numerical simulations can be deliberate from the refs. [8,27,37,38] and [39–42], respectively. T. Zubair reported in ref. [8] that the scheme is highly efficient and holds all the points discussed about numerical strategy in the start of this subsection. Therefore, we are going to extend this scheme to the higher dimensional problem along with partial-slip effects of boundaries in this paper. For this determination; we modify the Gegenbauer polynomials. Therefore, we are bounded to select or articulate the numerical strategy that covers all the above points. The idea to study VMS and FC collectively for higher dimensions is not reported yet in the literature except the ref. [8], in which the author of this current paper itself discussed the model, but it is a lower dimensional model.

The main idea of this paper is that we formulated an extended version of VMS using the concepts of fractional calculus and further numerical simulations of the problem, which is influenced by the concepts of partial slip boundaries [22–24], with the help of a modified algorithm. For this purpose, we strategized a specific geometry with partial slip boundaries and further verbalized the assumptions using the basic perceptions and theorems available in the literature [9,37,43,44]. We suggested the suitable modified version of a numerical algorithm. The numerical outcomes are validated using different approaches explained in the proceeding section of the paper. All the ideas discussed above, i.e., fractional concepts, modified numerical algorithm and partial slip, incorporated on Vlasov Maxwell system, which open new ways to study the problem which is not focused by researchers yet.

The presented study is separated into different sections. The first section consists of a detailed literature survey, creation of a problem, and applications. In the second portion, we have provided the knowledge regarding the modified numerical system. The third component consists of extensive investigation of the numerical outcome. In the last section, we concluded our offered study. This study developed new standards that can further motivate the readers to extend it to the Boltzmann approximations.

2. Formulation of Numerical Scheme

The function approximations are [8,20,21]:

$$\widetilde{f}(x,y,p_x,p_y) = \sum_{i=1}^{M_1}\sum_{j=1}^{M_2}\sum_{k=1}^{M_3}\sum_{l=1}^{M_4} \rho_{i,j,k,l} G^{\mu}_{i,j,k,l}(x,y,p_x,p_y) = {}^4K^T \Lambda(x,y,p_x,p_y),$$

and also:

$$^4K = [\rho_1, \rho_2, \rho_3, \dots, \rho_r]^T, \qquad \Lambda = \left[G^{\mu}_1, G^{\mu}_2, G^{\mu}_3, \dots, G^{\mu}_r \right]^T,$$
$$r = M_4(M_3(M_2(i_1 - 1) + j_1 - 1) + k_1 - 1) + l_1,$$

where the vectors ^4K and Λ are of $M_1 M_2 M_3 M_4 \times 1$ order. Some of the important results are listed below (see Figure 3) as:

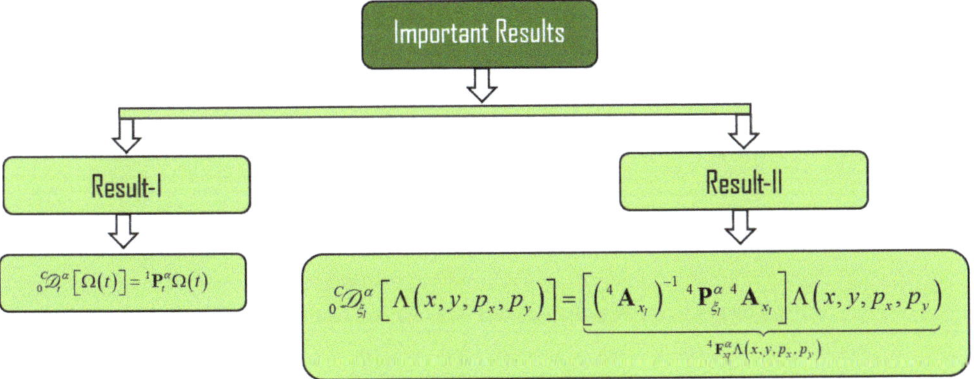

Figure 3. Important results.

In result-I, Caputo fractional differentiation with order $\gamma - 1 < \alpha < \gamma, \alpha \in \mathbb{R}^+$ is defined, and $^1\mathbf{P}_t^\alpha$ is square matrix with order $M \times M$ can written as [20,21,44]:

$$^1\mathbf{P}_t^\alpha = t^{-\alpha}
\begin{bmatrix}
0 & \cdots & 0 & 0 & \cdots & 0 & 0 \\
\vdots & \ddots & \vdots & \vdots & \cdots & \vdots & \vdots \\
U & \cdots & \frac{\gamma!}{\Gamma(\gamma-\alpha+1)} & 0 & \cdots & 0 & \vdots \\
0 & \cdots & 0 & \frac{(\gamma+1)!}{\Gamma(\gamma-\alpha+2)} & \cdots & 0 & \vdots \\
\vdots & \cdots & \vdots & \vdots & \ddots & \vdots & \vdots \\
0 & \cdots & \vdots & \vdots & \cdots & \frac{(M-2)!}{\Gamma(M-\alpha-1)} & 0 \\
0 & \cdots & \vdots & \vdots & \cdots & 0 & \frac{(M-1)!}{\Gamma(M-\alpha)}
\end{bmatrix}$$

The definition of Caputo fractional derivative is

$$\begin{smallmatrix}C\\0\end{smallmatrix}D_t^\alpha \varsigma_j = \frac{j!}{\Gamma(j-\alpha+1)}t^{j-\alpha}, \ j = \lceil\rho\rceil, \lceil\rho\rceil+1,\ldots,M-1,$$

Here, $\rho - 1 < \alpha < \rho$ and $\alpha \in \mathbb{R}^+$.

As we have the four-dimensional function, therefore the fractional order matrix form of differentiation is given in result-II. According to the variable we have,

$$\left.\begin{array}{l}
{}^4\mathbf{F}_{x_1}^\alpha = {}^4\mathbf{F}_x^\alpha, \ {}^4\mathbf{F}_{x_2}^\alpha = {}^4\mathbf{F}_y^\alpha, \ {}^4\mathbf{F}_{x_3}^\alpha = {}^4\mathbf{F}_{p_x}^\alpha, \ {}^4\mathbf{F}_{x_4}^\alpha = {}^4\mathbf{F}_{p_y}^\alpha, \ {}^4\mathbf{A}_{x_1} = {}^4\mathbf{A}_x, \ {}^4\mathbf{P}_{x_1}^\alpha = {}^4\mathbf{P}_x^\alpha, \\
{}^4\mathbf{F}_{x_1}^\alpha = {}^4\mathbf{F}_x^\alpha, \ {}^4\mathbf{F}_{x_2}^\alpha = {}^4\mathbf{F}_y^\alpha, \ {}^4\mathbf{F}_{x_3}^\alpha = {}^4\mathbf{F}_{p_x}^\alpha, \ {}^4\mathbf{F}_{x_4}^\alpha = {}^4\mathbf{F}_{p_y}^\alpha, \ {}^4\mathbf{A}_{x_1} = {}^4\mathbf{A}_x, \ {}^4\mathbf{P}_{x_1}^\alpha = {}^4\mathbf{P}_x^\alpha,
\end{array}\right\}$$

The above defined square matrices are of order $M_1 M_2 M_3 M_4 \times M_1 M_2 M_3 M_4$ and are explained in detail in refs. [8,20,21]. Finally, on the bases of results and ref. [8], methodology is defined in a flow chart that can be seen in Figure 4.

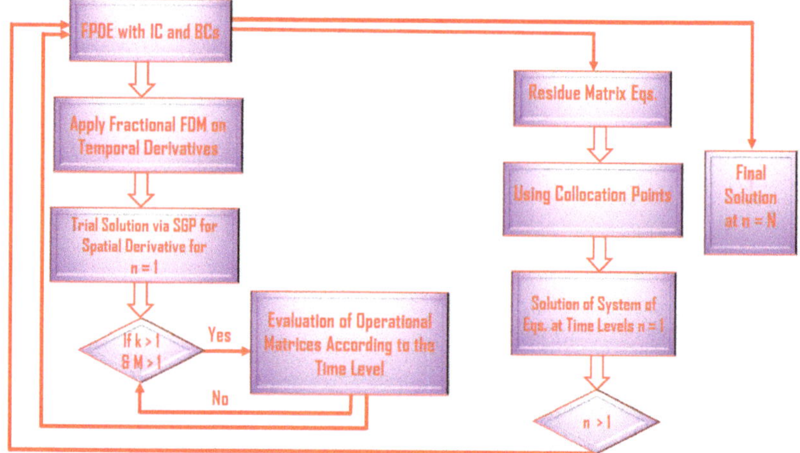

Figure 4. Flow chart of the methodology.

The above methodology will be initiated with the help of the following initial data as follows [45,46]:

$$f(0,x,y,p_x,p_x) = \frac{1}{\sqrt{2\pi}}e^{\frac{-(p_x^2+p_y^2)}{2}}(1+\varepsilon_1\cos(k_x x) + \varepsilon_2\cos(k_x y)),$$

The initial condition described above is referred to as a "2D Maxwellian two-stream cosine perturbation", and we consider $k_x = k_y = 0.5$ and $\varepsilon_1 = 0.25, \varepsilon_2 = 0.35$. This particular initial condition settled for the data is to boost up the complexities of the problem, so that with this complex initial perturbation, the presented scheme has been tested. The generic Python and MAPLE 13 codes have been developed, and several simulations have been run to obtain the necessary results. To demonstrate the numerical structure's competency, we presented and discussed the numerical convergence and stability of the projected technique using a distinct approach.

3. Discussion about Numerical Results

As we know that convergence and stability of the numerical strategy are very important, we are also aware that the exact and close form solution is not available in the literature yet. With this prospective, we choose two different methods to validate the numerical convergence of the formulated algorithm. The complete detail of both methods is given in ref. [8,20,21]. In this method, we use the concepts of norm of two consecutive iterations, which can be seen in ref. [8] in detail. There are two types of numerical values such as integer and fractional values of fractional parameter is used.

We can simply conclude from Figures 5 and 6 that the proposed algorithm is accurate and compatible with this model. The pivotal factor is that the technique is stable and numerical convergence increases as the computing domain increases for both NR and SR scenarios. In light of the intricacies of the problem stated in the preceding part, we can simply conclude that the recommended method is capable of covering the problem's physics better. In our current study, we have the following parameters.

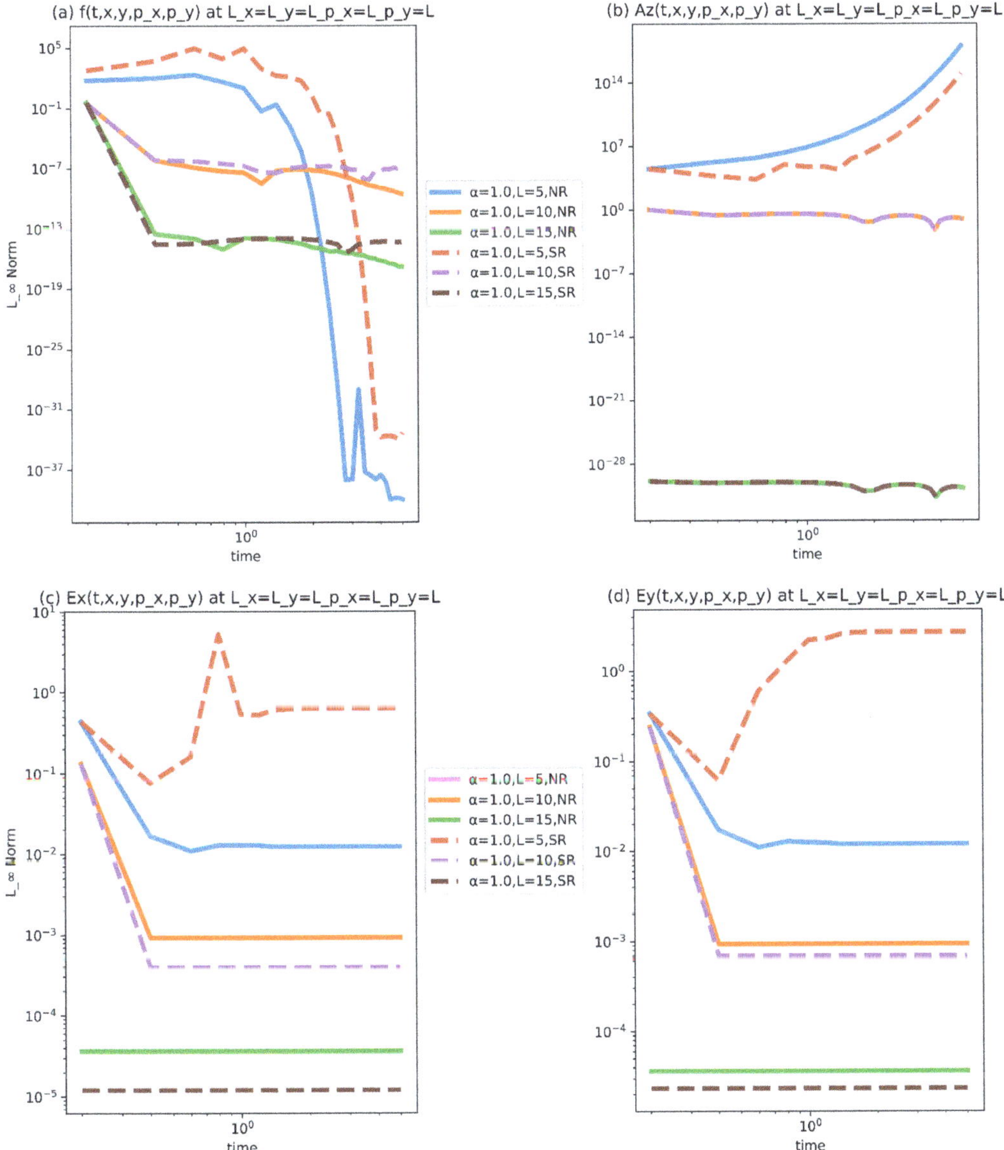

Figure 5. (**a–d**) Norm plots for the NR and SR case at $M_1 = M_2 = M_3 = M_4 = 5, N = 1000, \alpha = 1.0$.

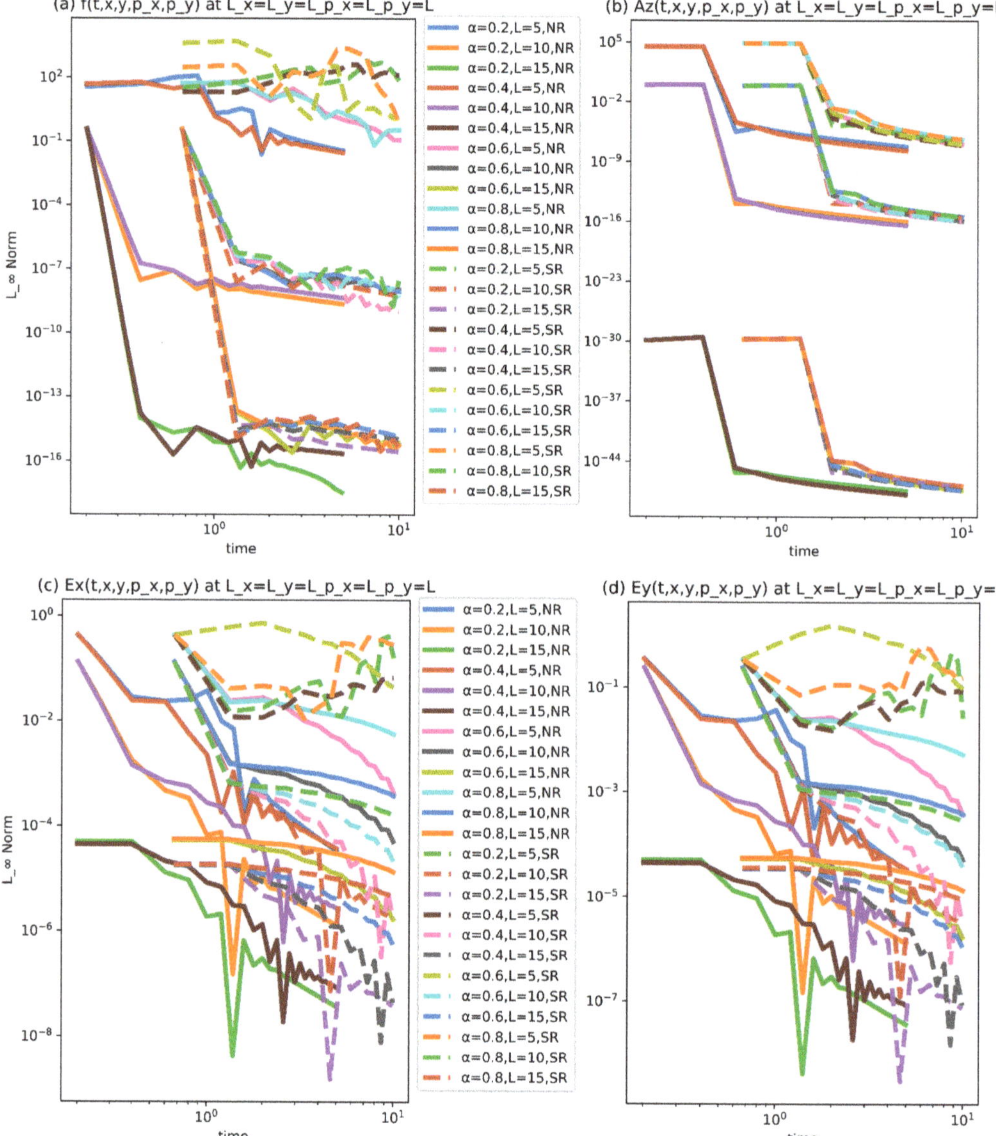

Figure 6. (**a–d**) Convergence of norm at $N = 1000$, $M_1 = 5$, $M_2 = M_3 = M_4 = M_1$.

- NR and SR
- DBCs and PSBCs
- Fractional parameter α

In order to study the numerical impact of the defined parameters, we have offered here higher-dimensional density plots at $t = 9.0$ in Figure 7. There are four different cases, we can see in the presented Figure 7.

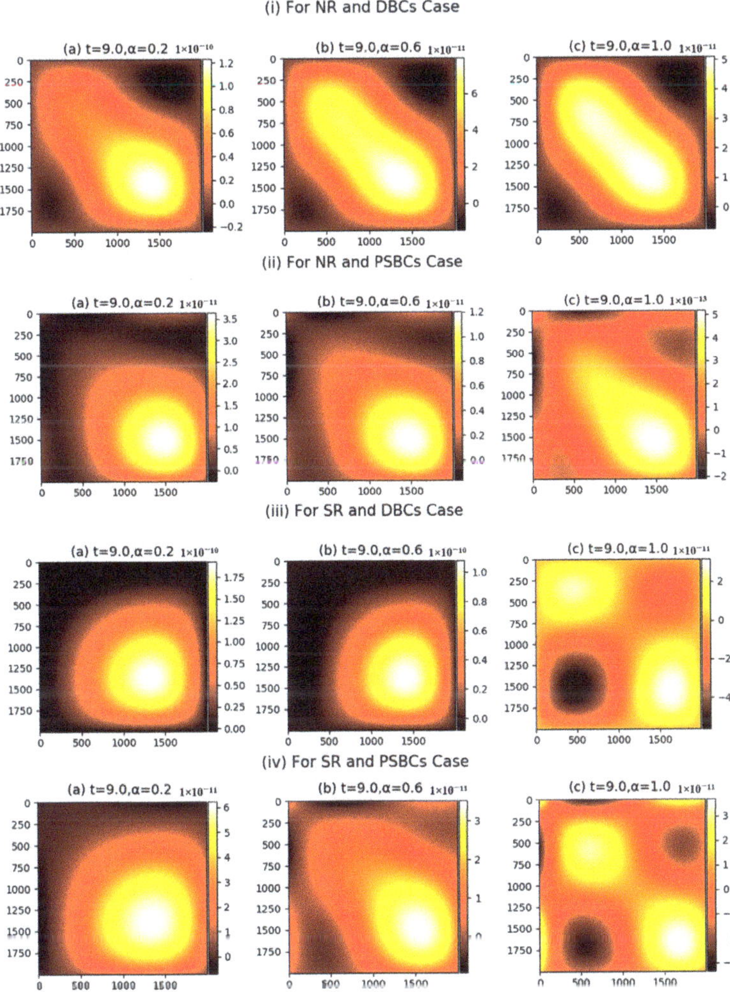

Figure 7. (**i–iv**) Density plots with $M_1 = M_2 = M_3 = M_4 = 5, N = 1000$.

When the initial data are applied, a considerable disturbance is observed. Thus, the energy (or momentum) transformation process began abruptly as a result of the massive burst of initial data. Additionally, the Dirichlet boundary conditions are included to demonstrate that boundaries have no flux. At specific time intervals, due to their identical energy levels, some plasma particles form clusters, which are articulated behind a thin layer (see Figure 7i(a)). As a result, the transformation of plasma particles from high density to low density occurs naturally until the equilibrium condition is reached.

The coated layer expands itself due to the transformation of plasma particles, which can be seen at $t = 9.0, \alpha = 0.6$ (see Figure 7i(b)). The dark portion of the domain shows that there are no excited plasma particles is found in this area, and it is reducing gradually because plasma particles inside the coated layer are trying to adjust itself. To acquire the equilibrium condition, movement of immense volume of plasma particles can be perceived from the upper part to the lower part with integer α (see Figure 7i(c)). Further in Figure 7ii, we can see the variations produced due to the partial slip boundary conditions.

At $t = 9.0, \alpha = 0.2$, particles formulate the cluster of plasma particles, which can be seen (see Figure 7ii(a)) between the mesh domains $1000 \leq x, y \leq 2000$.

Some quantity of volume of excited plasma particles (PP) can also be seen at the other position of the domain, especially at the boundaries of the computational domain, which is due to the slip effects. Two types of movements can be seen in further Figure 7ii(b,c), i.e., the transformation of plasma particles inside the layer and at the boundaries of the domain. In the final shape, i.e., at $t = 9.0, \alpha = 1.0$, the high density of excited plasma particles is placed at the boundaries. Figure 7iii,iv are strategized for different values of the fractional and integer values of the α. The SR parameter is provided to boost up the plasma particles on a large scale. Therefore, we can see the procedure to formulate the small bunches in Figure 7iii,iv.

Parametric study of current is explained in Figure 8, which shows that current density has high impact between $1000 \leq x, y \leq 2000$ (see Figure 8i(a)–iv(a). The reason behind this happening is that huge quantity of ions and electrons are accessible in the discussed domain. It is also witnessed that the current pattern is quite similar to the density profile. It is obvious because that current density is high, where high energy particles movement is recorded. Similarly, we can perceive the other parametric attitude of the current in Figure 8.

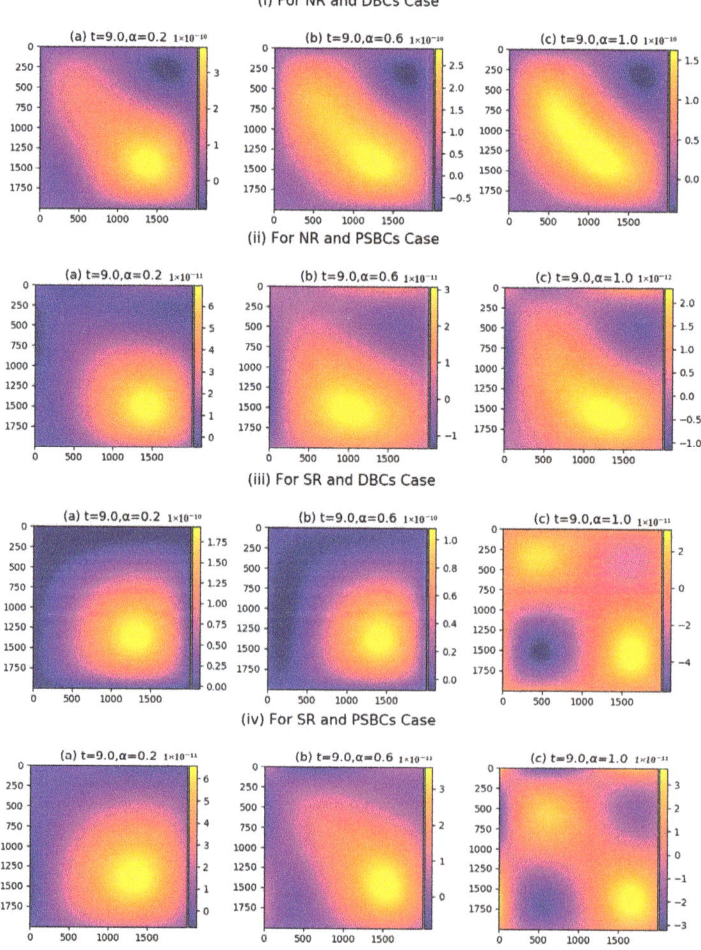

Figure 8. (**i–iv**) Current plot $M_1 = M_2 = M_3 = M_4 = 5, N = 1000$.

4. Concluding Remarks

We articulated an upgraded version of the Vlasov Maxwell system and then used a semi-spectral numerical method to define the numerical solution with partial slip boundaries. Additionally, a detailed discussion of the main aims and an analysis of the results are presented. The following are some closing points:

- Due to the SR parameter, plasma particles can withstand a high rate of destruction. As a result, the density performance of the SR case is completely different.
- Plasma particles scatter to different positions and are further arranged in a cluster form. With increasing values of α, this cluster expands itself under the coated layer.
- The fractional parameter established a new tradition for studying this subject in novel ways. It enables us to investigate the concealed figures of plasma particles.
- The PSBCs parameter assists the plasma particles to obtain more energy from the boundaries and further disperses it to the different positions of the computational domain.
- Although the specified problem contains numerous complications, the technique effectively handles them and produces extremely accurate and stable results that are demonstrated using dissimilar methodology.
- As described previously, this approach is expanded to higher dimensions in this article. As a result, we may conclude that the technique is also efficient, well-matched, and compatible with higher-dimensional problems.

Author Contributions: Conceptualization, T.Z. & M.U.; methodology, M.U.; software, T.Z. & M.U.; validation, T.Z. & M.U.; formal analysis, T.Z.; investigation, T.Z.; resources, T.Z. & M.U.; data curation, T.Z.; writing—original draft preparation, T.Z.; writing—review and editing, T.Z.; visualization, T.Z.; supervision, M.U.; project administration, M.U.; funding acquisition, M.I.A. and J.A. All authors have read and agreed to the published version of the manuscript.

Funding: This work was funded by the National Science Centre, Poland, under the grant OPUS 14 No. 2017/27/B/ST8/01330.

Data Availability Statement: Not applicable.

Conflicts of Interest: The authors declare no conflict of interest.

References

1. Bica, I. Nanoparticle production by plasma. *Mater. Sci. Eng. B* **1999**, *68*, 5–9. [CrossRef]
2. Bica, I. Some Mechanisms of SiO2 Micro-tubes Formation in Plasma Jet. *Plasma Chem. Plasma Process.* **2003**, *23*, 175–182. [CrossRef]
3. Bica, I. Pore formation in iron micro-spheres by plasma procedure. *Mater. Sci. Eng. A* **2005**, *393*, 191–195. [CrossRef]
4. Vatzulik, B.; Bica, I. Production of magnetizable microparticles from metallurgic slag in argon plasma jet. *J. Ind. Eng. Chem.* **2009**, *15*, 423–429. [CrossRef]
5. Bunoiu, M.; Bica, I.; Bunoiu, A.M.; Jugunaru, I.; Bica, I.; Balasoiu, M. Nonthermal Argon Plasma Generator and Some Potential Applications Electromagnetic properties of some composite nanostruc-tured materials View project Ferrofluid based elastomer microstructure View project Nonthermal Ar-gon Plasma Generator and Some Potential. *Ser. Fiz.* **2015**, LVIII.
6. Abbas, I.A.; Marin, M. Analytical Solutions of a Two Dimensional Generalized Thermoelastic Diffusions Problem Due to Laser Pulse. *Iran. J. Sci. Technol. Trans. Mech. Eng.* **2018**, *42*, 57–71. [CrossRef]
7. Marin, M.; Othman, M.I.A.; Seadawy, A.R.; Carstea, C. A domain of influence in the Moore–Gibson–Thompson theory of dipolar bodies. *J. Taibah Univ. Sci.* **2020**, *14*, 653–660. [CrossRef]
8. Zubair, T.; Lu, T.; Usman, M. A novel scheme for time-fractional semi relativistic Vlasov Maxwell system based on laser-plasma interaction with linear polarization and Landau damping insta-bility. *Numer. Methods Partial. Differ. Equ.* **2020**. [CrossRef]
9. Carrillo, J.A.; Labrunie, S. Global solutions for the one-dimensional vlasov–maxwell system for laser-plasma interaction. *Math. Model. Methods Appl. Sci.* **2006**, *16*, 19–57. [CrossRef]
10. Hamid, M.; Zubair, T.; Usman, M.; Huq, R.U. Numerical investigation of fractional-order un-steady natural convective radiating flow of nanofluid in a vertical channel. *AIMS Math.* **2019**, *4*, 1416–1429. [CrossRef]
11. Filbet, F.; Sonnendrücker, E. Numerical methods for the Vlasov equation. In *Numerical Mathematics and Advanced Applications*; Springer: Cham, Switzerland, 2003; pp. 459–468.
12. Birdsall, C.K.; Langdon, A.B. *Plasma Physics via Computer Simulation*; Taylor & Fracnics: Boca Raton, FL, USA, 2004. [CrossRef]
13. Glassey, R.T. The Cauchy Problem in Kinetic Theory. *SIAM* **1996**. [CrossRef]

14. Ross, B. The development of fractional calculus 1695–1900. *Hist. Math.* **1977**, *4*, 75–89. [CrossRef]
15. Barbosa, R.S.; Machado, J.A.T.; Ferreira, I.M. Tuning of PID Controllers Based on Bode?s Ideal Transfer Function. *Nonlinear Dyn.* **2004**, *38*, 305–321. [CrossRef]
16. Silva, M.F.; Machado, J.A.T.; Lopes, A.M. Comparison of Fractional and Integer Order Control of an Hexapod Robot. In Proceedings of the ASME Design Engineering Technical Conference, Chicago, IL, USA, 2–6 September 2003; Volume 5, pp. 667–676.
17. Silva, M.F.; Machado, J.A.T.; Jesus, I.S. Modeling and simulation of walking robots with 3 dof legs. In Proceedings of the IASTED International Conference on Modelling, Identification, and Control, MIC, Innsbruck, Austria, 20–21 February 2007.
18. Usman, M.; Hamid, M.; Haq, R.U.; Wang, W. An efficient algorithm based on Gegenbauer wave-lets for the solutions of variable-order fractional differential equations. *Eur. Phys. J. Plus* **2018**, *133*. [CrossRef]
19. Kilbas, A.A.; Srivastava, H.M.; Trujillo, J.J. *Theory and Applications of Fractional Differential Equations*; Elsevier: Amsterdam, The Netherlands, 2006.
20. Zubair, T.; Lu, T.; Usman, M. Higher dimensional semi-relativistic time-fractional Vlasov-Maxwell code for numerical simulation based on linear polarization and 2D Landau damping instability. *Appl. Math. Comput.* **2021**, *401*, 126100.
21. Zubair, T.; Lu, T.; Nisar, K.S.; Usman, M. A semi-relativistic time-fractional Vlasov-Maxwell code for numerical simulation based on circular polarization and symmetric two-stream in-stability. *Results Phys.* **2021**, *22*, 103932. [CrossRef]
22. Elazem, N.Y.A.; Ebaid, A. Effects of partial slip boundary condition and radiation on the heat and mass transfer of MHD-nanofluid flow. *Indian J. Phys.* **2017**, *91*, 1599–1608. [CrossRef]
23. Sharma, R.; Ishak, A.; Pop, I. Partial Slip Flow and Heat Transfer over a Stretching Sheet in a Nanofluid. *Math. Probl. Eng.* **2013**, *2013*, 1–7. [CrossRef]
24. Usman, M.; Zubair, T.; Hamid, M.; Haq, R.U.; Wang, W. Wavelets solution of MHD 3-D fluid flow in the presence of slip and thermal radiation effects. *Phys. Fluids* **2018**, *30*, 023104. [CrossRef]
25. Tskhakaya, D.; Matyash, K.; Schneider, R.; Taccogna, F. The Particle-In-Cell Method. *Contrib. Plasma Phys.* **2007**, *47*, 563–594. [CrossRef]
26. Denavit, J. Numerical simulation of plasmas with periodic smoothing in phase space. *J. Comput. Phys.* **1972**, *9*, 75–98. [CrossRef]
27. Wettervik, B.S.; DuBois, T.C.; Siminos, E.; Fülöp, T. Relativistic Vlasov-Maxwell modelling us-ing finite volumes and adaptive mesh refinement. *Eur. Phys. J. D* **2017**, *71*, 1–14. [CrossRef]
28. Zaki, S.; Gardner, L.; Boyd, T. A finite element code for the simulation of one-dimensional vlasov plasmas. I. Theory. *J. Comput. Phys.* **1988**, *79*, 184–199. [CrossRef]
29. Vlase, S.; Negrean, I.; Marin, M.; Năstac, S. Kane's Method-Based Simulation and Modeling Ro-bots with Elastic Elements, Using Finite Element Method. *Mathematics* **2020**, *8*, 805. [CrossRef]
30. Vlase, S.; Marin, M.; Öchsner, A. Gibbs–Appell method-based governing equations for one-dimensional finite elements used in flexible multibody systems. *Contin. Mech. Thermodyn.* **2020**, *33*, 357–368. [CrossRef]
31. Vlase, S.; Nicolescu, A.E.; Marin, M. New Analytical Model Used in Finite Element Analysis of Solids Mechanics. *Mathematics* **2020**, *8*, 1401. [CrossRef]
32. IQureshi, H.; Nawaz, M.; Rana, S.; Zubair, T. Galerkin Finite Element Study on the Effects of Variable Thermal Conductivity and Variable Mass Diffusion Conductance on Heat and Mass Trans-fer*. *Commun. Theor. Phys.* **2018**, *70*, 049. [CrossRef]
33. Nawaz, M.; Zubair, T. Finite element study of three dimensional radiative nano-plasma flow sub-ject to Hall and ion slip currents. *Results Phys.* **2017**, *7*, 4111–4122. [CrossRef]
34. Cheng, Y.; Gamba, I.M.; Li, F.; Morrison, P.J. Discontinuous Galerkin Methods for the Vlasov–Maxwell Equations. *SIAM J. Numer. Anal.* **2014**, *52*, 1017–1049. [CrossRef]
35. Juno, J.; Hakim, A.; TenBarge, J.; Shi, E.; Dorland, W. Discontinuous Galerkin algorithms for fully kinetic plasmas. *J. Comput. Phys.* **2018**, *353*, 110–147. [CrossRef]
36. Nunn, D. A Novel Technique for the Numerical Simulation of Hot Collision-Free Plasma; Vlasov Hybrid Simulation. *J. Comput. Phys.* **1993**, *108*, 180–196. [CrossRef]
37. Manzini, G.; Delzanno, G.; Vencels, J.; Markidis, S. A Legendre–Fourier spectral method with exact conservation laws for the Vlasov–Poisson system. *J. Comput. Phys.* **2016**, *317*, 82–107. [CrossRef]
38. Hoyos, J.H.; Valencia, J.; Ramirez, S. Landau damping in cilyndrical inhomogeneous plasmas. *J. Phys. Conf. Ser.* **2019**, *1247*, 012006. [CrossRef]
39. Kumar, S.; Pandey, P.; Das, S. Gegenbauer wavelet operational matrix method for solving varia-ble-order non-linear reaction-diffusion and Galilei invariant advection–diffusion equations. *Comput. Appl. Math.* **2019**, *38*, 1–22. [CrossRef]
40. Hamid, M.; Usman, M.; Zubair, T.; Haq, R.U.; Shafee, A. An efficient analysis for N-soliton, Lump and lump–kink solutions of time-fractional (2+1)-Kadomtsev–Petviashvili equation. *Phys. A Stat. Mech. Its Appl.* **2019**, *528*, 121320. [CrossRef]
41. Usman, M.; Hamid, M.; Zubair, T.; Haq, R.U.; Wang, W. Operational-matrix-based algorithm for differential equations of fractional order with Dirichlet boundary conditions. *Eur. Phys. J. Plus* **2019**, *134*, 279. [CrossRef]
42. Usman, M.; Hamid, M.; Zubair, T.; Haq, R.U.; Wang, W.; Liu, M.B. Novel operational matrices-based method for solving fractional-order delay differential equations via shifted Gegenbauer poly-nomials. *Appl. Math. Comput.* **2020**, *372*, 124985.

43. Yoshikawa, K.; Yoshida, N.; Umemura, M. Direct Integration of the Collisionless Boltzmann Equation in Six-dimensional Phase Space: Self-Gravitating Systems. *Astrophys. J.* **2013**, *762*, 116. [CrossRef]
44. Zubair, T.; Hamid, M.; Saleem, M.; Mohyud-Din, S.T. Numerical Solution of Infinite boundary integral equations. *Int. J. Mod. Appl. Phys.* **2015**, *5*, 18–25.
45. Fijalkow, E. A numerical solution to the Vlasov equation. *Comput. Phys. Commun.* **1999**, *116*, 319–328. [CrossRef]
46. Qiu, J.-M.; Shu, C.-W. Conservative Semi-Lagrangian Finite Difference WENO Formulations with Applications to the Vlasov Equation. *Commun. Comput. Phys.* **2011**, *10*, 979–1000. [CrossRef]

MDPI
St. Alban-Anlage 66
4052 Basel
Switzerland
www.mdpi.com

Nanomaterials Editorial Office
E-mail: nanomaterials@mdpi.com
www.mdpi.com/journal/nanomaterials